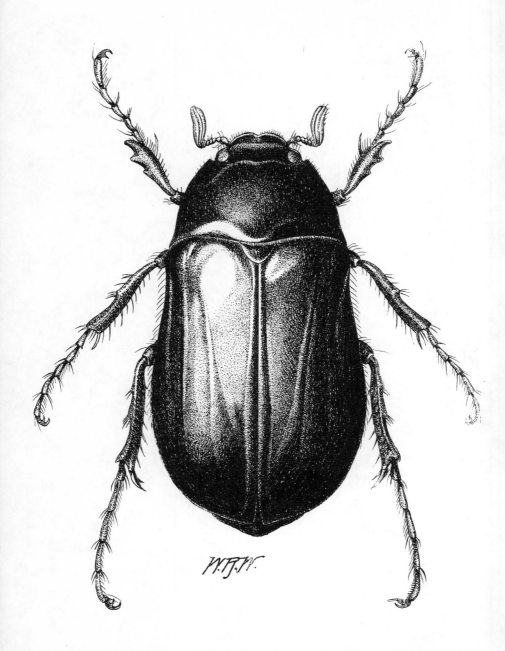

The common June beetle, *Phyllophaga fusca*, drawn by an outstanding entomological artist. (*W. R. Walton.*)

# INSECT LIFE

and

# INSECT NATURAL HISTORY

*(Formerly Entitled: General Entomology)*

By

## S. W. FROST

Professor Emeritus
The Pennsylvania State University
Honorary Member of the Entomological Society of America

## SECOND REVISED EDITION

DOVER PUBLICATIONS, INC.
NEW YORK

# PREFACE

This book is intended as a textbook for elementary college entomology and is written to serve as an introduction for more advanced insect studies. The book, as the name implies, is general in nature and attempts to cover all phases of the subject.

There are many excellent books dealing with the identification, classification, physiology, and morphology of insects and with the economics of entomology. The present book will minimize these subjects as much as possible and will emphasize the scope of the field. The immature forms and the habits of insects will be stressed. Insect orders will be dealt with in some detail, because a fundamental knowledge of the major groups is essential for an introduction to entomology.

During the last decade, the field of entomology has expanded enormously. Many new families, genera, and species have been added. The classification of insects, especially the immature forms, has been greatly improved. The practical phase of the subject has attracted considerable attention because of the general economic conditions of the country and because of the serious losses caused by insects. It therefore seems fitting to survey the subject and to summarize and generalize where the scope of the material demands it.

Insects are best observed in the field, and any conscientious study of entomology should be accompanied by frequent studies of living insects in their natural habitats. Dried, pinned insects should serve only for a detailed study of the species or for the observation of characters that cannot be seen in living specimens. It is far more satisfactory to observe a caterpillar roll a leaf, to watch a stinkbug suck the juices from its victim, or to study a colony of honeybees in an observation hive than to study a collection of dried insects, especially if there is no record of food preference or other ecological data. Many insects can be kept alive and studied in terraria, formicaria, or other suitable cages. Certain museums have featured living insects, and the possibilities along this line are unlimited. Grain insects, clothes moths, carpet beetles, and similar insects can be kept alive for many years with little effort and little feeding. Leaf-feeding caterpillars can be reared and studied without much difficulty. Even predacious species, such as the larvae of syrphus flies or the larvae of lacewings, can be fed if one is willing to gather the proper food.

Unfortunately, crowded classes, and courses of instruction given during the winter, make the study of insects in their natural habitats almost impossible. The alternative is a study of dried specimens supplemented by illustrative material, preferably motion pictures, and the consultation of some authoritative text. "General Entomology" is written to serve this purpose by presenting introductory material and by reviewing and summarizing the more outstanding habits of insects.

Our knowledge of the habits of insects is still too meager to make a book of this sort complete. For example, a few writers have commented upon the characteristic form and texture of the fecular pellets of insects, but no exhaustive study has been made of the subject. The burrows of only a comparatively few insects have been studied. The habits of perhaps only one-half of the leaf-mining species are well known. This book, therefore, will serve as a guide for future constructive work along similar lines.

Extensive lists of references accompany each chapter. No attempt has been made to compile exhaustive bibliographies, because this has been done by specialists in the various fields. The more important and the more recent references are listed. Texts are cited where possible, especially those with bibliographies covering these subjects. As a rule, only papers dealing with insects in general or with groups of insects, such as orders or families, are included. Exceptions are made where papers dealing with individual species seem to add to some important phase of the subject. General texts and other important references are cited in the Appendix.

In the preparation of this work the author has received much help from numerous entomologists, for which grateful acknowledgment is made, especially to the following, who read and criticized certain chapters: Dr. O. A. Johannsen, Dr. C. P. Alexander, Dr. E. P. Felt, Dr. J. Manson Valentine, Dr. V. R. Haber, Dr. E. H. Dusham, Dr. F. M. Carpenter, Dr. Wm. T. M. Forbes, Mr. H. S. Barber, Mr. C. C. Hill, and Mr. H. E. Hodgkiss.

The author is also deeply indebted to many for contributing excellent photographs and takes this opportunity to thank the Pennsylvania Department of Agriculture, Mr. A. B. Champlain, Mr. Charles Macnamara, Mr. E. J. Udine, Dr. F. M. Jones, Dr. B. A. Porter, Dr. B. B. Fulton, Dr. J. I. Hambleton, and others.

An attempt has been made to use new illustrations as far as possible, but it was frequently necessary to borrow from other sources. Many illustrations have been taken from "Destructive and Useful Insects" by C. L. Metcalf and W. P. Flint, "Insect Transmission of Plant Diseases" by J. G. Leach, "The World of Insects" by C. D. Duncan and G. Pickwell, and from numerous bulletins. The author is grateful for permission to use these. Illustrations for Chap. XVIII have been taken largely from William Beutenmuller's work on gall insects.

THE PENNSYLVANIA STATE COLLEGE,
STATE COLLEGE, PA.,                                        S. W. FROST.
    *April*, 1942.

# CONTENTS

# INSECT LIFE

and

# INSECT NATURAL HISTORY

# GENERAL ENTOMOLOGY

## CHAPTER I

### THE POSITION OF INSECTS IN THE ANIMAL WORLD

To understand fully the position that insects occupy in the animal kingdom, it is necessary to consider the larger groups of the world. Ultimately all matter, organic and inorganic, is divided into three great kingdoms: the mineral kingdom, the plant kingdom, and the animal

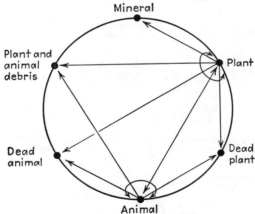

Fig. 1.—The plant and animal food relations.

kingdom. Modern scientists add a fourth, the synthetic kingdom, which may be developed from the animal, plant, or mineral kingdoms. To this kingdom belong many valuable products, such as rayon, nylon, ascorbic acid, sulfanilamide, and hundreds of valuable products. The three natural kingdoms are closely related and their functions are often inseparable. The mineral kingdom supplies the elements for the plant kingdom, which in turn directly or indirectly furnishes food for the animal kingdom. Although animals take mineral substances, which may be essential, they cannot depend entirely upon them for sustenance.

To the mineral kingdom belong all unorganized, lifeless substances and objects. The mineral kingdom is generally considered to be inanimate, immovable, and lacking in growth. The plant kingdom is characteristically sessile, takes inorganic food, grows, and reproduces. The

1

animal kingdom is generally free moving, takes food, grows, and reproduces.

The kingdoms are divided into subkingdoms, tribes, or *phyla*. *Phyla* is the plural of *phylum*, which is the Greek for tribe. In the animal kingdom, the phylum is the common basis of division. Some botanists divide the plant kingdom into two phyla, the Cryptogamia, or spore-

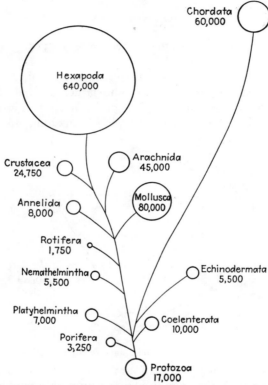

Fig. 2.—A phylogenetic tree of the animal kingdom. The areas of the circles represent approximately the number of species in the respective phyla.

bearing plants, and the Phanerogamia, or seed-bearing plants. Other specialists recognize six phyla; namely, Schizophyta, Thallophyta, Bryophyta, Pteridophyta, Spermatophyta, and Myxophyta. In the mineral kingdom the classification is obscure and greatly complicated by the antiquity of the subject. Rocks and minerals are classified by groups and classes rather than phyla.

Nineteen phyla are generally recognized in the animal kingdom. Only the major phyla are listed below. These are generally arranged in a definite order with the more primitive at the bottom and the more specialized at the top.

THE ANIMAL KINGDOM

| Phylum | Common name | Number of species |
|---|---|---|
| Chordata............... | Mammals, birds, fish, reptiles, amphibia | 60,000 |
| Arthropoda.............. | Spiders, mites, scorpions, insects, etc. | 713,500 |
| Mollusca................ | Clams, oysters, snails, squids, etc. | 80,000 |
| Annelida................ | Earthworms | 8,000 |
| Bryozoa................. | Moss animals, sea mats | 3,100 |
| Echinodermata........... | Starfishes, crinoids | 5,500 |
| Trochelmintha........... | Rotifers, animiculata | 1,750 |
| Nemathelmintha......... | Roundworms | 5,500 |
| Plathyhelmintha......... | Flatworms, flukes | 7,000 |
| Ctenophora............. | Marine animals related to jellyfish | 100 |
| Coelenterata............ | Jellyfish, coral hydra | 10,000 |
| Porifera................ | Sponges | 3,250 |
| Protozoa................ | One-celled animals | 17,000 |
| Other minor phyla....... |  | 1,300 |
| Total................ |  | 916,000 |

**Arthropoda.**—The Arthropoda is by far the largest phylum in the animal kingdom. The numbers given by different specialists vary considerably but it is generally conceded that there are approximately 713,500 species, or about 80 per cent of all the species in the animal kingdom.

The animals of the phylum Arthropoda are commonly called arthropods. The name is derived from the Greek terms *arthron*, joint, and *pous*, foot, having reference to the jointed feet. The Arthropoda include the insects and their near relatives the lobster, crayfish, crab, horseshoe crab, barnacle, shrimp, millipede, centipede, scorpion, spider, mite, peripatus, etc.

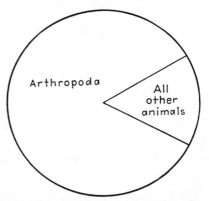

Fig. 3.—Diagram representing the relative abundance of the species of Arthropoda and other animals.

This phylum is distinguished by the following characters:

1. Bilateral symmetry.

2. Body composed of a linear series of rings, segments, or somites.

3. Each segment typically bears a pair of jointed appendages.

4. Body possesses an external skeleton of hard lime or sclerotinous material.

5. A dorsal heart and ventral nervous system.

These characteristics separate the Arthropoda from all other animals and naturally exclude certain small animals frequently but incorrectly mistaken for arthropods, such as the true slugs, which are snails, Mollusca. The larvae of certain Hymenoptera are called slugs and properly belong to the Arthropoda. The adults of these insects are known as sawflies.

**Classes of Arthropoda.**—All phyla are subdivided into classes. The phylum Arthropoda is generally divided into 13 classes which can be separated by the following key:

### KEY TO THE CLASSES OF ARTHROPODA*

*A.* Wormlike animals, with an unsegmented body, but with many unjointed legs, *Peripatus* .......................................... ONYCHOPHORA
*AA.* Body more or less distinctly segmented except in a few degenerate forms.
    *B.* With two pairs of antennae and at least five pairs of legs, sow bug, crayfish, shrimp, etc................................. CRUSTACEA
    *BB.* Without or apparently without antennae.
        *C.* With well-developed aquatic respiratory organs, horseshoe crab.... PALAEOSTRACHA
        *CC.* With well-developed aerial respiratory organs or without distinct respiratory organs.
            *D.* With well-developed aerial respiratory organs.
                *E.* Body not resembling that of Thysanura in form, spiders and mites......................... ARACHNIDA
                *EE.* Body resembling that of Thysanura in form............ MYRIENTOMATA
            *DD.* Without distinct respiratory organs.
                *E.* With distinctly segmented legs.
                    *F.* Body resembling that of Thysanura in form, but without antennae, and with three pairs of thoracic legs and three pairs of vestigial abdominal legs.... MYRIENTOMATA
                    *FF.* With four or five pairs of ambulatory legs; abdomen vestigial.................... PYCNOGONIDA
                *EE.* Legs not distinctly segmented.
                    *F.* With four pairs of legs in the adult instar.......... TARDIGRADA
                    *FF.* Larva with two pairs of legs, adult without legs.... PENTASTOMIDA
    *BBB.* With only one pair of feelerlike antennae, respiration aerial.
        *C.* With more than three pairs of legs, and without wings.
            *D.* With two pairs of legs on some of the body segments, millipedes DIPLOPODA
            *DD.* With only one pair of legs on each segment of the body.
                *E.* Antennae branched.................... PAUROPODA
                *EE.* Antennae not branched.
                    *F.* Head without a Y-shaped epicranial suture, tarsi of legs with a single claw each, opening of the repro-

* Adapted from J. H. Comstock, "An Introduction to Entomology."

ductive organs near the caudal end of the body, centipedes...................... CHILOPODA

   *FF.* Head with a Y-shaped epicranial suture, tarsi of legs with two claws each, openings of the reproductive organs near the head.............. SYMPHYLA

  *CC.* With only three pairs of legs, and usually with wings in the adult stage, insects................................ HEXAPODA

*Class Onychophora.*—The Onychophora include the lowest forms of the Arthropoda and are considered a connecting link between the earth-

Fig. 4.—Peripatus of authors, a representative of the class Onychophora.

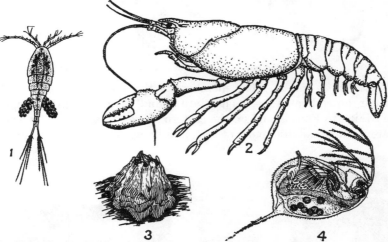

Fig. 5.—Examples of typical crustacea: (1) cyclops; (2) crayfish; (3) barnacle; (4) daphnia. ( (1), (3) and (4) are greatly enlarged.)

worms, (Annelida) and the Arthropoda. Fifty or more species are known from various parts of the world including Australia, Africa, Central America, and the West Indies. They are air-breathing animals and live beneath stones in dark places. They have no definite head but possess a pair of comparatively short ringed antennae and two pairs of jaws or mandibles. The body is cylindrical and externally unsegmented. Internally, the arrangement of the nerve ganglia and paired nephridia give evidence of segmentation. The legs are ringed but not distinctly segmented. Some writers consider Onychophora a separate phylum.

*Class Crustacea.*—The most familiar examples of the class Crustacea are the crayfish and sow bug. Those who dwell along the seashore are familiar with the lobster, shrimp, and barnacle. The members of this class are aquatic and breathe by means of true gills. The anterior segments are united to form a *cephalothorax*, *i.e.*, a head and thorax combined. They have two pairs of antennae and at least five pairs of legs.

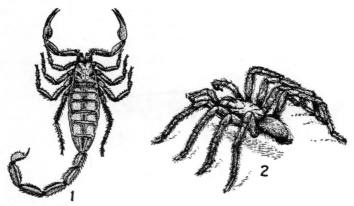

FIG. 6.—Examples of Arachnida: (1) scorpion; (2) spider.

*Class Palaeostracha.*—The members of this class are known as king crabs or horseshoe crabs. They are found along the Atlantic coast from Maine to Florida and on the eastern shores of Asia. They have no apparent antennae, five pairs of legs, and breathe by means of true gills.

 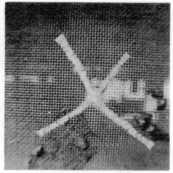

FIG. 7.—Orb web of *Argiope argentata* on water tank, Panama. Only the stabilimentum in the center is visible. The radii are exceedingly delicate.

FIG. 8.—Stabilimentum of the web of *Argiope argentata* on a window screen, Panama.

*Class Arachnida.*—The Arachnida form one of the largest of the classes of Arthropoda and are exceeded in number of species only by the Hexapoda. There are approximately 45,000 described species, which include spiders, scorpions, harvestmen, mites, and related forms. They are air-breathing arthropods with no apparent antennae, four pairs of legs in the adult; the head and thorax are united to form a cephalothorax. There is considerable diversity of body form within this class. The spiders (Araneida) have two body divisions, the cephalothorax and abdomen. The latter is unsegmented. The scorpions (Scorpionida) have two major body divisions but the abdomen is distinctly segmented. The mites (Acarina) have but one body division and this is distinctly unsegmented. Some authorities place the trilobites and horseshoe crabs in this class; others deem it wise to place them in separate classes.

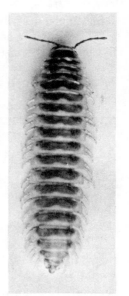

FIG. 9.—A millipede.

*Class Trilobita.*—The trilobites were fossil animals quite closely related to the Arachnida but they were aquatic in habits and possessed a pair of antennae. Their position is uncertain and they have been placed by some in the Crustacea.

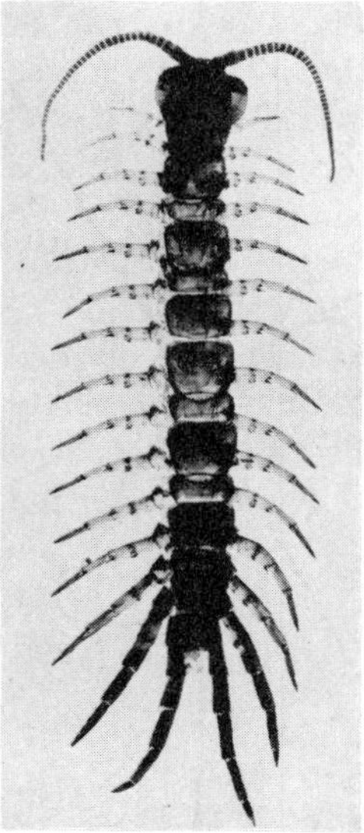

*Class Diplopoda.*—These are popularly known as millipedes. They are frequently mistaken for wireworms because of their shape and form. True wireworms are the larvae of Elateridae or click beetles. Although millipedes live in damp places and generally feed on decaying vegetable matter, a few species attack growing plants. The antennae are short and each body segment appears to bear two pairs of legs.

*Class Chilopoda.*—The animals constituting this class are commonly known as centipedes. They vary considerably in body form. The common house centipede, *Scutigera forceps*, is a harmless creature and as a matter of fact is beneficial for it feeds upon flies and other small insects about the house. On the other hand, some tropical species may attain a length of one foot and are exceedingly poisonous. Unlike the Diplopoda each segment of the body bears only one pair of legs, and the antennae are long and many-jointed.

FIG. 10.—A centipede, *Bothropolys multidentatus.* (*Courtesy of General Biological Supply House.*)

There are a few classes of Arthropoda that are rare or seldom collected; namely, Pentastomida, Tardigrada, Pycnogonida, Pauropoda, Symphyla, and Myrientomata.  The garden symphylid, *Scutigerella immaculata*, is often a pest in greenhouses and gardens.

*Class Hexapoda.*—This is the largest class of Arthropoda, with more than 640,000 described species.  The Hexapoda or Insecta will constitute the major subject of the discussion of this book.  There are three general divisions to the insect body: head, thorax, and abdomen.  One pair of antennae, three pairs of legs, and one or two pairs of wings characterize the adults of most species.

CHARACTERISTICS OF THE PRINCIPAL CLASSES OF ARTHROPODA

| Class | Major body divisions | Number of pairs of antennae | Number of legs |
|---|---|---|---|
| Onychophora | 1 | 1 | Numerous |
| Diplopoda* | 1 | 1 | 2 pairs per segment |
| Chilopoda* | 1 | 1 | 1 pair per segment |
| Crustacea | 2 | 2 | 5 or more pairs |
| Arachnida | 1 or 2 | none | 4 pairs |
| Hexapoda | 3 | 1 | 3 pairs |

\* These two classes are sometimes combined and known as Myriapoda.

**Relationships.**—The phylum Arthropoda is placed near the top of the phylogenetic tree (see Fig. 2) and near the top of the list of phyla (see page 3).  The position indicates that the Arthopoda are highly developed organisms and far removed from the lower forms of organisms such as the Protozoa.

Some sort of arrangement is necessary in order to organize and study insects systematically.  This arrangement is known as classification or taxonomy.  Insects, like other animals and plants, are therefore divided into various categories or related groups, which become successively smaller in regard to number of species until the species itself or variety is reached.  We may now classify a common insect such as the imported cabbage worm, *Ascia rapae* (L.), thus,

Kingdom: Animal
  Phylum: Arthropoda (arthropods)
    Class: Hexapoda (insects)
      Order: Lepidoptera (moths and butterflies)
        Family: Pieridae (the pierids or whites and yellows)
          Genus: *Ascia*
            Species: *rapae*

L. is the abbreviation for Linnaeus, the one who described the species.  Some insects have several forms or color varieties, in which case we may find classifications lower than the species, such as variety or subspecies.

Although insect names are essential as handles to discuss species intelligently, we must not forget that insects are living organisms. A collection of insects is not analogous to a stamp collection in which the total number of forms is known at a given time. Insects are subject to modifications by environmental conditions and, as a result, show remarkable variations in color and form. Each insect has a definite geographical distribution, a host or hosts, and definite biological habits. The life histories of insects are complicated by numerous developmental forms such as eggs, larvae, nymphs, pupae, and adults, not to mention color varieties and sexual forms. In a sense, it would be far better to know that a certain insect occurs in the southeastern part of the United States, that it is a sucking insect, that it is beautifully marked with red and black, and that its favorite food plant is cabbage, than simply to know that it is *Murgantia histrionica*, the harlequin cabbage bug. The beginner wants to know the scientific name of each species; the specialist is never satisfied until he learns everything known about the species. Such knowledge is essential to classify insects accurately.

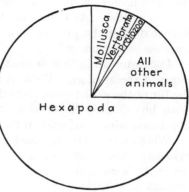

FIG. 11.—Diagram representing the relative abundance of insects (Hexapoda) and other animals.

FIG. 12.—The grasshopper represents a typical insect; there are three distinct divisions of the body, one pair of antennae, three pairs of legs, and two pairs of wings (the second pair of wings is obscured by the first pair of wings, which take the form of wing covers).

**Insect Defined.**—An insect is defined as an air-breathing arthropod with a distinct head, thorax, and abdomen, one pair of antennae, three pairs of legs, and usually one or two pairs of wings in the adult stage.

Only rarely do insects have reduced front legs. The Nymphalidae, or so-called four-footed butterflies, have very much reduced front legs which are useless for locomotion.

Wings may or may not be present in the adult insect. In the first two orders, Thysanura (the bristletails) and Collembola (the springtails), there are no wings and there is no evolutionary evidence that they ever had wings. Since they have not therefore been lost through the process of evolution, these insects are said to be primitively wingless. In the parasitic orders, such as Anoplura (lice) and Siphonaptera (fleas), wings have been lost. As these insects had no use for them, the wings disappeared during the process of evolution. This is known as an acquired wingless condition.

It is interesting to note that some insects living on animal hosts have retained their wings. There is a strange moth (*Bradypus*) that lives in the fur of the sloth; when disturbed, it moves about over the body of the sloth but quickly hides away again in the fur. This insect has well-developed wings but seldom flies away from the sloth.

Wings are often obscured or so modified that they are not at first evident. The wing covers, or elytra, of beetles are hard and horny and cover the second pair of wings. The wing covers of grasshoppers and roaches are leathery and likewise obscure the second pair of wings.

**Hexapoda.**—This class is sometimes known as Insecta. The term is derived from the Greek, *hex*, six, and *pous*, foot, relating to the six feet of most adult insects.

**Entomology.**—Entomology is the science that deals with insects or the Hexapoda. The term is derived from the Greek *entomon*, meaning an insect.

### BIBLIOGRAPHY

CARPENTER, G. H. (1903). On the relationships between the classes of Arthropoda. *Proc. Roy. Irish Acad.* **24.**
——— (1906). Notes on the segmentation and phylogeny of the Arthropoda. *Quart. Journ. Micros. Sci. Lond.* **49.**
CLAUS, C. (1887). On the relations of the groups of Arthropoda. *Ann. Mag. Nat. Hist.* **19.**
COMSTOCK, J. H. (1940). The spider book. Ed. by W. J. Gertsch. Doubleday, Doran & Company, Inc, New York.
COPELAND, H. F. (1938). The kingdoms of organisms. *Quart. Rev. Biol.* **13** (4).
CRAMPTON, G. C. (1919). The evolution of arthropods and their relatives with especial reference to insects. *Amer. Nat.* **53** (625). *References Cited.*
LANKESTER, E. R. (1904). The structure and classification of the Arthropoda. *Quart. Journ. Micros. Sci. Lond.* **47.**
PEARSE, A. S. (1936). Zoological names, a list of phyla, classes and orders. *Sect. F. Amer. Assoc. Adv. Sci.*, Duke University Press.
SNODGRASS, R. E. (1938). Evolution of the Annelida, Onychophora and Arthropoda. *Smiths. Misc.* **97** (6). *Bibliography.*
Consult any of the general texts on entomology.

# CHAPTER II

## ORIGIN AND DISTRIBUTION OF INSECTS

The ancestry of insects is not definitely known. They are primarily terrestrial animals and it is generally believed that they originated from some aquatic ancestor. Most entomologists agree that the Arthropoda have descended through a polychaete annelid. *Peripatus* stands as a connecting link between the Arthropoda and the Annelida. This genus shows relationship to the Annelida by the possession of paired nephridia in most of the body segments and cilia in the generative tract. It is also evident that the Arthropoda are closely related to the Mollusca. The barnacle is no doubt a connecting link between the Arthropoda and the Mollusca. The barnacle shows relationship to the Mollusca through the possession of a mantle in which calcareous plates are secreted. Further affinities are shown by the trochophore or free-swimming larval form of certain annelids and mollusks.

Various hypotheses have been offered to explain the origin of insects. Müller (1864) suggests that insects have descended from species resembling the zoea larvae of the Crustacea which have a cephalothorax, three pairs of mouth appendages, and three pairs of thoracic legs. Brauer (1868) suggests the origin of insects from a myriapodlike ancestor. Tillyard (1930) believes that the Hexapoda, Chilopoda, and the Diplopoda have descended from a common ancestor which he calls Protaptera.

A study of fossils proves that insects have an ancient origin. They are definitely known from the Carboniferous period, many millions of years ago. It is possible that insects may have occurred earlier than this, for a Collembolalike species has been recorded from the Devonian of Scotland.

We do not know where many of the different species of insects originated. No doubt each species developed at some central point. As individuals spread to other areas, they were affected by environmental conditions and during the course of centuries were modified to form new species.

The origin of the North American species has been studied considerably. Undoubtedly, many originated in this country; others migrated from distant places guided or forced by cataclysmic changes in the contour of the earth's surface. On page 14 we describe some of these factors. During comparatively recent times, four paths have been opened by

11

which insects have come to North America: from Asia via Alaska, from Asia via the Pacific Ocean, from South and Central America via Mexico, and from Europe via the Atlantic Ocean. These avenues of introduction are discussed in detail in Chap. IV.

**Ancient Distribution of Insects.**—At one time, as is evident from the table on page 13, insects were supposed to have existed during the Ordovician and Silurian. More recent studies, however, have led to the belief that, with the possible exception of a Collembolalike species, insects did not appear until the Carboniferous period. By the time of the Upper Carboniferous, insects appear to have become numerous and generally well distributed. Numerous groups of insects are recorded from Carboniferous rocks but many of these are extinct today. During the Carboniferous, as well as the Permian and the Triassic, certain groups reached the peak of their development and disappeared before the end of the Jurassic. These include the extinct Orthoptera (Protorthoptera), the extinct dragonflies (Protodonata), the extinct stone flies (Protoperlaria), and others. Many of these insects greatly exceeded in size their living relatives. The Carboniferous period is noted for the number and size of the roaches. Over 200 species have been described from Europe and North America. Other Carboniferous species have been taken from the coal mines of Nova Scotia, Rhode Island, Illinois, Pennsylvania, and Ohio. The Permian and Triassic

Fig. 13.—Fossil tsetse fly, *Glossina veterna* Cockerell. (*After U. S. National Museum, Vol. 54.*)

strata are relatively rich in fossil insects. Specimens have been collected largely from Australia, New South Wales, and Kansas. The Jurassic has contributed many fine specimens. One hundred and forty-three have been described from the limestone beds of Bavaria. The Tertiary strata supply the largest number of fossils. Eight hundred and forty-four have been described from the Miocene beds of Bavaria, 900 from the Baltic amber, 193 Rhynchophora from the Florissant of Colorado and the Green River of Wyoming, and 150 species from Europe. In addition, fossils have been found in New Brunswick and Ontario, Canada, Scotland, England, France, Switzerland, Germany, Spain, Russia, and Sweden. They have been preserved in amber, resin, copal, lignite, and occasionally in limestone. There can be no doubt that insects were widely distributed even at these early periods.

The relative percentages of species occurring during different geological periods vary considerably and lend a clue to their dominance. During the Permian, the Mecoptera constituted 9 per cent of the total number of insects. This fell to 3.7 per cent in the Mesozoic, to 0.16 per cent in the Tertiary, and to 0.035 in 1930. The Homoptera constituted 12.5 per cent of the species in the Permian, 9.0 per cent in the Mesozoic, 4.0 per

GEOLOGICAL APPEARANCES OF ARTHROPODS, DETERMINED BY FOSSILS

| Period | First appearance according to | | | | Other important appearances |
| --- | --- | --- | --- | --- | --- |
| | Scott, 1909 | Folsom, 1934 | Carpenter, 1930 | Carpenter,† 1941 | |
| Quaternary | | | | | Man Glaciation |
| Tertiary | Balance of insects | Balance of insects | Isoptera Lepidoptera | Thysanura Isoptera Siphonaptera Strepsiptera Lepidoptera | Larger mammals |
| Cretaceous | | | | Collembola | |
| Jurassic | Diptera Lepidoptera Hymenoptera | Diptera | Dermaptera Thysanoptera Trichoptera Hymenoptera | Trichoptera Diptera Hymenoptera | Archaeopteryx Birds |
| Triassic | Coleoptera | Coleoptera Hymenoptera | Heteroptera | Heteroptera | Mammals |
| Permian | | Trichoptera | Orthoptera Neuroptera Homoptera Odonata Plecoptera Coleoptera Diptera Psocoptera Mecoptera | Embiodea Raphidiodea Neuroptera Corrodentia Thysanoptera Homoptera Odonata Plecoptera Perlaria Mecoptera Coleoptera | Reptiles |
| Carboniferous | Centipedes | Roaches | Blattaria | Blattaria Orthoptera | |
| Devonian | Orthoptera* Neuroptera* | Odonata* Ephemerida* | | Collembola‡ | Amphibia |
| Silurian | Scorpion | Blattidae* Protocimex* | | | Vertebrates |

* Protocimex was subsequently determined not to be an insect, the Silurian and Devonian species were placed by incorrect determination of the rocks; they properly belong to the Carboniferous.
† Unpublished.
‡ Doubtful.

cent in the Tertiary, and 3.4 per cent in 1930.  The Blattaria show the most remarkable contrast, 34 per cent during the Permian as against 0.42 per cent at present.  These figures show that the Mecoptera, Homoptera, and Blattaria were relatively more abundant during the early geologic periods.  The Blattaria apparently reached the peak of development during the late Carboniferous or early Permian.

On the other hand, the Coleoptera show increased domination with the advance of geologic periods.  In the Permian they constituted only 1 per cent of the insects.  This percentage increased to 37 in the Mesozoic, remained the same during the Tertiary, and at present is 41.5 per cent of the known insects.

The preceding figures are based on Carpenter's observations in 1930, at which time 10,400 fossil insects had been described.  Although new fossil species are continually being described, the increase during recent years is not sufficient to alter the percentages materially.  The number of present-day species is likewise increasing.  Approximately 640,000 have been described and it is estimated that there may be 2,000,000 species when all are described.  This may alter the relative percentages, especially if a large number of species of one order is described.  However, it appears that the descriptions of new species of the various orders remain relatively constant, except perhaps the Homoptera, which are being described at an unusual rate.

**Ancient Conditions Affecting the Distribution of Insects.**—Many writers believe that there were, in ancient times, certain avenues and certain barriers that played an important part in the distribution of insects.  Outstanding among these were the effects of glaciation.  Several advancing ice sheets during the Devonian, Permian, and Pleistocene periods may have forced insects toward the equator.  Some of the species were probably modified by the new environmental factors; others were eliminated.  As the ice sheets receded, it is supposed that insects advanced to occupy the new areas freed from ice.  Many of the arctic species sought the mountain tops where they found an agreeable climate.  They remained there while the temperature at the bases of the mountains became milder and, in turn, became stranded on these lofty islands.  Today, a small butterfly, *Oeneis semidea*, remains on two widely separated mountains, Pikes Peak, Colo. and Mt. Washington, N. H., as evidence of what undoubtedly happened to other species.  The two colonies are the remnants of a cold-loving species which occurs nowhere else.  Other individuals of the genus *Oeneis* continued northward and reached Maine and Labrador where several new species were apparently developed.  The unusual fauna and flora that are found in the comparatively high bogs of the mountains of the Eastern United States are believed to be similar remnants.

The proximity of land masses often formed ancient highways for the distribution of insects. The nearness of Asia to North America still forms a means by which species may migrate from one of these continents to the other.

The belief that land bridges existed across the Atlantic during ancient times is popular in Europe but is not so generally accepted in America. Scharff (1911) describes numerous ancient land bridges connecting the Western Hemisphere with the Old World. Although some of these bridges are questionable, they present some interesting studies of distri-

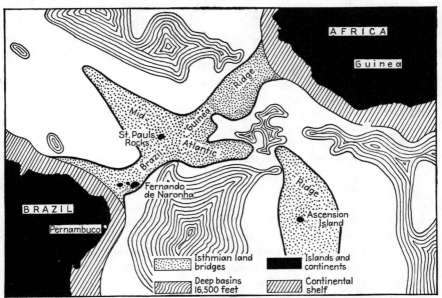

Fig. 14.—The Brazil-Guinea isthmus or land bridge in the late Carboniferous and early Permian. (*After Willis*, 1932.)

bution. The Cretaceous land bridge connected Northern South America with Northern Africa. The West Indies and the Canary Islands were included. This bridge coincides in part with the Gondwana continent mentioned below. The mid-Atlantic bridge of the Tertiary era, according to Scharff, connected the West Indies and the American continents with Europe. The Western Hemisphere was then narrow and much of South America was submerged, with the exception of Patagonia, a small area now Ecuador, and an area to the eastward, now Brazil. At this time, European species could not reach Patagonia or eastern South America although they had access to the West Indies, Central America and western South America. These isolated areas formed centers from which the species supposedly migrated after South America emerged from the sea.

Some sort of land connection between South America and Africa seems to be quite generally accepted by numerous geologists. Neumayr (1886–1887) called this the Brazil Ethiopian continent, which according to his conception included a broad area in the southern Atlantic, most of South America, Africa, and India. This is apparently the same continent that Suess (1893) calls Gondwana. Schuchert (1932) believes this land connection was much narrower. He substantiates his ideas by geologic and biogeographic evidence and places this land bridge in the same general location as the Brazil-Guinea bridge described by Willis (1932). Willis locates this bridge by means of the basins at the bottom of the ocean, by the pressures exerted in this area, and by means of the nature of the rock formation. Four deep basins occur in the Atlantic, one pair to the north of this bridge and one pair to the south. The bridge includes the islands of Fernando de Naronha, 200 miles off the coast of Brazil and St. Paul's Rocks, about 540 miles off the Brazilian coast.

Rising and falling land masses had much to do with the ancient distribution of insects. Undoubtedly, North and South America were united and separated many times during the formation of the continents. The West Indies had a varied and complicated geological history. During the Cretaceous period, all the West Indies, with the exception of the Bahamas, were submerged. During the Eocene, the greater part of Cuba was above the sea. During the Oligocene, all the West Indies except the Bahamas were again submerged. During the late Miocene and Pliocene periods, the West Indies assumed somewhat their present form. Many species were destroyed during the periods of submersion as the meager fauna of Cuba bears testimony. The redistribution of the species to other islands of the group is an involved problem.

**Present Distribution of Insects.**—Insects are widely distributed over the earth's surface. They are found nearly everywhere that green plants grow; that is, wherever food is available. Green plants serve as food for insects and other animals. Insects in turn may feed upon other insects or animals. The decayed remains of plants and animals nourish many species. Insects occur from the equator to the polar regions and are more widely distributed than any other group of animals. They are found on mountain tops, in the caves of the earth, at great altitudes in the air, and a few are found far at sea. In ancient times, before the ice cap crept down from the poles, insects undoubtedly encircled the globe from east to west and from north to south. Today the major orders of insects are world-wide in their distribution. The Thysanura, Collembola, and many of the Coleoptera feed largely upon decayed organic material, a type of food that is abundant and available nearly everywhere. Hence these insects find favorable food conditions in many parts of the world. A few species of insects are cosmopolitan in their distribution.    Among the Lepidoptera, the painted lady, *Pyrameis cardui*, and the monarch, *Danaus menippe*, occur in nearly every part of the world.    Some beetles, especially those that live in domestication, have a world-wide distribution.    The larder beetle, *Dermestes lardarius*, and the black carpet beetle, *Attagenus piceus*, are excellent examples.    Our beautiful mourning-cloak butterfly, *Hamadryas antiopa*, is apparently distributed over the entire breadth of North America below the Arctic circle.    Today the polar ice caps exclude insects from certain comparatively small areas.    Some insects find the humid hot climate of the Tropics more favorable, while others do better in the cool, arid, temperate areas.

Fig. 15.—The mourning-cloak butterfly, *Hamadryas antiopa*.    A common butterfly with a wide distribution in North America.

**Faunal Areas.**—For many years naturalists have recognized that the surface of the earth can be divided into several areas or realms each with its characteristic fauna and flora.    The voyages of the early naturalists such as Darwin, Humboldt, Wallace, Bates, and Belt were made largely to gather information on new and unexplored realms.    Wallace was one of the first to describe the areas visited (see Fig. 17).    Zoologists do not agree as to the precise boundaries of these realms.    The Holarctic is sometimes divided.    The portion including the whole of Europe, Northern Africa as far as the Sahara, and Asia down to the Himalayas, is termed the Palaearctic.    The portion in North America as far south as Mexico is termed the Nearctic.

To the student of North America, the life zones of the United States will be found more useful.    These areas were worked out by C. Hart Merriam in 1894 and are based upon temperatures.    He calculated these

Fig. 16.—Life zones of the United States.    (*After Merriam.*)

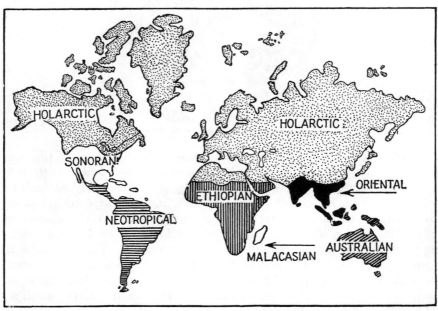

Fig. 17.—Faunal realms.    (*After Sclater and Wallace.*)

zones from the fact that animals are restricted in their northern distribution by the total quantity of heat during the season of their growth and that they are likewise restricted in their southern range by the mean temperatures of a brief period during the hottest part of the year. Four life zones are recognized: *Boreal, Transition, Austral,* and *Tropical.* The Boreal covers Canada, a small part of Northern United States, and the higher elevations of the Rocky Mountains and the Appalachians. It is sometimes divided into the Arctic, Hudsonian, and Canadian. The Transition includes the northern part of the United States and the lower altitudes of the mountainous areas of the United States. The Austral region includes the greater part of the United States with the exception of the areas mentioned above. The Austral is divided into the Upper and the Lower Austral. The Upper Austral includes the Plains and a large part of the central portion of the United States. The Lower Austral includes most of the southern part of the United States with the exception of the Gulf strip and the tropical areas. The tropical region embraces the southern portion of Florida, a small area in Texas in the vicinity of Brownsville, and southern California.

A COMPARISON OF FAUNAL AREAS

| According to Wallace, 1876 | According to Newbigin, 1936 |
|---|---|
| Nearctic | Holarctic and Sonoran |
| Palaearctic | Holarctic of the Old World |
| Neotropical | Neotropical |
| Oriental | Oriental |
| Australian | Australian |
| Ethiopian | Ethiopian of Wallace without Madagascar |
| Madagascar, a part of the Ethiopian | Malacasian |

It seems that the Holarctic of Newbigin and the Nearctic and Palaearctic of Wallace are entirely too large and include animal forms of great diversity of form and habits. A division of this realm into the more northern forms and the more southern forms would aid in placing the fauna and flora of Europe, Asia, and North America.

We may consider the distribution of insects as to latitude and as to altitude. It should be remembered that the change in temperature in traveling 910 miles of latitude is the equivalent to the change in temperature in ascending or descending one mile of altitude. Thus, we might expect to find the same insects on a mountain one mile high as we would at a point 910 miles northward from the foot of this mountain. At 40° north latitude, snow occurs at approximately 8,300 feet; in Peru snow does not occur until we reach 15,800 feet. In other words, cold temperate conditions exist on the high mountains of the Tropics, and here we may find insect forms similar to those found in the Far North or the Far South. Altitude, therefore, has much to do with the distribution of species.

In general, the distribution of insect life is dependent upon three factors: temperature, moisture, and food. Insects are more abundant, in number of species, in the Tropics. As we proceed toward the poles, the number of species becomes less. On the other hand, the number of individuals of a species may be more numerous in the temperate climate. This is true with the exception of ants and termites. We must also bear

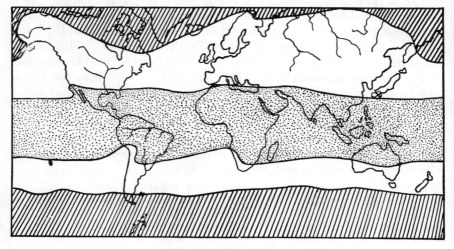

TEMPERATURES ZONES OF THE WORLD

Zone limited by the mean annual temperature of 68°F.

Zone limited by the mean temperatures between 68°F. and 50°F.

Zone limited by the mean annual temperatures between 50°F. and polar temperatures

FIG. 18.—Temperature zones of the world. (*After Hopkins, U. S. Department of Agriculture.*)

in mind that aquatic life, especially in the Atlantic Ocean, is more abundant in the cooler waters near the poles than in the warmer waters at the equator. This is particularly true of microscopic plants and animals. Lohmann records 2,500 organisms per liter in the colder waters and 700 organisms per liter in tropical waters. James Orton (1870), in his classical work "The Andes of the Amazon," gives a few figures which, although old, illustrate the general trend in insects.

DISTRIBUTION OF INSECTS ACCORDING TO ORTON, 1870
11 species of insects occur in Greenland
2,500 species of insects occur in England
8,000 species of insects occur in Brazil

Wallace gives us a better idea of the distribution of species.

DISTRIBUTION OF INSECTS ACCORDING TO WALLACE, 1876

| Group of insects | Tropical and sub-tropical species | Nontropical species |
|---|---|---|
| Locustidae | 2,726 | 1,120 |
| Acrididae | 2,811 | 1,842 |
| Odonata | 2,096 | 921 |
| Pentatomidae | 3,675 | 1,560 |
| Formicidae | 2,888 | 1,055 |
| Totals | 14,196 | 6,498 |

Confining our interests to one of the larger orders of insects, we find that certain families are distinctly tropical and that others have a more general distribution. The largest number of genera are tropical.

DISTRIBUTION OF THE GENERA OF LEPIDOPTERA ACCORDING TO WALLACE, 1876

| Family | Number of genera* | | | | | | Number of genera |
|---|---|---|---|---|---|---|---|
| | Neotropical | Nearctic | Palaearctic | Ethiopian | Oriental | Australian | |
| Danaidae | 20 | 1 | 1 | .. | 3 | 3 | 24 |
| Satyridae | 25 | 1 | 6 | 6 | 11 | 7 | 60 |
| Elymniidae | .. | .. | .. | 1 | 1 | 1 | 1 |
| Morphidae | 1 | .. | .. | .. | 9 | 1 | 10 |
| Brassolidae | 7 | .. | .. | .. | .. | .. | 7 |
| Acraeidae | 1 | .. | .. | 1 | 1 | 1 | 1 |
| Heliconiidae | 2 | 1 | .. | .. | .. | .. | 2 |
| Nymphalidae | 74 | 6 | 8 | 21 | 30 | 7 | 113 |
| Libytheidae | 1 | 1 | 1 | 1 | 1 | 1 | 1 |
| Nemeobiidae | 6 | .. | 2 | 1 | 4 | 1 | 12 |
| Eurygonidae | 2 | .. | .. | .. | .. | .. | 2 |
| Erycinidae | 59 | 2 | .. | .. | .. | .. | 59 |
| Lycaenidae | 12 | 5 | 6 | 20 | 18 | 14 | 39 |
| Pieridae | 17 | 10 | 6 | 11 | 14 | 9 | 35 |
| Papilionidae | 2 | 1 | 6 | 1 | 4 | 2 | 13 |
| Hesperidae | 33 | 9 | 6 | 13 | 18 | 12 | 52 |
| Zygaenidae | 4 | 2 | 5 | 4 | 4 | 3 | 46 |
| Castniidae | 3 | .. | .. | .. | .. | 4 | 7 |
| Agaristidae | .. | .. | .. | 2 | 10 | 9 | 13 |
| Uraniidae | 2 | .. | .. | .. | 2 | 1 | 2 |
| Stygiidae | 1 | .. | 2 | .. | .. | .. | 3 |
| Sphingidae | 5 | 7 | 7 | 4 | 6 | 3 | 7 |
| Totals | 277 | 46 | 56 | 86 | 136 | 79 | 509 |

* Only the genera peculiar to the areas indicated are included.

Although Wallace's figures are old and many changes have been made in the classification and many new genera have been described, they still show the trend in distribution. Some of the genera characteristically belong to one area but spread into others. One Neotropical genus of the *Danaidae*, for example, occurs also in the Nearctic. The same is true of a genus of the *Satyridae*. The Oriental genera of Elymniidae spread into the Ethiopian and Australian areas. The Nearctic genus *Heliconia*, containing but a few species in the Southern United States, has spread from the Neotropical. Two Nearctic genera of Erycinidae occur also in the Neotropical area. Since a single genus may occur in several areas, it is difficult to give the total number of genera for each life zone. Five hundred and seventy-eight genera are tropical or subtropical; and only 102 are Nearctic or Palaearctic. In the case of the Coleoptera, 2,230 genera are tropical or subtropical and 515 genera are Nearctic or Palaearctic.

DISTRIBUTION OF SPECIES OF LEPIDOPTERA OF THE SUPERFAMILIES HESPERIOIDEA AND PAPILIONOIDEA*

| Family | Neotropical | Nearctic |
|---|---|---|
| Papilionidae | 160 | 20 |
| Pieridae | 300 | 50 |
| Nymphalidae | 2,130 | 210 |
| Erycinidae | 800 | 12 |
| Lycaenidae | 750 | 110 |
| Hesperidae | 1,600 | 170 |
| Megathymidae | 3† | 9 |
| Totals | 5,743 | 581 |

* An estimate by Dr. Wm. T. M. Forbes based on Barnes and McDunnough, "Check List of the Lepidoptera of Boreal America" and Seitz, "Macrolepidoptera of the World."
† Not truly Neotropical for they occur on the highlands of Central America.

DISTRIBUTION OF THE GENERA OF COLEOPTERA ACCORDING TO WALLACE, 1876

| Family | Number of genera | | | | | | Number of genera |
|---|---|---|---|---|---|---|---|
| | Neotropical | Nearctic | Palaearctic | Ethiopian | Oriental | Australian | |
| Cicindelidae | 15 | 5 | 2 | 13 | 8 | 9 | 35 |
| Carabidae | 100 | 53 | 64 | 75* | 80* | 95 | 620 |
| Lucanidae | 12 | 5 | 8 | 10 | 16 | 15 | 45 |
| Cetoniidae | 14 | 7 | 13 | 76 | 29 | 11 | 120 |
| Buprestidae | 39 | 24 | 27 | 27 | 41 | 47 | 109 |
| Longicornia | 516 | 111 | 196 | 262 | 360 | 360 | 1,488 |
| Totals | 696 | 205 | 310 | 463 | 534 | 537 | 2,417 |

* These figures show the number of genera peculiar to the areas; the total number of genera would be much larger.

Even the distribution of insects in Alaska shows more species to the southward. Hatch (1937) summarizes the collections of Coleoptera from Alaska as follows:

DISTRIBUTION OF COLEOPTERA IN ALASKA

| Area | Number of species | Per cent of total* of 712 species |
| --- | --- | --- |
| Southern Alaska.................. | 403 | 57 |
| Only southern Alaska.............. | 256 | 36 |
| | | |
| South of Alaska Mts.............. | 628 | 88 |
| Only south of Alaska Mts.......... | 566 | 79 |
| | | |
| Central Alaska................... | 156 | 22 |
| Only central Alaska.............. | 83 | 12 |
| | | |
| Aleutian and Pribilof Islands....... | 104 | 15 |
| Only Aleutian and Pribilof Islands... | 18 | 3 |
| | | |
| North of Baird Mts.............. | 3 | 0.4 |
| Only north of Baird Mts.......... | 0 | 0.0 |

\* Only a portion of the table is included.

**Extreme Limits of Insect Distribution.**—In the north-polar region, Thysanura and Panorpidae are absent and Ephemerida, Neuroptera and Dermaptera are rare. Butterflies have been reported at 80° N., that is, within the Arctic circle, or approximately 800 miles from the North Pole. Mosquitoes have been found at 70° N. They can withstand temperatures as low as −46°C. As a matter of fact, insects live farther north than man. Since insects are closely associated with plants, it is interesting to note that 400 species of plants grow north of the Arctic circle and 120 flowering plants are known from Greenland, including the well-known dandelion.

The antarctic fauna is more impoverished than the arctic. Plant life is almost nonexistent as one approaches the South Pole. Land vertebrates are absent in Antarctica and the birds and mammals are wholly dependent upon the wealth of life in the sea. To quote from Byrd's " Discovery," 1935, "Aside from a few known penguin rookeries on the Antarctic coast, in the vicinity of which occur mites, such lowly organized insects as collembola and a wingless chironomid fly, together with a few rotifers . . ., a Tardigrada (*sic*) and two or three Protozoa in the moss—life is almost unknown."

On Scott's first expedition to the South Pole, only one insect was found. On the second expedition, another species was collected. Priestly found tiny red insects on the underside of boulders of Cape

Adare, 1,200 miles from the South Pole.   One of Mawson's party found myriads of mites attached to upturned stones on the bleak windswept coast of King George V Land, 1,000 miles from the South Pole.   According to Enderlein, the only species with any sort of wide distribution appears to be the parasites of the seals.

We can better understand the scarcity of insects in Antarctica when we consider the rarity of the plants.   Ferns do not exist there.   There are about 100 species of lichens and about 63 species of mosses.   The only flowering plants are a species of grass and a small caryophyllaceous plant; both are dwarfed and rare.

**Abundance of Insect Life in the Tropics.**—In sharp contrast to the paucity of life in the Arctic and Antarctic, life is strikingly abundant in the Tropics.   When life in the temperate areas is compared with that of the Tropics, the contrast is not always so impressive.   The high moisture and the prolonged sunshine of the Tropics and the abundant vegetable food, resulting from these two factors, favor insect development.   On the other hand, the deep shade of the forests and the excessive sunshine of the open places are unfavorable for the existence of some species.

The number of individuals is frequently larger in the temperate area than in the Tropics.   The abundance of insectivorous birds and animals makes unusual inroads into the populations of tropical insects.   Although plant food is abundant in the Tropics, the competition between predacious insects is more acute.   Among the social insects there is a tendency toward small colonies.   This is well illustrated by the wasps of the genus *Polybia* that build small nests of paper and mud.   The number of individuals in these nests seldom exceeds 600 or 700 and the colonies are frequently much smaller.   Compare this with our paper wasp *Vespula maculata*, which may have a population of 2,000 or 3,000 individuals.   Of course there are tropical wasps that have large colonies.   The African *Polybioides melaena* would qualify here.   These wasps have painful stings and violent tempers and are much dreaded by both natives and Europeans.   One must also remember that insects frequently appear in large numbers in the temperate areas.   Consider the Japanese beetle at the time of its emergence in a heavily infested area or the migratory locust on its annual pilgrimage.

Tropical life often presents false impressions.   Large insects, such as the rhinoceros beetle, *Megasoma elephas*, tend to focus attention on such species and to emphasize their abundance.   The morphos are conspicuous and not easily overlooked except when at rest with the brilliant blue surfaces of the wings folded inward.   Since one of these butterflies may be seen several times during a short period, it is difficult to determine the abundance of these striking insects by actual count.   The shrill call of the Panama cicada, far eclipsing the murmur of our seventeen-year cicada, attracts considerable attention and produces a lasting impression. Pestiferous insects, such as the mosquitoes and predacious bugs, often seem more numerous than they actually are.

The season of the year has much to do with the abundance of insects. There is, of course, no true dormant period in the Tropics corresponding to winter of the temperate zone. There are, however, periods when trees, insects, and other animals estivate or become inactive. The dry season promotes inactivity and insects are usually scarce during this period. A tropical downpour stimulates the growth of new leaves and shoots. Under such conditions bamboo may shoot nine inches a day and the banana leaf may grow about four inches during the same period. Leaf miners, leafhoppers, aphids, and other insects that feed upon tender shoots find utopia when the rainy season begins. Ants and termites emerge in great numbers and insect life in general is noticeably more abundant.

Ants and termites are dominant animals of the warm countries. Wallace lists 2,888 species of ants from the Tropics and 1,055 species from the temperate areas. Individuals are abundant as well as species. They often play an important part in the destruction and decay of trees. The anteater and the armadillo depend largely upon these insects for food. The nests of ants and termites are conspicuous in trees and may be found in dead wood and in the soil. Ants are always in evidence during favorable weather. The leaf-cutting ants can usually be seen carrying small pieces of leaves to their nests. The small azteca ants rush forth from their arboreal, pendant nests to attack the intruder. The army or legionary ants are frequently seen on the march. A closer examination of the forests reveals that the nodes of many trees, the thorns and other parts of plants are often populated by colonies of ants.

Something of the abundance of insect life and the struggle that goes on in a tropical forest can be gleaned from the nightly operation of light traps. Every night the sequence is somewhat the same. Darkness settles about seven o'clock and the lights are turned on. Within a short time insects are attracted in vast numbers. The large beetles and Heteroptera come with a tremendous hum. Termites, leafhoppers, gnats, and small moths envelop the light in a fog. The enormous grasshoppers hit the light with a thud. Many insects are stunned by the impact and fall to the ground. Before they cease struggling, hordes of ants are ready to drag them off. The roaches come in for their share. The larger species keep to the shadows and nibble at the dead or dying insects. Some of the smaller species are attracted to the light and perish like the others. When the lights are turned off, bats and certain birds make their appearance and clean the table. By morning not a trace of an insect can be found. Some of the visitors have gone away well fed; others have perished for their sake.

There is no doubt that the number of tropical species far exceeds that of the temperate areas. It is also evident that a large proportion of the tropical species still remain undescribed. This is partly due to the fact that collecting has been more limited in the Tropics and that the demands for work on the temperate species precludes the study and identification of the tropical species.

The most striking feature of tropical life is the abundance of certain groups that are comparatively scarce or absent in the temperate areas. Hundreds of species of termites are known to occur in Central America, while only a single species occurs in Northeastern United States. Mantispids are conspicuous in the Tropics but only two species occur in the northern part of our country. Wasps of the genus *Polybia* are numerous in Central America but not a single species occurs in North America. Among the butterflies, we find the Morphidae and the Heliconiidae conspicuous in Central America. With the exception of a few species of heliconias, they do not occur in North America. On the other hand, species of bees, caddis flies, and May flies are more numerous in temperate areas.

**Modes of Distribution.**—Insects, like other animals, are characterized by their ability to move about. Their movements help to avoid competition for space, for food, for oxygen (in the case of aquatic insects), and to escape parasitic and predacious enemies. The change of location may be accidental or may be the result of certain impelling conditions. They are attracted or urged by favorable temperature, humidity, or food and are impeded by mountain ranges, deserts, or watercourses.

There are many degrees of insect migration. Movements may be local, such as the emergence of insects from their resting places, their wanderings in search of hibernating quarters, or the migration in search of new feeding grounds. Certain factors may compel them to move a considerable distance. When a pond dries up, aquatic insects must seek a new location. If the hive becomes crowded, bees must find new quarters. A change of wind direction may force a swarm of Chironomidae to find a more sheltered spot.

If an insect is introduced into a new country, its movements may be exceedingly rapid. The Japanese beetle was accidentally brought into Riverton, N. J., in 1916. Conditions were favorable for its development and it soon spread throughout the state and crossed the Delaware River to adjacent areas. The center of the heaviest infestation still remains about Riverton, although the pest has spread over thousands of square miles and now occurs in all the states north of Georgia and east of Indiana.

Migration is a special form of distribution and is usually a seasonal affair. Armyworms, chinch bugs, and grasshoppers migrate to new fields when the crop upon which they are feeding is cut or when the food otherwise becomes scarce. Some insects have a definite annual migration as remarkable as the migration of birds. The common milkweed butterfly, *Danaus menippe*, gathers by the thousands in the fall and travels

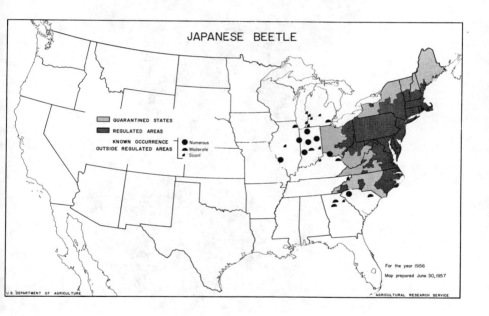

FIG. 19.—Regional concentration of the Japanese beetle in the area of general distribution, 1957. (From insect pest survey.)

southward, where it passes the winter. In the spring it returns to the north. The painted lady, *Pyrameis cardui*, is another species that migrates southward in the fall. On the other hand, certain insects migrate northward. Each fall the moths of the cotton leaf worm, *Alabama argillacea*, fly northward from the Gulf States. They are caught by cold weather and perish during the winter. Unusual storms may drive them northward in enormous numbers. In the cities they are attracted by lights and their abundance frequently draws the attention of the public.

Tropical migrations are frequently spectacular. Thomas Belt, in "The Naturalist in Nicaragua," describes a flight of the brown-tailed butterflies, *Timetes chiron*, which at the time of his observation were flying to the southeast. "They occurred, as it were, in columns. The

Fig. 20.—The black witch, *Erebus odora*, a moth that frequently migrates northward in the fall from the Gulf States; above, a fresh specimen; below, a specimen after the northward flight. (*Champlain and Kirk.*)

air would be comparatively clear of them for a few hundred yards; then, we would pass through a band, perhaps fifty feet in width, where hundreds were always in sight and all traveling one way. . . . In some seasons these migratory swarms of butterflies continue passing over to the southeast for three to five weeks and must consist of millions upon millons of individuals comprising many different species and genera." Those who have traveled in the Tropics have witnessed migrations of this sort.

**Manner of Distribution.**—Insects possess many structures and devices by which they propel themselves. Wings, the most important organs of propulsion, do not occur in all orders. Some insects, such as the bees, wasps, dragonflies, and grasshoppers, have two pairs of wings; others, such as the true flies and certain May flies, have but a single pair of wings. The number or size of wings does not entirely determine the ability to fly. The dobson fly, *Corydalis cornuta*, has two large pairs of wings but is a clumsy flier. The aphis lion or golden eye, *Chrysopa*, has comparatively large wings but is a poor flier. On the other hand, the bees and wasps with comparatively small wings are powerful fliers. Their wings are well braced with strong veins and provided with strong muscles which make rapid motion possible. The dragonfly is one of the most powerful and skillful fliers. Dr. Tillyard showed that one of the large dragonflies could travel approximately 60 miles an hour. On the other hand this insect can stop quickly, alight gracefully upon a blade of grass, or hover over the water with ease. It captures its prey in flight and depends upon its wings to get a living. The sphingid moths are also powerful fliers. Their wings are narrow and their bodies are slender. Grasshoppers have been

Fig. 21.—A springtail, *Neosminthurus curvisetis.*

found 1,200 miles at sea, where they were probably driven by the wind. Grasshoppers are known to fly 15 miles an hour. *Schistocerca peregrina* has been found 500 miles at sea. The most remarkable speed is exhibited by the male of the deer botfly, *Cephanomyia pratti*, which according to some authorities has a velocity of 818 miles an hour. Some physicists believe that such speed is impossible and that an insect traveling at that rate would be crushed. Nevertheless, this insect holds the fastest record in the animal world. By comparison, a modern shell travels at the rate of 2,000 miles per hour.

Powerful legs aid many insects in moving when disturbed. The grasshopper is the best example. The femora of the hind legs are enlarged for the attachment of powerful muscles. Flea beetles, leafhoppers, tree hoppers, fleas, and other insects have similar legs for jumping.

The springtail is provided with a springlike attachment to the underside of the body. When the insect is ready to jump, this spring is thrown backward and propels the insect forward.

Air currents are a considerable factor in the distribution of many insects. Numerous insects are carried upon the prevailing winds. Experiments have shown that many of our pests are scattered in this manner.

When the air current reaches cyclonic proportions, insects may be carried remarkable distances. A consideration of man's attempts to conquer the air gives some idea of what might be expected of insects under similar circumstances. The average velocity of balloons is 18 miles an hour. At times they attain a speed of 100 miles an hour. The longest balloon flight on record was 400 miles in 1923. In 1937, a glider traveled 200 miles in six hours and forty-five minutes, attaining an altitude of 3,800 feet. In 1939, a stratosphere balloon, with robot, reached an altitude of 14 to 16½ miles. Thus, we might expect to find insects at great altitudes or far out at sea. As a matter of fact Cook (1890) records a spider 2,000 miles at sea. Whitefield (1940) summarizes the insects of the air.

Spiders and small caterpillars readily drop from higher to lower elevations by spinning strands of silk. They can ascend by gathering

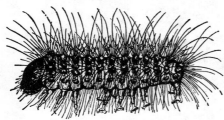

up the silk in a ball and discarding it. The young of several species of spiders use silk as a means of transportation. A spider climbs to some elevated position. With its front feet grasping a support, it elevates its abdomen and spins several strands of silk, which are carried upward by the air currents. When sufficient silk has been spun to buoy

FIG. 22.—First instar gypsy moth larva showing vesicular and acuminate hairs.

up the spider, it lets go its hold and drifts through the air. Sometimes open fields become strewn with the silk of these spiders as they make unsuccessful attempts to take off.

The newly hatched gypsy moth larvae have a unique method of transportation. The young larvae have hairs with small air pockets that make them buoyant and, when the wind is strong, carry them a long distance. They have been known to travel five miles a day and have been found 300 feet in the air.

In many insects there are expansions of the tracheae, or breathing tubes, which are termed *air sacs*. They vary in number and position in different species. In the honeybee there are two large sacs which occupy a considerable part of the abdominal cavity. In the May beetle there are hundreds of small sacs. These air sacs help to lighten the insects and make flight easier. They are especially well developed in the May fly. The adults take no food and the alimentary canal acts as a balloon which lessens the specific gravity of the insects.

Many insects naturally live at high altitudes or are found only in the colder parts of the country. In Alpine fauna, insects take first place among animals. According to Whymper (1892) Hemiptera are found at 16,500 feet, Orthoptera at 16,000 feet, Lepidoptera at 15,000 feet, and Hymenoptera at 12,000 feet. In comparison, man lives at 15,900 feet in West Tibet. He climbs to 28,000 feet without artificial oxygen and ascends much higher in planes with the aid of oxygen. Vultures are known to soar to 35,000 feet. Honeybees have been seen about balloons at an altitude of 35,000 feet. The most remarkable range of insect distribution is exhibited by a large moth, *Erebus*, found in the lowlands of Brazil and up to 10,000 feet in Ecuador.

Oceans and streams may carry insects long distances. The effect of ocean currents is well illustrated by some recent investigations. In 1937, Mr. E. Thomas Gilliard of the American Museum of Natural History released 100 bottles from Funk Island off the coast of Newfoundland. These bottles were properly weighted so that they would float but would not be driven by the wind. Of the first 17 bottles recovered, 7 were taken off the coast of Ireland, 5 off the coast of England, 4 off the coast of France, and 1 off the coast of Spain. One bottle, driven by the prevailing easterly current, crossed the ocean in six months and twenty-four days. The average time for the crossing was eight and a half months.

Glass buoys used to support fishing nets in Portugal have been found off the coast of Florida. In this case the prevailing westerly currents carried them. Similar buoys have drifted from the Japanese coast to the California coast. The abrasion of the glass indicated that the buoys had been in the water for a considerable time.

Birds may be responsible for carrying certain insects, especially those like the chalcids that infest seeds.

Man is one of the most important factors in disseminating insects. They are carried by boats, cars, trains, and airplanes. The recent fear of the introduction of mosquitoes or similar insects from various parts of the world by airplane is a problem. Introductions such as the honeybee, silkworm, parasites, and predacious insects were intentional but too many species have been introduced accidentally.

### BIBLIOGRAPHY

There are hundreds of references to the subjects of this chapter. Only the more important and more recent papers are given here.

I. Origin and Ancient Distribution of Insects.

BEQUAERT, J. C., and CARPENTER, F. M. (1941). The antiquity of social insects. *Psyche* **47** (1).

BRUES, C. T. (1933). Progressive change in the insect population of forests since the early Tertiary. *Amer. Nat.* **67** (712). *Literature.*

———— (1940). Is ours the "Age of insects"? *Scientific Mo.* **50** (5).

CARPENTER, F. M. (1930). A review of our present knowledge of the geological history of the insects. *Psyche* **37** (1).

CLARK, A. H. (1927). The biological relationships of the land, the sea and man. *Science* **65** (1680).

COCKERELL, T. D. A. (1931).    The antiquity of insect structures.    *Amer. Nat.* **65** (699).

CRAMPTON, G. C. (1919).    The evolution of arthropods and their relatives with especial reference to insects.    *Amer. Nat.* **53** (625).    *Reference cited.*

———— (1920).    Remarks on the ancestry of insects and their allies.    *50th Ann. Rept. Ent. Soc. Ontario* **36.**

FOLSOM, J. W. (1934).    Entomology with special reference to its ecological aspects. Revised by R. A. Wardle.    The Blakiston Company, Philadelphia.

GOSS, H. (1900).    The geological antiquity of insects.    2d ed.    Gurney & Jackson, London.

HANDLIRSCH, A. (1906–1908).    Die fossilen Insekten und die Phylogenie der resenten Formen.    Leipzig.

———— (1925).    Paläontologie.    *In,* Schröder's Handbuch der Entomology, 3. Gustav Fischer, Jena.

IMMS, A. D. (1937).    Palaeontology.    *In,* Recent advances in entomology.    2d ed. The Blakiston Company, Philadelphia.

RAYMOND, P. E. (1920).    Phylogeny of the Arthropoda with special reference to the Trilobites.    *Amer. Nat.* **54** (534).

———— (1939).    Prehistoric life.    Harvard University Press, Cambridge, Mass.

SCHARFF, R. F. (1911).    Distribution and origin of life in America.    Constable & Company, Ltd., London.    *Bibliography.*

SCHUCHERT, C. (1932).    Gondwana land bridges.    *Bull. Geol. Soc. Amer.* **43.**    *Bibliography.*

SCOURFIELD, (1939–1940).    The oldest known fossil insect.    *Proc. Linn. Soc. Lond. Sessions* **152.**

SCUDDER, S. H. (1875–1901).    Twenty-eight papers on fossil insects.

SNODGRASS, R. E. (1938).    Evolution of the Annelida, Onychophora and Arthropoda. *Smiths. Misc. Coll.* **97** (6).    *Bibliography.*

TILLYARD, R. J. (1930).    The evolution of the class Insecta.    *Pap. Roy. Soc. Tasmania.*

———— (1930).    A new theory of the evolution of insects.    *Nat. Lond.* **126.**

TOTHILL, J. D. (1916).    The ancestry of insects, with particular reference to Chilopods and Trilobites.    *Amer. Journ. Sci.* **42** (251).    4th ser.

WALTON, L. B. (1927).    The polychaete ancestry of the insects, I, the external structure.    *Amer. Nat.* **61** (647).

WILLIS, B. (1932).    Isthmian links.    *Bull. Geol. Soc. Amer.* **43.**

## II. Recent Distribution of Insects.

ALLEE, W. C. (1926).    Distribution of animals in a tropical rain forest with relation to environmental factors.    *Ecology* **7** (4).

BODENHEIMER, F. S. (1927).    Factors that determine distribution.    Ueber die für Verbreitungsgebiet einer Art bestimmenden Faktoren.    *Biol. Zbl. Leipzig* **47.**

CALVERT, P. P. (1923).    The geographical distribution of insects and the age and area hypothesis of Dr. J. C. Willis.    *Amer. Nat.* **57** (650).

COLE, L. J., *et al.* (1940).    The relation of genetics to geographical distribution and speciation.    *Amer. Nat.* **74** (752).

COMON, R. (1937).    Dispersion des espèces.    Des causes qui réglent la répartition des espèces.    *Misc. Ent. Castanet-Tolosan* **38.**

COOK, W. C. (1931).    Notes on predicting the probable future distribution of introduced insects.    *Ecology* **12** (2).    *Literature.*

CRAMPTON, C. B. (1913).    Ecology the best method of studying the distribution of species.    *Proc. Roy. Physic. Soc. Edinburgh* **19.**

DAUBENMIRE, R. F. (1938).    Merriam's life zones of North America.    *Quart. Rev. Biol.* **13** (3).    *Literature cited.*

Fox, H. (1939). The probable future distribution of the Japanese beetle in North America. *Journ. N. Y. Ent. Soc.* **47** (2). *Bibliography.*

Hesse, R., *et al.* (1937). Ecological animal geography. John Wiley & Sons, Inc., New York.

Howell, A. B. (1922). Agencies which govern the distribution of life. *Amer. Nat.* **56** (646). *Literature.*

Merriam, C. H. (1892). The geographical distribution of life in North America. *Proc. Biol. Soc. Wash.* **7.**

——— (1898). Life zones and crop zones of the United States. *U. S. Biological Survey* **10.**

Newbigin, M. I. (1936). Plant and animal geography. Methuen & Co., Ltd., London.

Pearse, A. S. (1939). Animal ecology. McGraw-Hill Book Company, Inc., New York. *Bibliography.*

Sanderson, E. D. (1908). The influence of minimum temperatures in limiting the northern distribution of insects. *Journ. Econ. Ent.* **1** (4).

Scharff, R. F. (1911). Distribution and origin of life in America. Constable & Co., Ltd., London. *Bibliography.*

Stiasny, G. (1934). Das Bipolaritatsproblem. *Arch. Neerland Zool.* **1.**

Sweetman, H. L. (1933). Ecological studies in relation to the distribution and abundance of economic pests. *Journ. Econ. Ent.* **26** (2).

Van Dyke, E. C. (1919). The distribution of insects in western North America. *Ann. Ent. Soc. Amer.* **12** (1).

——— (1929). The influence which distribution has had in the production of the insect fauna of North America. *Trans. 4th Intern. Congress Ent.*

Wallace, A. R. (1876). Geographical distribution of animals. Macmillan & Company, Ltd., London.

Warnecke, G. (1935). Principles of geography applied to entomology. Grundsatzliches zur Methodik zoographischer Untersuchungen in der Entomologie. *Int. Ent. Zeit. Guben* **28.**

Webster, F. M. (1903). The diffusion of insects in North America. *Psyche* **10.**

Webster, R. L., and Smith, L. G. (1937). Insect immigrants in Washington. *Northwest Sci. Wash.* **11.**

Willis, J. C. (1922). Age and area study in geographical distribution and origin of species. Cambridge University Press, London.

### III. Wind and Air Currents as Factors in the Distribution of Insects.

Felt, E. P. (1928). Dispersal of insects by air currents. *N. Y. State Mus. Bull.* **274,** *Bibliography.*

——— (1938). Wind drift and dissemination of insects. *Can. Ent.* **70** (11).

———, and Chamberlain, K. F. (1935). The occurrence of insects at some height in the air, especially on the roofs of high buildings. *N. Y. State Mus. Circ.* **17.**

Glick, P. A. (1939). The distribution of insects, spiders and mites in the air. *U. S. Dept. Agric. Tech. Bull.* **673.**

Hardy, A. C., and Milne, P. S. (1937). Studies in the distribution of insects by aerial currents. *Journ. Animal Ecol.* **7** (2).

——— and ——— (1937). Insect drift over the North Sea. *Nat. Lond.* **139** (3516).

——— and ——— (1938). Studies in the distribution of insects by aerial currents. *Journ. Animal Ecol.* **7** (2).

Imms, A. D. (1939). Insects in the upper air. *Nat. Lond.* **144** (3644).

McClure, H. E. (1938). Insect aerial populations. *Ann. Ent. Soc. Amer.* **31** (4).

Webster, F. M. (1902). Winds and storms as agents in the diffusion of insects. *Amer. Nat.* **36** (430).

WHITEFIELD, F. G. S. (1940). Air transport, insects and disease. *Bull. Ent. Res.* **30** (3). *Extensive Bibliography.*

**IV. Flight and Migration.**

ANDREWS, R. C. (1937). Wings win. *Natural History,* October.

BIRD, R. D. (1937). Records of northward migration of southern insects during drought years. *Can. Ent.* **69** (5).

BISHOPP, F. C., and LAAKE, E. W. (1921). Dispersion of flies by flight. *Journ. Agric. Res.* **21** (10).

ECKERT, J. E. (1933).. The flight range of the honeybee. *Journ. Agric. Res.* **47** (5).

ENGELHARDT, P. (1934). Tornados and butterfly migration in Texas. *Bull. Brooklyn Ent. Soc.* **29** (1).

EWING, H. E. (1938). The speed of insect flight. *Science* **87** (2262).

FLETCHER, T. B. (1936). Marked migrant butterflies. *Ent. Rec.* **48** (10).

GRANT, K. J. (1936). The collection and analysis of records of migrating insects, British Isles. *Entom. Lond.* **69.**

KNIGHT, H. H. (1936). Records of southern insect species moving northward during the drought years of 1930 and 1934. *Ann. Ent. Soc. Amer.* **29** (4).

LANGMUIR, I. (1938). The speed of the deer-fly. *Science* **87** (2254).

MUNRO, J. A., and SAUGSTAD, S. (1938). A measure of the flight capacity of grass-hoppers. *Science* **88** (2290).

SHANNON, H. J. (1926). A preliminary report on the seasonal migrations of insects. *Journ. N. Y. Ent. Soc.* **34** (1).

SMITH, S. (1938). Alabama records of butterfly migration. *Ala. Acad. Sci.* **10.**

SNODGRASS, R. E. (1930). How insects fly. *Ann. Rept. Smiths. Inst.* 1929.

TOWNSEND, C. H. T. (1927). On the Cephenemyia mechanism and the daylight-day circuit of the earth by flight. *Journ. N. Y. Ent. Soc.* **35** (3).

———— (1939). Speed of Cephenemyia. *Journ. N. Y. Ent. Soc.* **47** (1).

TURNER, H. J. (1937). A few random thoughts on mass movement of Lepidoptera or pseudo-migration. *Ent. Rec. Lond.* **49.**

WILLIAMS, C. B. (1930). Migrations of butterflies. Oliver and Boyd, Edinburgh.

———— (1936). Collected records relating to insect migrations. *Proc. R. Ent. Soc. Lond.* **11.**

———— (1937). The migrations of day-flying moths of the genus Urania in Tropical America. *Proc. R. Ent. Soc. Lond.* **12.**

———— (1938). Recent progress in the study of some North American migrant butterflies. *Ann. Ent. Soc. Amer.* **31** (2).

———— (1939). Some records of butterfly migrations in America. *Proc. Roy. Soc. Lond. Ser. A* **14.**

**V. Local Movements of Insects.**

CROZIER, W. J. (1923). On the locomotion of the larvae of the slug-moths (Cochlidiidae). *Journ. Exp. Zool.* **39** (2). *Literature cited.*

HALLOCK, H. C. (1935). Movements of larvae of the oriental beetle through soil. *Journ. N. Y. Ent. Soc.* **43.**

HARTZELL, A., and McKENNA, G. F. (1939). Vertical movements of the Japanese beetle larvae. *Contrib. Boyce Thompson Inst.* **11** (1). *Literature cited.*

KALMUS, H. (1938). Die Orientierung der Bienen im Stock. *Forsch. u. Fortsch. Berlin* **14.**

MALENOTTI, E. (1940). Nocturnal migrations, on the surface of the ground, of subterranean larvae of Melonthids. *Ath. Accad. Agric. Verona* **5** (18).

TRAVIS, B. V. (1939). Migrations and bionomics of white grubs in Iowa. *Journ. Econ. Ent.* **32** (5). *Literature cited.*

See also references for Chap. XI, Insect Behavior.

# CHAPTER III

## ABUNDANCE, SIZE, AND REPRODUCTIVE CAPACITY OF INSECTS

The Hexapoda, or insects, comprise the largest number of species of the animal world. There are about six times as many species of insects as of all other kinds of animals combined. Plants may seem numerous but the flowering plants of the United States can be adequately listed and briefly described in a single volume. On the other hand, insects are too numerous to be listed conveniently. If the names of all insects were arranged in two columns and printed on an average size page, they would fill 10 volumes and still some names would not be included. There are approximately 640,000 described species of insects. They are being described at the rate of 6,000 a year and it is estimated that there may be approximately 2,000,000 species in all.

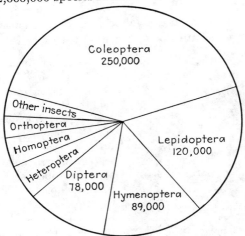

Fig. 23.—Diagram illustrating the relative number of species of the orders of insects.

When we consider that each insect has several forms—egg, larva or nymph, pupa, and adult—we can readily understand that entomology is a tremendous subject. It would be impossible for one to know the names of even a small portion of the species, not to mention their hosts, habits, and other characteristics.

The number of species common to a particular locality is far smaller. Only 15,449 species are given in "A List of the Insects of New York." It is probable that 20,000 would be a high figure for any locality.

35

THE MORE IMPORTANT ORDERS OF INSECTS

| Order | Common name | Approximate number of species |
|---|---|---|
| Orthoptera | Grasshoppers, crickets, roaches | 18,000 |
| Heteroptera | True bugs, water boatmen, etc. | 25,000 |
| Homoptera | Cicadas, scale insects, aphids | 80,000 |
| Coleoptera | Beetles | 250,000 |
| Lepidoptera | Butterflies and moths | 120,000 |
| Diptera | True flies | 78,000 |
| Hymenoptera | Bees, wasps, and ants | 89,000 |

**Age of Insects.**—Insects have lived in this world for many millions of years. Their persistence to the present day indicates that they are able to exist under the most trying conditions. There is geologic evidence that certain groups of insects, such as the roaches, reached the peak of their development centuries ago during the late Carboniferous or early Permian period. Other groups, such as the Coleoptera, are coming into prominence today. It is sometimes stated that insects may supersede man. Dr. W. J. Holland, in his classic work on the moths, expressed his views on the dominance and persistence of insects as follows:

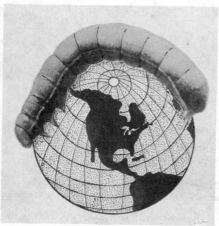

FIG. 24.—If it were not for enemies and adverse conditions, insects might overpopulate the world.

When the moon shall have faded out from the sky and the sun shall shine at noonday a dull cherry red, and the seas shall be frozen over, and the ice cap shall have crept downward to the Equator from either pole, and no keels shall cut the waters, nor wheels turn in mills, when all cities shall have long been dead and crumbled into dust, and all life shall be on the verge of extinction on this globe— then, on a bit of lichen, growing on the bald rocks besides the eternal snows of Panama, shall be seated a tiny insect, preening its antennae in the glow of the worn out sun, representing the sole survival of animal life on this our earth, a melancholy "bug."

There is no doubt that insects greatly exceed the number of other living animals of the world today. During the summer, we are overwhelmed by their abundance. Everyone has had some experience with insects, black flies when hunting in the woods, worms in juicy fruit, or moths in his clothes. The enormous flights of May flies along the shores of lakes, the swarms of Japanese beetles in heavily infested areas, or the inroads of thrips during the middle of the summer are sufficient to remind us of the abundance and prevalence of insects at certain seasons of the year. Some cut leaves, roots, or stems into shreds, some attack meat, some feed on filth, some fly through the air with a buzzing sound, some sing monotonous songs, and some even press their beaks relentlessly into our tender skin. During the winter, insects appear to be scarce but, if one looks beneath stones or bark, they may be found congregated in enormous numbers. The spotted ladybird beetles, *Ceratomegilla fucilabris*, for example, assemble in large numbers in the late fall in some secluded spot where they pass the winter often deeply buried beneath the snow. It is not unusual to find thousands of them in their winter quarters. Indoors, household insects, such as clothes moths, larder beetles, cereal moths, and numerous other species, continue to reproduce during the entire winter. Insects are conspicuous because of their abundance, their enormous capacity for reproduction, their extremely wide distribution, and the economic losses resulting from their depredations.

To express the abundance of insects more concretely, we cite a few examples. Insects are plentiful in number of individuals as well as in number of species. The normal method of reproduction may lead to large numbers of individuals, for the reproductive capacity of insects is often phenomenal. This is particularly true if reproduction is allowed to continue without interruption, *i.e.*, if not checked by parasitic or predacious enemies, if not restrained by adverse temperature or moisture conditions, and if not affected by lack of food.

Many of our social insects have large families. A mature colony of bald-faced hornets, *Vespula maculata*, may contain 3,000 individuals, an organization surely to be avoided. The queen honeybee is capable of laying 4,000 eggs a day and a strong hive in spring may contain 40,000 to 50,000 individuals. Large ant colonies are said to comprise 500,000 individuals. Millions of grasshoppers are included in a "locust plague." Such plagues have been more serious in Europe and Asia than in this country. Some grasshopper infestations are reported to cover 15,000 acres and to consume the crops of such an area in seven or eight hours. On one occasion, 300,000 pounds of grasshoppers were gathered and destroyed. Millions of emerging May flies or other insects are often washed upon the beaches of lakes and become offensive when they decay. Moths sometimes become so numerous about beacons or light-houses that they interfere with the operation of the lights. Thrips may become so numerous over fields that they appear as fog or mist arising from the ground. The farmer is in a position to see insects at the height of their abundance. Incrustations of scale insects and aphids are common on many plants; 24,688 aphids have been counted on a single tomato plant. It is estimated that there may be from 1,000,000 to 65,000,000 insects per acre in some soil.

Insects appear more abundant and more conspicuous under certain circumstances. When they emerge from their hibernating quarters, when they come forth in swarms from the water, or when they congregate

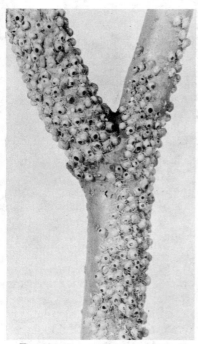

for their southern migration, they are generally more noticeable. The adults of the Japanese beetle emerge late in July or early in August and spend a comparatively short part of their existence above ground; nevertheless, all people living in the infested area are familiar with the tremendous flights of these beetles. The seventeen-year cicadas emerge more or less simultaneously over a given area. These conditions tend to increase the apparent number and to focus attention upon the species. The resulting sudden injury may likewise attract attention. If the same number of insects emerged over a period of six or eight months, they might not be noticed. When the milkweed butterflies, painted ladies, and other species of butterflies migrate southward in the fall, they are very conspicuous. These large well-marked species naturally attract attention. During the summer, however, the larvae feed on widely

Fig. 25.—Incrustation of aphids on a twig. Hymenopterous parasites have emerged from many of the aphids.

separated plants and appear to be scarce. Try to collect 100 milkweed butterfly caterpillars from an area of an acre and you will find it somewhat difficult but, at the migration season, thousands of adults may be seen resting on a single tree.

The abundance of insects is dependent upon many factors, of which only a few can be considered. On the whole, insects are small, but insignificance, at times, is advantageous. Small insects obviously require less food for their development. This is striking especially when two insects are competing for the same supply of food. Leaf-mining larvae may feed within a leaf serving also as food for a larger caterpillar.

Even though the caterpillar may nearly skeletonize the leaf or reduce it to mere veins, a species like *Nepticula* finds enough leaf material to complete its small mine. The amoeba, an animal with an ancient historic background, still exists today largely because of its small size. The dinosaur and mammoth, on the other hand, disappeared long ago. The largest insects of which we know are represented only by fossil forms.

**Reproductive Capacity of Insects.**—The reproductive capacity of insects is dependent upon many factors, chiefly the number of eggs or young produced, the short life cycle which is often coupled with numerous egg-laying periods following in rapid succession, and special provisions to speed up reproduction.

Insects generally lay large numbers of eggs. The San José scale produces 400 to 500 young, and the dobson fly, *Corydalis cornuta*, may lay 2,000 or 3,000 eggs. Social insects are heavy egg layers. The queen honeybee may lay 2,000 or 3,000 eggs a day. The queen termite is able to lay 60 eggs per second until several million are laid. These are extreme cases; the average number of eggs laid by insects is from 100 to 150.

Fig. 26.—Egg mass of *Corydalis cornuta*, the dobson fly. Several thousand eggs may be contained in a single egg mass.

The results of such prolific tendencies has led to much speculation as to what would happen if insects were permitted to reproduce without any check. Hodge figured that a pair of flies beginning in April might be the progenitors, if all were to live, of 191,010,-000,000,000,000,000 individuals by August. Allowing ⅛ cubic inch per fly, there would be enough of them to cover the earth 47 feet deep. Herrick determined that the cabbage aphid might have 12 generations between March 31 and August 15. During this time a female would be able to produce 564,087,-257,509,154,652 individuals or approximately 1,655,254,501,086 pounds. This would be 822,627,250 tons a year. Huxley figured that a single female aphid might produce during a year a progeny whose total protoplasm would equal that represented by the inhabitants of China. Of course, no insect ever reaches its maximum reproductive capacity. There are many adverse conditions, parasitic and predacious enemies that prevent such things from happening and thus these insects are held in check.

The eggs of many insects are remarkably well protected. Those of the tent caterpillar and the fall canker worm have a hard coating which protects them from severe winter temperatures. An interesting comparison exists between the eggs of the fruit tree leaf roller, *Cacoecia argyrospila*, and the eggs of the red-banded leaf roller, *Argyrotaenia velutinana*. The eggs of the fruit tree leaf roller are covered by a thick varnishlike material which protects them from severe winter conditions. The eggs of the red-banded leaf roller are laid only during the summer and are unprotected. They are so thin and transparent that they are scarcely visible upon the leaf surface. The eggs of other kinds of insects are laid deep in the wood or similar substrata where they receive some protection. Most of the grasshoppers lay their eggs in the ground where they are protected. Many of the Buprestidae and Curculionidae cover their eggs with fecula or secretions from the mouth. In this manner they are obscured from parasitic and predacious enemies. The eggs of the aphis lion are the most remarkably developed. Each egg is placed on a stiff stalk which elevates it above prowling enemies.

The short life cycle of most insects helps to build up infestations rapidly. The majority of insects mature within a year and many have several generations during a single season. Insects with more than one generation a year are quite common. The codling moth may have two or three generations and the oriental fruit moth may have five to seven generations, depending upon the latitude in which they live. Red mites frequently go through eight or nine generations during a summer; aphids may have 30 generations during a single season. Each generation is a means of stepping up the insect population. There are certain exceptions, such as the periodical cicada, which requires a number of years in which to mature. Some of the wood-boring beetles, likewise, require two and sometimes ten or more years to mature.

The wide range of substances upon which insects feed and the great abundance of these food materials are decided advantages to the development of insect life. Slightly more than half of the insects are plant feeders; the rest feed upon animals and upon plant or animal debris.

Insects are favored by many unique methods of defense. The hard sclerotinous exoskeleton, stinging hairs, poison glands, and other features help to ward off enemies. In spite of all these protective devices, parasitic and predacious insects succeed in attacking their hosts. These in turn are struggling for existence.

Special methods of reproduction such as *polyembryony* and *parthenogenesis* aid in building up a great insect population. In the case of *polyembryony*, hundreds of individuals may be produced from a single egg. This form of reproduction occurs in comparatively few insects and is especially well adapted for the parasitic habit. An insect that might have difficulty in maintaining its position on an active caterpillar long enough to deposit several eggs has only to remain long enough to thrust its ovipositor once into the body of the host. A single egg, deposited, splits at an early stage of development into a number of embryos. In some cases as many as 2,500 individuals may develop from one egg. As a rule the number is much smaller. *Platygaster hiemalis*, a well-known polyembryonic form, has been studied considerably. It is an enemy of the hessian fly. Only six or seven young develop from a single egg although this insect may lay several eggs in its host.

Some insects produce living young without the necessity of mating. This method of reproduction is known as *parthenogenesis*. Slingerland reared aphids through 98 generations over a period of four years and three months and stopped only because the problem became monotonous and other subjects needed attention. Subsequent workers have produced successive generations over a period of many years without a single fertilization. This evidently happens commonly in greenhouses or in tropical areas where the laying of winter eggs is not induced by low temperatures or other unfavorable conditions. Among the aphids, parthenogenesis is the common form of reproduction during the summer. The same method is also found among some Hymenoptera, Diptera, and a few other groups. It increases the reproductive possibilities because there are no casualties or loss of time in mating. The process continues almost uninterrupted. An unusual form of parthenogenesis, occurring chiefly in the Cecidomyiidae, is known as *paedogenesis*. The larvae, or rarely the pupae, give birth to living young. Other insects speed up reproduction by depositing their young in an advanced stage of development. This occurs chiefly in the parasitic Diptera, the louse flies, Hippoboscidae, the bat ticks, Streblidae and Nycteribiidae, and in a

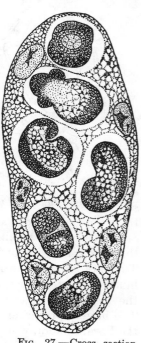

Fig. 27.—Cross section of a 23-day-old embryonic mass of *Platygaster zosine*, a parasite of the hessian fly, showing six embryos that have developed from a single egg laid by the parasite. (*After Leiby and Hill*).

few species of Sarcophagidae. The fully grown larvae or puparia are deposited by the adults of these species.

Finally we may say that insects have neither birth control nor old-age problems. When work is over or their functions cease, they die. They can eke out a living where man would fail miserably. They can feed on materials most unpalatable to human beings, withstand extreme temperatures, and in some cases endure severe desiccation. They are, in short, not easily restrained when trying to get a living under the most unfavorable circumstances. Darwin, in his "Voyage of the Beagle," has shown how insects live on the barren St. Paul's Rocks, 540 miles from the coast of South America.

Not a single plant, not even a lichen, grows on this islet; yet it is inhabited by several insects and spiders. The following list completes, I believe, the terrestrial fauna: a fly (Olfersia) living on the booby, and a tick which must come here as a parasite on the birds; a small brown moth, belonging to a genus that feeds on feathers; a beetle (Quedius) and a woodlouse from beneath dung; and lastly, numerous spiders, which I suppose prey on these small attendants and scavengers of the waterfowl

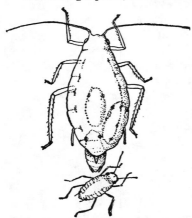

Fig. 28.—A wingless parthenogenetic aphid giving birth to living young. (*After Hunter and Glenn*).

**Size of Insects.**—Insects are generally small and attract attention more often because of their abundance and grotesque form than because of their large size. The great bulk of them, those that we see most frequently—the ant, the fly, the mosquito, the bee, and the wasp—are indeed small when compared with the average vertebrate. Some insects are smaller than the largest protozoan; others are larger than the smallest chordate. The smallest chordates belong to the group Tunicata, some of which are only one one-hundredth of an inch long. The majority of the insects, of course, exceed these in size. The largest protozoan is a *Nummulites*, one of the Foraminifera, which is about the size of a 25-cent piece.

In the Tropics, one meets many surprises. On the one hand, certain scarab beetles, wood-boring Buprestidae, and grasshoppers have developed great size. The rhinoceros beetle, *Megasoma elephas*, from Central America is about five inches long and the volume of its body is about 20 times that of the related June beetle of our country. The occasional observer of nature is often impressed by the size of insects. A traveler in Panama saw a cat playing with a roach and he mistook the insect for a mouse. A bird collector in the same country shot a buprestid beetle measuring two and a quarter inches, believing it to be a bird. On the other hand, some tropical insects are remarkably small.

Fig. 29.—*Megasoma elephas*, one of the largest living beetles. It measures approximately five inches long and is somewhat common in Central America.

The stingless bees, *Melipona*, are much smaller than our honeybees and are among the smallest of the bees that occur in the Tropics. Some of the tropical species of wasps that build paper nests are minute. Forty or fifty of them can congregate conveniently on a nest scarcely two inches in diameter. The colonies are likewise small; *Protopolybia sedula* may contain only 150 or 200 individuals. Compare this with the nest of our bald-faced hornet, *Vespula maculata*, which may contain 2,000 or 3,000 individuals. Although some of the dragonflies of the Tropics are enormous, the genera allied to *Agrion* contain insects of extraordinary fragility and delicacy.

The largest living insect is *Erebus agrippina*, a moth from Brazil which has a wing expanse of 11 inches. A moth from Indo-China, *Attacus atlas*, has a spread of nine and a half inches. *Attacus caesar* from the Philippines has an expanse of 10 inches.

Many grasshoppers are large. As a matter of fact, there are very few small insects among the Orthoptera. Aside from the pigmy crickets and possibly some roaches, the Orthoptera are comparatively large and showy. The lubber grasshopper of the southwestern plains is nearly three inches long. The Central and South American *Tropidacris latreillei* is six and a half inches long and has a wing expanse of nine and a half inches.

The largest living beetle is a tropical species of longhorn (Cerambycidae), *Acrocinus longimanus*. It measures from six to eight inches with the legs and antennae outstreched. The front legs are long and almost exceed the length of the long antennae. Another large species is the giant scarab beetle from Africa, which is six and a half inches long. A Venezuelan beetle, *Dynastes hercules*, is slightly more than six inches long. In rating the size of insects, bulk must be considered as well as length. Some three-inch beetles are greater in weight than those that are six inches long. The rhinoceros beetle, previously referred to, probably rates first as to bulk.

Parts of insects are often developed enormously. The growth of antlerlike processes on the thorax of the males of exotic rhinoceros beetles have been described by entomologists. In some cases these have developed to such an extent that they are a hindrance to the beetle and threaten the destruction of the species. Mandibles occasionally develop grotesquely. The jaws of the males of the dobson fly, *Corydalis cornuta*, are exceedingly long and cruciate. The pupae of *Phryganea* (Trichoptera) and of the Micropterygidae (Lepidoptera) also have comparatively conspicuous cruciate mandibles. The mandibles of the males of the European *Chiasognathus grantii* exceed the length of the body but those of the female are normal. In other insects the head or the sucking mouth parts may be considerably elongated. In Balaninus and in Mecoptera the head is elongated. The Brazilian sphinx moth, *Macrosila cruentis*, has a proboscis that is nine and a half inches long. An East Indian horsefly, *Pangonia longirostris* has a proboscis three or four times the length of the body. A tropical bee, *Euglossa cordata*, has a tongue considerably longer than the rest of the body.

Some spiders also come in large sizes. There is a South American species belonging to the genus *Avicularia*, the body of which is two inches long and measures seven inches with its legs outstreched. This species feeds on birds and spiders and is capable of swift movement. It was described by Linnaeus over 200 years ago. He called it *Avicularia* because at that time its food was supposed to consist entirely of birds.

Some of the neotropical centipedes attain a large size, *Scolopendra gigantea* reaching a length of nearly a foot. This species, incidentally, is exceedingly poisonous and its bite often causes death.

Among the fossil or extinct insects, there are many that exceed the size of living insects. A phasmid, *Titanophasma*, from the Carboniferous is a quarter of a meter long; a species of dragonfly from the Carboniferous of France has a wing expanse of 28 inches, and a Permian species of the genus *Meganeuropsis* from Kansas and Oklahoma has a wing expanse of 30 inches. A May fly, *Platephemera antiqua*, is of gigantic size, with a wing expanse of approximately 5½ inches. This is exceedingly large when we consider that species existing today have a wing expanse of scarcely three quarters of an inch.

On the other hand, some insects are exceedingly small and can scarcely be seen without the aid of a microscope or hand lens. The smallest beetles belong to the family Trichopterygidae. The smallest North American species is *Nanosella fungi*, which occurs in powdery fungi. It is only one one-hundredth of an inch long. Certain Hymenoptera of the genus *Trichogramma* are also small, measuring less than one thirty-second of an inch. Upon a microscope slide, these insects appear to the unaided eye like specks of dust. They are parasitic upon the eggs of other insects and are naturally small in size. Other Hymenoptera which are extremely small are the Proctotrupidae. These are likewise parasitic upon the eggs of insects.

## BIBLIOGRAPHY

I. Abundance.

BEALL, G. (1935). Study of arthropod populations by the method of sweeping. *Ecology* 16 (2). *Literature cited.*

BETZ, B. J. (1932). The population of a nest of the hornet *Vespa maculata*. *Quart. Rev. Biol.* 7 (2).

BODENHEIMER, F. S. (1938). Biological equilibrium in nature. *In*, Problems of animal ecology. Oxford University Press, New York.

BRUES, C. T. (1923). Choice of food and numerical abundance among insects. *Journ. Econ. Ent.* 16 (1).

CARPENTER, J. R. (1939). European insect outbreaks as population fluctuations. *Proc. Okla. Acad. Sci.* 19.

——— (1939). Fluctuations in biotic communities, aspection in a mixed-grass prairie in central Oklahoma. *Amer. Midl. Nat.* 22 (2).

CHAPMAN, ROYAL N. (1931). Population equilibrium and trends. *In,* Animal ecology. McGraw-Hill Book Company, Inc., New York.

———, and BAIRD, L. (1934). The biotic constants of *Tribolium confusum* Duval. *Journ. Exp. Zool.* 68 (2). *Literature cited.*

COWLEY, J. (1938). Quantitative methods of local entomofaunistic survey. *Entom. Lond.* 71. *References.*

CREIGHTON, J. T. (1938). Factors influencing insect abundance. *Journ. Econ. Ent.* 31 (6).

DAVIES, WM. M. (1939). Studies on aphides infesting the potato crop. VII Report on a survey of the aphis population. *Ann. Appl. Biol.* 26 (1). *References.*

GAUSE, G. F. (1932). Ecology of populations. *Quart. Rev. Biol.* 7 (1). *Literature.*

GINSBURG, J. (1939). The measure of population divergence and multiplicity of characters. *Journ. Wash. Acad. Sci.* 29 (8).

GRAY, N. E., and TRELOAR, A. E. (1933). On the enumeration of insect populations by the method of net collection. *Ecology* 14 (4). *Literature cited.*

HAMMOND, E. C. (1938–1939). Biological effects of population density in lower organisms. *Quart. Rev. Biol.* 13, 14 (1). *Literature cited.*

HANDLIRSCH, ANTON (1913). The number of living insects. *In,* Die Handbuch der Entomologie. Ed. by Schröder. *Palaeontology* 7.

JOHNSON, M. S., and MUNGER, F. (1930). Observations on excessive abundance of the midge *Chironomus plumosus* at Lake Pepin. *Ecology* 11 (1). *Literature.*

KRECKER, F. H. (1939). A comparative study of the animal population of certain submerged aquatic plants. *Ecology* 20 (4).

McCLURE, H. E. (1938). Insect aerial populations. *Ann. Ent. Soc. Amer.* 31 (4).

METCALF, Z. P. (1940). How many insects are there in the world? *Ent. News* 41 (8).

MILLER, D. F. (1940). Insect cultures inbred for 200 generations. *Science* 92 (2381).

PARK, T. (1937). Experimental studies on insect populations. *Amer. Nat.* 71 (732).

——— (1939). Analytical population studies in relation to general ecology. *Amer. Midl. Nat.* 21. *References cited.*

PHILLIPS, J. F. V. (1931). Quantitative methods in the study of numbers of terrestrial animals in biotic communities; a review, with suggestions. *Ecology* 12 (4). *Literature cited.*

SMITH, R. C. (1934). Hallucinations of insect infestations causing annoyance to man. *Bull. Brooklyn Ent. Soc.* 29 (5).

——— (1931–1938). Summary of the population of injurious insects in Kansas. *Journ. Kan. Ent. Soc.* 5, 6, 7, 10, and 11. *Trans. Kan. Acad. Sci.* 39.

——— (1938). Annual insect population records, with special reference to the Kansas summary. *Journ. Econ. Ent.* 31 (5).

WEBSTER, F. M. (1903). The price of dairy products as influencing the abundance of some insects. *Journ. N. Y. Ent. Soc.* 11 (1).

WILBUR, D. A., and FRITZ, R. A. (1940). Grasshopper populations (Orthoptera, Acrididae). *Journ. Kan. Ent. Soc.* 13 (3).

**II. Size of Insects.**

BODENHEIMER, F. S. (1932). Ueber Regelmassigkeiten im Wachstum der Insekten. Das Gevichtswachstum. *Arch. Entwmech. Berlin* 126.

BRUES, C. T. (1934). Growth and determination of size in insects. *Psyche* 41 (1).

FOLSOM, J. W. (1934). Number and size of insects. *In,* Entomology with special reference to its ecological aspects. The Blakiston Company, Philadelphia.

METCALF, C. L., and FLINT, W. P. (1939).   The size of insects.   *In,* Destructive and useful insects.   McGraw-Hill Book Company, Inc., New York.

SCUDDER, S. H. (1891).   A decade of monstrous beetles.   *Psyche* **6.**

See also articles on the effect of temperature, moisture, and nutrition on the growth and size of insects.

III. Reproductive Capacity of Insects.

AZEKESSY, W. (1937).   Ueber parthenogenese bei Kolopteren.   *Biol. gen. Vienna* **12.**

EDWARDS, F. W. (1919).   Some parthenogenetic Chironomidae.   *Ann. Mag. Nat. Hist.* **9** (3).

EWING, H. E. (1914).   Pure line inheritance and parthenogenesis.   *Marine Biol. Lab. Woods Hole, Bull.* **26** (1).   *Bibliography.*

———— (1916).   Eighty-seven generations in a parthenogenetic pure line of *Aphis avenae. Marine Biol. Lab. Woods Hole, Bull.* **31** (2).   *Literature cited.*

FELT, E. P. (1911).   *Miastor americana* Felt, an account of Pedogenesis.   *26th Ann. Rept. N. Y. State Entom. Mus. Bull.* **147.**

GATENBY, J. B. (1918).   Polyembryony in parasitic Hymenoptera.   *Quart. Journ. Micros. Sci. Lond.* **63** (2).   New series.

HERRICK, G. W. (1926).   The "ponderable" substance of aphids (Homoptera).   *Ent. News* **37** (7).

HOLMGREN, N. (1903).   Ueber vivipare Insekten.   *Zool. Jahrb. Syst.* **19.**

HOWARD, L. O. (1906).   Polyembryony and the fixation of sex.   *Proc. Ent. Soc. Wash.* **8** (3–4).

HUXLEY, T. H. (1858).   On the agamic reproduction and morphology of aphis.   *Trans. Linn. Soc. Lond.* **22.**

LAWRENCE, P. S. (1941).   The sex ratio, fertility and ancestral longevity.   *Quart. Rev. Biol.* **16** (1).   *Literature.*

LEIBY, R. W. (1929).   Polyembryony in insects.   *Trans. 4th Intern. Congress Ent.* **2.**   *Literature.*

PARKER, G. H. (1922).   Possible Pedogenesis in the blow-fly, *Calliphora erythrocephala* Meigen.   *Psyche* **29** (4) *References.*

PEACOCK, A. D. (1939).   Recent work on experimental parthenogenesis.   *Nat. Lond.* **144** (3660).   *Literature cited.*

PETRUNKEVITCH, A. (1905).   Natural and artificial parthenogenesis.   *Amer. Nat.* **39** (458).   *Bibliography.*

PHILLIPS, E. F. (1903).   A review of parthenogenesis.   *Amer. Philos. Soc. Proc.* **42** (174).

RICHARDSON, C. H. (1925).   The oviposition response of insects.   *U. S. Dept. Bull.* **1324.**

SILVESTRI, F. (1937).   Insect polyembryony and its general biological aspects.   *Bull. Mus. Comp. Zool. Harvard* **71** (4).   *References.*

SNELL, G. D. (1932).   The rôle of male parthenogenesis in the evolution of the social Hymenoptera.   *Amer. Nat.* **66** (705).

WILLIAMS, C. B. (1917).   Some problems of sex ratios and parthenogenesis.   *Journ. Genetics,* **6** (4).

# CHAPTER IV

## BENEFICIAL AND INJURIOUS INSECTS

The study of beneficial and injurious insects is popularly known as economic entomology. The subject is treated in numerous texts and in various books dealing with groups, such as fruit insects, vegetable insects, forest insects, etc. It includes a study of the control, reduction, or rarely the elimination of the injurious species, the propagation of the beneficial, and the appreciation of the beautiful and the harmless kinds.

**Brief History of Economic Entomology.**—The cultivation of plants and the control of insects and diseases go hand in hand. It is true that our forefathers were little concerned with some of these pests. As soon as white man set foot upon the American continent, he started the cultivation of plants. He brought with him many fruits and grains. He also unwittingly introduced many of the injurious insects with which we are confronted today. The codling moth made its appearance in this way. It has developed as one of the most serious pests of apple and will probably continue to be an important problem for years. The early settlers soon noticed these pests and made mention of them in their writings. In 1588, Thomas Harriot, published "A Briefe and True Report of the New Found Land of Virginia," in which he described the habits of many insects and mentioned the silkworm and the honeybee in particular. In 1666, John Esquemeling, a Dutch buccaneer, described the pestiferous insects of this period. Many other reports on injurious insects were sent back to Europe.

Economic entomology was still undeveloped at the beginning of the eighteenth century. This is evident from the ridiculous recommendations suggested for the control of insects. Materials such as salt, lime, alcohol, lampblack, cayenne pepper, ashes, and dung were used. Most of these materials are recognized today as of little or no insecticidal value. Toward the middle of the century, more reasonable and practical methods of control came into use. It is interesting to note that George Washington experimented with the hessian fly in 1760. In 1763, a Frenchman suggested the use of tobacco water against plant lice. Nicotine sulphate, the standard spray for aphids today, was not discovered until a much later date.

From these meager beginnings, the subject of economic entomology was developed. The science of insect control did not become very

47

important until about 1800.   The years 1800 to 1900 mark a period of pioneering among the early enthusiasts.   Many remarkable entomologists came to the front during this period.   The field was broad and there was much to be done.   We shall long remember many of the early economic workers, such as Asa Fitch, the first New York State entomologist, B. D. Walsh, state entomologist of Illinois, Townend Glover,

Fig. 30.—Spraying by means of modern orchard equipment.   The pump mounted in front of the spray tank is operated by a power take-off from the tractor.   (*After John Bean Manufacturing Company.*)

the first United States entomologist, J. A. Lintner of New York State, A. S. Packard of the United States Entomological Commission, C. V. Riley, and many others.

In 1807, an impetus was given to spraying through the discovery of the value of copper salts as a control for certain fungus diseases.   The discovery of sulphur as a fungicide for peach mildew in 1821 also advanced the science.   Lime sulphur was developed in 1886 as a stock dip.   Later it was found to be a valuable contact spray against scale insects.   Paris green, one of the first arsenicals, was used against the Colorado potato beetle in 1860. Arsenate of lead was developed in 1892 as a spray against the gypsy moth. It still remains a standard spray against chewing insects, more than 3,000,000 pounds being used annually.

The years following 1862 mark considerable growth in agriculture in general, and entomology developed in proportion. The economic problems associated with the large agricultural enterprises of the day, such as the fruit industry, cotton growing, and tobacco production, woke the country to the need of better cultural and insect pest control methods. In 1862, the first Morrill Act was passed by Congress providing for the

FIG. 31.—Dusting cotton by means of an airplane. The cloud of dust, discharged at a height of about 10 feet above the ground, covers a swath about 150 feet in width. (*After U. S. Department of Agriculture.*)

foundation and maintenance of state colleges, "where the leading subject shall be, without excluding other scientific and classic studies and including military tactics, to teach such branches of learning as are related to agriculture and mechanic arts." In 1890, the second Morrill Act was passed giving $25,000 annually to each land-grant state college throughout the country. The Hatch Experiment Station Act, passed in 1887, supplemented Federal aid to state experiment stations. In 1925, the Purnell Act provided further aid and, in 1935, the Bankhead-Jones Act again voted additional funds for agricultural research.

**Centers of Entomological Investigation.**—As we have noted, various Federal funds were voted by Congress to carry on experimental work at the state colleges. The first state experiment station was organized at

New Haven, Conn. Shortly afterward, stations were set up in other states. Those at Geneva, N. Y., New Brunswick, N. J., Wooster, Ohio, State College, Pa., and Sacramento, Calif., have made outstanding contributions to entomological research. Excellent work has also been produced by some of the privately endowed colleges of the country. Certain types of entomological work are also conducted by the state departments. Private organizations, such as the Boyce Thompson Institute at Yonkers, N. Y., and the Rockefeller Foundation, New York, have contributed much entomological research. Many commercial firms likewise produce scholarly work in entomology. The Crop Protection

Fig. 32.—Mechanical injury produced on an apple by a circular metal tool when the fruit was the size of a marble. The growth of this injury is similar to that of injuries produced by many chewing insects.

Institute has the financial backing of outstanding commercial companies, and research is conducted by trained scientists from various colleges or experiment stations. Finally, Washington, D. C., is an important center of entomological activity, which correlates and contributes to the work of the various state experiment stations.

**Control of Injurious Insects.**—It is unfortunate that we must consider the harmful insects first. There is no doubt that there are many injurious species that annually cause serious losses; nevertheless, the number is comparatively small. Scarcely 10,000 of the 640,000 species of insects are injurious and of these not more than 500 can be considered as major pests in this country. Many plant- or animal-feeding species of insects may, under the proper conditions, become pests. When Dr. A. L. Quaintance reported the oriental fruit moth, *Grapholitha molesta*, for the first time in this country in 1916, he intimated that this species might become a serious pest. As a matter of fact, we recognize it as one of the most serious pests of peaches today.

We must face the fact that some insects are extremely injurious and that their control is necessary and often difficult and expensive. Economic entomology resolves itself into the identification of the species causing the injury, a study of the habits of the species in order that some vulnerable point in the life history may be detected, the selection of a means of control, and the dissemination of this information to fruit growers and farmers.

**Extent of Insect Injury.**—Little that man grows, manufactures, or borrows from nature is free from the ravages of insects. They invade

FIG. 33.—Japanese beetle injury to foliage and fruit of peach.

the animal, the plant, and the mineral kingdoms. Approximately one-half of them feed upon plants. Not a small number attack man or other animals. Some can penetrate exceedingly hard substances. The short-circuit beetle, Bostrichidae, can bore through lead cables. A Central American curculionid bores into the tagua, a nut which is as hard as ivory. Insects attack leaves, stems, buds, flowers, seeds, fruits, bark, wood, and roots; in fact, few parts of a plant are free from their feeding. Fruits and vegetables, grains and stored products, flowers and shade trees suffer.   About 1,000 species attack oaks, 400 species are injurious to apple, 200 species affect clover, and 200 species attack corn.   In the home, cereals, dried fruits, meats, clothing, stuffed furniture, woodwork, and many other products may be attacked.   Many products derived from plant and animal sources, such as wool, feathers, cigars, and tool handles, are also attacked by insects.

The damage by insects in 1938 amounted to approximately $1,601,-527,000.   This figure is calculated upon the actual loss by the insects plus the cost of control.   It has been said that it costs the farmer more to feed his insects than it does to educate his children.   The losses by insects can best be illustrated by some comparisons.   The yearly loss by insects is equal to two times the annual loss by fire, or about two times the capital invested in manufactured farm machinery, or about three times the value of all fruit orchards, vineyards, and small fruit farms in the country.   It is believed that about 10 per cent of every crop is lost through insects.   Often this loss is unobserved.   The growth of crops such as wheat or corn may compensate for the loss so that the farmer is not aware of the injury.   Cicada nymphs may feed for 17 years upon the roots of the same tree; still there is no sign of injury because the feeding is prolonged over a considerable period and new rootlets grow to replace those killed.   However, when the adults emerge, they produce serious injury by their egg-laying habits.

LOSSES RESULTING FROM SOME OUTSTANDING INSECTS*

| Insect | Annual insect loss | Cost of control | Total loss |
|---|---|---|---|
| Codling moth.................... | $ 13,500,000 | $17,500,000 | $ 31,000,000 |
| Plum curculio................... | 7,000,000 | 3,000,000 | 10,000,000 |
| Peach tree borer................ | .......... | 1,000,000 | 1,000,000 |
| Japanese beetle................. | 2,739,000 | 751,000 | 3,490,000 |
| Colorado potato beetle.......... | 16,500,000 | 2,000,000 | 18,500,000 |
| Mexican bean beetle............. | 1,276,000 | 730,000 | 2,006,000 |
| Stored grain insects............ | 300,000,000 | 350,000 | 300,350,000 |
| Forest insects.................. | 100,000,000 | .......... | 100,000,000 |
| Shade tree insects.............. | 62,000,000 | 25,000,000 | 87,000,000 |
| Greenhouse insects.............. | 33,000,000 | 9,362,000 | 42,362,000 |
| Termites....................... | 40,000,000 | .......... | 40,000,000 |
| Grasshoppers.................... | 25,243,000 | 458,000 | 25,701,000 |
| Clothes moths................... | 20,800,000 | 1,914,000 | 22,714,000 |

* Figures for 1938.

**Rapid Spread of Insects.**—The seriousness of insect damage often depends upon the rapidity with which pests spread into new areas. Insects often increase tremendously because their natural enemies are absent.   Extended distribution usually follows heavy infestations.   The majority of our pests have been introduced through eastern ports such as Boston, New York, and Philadelphia.   The imported cabbage worm, *Ascia rapae*, for example, was introduced into Quebec in 1860.   It was reported from New York and New Jersey in 1875, from Iowa in 1878,

POINTS AT WHICH JAPANESE BEETLES
HAVE BEEN RECOVERED

▨▨ Area continuously infested by natural spread

▦ Localized colonies or points of minor occurrence

FIG. 34.—Distribution of the Japanese beetle, showing the heaviest infestation in New Jersey, Pennsylvania, Maryland, and Delaware in the vicinity of the original infestation at Riverton, N. J.   (*Courtesy of U. S. Department of Agriculture.*)

from Minnesota in 1880, and from Colorado in 1886.   This insect crossed the continent in 26 years.

The gypsy moth and the Japanese beetle likewise entered through eastern ports of the United States.   The gypsy moth was accidentally introduced at Medford, Mass., by a French entomologist conducting experiments on the silkworm.   It was immediately recognized as a pest but the public did not take the matter seriously.   For 20 years this moth increased in numbers, unmolested.   In 1905, a serious outbreak in

Massachusetts attracted sufficient attention to warrant control measures. It had already spread over such a wide area that elimination was impossible. By means of scouting, quarantine, and control measures, the pest was brought within reasonable control but it continued to spread until today it is found all over the New England states as well as in small adjoining portions of New York, New Jersey, Pennsylvania, and southeastern Canada.

The Japanese beetle, *Popillia japonica*, has also had a remarkable development in this country. It was introduced at Riverton, N. J., in 1916, upon ornamental trees from Japan. With the expenditure of a comparatively small sum, the original infestation could have been cleaned up. However, the insect spread rapidly into Pennsylvania, New York, the New England states, Virginia, West Virginia, Ohio, North and South Carolina. The infestation still remains heaviest in New Jersey in the vicinity of the original introduction. More than 260 kinds of trees, shrubs, and plants are affected and this pest will probably continue to spread and cause destruction until its parasites are well established.

Many of the North American insects evidently came from South and Central America through Mexico. Among the economic species that came by this route we may mention the harlequin bug, *Murgantia histrionica*, the sugarcane borer, *Diatraea saccharalis*, the potato tuber worm, *Gnorimoschema operculella*, the cotton boll weevil, *Anthonomus grandis*, and others. As a rule, these insects are subtropical in nature and are unable to breed beyond a certain latitude in the United States. The harlequin bug, for example, seldom breeds in Pennsylavnia. It has been taken as far north as Elmira, N. Y., but apparently does not hibernate at this latitude. The cotton boll weevil was introduced into the United States about 1892. It was reported from Louisiana in 1903, from Mississippi in 1910, from Georgia in 1915, from South Carolina in 1917, and from North Carolina in 1919. Its distribution is limited by its food plant and by temperature. The insect is now occupying the total area satisfactory for its development.

Other pests have been introduced from the west. The San José scale, *Aspidiotus perniciosus*, is a notable example. It is a native of the Hawaiian Islands, Japan, China, Australia, and Chile. No one knows definitely from which of these countries it came. It was introduced into the San José Valley, Calif., in 1870, reported from New Jersey in 1887 and from Virginia in 1893, thus making its journey across the country in approximately 23 years.

The Colorado potato beetle, *Leptinotarsa decemlineata*, has a unique history. It is a native of the Rocky Mountains, feeding upon wild species of Solanaceae, chiefly buffalo bur. As the early settlers migrated westward, they planted potatoes, which eventually formed steppingstones, and the beetles took advantage of these plantings and gradually moved eastward. It was reported from Nebraska in 1859, from Missouri in 1868, from Ohio in 1869, and from New York in 1872. It crossed the United States from west to east in approximately 15 years.

A few serious pests have been introduced from Europe. The European corn borer, *Pyrausta nubilalis*, the hessian fly, *Phytophaga destructor*, the codling moth, *Carpocapsa pomonella*, and the imported cabbage worm, *Ascia rapae*, are the most injurious species that we have from the Old World.

Approximately 100 species of injurious insects have been introduced into the United States up to 1920, but the number is gradually being reduced by careful inspection of plant material brought into this country. Among the more recent introductions are the satin moth, *Stilpnotia salicis*, an enemy of poplars, willows, and oaks, introduced into Massachusetts in 1920; the oriental or Asiatic beetle, *Anomala orientalis*, introduced into Connecticut in 1920; the Asiatic garden beetle, *Autoserica castanea*, introduced into New Jersey in 1921; the Mediterranean fruitfly, *Ceratitis capitata*, introduced into Florida in 1927, and the white fringed beetle, *Pantomorus (Naupactis) leucoloma*, reported from Florida in 1936 as a serious pest on cotton, corn, peanut, sweet potato, and cow pea.

Fig. 35.—Injury caused by the silver leaf mite, *Phyllocoptes cornutus*, to peach foliage: (left) infested; (right) normal leaf.

**Insect Transmitters of Human Disease.**—Many insects act as the primary or intermediate hosts or carriers of human diseases. A wide range of organisms, such as protozoa, nematodes, and viruses, may be transmitted by them. Malaria is one of the most serious insect-borne diseases of man. It is a protozoan (*Plasmodium*) which is transmitted by several species of *Anopheles* mosquitoes. There are three types of malaria resulting from different species of parasites. The *Plasmodium* undergoes a complicated sexual cycle in the body of the mosquito and is injected into man with the saliva at the time the mosquito bites. In India, the mean annual death rate from malaria in 1910 was five per thousand. In Alabama, in 1911, there were 70,000 cases and 770 deaths.

The plague, due to *Bacillus pestis*, infects man as well as rats and other rodents. It is transmitted by means of fleas. The flea shoots blood from the anus or regurgitates drops of blood which are enriched with the bacillus. It is through this means that man contracts the disease. India is the hotbed of this disease. In 1904, there were 1,040,429 deaths; in 1914, there were 198,875 deaths. In areas where rats have been eradicated, there is a corresponding decrease in the plague.

Typhoid, another bacterial disease, may be contracted in many different ways. Houseflies are considered an important agent in distributing this disease because they commonly frequent human excrement to feed and oviposit, then visit man's home, where they crawl over food and spread the disease.

A number of tapeworms are known to pass a part of their life history in the bodies of insects. A species attacking dogs and cats is frequently carried by the flea. The eggs of the tapeworm are discharged in the feces of the animal and are taken up by the flea. The cat or dog in turn, in trying to dislodge the fleas, takes them into its mouth, and eventually the tapeworm eggs reach the alimentary canals of these animals. Children may contract tapeworms by close contact with cats and dogs, or by feeding from dishes from which these animals have eaten.

For a long time it has been suspected that insects are associated with the transmission of infantile paralysis, acute anterior poliomyelitis. Howard and Clark (1912) demonstrated that the housefly can carry the virus for several days on the surface of the body. The bedbug was found to take the virus from infected monkeys and maintain it within the intestinal tract in a living condition for a period of seven days. However, the experimenters doubted that the bedbug was a natural carrier of the disease. Although the stablefly, *Stomoxys calcitrans*, has been suspected as a carrier of infantile paralysis, early experiments do not prove the point conclusively. Sabin and Ward (1941) state that the virus has been isolated from numerous crushed flies, especially those of the family Metopidae. Positive reactions have been secured by injecting filtrates into monkeys. Flies were collected in the vicinity of infantile epidemics at a camp in Connecticut and in several urban areas in Alabama. Ether extracts were injected intraperitoneally and unetherized extracts were given both intranasally and by mouth. The monkeys so treated succumbed to the effects of poliomyelitis in from six to fifteen days. Unfortunately the workers did not give specific identification of the flies concerned. The flies appear to carry the disease in a manner similar to that in which the housefly carries typhoid.

Many other diseases are transmitted by insects. In the Tropics, sleeping sickness is transmitted by the tsetse fly; elephantiasis is transmitted by mosquitoes, etc.

**Poisonous Insects.**—There is a large class of insects that sting or irritate the skin of man. The sting of the honeybee is generally transitory and not serious. It may cause severe swelling and in rare cases, owing to other complications, cause death. The sting is a sharp lancelike organ located at the posterior end of the body. The sensation is caused by an acid and an alkali which are forced into the wound made by the sting. When the bee attacks, the sting, including the tip of the abdomen, is torn loose. The muscles attached to the sting continue to contract and force the sting and the poisons into the victim's flesh.

Many other Hymenoptera sting. The bald-faced hornet and the yellow jacket are perhaps the most ferocious. A colony of several thousand wasps is an organization to be avoided. In the Tropics, one meets even more ferocious wasps. The colonies are often larger and the wasps more aggressive. Some of the large wasps are said to attack without provocation and to follow man for miles through the jungles. *Polistes*, the North American maker of the small, suspended paper nest, gives a ferocious sting but is not easily provoked. The mud daubers may likewise sting but generally go about their business and seldom attack man. Stings

Fig. 36.—A nettling insect, the saddleback, *Sibine stimulea*. When the barbed hairs break, a poison is released. (*Champlain and Kirk.*)

are intended to paralyze their prey. The same may be said of all wasps that provision their nests with insects. Ants may inflict serious pain by means of their jaws. Other ants possess stings that inject formic acid directly into the victim. The fire ant, *Solenopsis geminata*, occurs only in the Tropics and Southern United States. It is extremely pugnacious and causes serious irritation.

Many of the Heteroptera are poisonous. The bedbug is best known of these. A number of species of the family Reduviidae, popularly known as assassin bugs, are capable of inflicting very painful wounds. In the Tropics one of these species carries chagas fever. The back swimmers, Notonectidae, and the giant water bugs, Belostomidae, also cause severe pain.

Certain flies are exceedingly pestiferous. The bite of the mosquito is rather well known. The horseflies, Tabanidae, and the stableflies,

Muscidae, have sharp stiff mouth parts which often inflict severe pain. *Pangonia longirostris* of India has a proboscis three or four times the length of the body and can pierce the skin of the human body even through clothing of considerable thickness. Black flies and a few Chironomidae also bite severely.

**Nettling Insects.**—There are a considerable number of insects that possess poisonous hairs. These are found among the Lepidoptera.

Fig. 37.—The io moth, male, female, and larva. The larva bears barbed hairs which are poisonous. (*Champlain and Kirk.*)

The larvae of the saddleback, *Sibine stimulea*, the io moth, *Automeris io*, the flannel moth, *Lagoa crispata*, and the brown-tail moth, *Nygmia phaeorrhoea*, possess such hairs. These hairs are shorter than the regular hairs of the larvae and are barbed. They connect with underlying hypodermal poison glands. Nettling hairs are especially dangerous when they lodge on the cornea of the eye, where they cause considerable irritation.

**Beneficial Insects.**—Beneficial insects may be divided into five general groups: those yielding useful products, those used in the medical profession, those used as natural control of injurious species, those that pollinate flowers, and those used in art.

Many useful products are obtained from insects. From the bees we obtain honey and wax. Virgil, Aristotle, and Pliny were familiar with the habits of the bees and mention the value of their honey and wax. When we consider that 20 pounds of honey are necessary for the bees to make one pound of wax, we readily see that the wax is really more valuable than the honey. Incidentally wax is obtained from the plant and the mineral kingdoms also. Carnauba is obtained from the wax palm of Brazil, and Japanese wax is obtained from the Japanese ash. Paraffin, of course, is the most abundant wax from the mineral kingdom.

Fig. 38.—A harmless but ferocious-looking caterpillar and its pupa, the hickory horned devil, *Citheronia regalis*. (*Champlain and Kirk.*)

Many insects secrete a waxy material which man puts to various uses. China wax is secreted by a scale insect, *Ericerus pele*. The lac insect *Laccifer* (*Tachardia*) *lacca*, of India produces a secretion from which stick lac of commerce is manufactured. Nearly 40,000,000 pounds are gathered annually. The San José scale, woolly aphis, and other insects produce considerable wax but not in sufficient quantities to be of commercial value. A scale insect, *Coccus mannifera*, which occurs on a tree on Mt. Sinai, produces a flaky secretion which is edible and is said to have been the manna of the Israelites.

Cochineal is a beautiful red pigment obtained from the dried pulverized bodies of a scale insect, *Coccus cacti*, which lives upon the prickly pear and other cacti. They are cultivated in Hondurus, Mexico, Peru, Algeria, Spain, and the Canary Islands. It takes 70,000 insects to make a pound of the dye which sells as high as $3 a pound and is used largely as a cosmetic, in rouge, and in coloring fancy cakes, beverages, and medicines.

Silk is one of the most important products from insects. It is spun by a large number of caterpillars, by the larvae of many other insects, and by spiders. The insect uses silk for many purposes but chiefly to make cocoons. None except that of the silkworm, *Bombyx mori*, has been found of commercial value. Sericulture, the culture of silkworms, has an ancient and interesting history. About 1800 B.C., Hoang-ti encouraged his wife, Si-ling-Chi, to study the silkworm. She discovered methods of rearing the insects on mulberry leaves and of winding the

Fig. 39.—*Icerya brasiliensis*, an exotic scale insect that produces an abundance of wax. Many scale insects and aphids produce wax to a lesser extent. (*Pennsylvania Department of Agriculture.*)

silk from the cocoons. The secrets of silk culture were jealously guarded by the Chinese, but finally a few silkworms were smuggled into Constantinople by some monks. The early colonists brought silkworms to America and endeavored to produce silk here but their attempts were unsuccessful. The larvae naturally feed upon mulberry leaves. Many old mulberry trees still persist throughout the eastern part of our country as evidence of early attempts to propagate this insect. Today silk has given way largely to the cheaper rayon; more recent synthetic products promise to produce cheaper and stronger fibers.

Attempts have been made to utilize the silk of large spiders belonging to the genus *Nephila* which are common in the Tropics and the warmer portions of subtropical regions. *Nephila clavipes* builds large webs frequently two or three feet in diameter. The supporting lines of these webs are strong and tough but the silk of this or related species has never been successfully utilized. A set of bed hangings was made from the silk of these spiders and shown at the Paris Exposition some years ago. Fabrics from spiders' silk still remain a curiosity. It is used, however, as reticles in optical instruments.

Tannin is obtained from insect galls and is used in the process of tanning hides and in the manufacture of permanent durable inks. The natives of certain countries use tannin in tattooing. Turkey red is another product obtained from certain insect galls known as "mad apple." This dye comes from Asia Minor.

**Insects Used in the Medical Profession.**—Many kinds of insects are reputed to possess medical properties. Cantharidin is extracted from the bodies of blister beetles, Meloidae. There are many different species that yield this material but a species from India, *Mylabris cichorii*, supplies fully twice as much as the majority of the species. Cantharidin is a blistering agent having various uses in the medical profession. It is employed chiefly in the internal treatment of certain diseases of the urinogenital system. It has been used in hair washes but has been pronounced unsafe and injurious. Another preparation, known as "specific medicine apis," is extracted by means of alcohol from the bodies of bees. It is used in treating "hives," diphtheria, scarlet fever, dropsy, and erysipelas. Dipterous maggots of the genus *Wohlfahrtia* have been used to clean up decayed tissue and bacteria present in wounds. Recently, however, substitutes have been discovered in the form of synthetic products derived from urea which are used in the place of maggots and are considered by some to be much better and safer.

**Parasitic and Predacious Insects.**—The insect parasites* and predators are exceedingly valuable in controlling injurious species. Insects may have natural enemies that tend to reduce their numbers and hold them in check. When these forms are absent or have been reduced, the insects may increase in numbers and develop serious outbreaks. The science of breeding and releasing parasitic and predacious species has received considerable attention during the past few years. Hundreds of parasites of our outstanding pests, such as the Japanese beetle, the oriental fruit moth, and the European corn borer, have been introduced from Europe and Asia. Some have become established in this country and are quite efficient in holding these pests under control. Many years

---

* See Chap. XIV, Scavengers, Predators, and Parasites.

are often necessary to develop parasite populations sufficient to hold injurious insects under control.

Fifteen of the twenty-six orders of insects contain predacious or parasitic insects. It is difficult to generalize, especially in such an enormous field, but it is safe to say that the Hymenoptera and the Diptera contain the largest number of parasitic species. We find the principal parasites among the Braconidae and the Ichneumonidae, although many other families of the Hymenoptera are parasitic. *Apanteles* is a common genus and various species attack caterpillars. The cocoons are formed in great numbers on the body of the caterpillar or adjacent to the caterpillar that has succumbed to the parasite. The common tobacco worm,

Fig. 40.—A sphinx caterpillar bearing the cocoons of a braconid parasite.    (*Champlain and Kirk.*)

*Protoparce quinquemaculata*, is frequently attacked by one of these species. In the Diptera, the Tachinidae are the predominating forms.

The Hemiptera, Coleoptera, Neuroptera, and Diptera contain the largest number of predacious forms. Common predacious species are the larvae of the syrphus flies, aphis lions, and Hemerobiidae, and the larvae and adults of the ladybird beetles (see Chap. XIV).

**Insect Pollination.**—The value of insects in pollenizing flowers is unquestionable. Cross pollination is necessary for the fertility and vigor of plants. In many cases the transportation of the pollen is a simple matter, since it merely adheres to the hairs of the insect and is taken to the next flower. In other cases, the flower development favors insect visitation, or the insect may be developed in favor of flower visitation. These are subjects that will be discussed in the chapter on the interrelation of insects and flowers. We shall also discover that *Blastophaga* is essential for the growth of the Smyrna fig.

**Insects as Food.**—Insects form an abundant supply of food for birds, lizards, frogs, snakes, fish, and other animals. The food of bluebirds, meadow larks, chickadees, and house wrens consists largely of insects or their near relatives. Moles feed on insects and other small animals. Many tropical animals, such as armadillos and anteaters, feed almost wholly on small insects. The smaller species of frogs are largely insectivorous. Fish consume a great quantity of insects, especially caddis flies, May flies, gnats, and mosquitoes. It is evident, therefore, that indirectly insects are of great benefit to man.

Even man consumes many insects, generally only by accident along with apples, spinach, asparagus, or other food. Primitive races, however, consume insects in great quantities and often consider them a great luxury. The products of insects, for example, honey, is quite generally accepted as a delicacy. The secretion from the scale insect, *Coccus mannifera*, it is reputed, was eaten by the natives of ancient Jerusalem.

The selection of insects as food by the natives of many countries undoubtedly has a logical basis. Insects are frequently abundant and easily gathered. The primitive idea of economy might therefore suggest their use. When insects emerge in great numbers, such as cicadas from the ground or May flies from the water, they may be gathered with little effort. The relatively high salt content of the blood of insects probably makes them more tempting than other morsels of food when one has overcome the prejudice against such food. In portions of the world where salt is scarce, insects may supply this necessary food requirement. The scarcity of food at times might lead certain populations to select insect food. In the Eastern United States, insects were not frequently eaten, chiefly because the rainfall was generally sufficient to produce good crops and insect food was not essential. In the West, where famines were more common, the Indians resorted to any kind of food that was available. Insects were thus regarded by some tribes as a staple.

Since the list of insects eaten by primitive people is long, we can select only a few for discussion. The natives of the Amazon Valley eat the sauba ants, *Atta cephalotes*. To strangers it is a peculiar sight to see a native make his breakfast of the great winged ants with alternate handfuls of cassava meal. Termites are a favorite food of the peoples of the Tropics. The large queens, sometimes measuring three inches, are sought in particular. In Mexico, the eggs of *Corixa femorata* are considered a delicate dish. They are commonly fried in fat. The grub of the goliath beetle of Africa, one of the largest of the Coleoptera, is an especially fine food morsel. It lives upon the roots of the banana tree and measures five and a half inches long when fully grown. Young fat larvae of various wood-boring insects, bees, and wasps are especially tempting. In the Western United States, *Prionus californicus* was a favorite because of its size. Another insect often gathered and eaten in the West is the puparia of a small fly, *Ephydra hians.* This insect was known to the Indians as Koo-tsabe. At certain seasons of the year the puparia of these flies are blown in great windrows along the shores of brackish lakes, where they can be gathered easily.

Insects are sometimes eaten raw but more often they are prepared in various ways. The Greeks ground locusts in stone mortars and made flour of them. The American Indians dried or smoked the larger caterpillars and preserved them for later use. The Koo-tsabe were separated from their shells; that is, the pupae were removed from the puparia.

**Insects in Music.**—Many insects produce noises in various ways: by rubbing their wings together, by rubbing the wings against the legs, by vibrating certain membranes, by rapidly vibrating their wings, and in other ways. It is questionable whether these sounds can be called music. The entomologist avoids criticism by calling such noises sonification,* stridulation, or phonation. What is music to one may be an unpleasant sound to another. Virgil, like many of us, disliked the harsh calls of the cicada for he remarks, "They burst the very shrubs with their noise." On the other hand, Shelley obtained great joy from the songs of the grasshoppers for he writes, "Merrily one joyous thing in a world of sorrowing."

That the songs of insects have inspired many is unquestionable. The chirps of crickets or the songs of the katydids are a pleasure to some, or at least, bring pleasant recollections of bygone days. Dickens was surely inspired by the cricket on the hearth. Nicolas Rimsky-Korsakov, in his delightful "Flight of the Bumblebee," reproduces the familiar hum of the bees interposed with the high-pitched song of the queen. Joseph Strauss caught an inspiration revealed in "Dragonfly," which typifies the zigzag flight and swift movements of the dragonfly. Anatol Liadov, a pupil of Korsakov, adds to our repertoire another, "Dance of the Mosquito," in which the violin represents the singing of the mosquito and is picked up in a sort of echo by the flute and the piccolo. Edvard Grieg has taken a motif from the butterfly in "Papillon." Few but a poet or a musician could sense the light, even flight of the butterfly. "The Beetles' Dance" by Gustav Holtz, "Minuet of a Fly" by Alphonse Cibulka, and "Song of the Flea" by Moritz Moszkowski, probably have little reference to the habits of these insects.

In Japan, cicadas and crickets are placed in small cages, like birds, and their songs are considered agreeable. In Tokyo, it is said, there are over 50 markets dealing with these commodities. Fancy goldfish and musical insects occupy a large place upon their shelves. So important a place do these insects hold that a week of celebration is held each year, "The festival of the singing insects." At this time, all caged insects

* See Chap. X, Sonification.

must be released so that they can have a few weeks of freedom before the first frost comes.

**Insects in Art.**—One first thinks of the beautiful iridescent blue *Morphos* that are so frequently used in art. These are known in Central America as royal blues. Jewelry and pictures made from pieces of the wings of these butterflies are especially durable because the colors do not fade. Some of the beetles are equally beautiful but frequently lose their brilliancy when preserved. The Indians of America used portions of insects much like feathers in their crafts. The Jivaros of Ecuador make earrings of the shiny green elytra of buprestid beetles.

The insect motif has likewise been used by many races. In spite of

Fig. 41.—Bee on a silver Greek coin of Ephesus milled about 400 B.C. (*Museum of Fine Arts, Boston.*)

their comparatively small size, we find insects mentioned frequently in the writings of poets and philosophers, and the folklore of nearly every country refer to them. In the Bible, for example, mention is made of

Fig. 42.—Grasshopper motif on an ancient piece of Mimbres pottery from New Mexico. (*Restored, after Cosgrove*).

bees, beetles, fleas, flies, gnats, grasshoppers, hornets, lice, locusts, moths, and palmerworms.

The scarab is the most popular insect in Egypt. The Egyptians were familiar with the habits of this beetle and selected the scarab as a symbol of their sun God, Khepera. It represented the soul emerging from the body and was always figured on the mummy case.

In Greece, the insect took a prominent place on coins, especially those of Ephesus. A silver tetradrachm, in use between 400 and 336 B.C., carries a bas-relief of a bee with the letters $E \phi$ (of the Ephesians). On the reverse are the foreparts of a stag and other decorations. The bee and the stag are symbols of Diana. The bee is found on many other Greek coins.

In Japanese art, one finds the insect frequently upon inros, netsukes, and carvings of ivory, jade, rock crystal, wood, and other materials. One of the finest examples is a highly polished netsuke in the form of a chestnut with a curculio larva emerging. Cicadas, dragonflies, and crickets are often featured. Exquisite reproductions of insects are often found upon Japanese sword guards and present the most refined and elaborate technique found anywhere in the realm of art. They are worked in tin, copper, iron, zinc, gold, silver, and numerous remarkable alloys.

Among the American Indians the insect was used frequently to adorn pottery, baskets, and other objects. The insect is often highly conventionalized. The Sioux, for example, use a symbol similar to the patriarchal cross to represent a dragonfly with its wings outstretched. The Hopi and the Arapaho use the triangle to represent a butterfly.

There is scarcely a branch of art that does not exhibit the insect in some form. We find them as marks on china and pewter, as printer's marks, as watermarks, and even on postage stamps.

## BIBLIOGRAPHY

I. History of Economic Entomology.

ESSIG, E. O. (1931). A history of applied entomology. The Macmillan Company, New York.

HOWARD, L. O. (1930). A history of applied entomology. *Smiths. Misc. Coll.* **84.**

—— (1937). Fighting the insects. The Macmillan Company, New York.

IMMS, A. D. (1938). Fifty years of entomology. *Ent. Rec. Lond.* **50.**

OSBORN, HERBERT (1937). Fragments of entomological history. The author, Columbus, Ohio.

WEISS, H. B. (1936). The pioneer century of American entomology. H. B. Weiss, New Brunswick.

II. General References to Economic Literature.

HERRICK, G. W. (1925). Manual of injurious insects. Henry Holt and Company, Inc., New York.

METCALF, C. L. and FLINT, W. P. (1951). Destructive and useful insects. 3d ed. McGraw-Hill Book Company, Inc., New York.

PEAIRS, L. M. and DAVIDSON, R. H. (1956). Insect pests of farm, garden and orchard. 5th ed. John Wiley & Sons, Inc., New York.

WEISS, H. B. (1940). Money losses due to destructive insects. *Journ. N. Y. Ent. Soc.* **48.**

III. References to Insecticides and Their Application.

BOUCART, E. (1913). Insecticides, fungicides and weed killers. Scott, Greenwood and Sons, London.

CUNNINGHAM, G. H. (1935). Plant protection by the aid of therapeutants. John McIndoe, New Zealand.

FREAR, D. E. (1942). Chemistry of insecticides and fungicides. D. Van Nostrand Company, Inc., New York.

LODEMAN, E. G. (1916). The spraying of plants. The Macmillan Company, New York.

MARTIN, H. (1940). The scientific principles of plant protection. 3d. ed. Edward Arnold & Co., London.

MASON, A. F. (1929). Spraying, dusting and fumigating plants. The Macmillan Company, New York.

SHEPARD, H. H. (1939). The chemistry and toxicology of insecticides. Burgess Publishing Co, Minneapolis.

WARDLE, R. A. (1929). The problems of applied entomology. Manchester Univ. Press, England. McGraw-Hill Book Company, Inc., New York.

IV. References to Forest and Shade Tree Insects.

DOANE, R. W., et al. (1937). Forest insects. McGraw-Hill Book Company, Inc., New York.

FELT, E.. P. (1905–1906). Insects affecting park and woodland trees. *N. Y. State Mus. Memoir* **8.**

———— (1. 24). Manual of tree and shrub insects. The Macmillan Company, New York.

———— and RANKIN, W. H. (1932). Insects and diseases of ornamental trees and shrubs The Macmillan Company, New York.

GRAHAM, SAMUEL A. (1939). Principles of forest entomology. 2d ed. McGraw-Hill Book Company, Inc., New York.

HERRICK, G. W. (1935). Insects of shade trees. Comstock Publishing Company, Inc., Ithaca, New York.

KOFOID, C. A. (1934). Termites and their control. University of California Press Berkeley.

SNYDER, T. E. (1935). Our enemy the termite. Comstock Publishing Company, Inc., Ithaca, New York.

V. References to Fruit and Vegetable Insects.

CROSBY, C. R., and LEONARD, M. D. (1918). Manual of vegetable garden insects. The Macmillan Company, New York.

QUALE, H. J. (1938). Insects of citrus and other subtropical fruits. Comstock Publishing Company, Inc., Ithaca, New York.

SLINGERLAND, M. V., and CROSBY, C. R. (1914). Manual of fruit insects. The Macmillan Company, New York.

VI. References to Medical Entomology.

HERMES, W. B. (1950). Medical entomology. 4th ed. Macmillan Company, New York.

MATHESON, R. (1950). Medical entomology. 2d ed. Charles C. Thomas, Springfield, Ill.

PATTON, W. S., and CRAGG, F. W. (1913). A text book of medical entomology. Christian Lit. Soc. for India.

PIERCE, W. D. (1921). Sanitary entomology. Richard G. Badger, Boston.

RILEY, WILLIAM A., and JOHANNSEN, OSKAR A. (1915). Handbook of medical entomology. Comstock Publishing Company, Inc., Ithaca, New York.

———— and ———— (1932). Medical entomology. McGraw-Hill Book Company, Inc., New York.

SABIN, A. B., and WARD, R. (1941).   Flies as carriers of Poliomyelitis virus in urban districts.   *Science* **94** (2451).

**VII. References to Meadow and Pasture Insects.**

OSBORN, H. (1939).   Meadow and pasture insects.   Educator's Press, Columbus, Ohio.

**VIII. References to Household Insects.**

HARTNACK, HUGO (1939).   202 common household pests of North America.   Hart-nack Pub. Co., Chicago.

HERRICK, G. W. (1914).   Insects injurious to the household and annoying to man. The Macmillan Company, New York.

**IX. References to the Biological Control of Insects.**

CLAUSEN, CURTIS P. (1940).   Entomophagous insects.   McGraw-Hill Book Company, Inc., New York.   *References.*

RILEY, C. V. (1893).   Parasitism in insects.   *Proc. Ent. Soc. Wash.* **2** (4).

SWEETMAN, H. L. (1936).   The biological control of insects.   Comstock Publishing Company, Inc., Ithaca, New York.

**X. References to Insects Used in Medicine.**

BECK, B. F. (1935).   Bee venom, its nature and effects on arthric and rheumatoid conditions.   D. Appleton-Century Company, Inc., New York.

HAMILTON, JOHN (1893).   Medico-entomology.   *Ent. News* **4**.

HOLLANDER, J. L. (1941).   Bee venom in the treatment of chronic arthritis.   *Amer. Journ. Med. Sci.* **20** (6).   *References.*

HORIKAWA, Y. (1929).   Animals, including numerous insects, used as medicine in Formosa.   (Japanese) *Dept. Hyg. Res. Inst. Formosa. Cont.* **84.**

IMMS, A. D. (1939).   Dipterous larvae and wound treatment.   *Nat. Lond.* **144** (3646).

MIYAKE, T. (1920).   Investigations on the insects used as food and in medicine. (In Japanese, a review in *Rev. Appl. Ent.* **8**.)

PIERCE, W. D. (1915).   The uses of certain weevils and weevil products in food and medicine.   *Proc. Ent. Soc. Wash.* **17** (3).

**XI. References to Products Obtained from Insects (Honey and Wax Excluded).**

ANONYMOUS.   Silk, its origin, culture and manufacture.   Belding-Heminway-Corticelli, New York.

BANKS, C. S. (1912).   A manual of Philippine silk culture.   Wesley, England.

CLARK, C. A. (1932).   A commercial use for a destructive insect.   *Journ. Econ. Ent.* **25** (1).

FAGAN, M. M. (1918).   The uses of insect galls.   *Amer. Nat.* **52** (614).

GLOVER, P. M. (1937).   A practical manual of lac cultivation.   2d ed.   Criterion Printing Works.

GOSSARD, H. A. (1909).   Relation of insects to human welfare.   *Journ. Econ. Ent.* **2** (5).

HOOPER, L. (1927).   Silk, its production and manufacture.   Pitman Publishing Corporation, New York.

KELLY, H. A. (1903).   The culture of the mulberry silkworm.   New ser. *U. S. Div. Ent. Bull.* **39**.

KIRKALDY, G. W. (1898).   An economic use of waterbugs.   *Ent. Mo. Mag.* **34.**

LEONARD, D. D. (1935).   The story of silk.   *Nat. Hist.* **35** (3).

MINNS, SUSAN (1929).   Book of the silkworm.   *National Americana Soc. N. Y.*

PARRY, E. J. (1935).   Shellac, its production, manufacture, chemistry, analysis, commerce and uses.   Pitman Publishing Corporation, New York.

——— (1937).   Lac cultivation in India.   *Indian Lac. Res. Inst.*

PETRUNKEVITCH, ALEX (1921).   Spider silk and its use.   *Nat. Hist.* **21.**

RAWLLEY, R. C. (1919). Economics of the silk industry. P. S. King, London.

RILEY, C. V. (1878). A new source of wealth to the United States. *Proc. Amer. Assoc. Adv. Sci.* **27.**

SMITH, D. K. (1937). Curing gourds by aid of insects. *Intern. Gourd Soc. Bull.* **6** (5). Los Angeles, Calif.

**XII. References to Insects Used as Human Food.**

BEQUAERT, J. (1921). Insects as food. *Nat. Hist.* **21** (2).

BRISTOWE, W. S. (1932). Insects and other invertebrates for human consumption in Siam. *Trans. Ent. Soc. Lond.* **80** (2).

CURRAN, C. H. (1939). On eating insects. *Nat. Hist.* **43** (2).

DECARY, R. (1937). Insects eaten by natives in Madagascar. L'entomologie chez les indigènes de Madagascar. *Bull. Soc. Ent. Fr. Paris,* **42.**

EALAND, C. A. (1915). Insects and man. D. Appleton-Century Company, Inc., New York.

ELTRINGHAM, H. (1909). An account of some experiments on the edibility of certain lepidopterous larvae. *Trans. Ent. Soc. Lond.* **4.**

ESSIG, E. O. (1934). The value of insects to the California Indians. *Scientific Mo.* **38** (2).

HOWARD, L. O. (1915). The edibility of insects. *Journ. Econ. Ent.* **8** (6).

MARLATT, C. L. (1907). The cicadas as an article of food. *In,* The periodical cicada. *U. S. Dept. Agric. Bur. Ent. Bull.* **71.**

PACKARD, A. S. (1885). Edible Mexican insects. *Amer. Nat.* **19.**

SIMMONDS, P. L. (1885). The animal resources of different nations. E. and F. N. Spon, London.

SKINNER, A. (1910). The use of insects and other invertebrates as food by the North American Indians. *Journ. N. Y. Ent. Soc.* **18** (4).

**XIII. References to Insects Used as Food by Birds and Other Animals.**

ALLEN, A. A. (1914). Birds in their relation to agriculture in New York State *Cornell Reading Course* **4** (76).

BEAL, F. E. L. (1926). Some common birds useful to the farmer. *U. S. Dept. Agric. Farmers' Bull.* **630.**

BURT, C. E. (1928). Insect food of Kansas lizards with notes on feeding habits. *Journ. Kan. Ent. Soc.* **1** (3).

FORBES, S. A. (1888). On the food relations of fresh-water fishes. *Bull. Ill. State Nat. Hist. Lab.* **2.**

FROST, S. W. (1924). Frogs as insect collectors. *Journ. N. Y. Ent. Soc.* **32.**

——— (1932). Notes on feeding and molting in frogs. *Amer. Nat.* **66** (707). *Literature.*

GABRIELSON, J. N. (1924). Food habits of some winter bird visitants. *U. S. Dept. Agric. Dept. Bull.* **1249.**

HAMILTON, W. J. (1932). The food and feeding habits of some eastern salamanders. *Copeia* **2.**

KIRKLAND, A. H. (1915). The usefulness of the American toad. *U. S. Dept. Agric. Farmers' Bull.* **196.**

KNOWLTON, G. F. (1938). Lizards in insect control. *Ohio Journ. Sci.* **38** (5).

———, and JANES, M. J. (1932). Studies of the food habits of Utah lizards. *Ohio Journ. Sci.* **32** (5).

———, and THOMAS, W. L. (1935). Insect food of Trout Creek lizards. *Proc. Utah Acad. Sci.* **12.**

McATEE, W. L. (1925). Notes on drift, vegetable balls, and aquatic insects as a food product of inland waters. *Ecology* **6** (3). *References.*

Munz, P. A. (1920).   A study of the food habits of the Ithacan species of Anura during transformation.   *Pomona Col. Journ. Ent. and Zool.* **12** (2).

Pearse, A. S. (1918).   The food of the shore fishes of certain Wisconsin lakes.   *U. S. Bur. Fish. Bull.* **35,** Document **856.**

Uhler, F. M., and Clarke, T. E. (1939).   Food of snakes of the George Washington National Forest, Va.   *Trans. 4th N. Amer. Wild Life Conference* 1939.

Weed, C. M. (1884).   The food relations of birds, frogs and toads.   *Mich. Hortic. Soc.*

————, and Dearborn, N. (1924).   Birds in their relations to man.   J. B. Lippincott Company, Philadelphia.

**XIV.   References to Insects Used in Art and Literature.**

Ball, K. M. (1937).   Decorative motives of oriental art.   Dodd, Mead & Company, Inc., New York.

Barber, G. W., and Wade, J. S. (1933).   Note on a collection of old entomological paintings.   *Journ. N. Y. Ent. Soc.* **41** (1–2).

Birdsong, R. E. (1934).   Insects of the Bible.   *Bull. Brooklyn Ent. Soc.* **29** (3).

Crook, A. H. (1937).   Browning and natural science.   *Hong Kong Nat.* **8.**

Curran, C. H. (1937).   Insect lore of the Aztecs revealing early acquaintance with many of our pests.   *Natural History* **39.**

Dow, R. P. (1917).   Studies in the Old Testament.   *Bull. Brooklyn Ent. Soc.* **12** (1).

———— (1916).   The testimony of the tombs.   *Bull. Brooklyn Ent. Soc.* **11** (2).

———— (1918).   The grasshoppers of the Old Testament.   *Bull. Brooklyn Ent. Soc.* **13** (2).

Faulkner, Pearl (1931).   Insects in English poetry.   *Scientific Mo.* **33.**

Frost, S. W. (1937).   The insect motif in art.   *Scientific Mo.* **44.**

Montgomery, B. E. (1937).   Insect stamps.   *Ent. News* **48** (7).

Patterson, R. (1841).   The natural history of insects mentioned in Shakespeare's plays.

Rodeck, H. G. (1932).   Arthropod designs on prehistoric Mimbres pottery.   *Ann. Ent. Soc. Amer.* **25** (4).

Ross, E. (1937).   Beetles used for personal adornment in various countries.   Ueber Schmeckkäfer und deren Verwendung bei verschiedenen Volkern.   *Ent.* 3 *Frankfurt* 50.

Slosson, A. T. (1916).   Entomology in literature.   *Bull. Brooklyn Ent. Soc.* **11** (3).

Tozzer, A. M., and Allen, G. M. (1910).   Animal figures in Maya codices.   *Peabody Museum* **4** (3).

Wade, J. S. (1927).   Some insects of Thoreau's writings.   *Journ. N. Y. Ent. Soc.* **35** (1).

Walton, W. R. (1922).   The entomology of English poetry.   *Proc. Ent. Soc. Wash.* **24** (7–8).

Watson, J. R. (1932).   Insects in art.   *Guide to Nature* **24.**

Weigall, Arthur (1924).   Ancient Egyptian works of art.   George G. Harrap & Co., Ltd., London.

Weiss, H. B. (1925).   The bee, the wasp, the ant, insects of the "Physiologus."   *Journ. N. Y. Ent. Soc.* **38.**

———— (1926).   The entomology of Pliny and Elder.   *Journ. N. Y. Ent. Soc.* **34** (4).

———— (1927).   The scarabaeus of the ancient Egyptians.   *Amer. Nat.* **61** (675).

———— (1929).   The entomology of Aristotle.   *Journ. N. Y. Ent. Soc.* **37** (1).

———— (1930).   Insects and witchcraft.   *Journ. N. Y. Ent. Soc.* **38.**

Wood, J. G. (1876).   Bible animals.   Longmans, Green and Company, New York.

# CHAPTER V

## ORDERS OF INSECTS

The comprehension of the orders of insects may seem simple as there are only about a score of orders of present-day insects; however, the study is considerably involved. Orders are not merely names under which families, genera, and species are grouped; they are categories that represent related groups of insects. It is more difficult to lump genera and families than it is to consider a single species. Often it is more important to know the order to which an insect belongs than to know the species. This is particularly true when dealing with the eggs, larvae, nymphs, or pupae of insects. Even a well-trained entomologist may find it difficult, on certain occasions, to distinguish the larva of a coleopteron from the larva of a lepidopteron. It is even more difficult when leaf-mining species are concerned, where the bodies tend to become similar in form. Newly hatched nymphs of different species of Pentatomidae are so similar that we, at present, make little attempt to determine the species. To recognize that certain eggs belong to the Pentatomidae is often all we can expect in classifying them.

ANALYSIS OF THE LINNAEAN INSECT ORDERS

| Linnaean orders | Present-day orders probably included by Linnaeus | Number of orders |
|---|---|---|
| Coleoptera....... | Coleoptera, Euplexoptera, Orthoptera | 3 |
| Hemiptera........ | Heteroptera, Homoptera, Corrodentia, Thysanoptera | 4 |
| Lepidoptera...... | Lepidoptera | 1 |
| Neuroptera....... | Neuroptera, Mecoptera, Ephemerida, Odonata, Plecoptera, Trichoptera | 6 |
| Diptera.......... | Diptera | 1 |
| Hymenoptera..... | Hymenoptera | 1 |
| Aptera........... | Collembola, Thysanura, Dicellura, Siphonaptera, Isoptera, Anoplura, Mallophaga. (The Arachnida, Diplopoda, Chilopoda, Crustacea, and animals of other phyla are included) | 7 |
| Total insect orders including Embiidina, Zoraptera, and Strepsiptera which are of more recent origin..................................... | | 26 |

The idea of separating insects into orders was suggested by Linnaeus in "Systema Naturae" (1735–1768). He divided the Hexapoda into seven orders: Coleoptera, Hemiptera, Lepidoptera, Neuroptera, Hymenoptera, Diptera, and Aptera. With the exception of the Aptera, these orders are recognized to the present day. About 342 B.C., Aristotle in his "Historia Animalium" used some of the names now employed for orders such as Coleoptera and Diptera but apparently had in mind no definite system of classification.

The number of orders of insects is determined by the natural grouping of families, genera, and species, by ecological and structural characteristics, and by the interpretation of these groups by various workers. The conceptions of the orders Lepidoptera, Diptera, and Hymenoptera have remained unchanged since the time of Linnaeus, with the exception that their comprehensiveness has been increased by the addition of new families, genera, and species.

The largest number of orders has been created by splitting older orders. The Neuroptera have been repeatedly divided and at least six orders have been derived. The Linnaean order Aptera, a heterogenous group of unrelated forms, has yielded seven insect orders. The order Hemiptera was split to form Heteroptera and Homoptera. The Euplexoptera were taken from the Coleoptera at an early date. Dicellura was created by recognizing the suborder, Entotrophi, of the Thysanura. The Dictyoptera and Phylloptera of Wardle have been split off from the Orthoptera. The division is not always so simple as this. An idea of the complexity of the problem is gained when we consider the obsolete order Platyptera. (See Appendix.)

Certain orders have occasionally been thrown together by some workers. The Psocoptera, of Wardle, includes the Zoraptera and Corrodentia of other workers. The Anoplura, of Wardle, includes the Mallophaga and Anoplura (Siphunculata).

Several comparatively new orders of insects have been added, namely, Strepsiptera, Zoraptera, and Embiidina. The addition of numerous fossil orders has increased the list considerably. Thus the seven original orders of Linnaeus have been increased by some writers to 27 orders of present-day insects and 13 orders of fossil insects, making a total of 40. There is no objection to increasing the number of orders if it seems wise to do so, for a larger number of well-defined groups make it easier to classify insects than a smaller number of obscurely or poorly defined groups.

**Arrangement of Orders.**—It is customary to place the more primitive orders at the beginning and the more specialized ones toward the end of the list. This is a logical, evolutionary arrangement. It is not difficult to place such orders as Thysanura, Dicellura, and Collembola, which are wingless and decidedly primitive in habits and structure. They naturally come at the beginning of the list. The Lepidoptera, Siphonaptera, Diptera, and Hymenoptera are all highly developed, and it is difficult to say which should be considered the most highly specialized. Hymenoptera are usually placed at the end of the list. The social habit, common in so many species, helps to determine the position of this order. Many changes have taken place in the position of the orders. In 1912, Comstock considered the Trichoptera closely related to the Coleoptera; in 1933, he indicated by their position that they were more closely related to the Lepidoptera. Wardle places the Dermaptera next to the Orthoptera largely because the species have incomplete metamorphosis. Comstock places them far down the list because of their affinities with the Coleoptera.

DERIVATION OF ORDER NAMES

| Order | Prefix | Stem | Significance |
|---|---|---|---|
| Thysanura | Thysanos | oura | Fringe tail |
| Collembola | Colla | embolon | Glue bolt |
| Orthoptera | Orthos | ptera | Straight wings |
| Zoraptera | Zoros | apteros | Purely wingless |
| Isoptera | Isos | ptera | Equal wings |
| Neuroptera | Neuron | ptera | Nerve wings |
| Ephemerida | Ephemeros | ....... | Lasting but a day |
| Odonata | Odous | ....... | Tooth |
| Plecoptera | Plecos | ptera | Plaited wings |
| Corrodentia | Corrodere | ....... | To gnaw |
| Mallophaga | Mallos | phagein | To eat wool |
| Embiidina | Embios | ....... | Lively |
| Physopoda | Physao | pous | To blow up foot |
| Thysanoptera | Thysanos | ptera | Fringe wings |
| Anoplura | Anoplos | oura | Unarmed tail |
| Homoptera | Homos | ptera | Same, homogenous wings |
| Heteroptera | Heteros | ptera | Diverse wings |
| Euplexoptera | Eupleko | ptera | Well-folded wings |
| Dermaptera | Derma | ptera | Skin wings |
| Coleoptera | Coleos | ptera | Sheath wings |
| Strepsiptera | Strepsis | ptera | Twisted wings |
| Mecoptera | Mecos | ptera | Length, long wings |
| Trichoptera | Trichos | ptera | Hair wings |
| Lepidoptera | Lepido | ptera | Scale wings |
| Diptera | Dis | ptera | Two wings |
| Siphonaptera | Siphon | aptera | Siphon wingless |
| Hymenoptera | Hymen | ptera | Membrane wings |

**Order Names.**—Pearse (1936) suggested that the ending "ida" be used for all order names. This seems undesirable in entomology because names such as Coleoptera, Diptera, and Hymenoptera have been used since the time of Linnaeus and are well established. Wardle suggested changing the names Embiidina and Ephemerida to conform with the names of other orders of winged insects. Thus he proposes the names Embioptera and Ephemeroptera. On such a basis, the Odonata might be called Odonoptera.

A few synonyms in common use deserve special mention: Thysanoptera is equivalent to Physopoda; Siphonaptera is synonymous with Aphaniptera; Dermaptera is the same as Euplexoptera; and Hemiptera of some workers is equivalent to Heteroptera. The list of synonyms is long and complicated and is given in the Appendix for the benefit of the advanced student.

**Derivation of Order Names.**—Most of the order names are derived from the Greek. Corrodentia alone is derived from the Latin. The majority include winged insects and their names are built up from some descriptive term designating the character of the wings. These are combined with the Greek stem, *ptera*, designating wings. (See table on page 73.)

**Classification of Orders of Insects.**—Orders were proposed primarily to aid in classifying insects. Numerous characteristics are used: the presence or absence of wings, the texture of the wings or wing covers, the type of mouth parts, the presence or absence of cerci, the type of metamorphosis, and certain outstanding habits. The wing characteristics are summarized in the following table.

CLASSIFICATION OF INSECT ORDERS ACCORDING TO THE PRESENCE OR ABSENCE OF WINGS

| Entirely wingless | Winged | Winged and wingless forms | Females, wingless |
|---|---|---|---|
| Collembola | Neuroptera | Orthoptera | Homoptera |
| Dicellura | Ephemerida | Zoraptera | (Coccidae) |
| Thysanura | Odonata | Isoptera§ | Lepidoptera† |
| Mallophaga | Plecoptera | Corrodentia | Strepsiptera |
| Siphonaptera | Trichoptera | Embiidina | |
| Anoplura | Coleoptera* | Thysanoptera | |
| | Lepidoptera† | Homoptera | |
| | Diptera‡ | Heteroptera | |
| | Hymenoptera‖ | Dermaptera | |
| | | Mecoptera | |
| | | Hymenoptera | |

* A few Lampyridae are wingless.
† The females of Psychidae and some Liparidae and Geometridae are wingless.
‡ Most of the Pupipara are wingless.
§ Soldiers and workers are wingless.
‖ The workers and soldiers of certain social insects are wingless; the females of some ants and Mutillidae are wingless; in Blastophaga, the male is wingless.

**Thysanura,** Lubbock, 1869 (bristletails and fish moths). These are small primitively wingless insects, with mandibulate mouth parts, no metamorphosis, a pair of long cerci and a median caudal filament (*pseudocercus*), elongate antennae, and compound eyes. The body is clothed with scales. Common forms are *Machilis*, which is found under stones, and *Lepisma*, which is often found in houses.

**Dicellura,** Holliday, 1896 (*Rhabdura*) (also known as bristletails). These small primitively wingless insects have mandibulate mouth parts, no metamorphosis, moderately long antennae, and a pair of cerci. In *Japyx*, the cerci are unjointed and resemble short forceps. In *Campodea*, the cerci are elongate and many jointed. The median caudal filament is

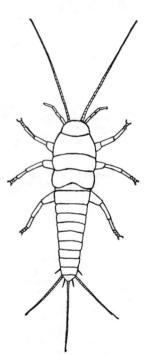

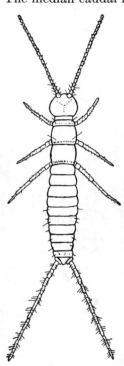

Fig. 43.—*Lepisma saccharina*, a silverfish (Thysanura).

Fig. 44.—*Campodea staphylinus*, an example of the order Dicellura.

absent and the body is not clothed with scales. Comstock, also Brues and Melander, place these insects in Entotrophi, a suborder of the Thysanura.

**Collembola,** Lubbock, 1870 (springtails). These insects are small primitively wingless species. They have mandibulate mouth parts and no metamorphosis, and the number of abdominal segments is reduced to six.

The first abdominal segment bears a ventral tube (the *collophore*), the fourth bears a forked *furcula* or spring, and the third is equipped with a *tenaculum* or triggerlike catch. Most of the species feed upon disintegrated organic matter; a few species are injurious to plants.

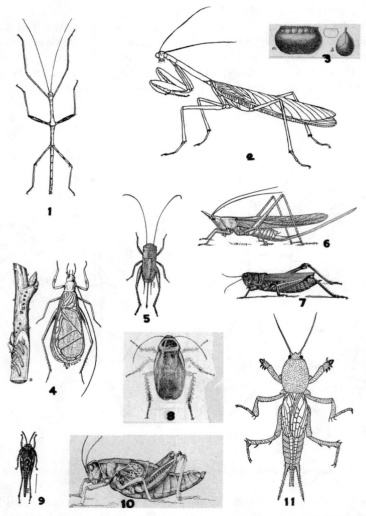

Fig. 45.—Examples of common Orthoptera: (1) walking stick, *Diapheromera femorata;* (2) Praying mantis, *Mantis religiosus;* (3) egg capsule American roach, *Periplaneta americana;* (4) snowy tree cricket, *Oecanthus niveus;* (5) common field cricket, *Gryllus assimilis;* (6) coneheaded grasshopper, *Neoconocephalus exiliscanorus;* (7) short-horned grasshopper, *Melanoplus differentialis;* (8) Australian roach, *Periplaneta australasiae;* (9) pigmy locust, *Paratettix cucullatus;* (10) south-western lubber grasshopper, *Brachystola magna;* (11) mole cricket, *Gryllotalpa hexadactyla.*

**Orthoptera,** Latreille, 1796 (grasshoppers, crickets, roaches, mantids, and others). The winged members of this order usually have two pairs of wings. The fore wings are more or less thickened, the hind wings are often folded in plaits like a fan. The mouth parts are mandibulate, the metamorphosis is incomplete, and the cerci are present but usually short.

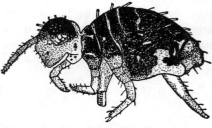

The order Dictyoptera, established by Bolivar, includes the Blattidae and the Mantidae. Wardle includes only the Mantidae in his conception of the Dictyoptera, leaving the Blattidae in the Orthoptera.

FIG. 46.—A spring tail, *Neosminthurus curvisetis* (Collembola).

The order Phylloptera, of Wardle, embraces two families of present-day insects, the Phyllidae and the Phasmidae, which are included by most workers in the Orthoptera.

Hemimeridae, usually placed in the Orthoptera, are considered under the Dermaptera by Wardle.

The Grylloblatta are of uncertain kinship. Some place them in the Orthoptera, others prefer to include them in a separate order.

A wide variety of insects belong to this order including roaches, mantids, walking sticks, grasshoppers and crickets. With the exception of the mantids and roaches, they are primarily plant feeders.

Roaches occur throughout the world, with the exception of the polar regions and reach their greatest development in the Tropics. Approximately 2,300 species are known from the entire world. They are omniverous, feeding upon dead plant and animal material.

Although 1,500 species of mantids are known, only one is indigenous to Eastern North America. Two species have been introduced from Asia.

Walking sticks are numerous in many parts of the world, but only one species is common in Eastern North America.

The grasshoppers constitute a large group generally divided into the short-horned species, *Acrididae* (*Locustidae*), and the long-horned species, the *Tettigoniidae*.

Crickets, *Gryllidae*, include the common black field species and many species of tree crickets.

**Zoraptera,** Silvestri, 1913. These small active insects are easily mistaken for termites. Both winged and wingless forms occur. The winged forms shed their wings much as do the termites. These insects have mandibulate mouth parts, incomplete metamorphosis, and the cerci are short and fleshy. Colonies are found under stones and under the bark of trees.  Wardle places the Zoraptera in the order Psocoptera.  Too little is known about these insects to place them definitely in the scheme of classification.  A. B. Gurney, who has studied the Zoraptera recently, remarks that they occur in four continents including the Old and the New Worlds.  They are represented by 16 species.  *Zorotypus hubbardi* has been found about Washington, D. C.  Gurney gives a key to the species of Zoraptera.  Three of the species occur in the West Indies but only one is known from the United States.  They show marked affinities to the Corrodentia.

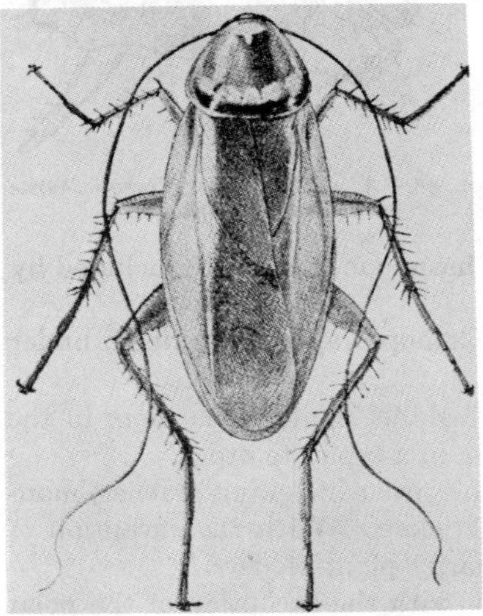

**Isoptera,** Brullé, 1832 (termites). These social insects live in colonies beneath stones or in wood. The abdomen is broadly joined to the thorax; the antennae are composed of

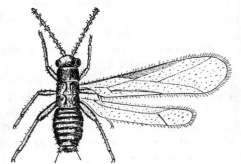

Fig. 47.—The American cockroach, *Periplaneta americana* (Orthoptera).

Fig. 48.—*Zorotypus hubbardi*, winged adult female (Zoraptera), left pair of wings removed.

a number of segments and are not elbowed as is the case in the ants. The cerci are present, though small.  The mouth parts are mandibulate; the metamorphosis is incomplete.  There are at least three distinct castes: the reproductive caste, the worker caste, and the soldier caste. Each caste includes both males and females.

**Neuroptera,** Linnaeus, 1758 (lacewings, ant lions, and others). The members of this order have two pairs of membranous wings with many longitudinal and cross veins.  The mouth parts are mandibulate, the metamorphosis is complete.  The Mantispidae have hypermetamorphosis.  Cerci are absent in the Neuroptera.  The form and habits of these species differ greatly.  Many are aquatic, some are terrestrial, and all are predacious, especially in the larval stage.  The larvae generally have strong mandibles and thoracic legs.  Lateral filaments frequently occur on the sides of the abdomen in the aquatic species.  The pupae are exarate and some form of silken cocoon is generally made.

**Ephemerida,** Krausse, 1906.  Ephemeridae, Leach, 1817 (May flies). The members of this order are medium in size with delicate membranous wings of which there are usually two pairs; in a few species there is but one pair.  The front wings are usually larger than the hind wings.  Additional longitudinal and cross veins are numerous.  The mouth parts are vestigial in the adults.  The cerci are long and many jointed, and a median filament is sometimes present.  About 500 species are known.

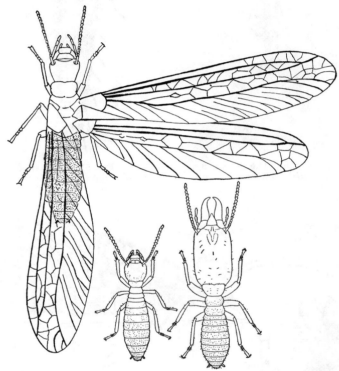

Fig. 49.—White ant or termite, *Reticulitermes flavipes*.  (Left) winged female; (middle) worker; (right) soldier (Isoptera).

from North America.  The adults are short-lived.  The naiads live in the water and are herbivorous.  Wardle suggested the name Ephemeroptera for this order.

**Odonata,** Fabricius, 1793 (dragonflies, damsel flies).  These are large insects with two pairs of membranous wings which are approximately equal in size.  They have numerous extra longitudinal and cross veins.  The antennae are minute.  The metamorphosis is incomplete. The mouth parts are mandibulate in both the naiads and the adults, and both forms are predacious.  The naiads are aquatic, the adults generally

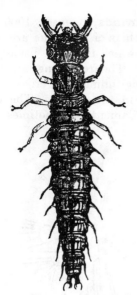

FIG. 50.—The dobson fly, *Corydalis cornuta* (Neuroptera).

FIG. 51.—The larva of a dobson fly (Neuroptera).

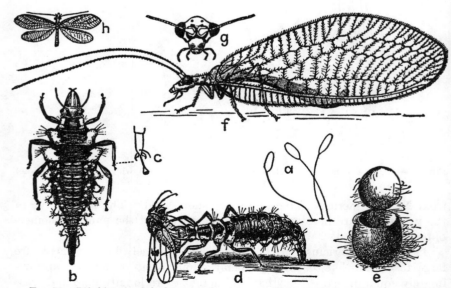

FIG. 52.—Life history of the lacewing fly; (*a*) eggs; (*b*) larva; (*c*) leg; (*d*) larva attacking a psyllid; (*e*) cocoon; (*f*) adult; (*g*) head of adult; (*h*) adult (Neuroptera). (*After Webster and Phillips, U. S. Department of Agriculture.*)

live near the water and are powerful fliers. There are two suborders, Zygoptera, the damsel flies, and Anisoptera, the dragonflies.

FIG. 53.—A May fly, *Callibaetis*, naiad and adult (Ephemerida).

**Plecoptera,** Burmeister, 1839 (stone flies). The members of this group have membranous wings. The hind pair is generally larger than the front pair. In some species, the branches of the principal veins are

FIG. 54.—Dragonfly (Odonata).

reduced in number and there are comparatively few cross veins. In others, the accessory veins are well developed and there are many cross veins. The cerci are long and many jointed, the antennae are long and slender; the metamorphosis is incomplete. The mouth parts of the naiads are mandibulate. Those of the adults are occasionally vestigial.

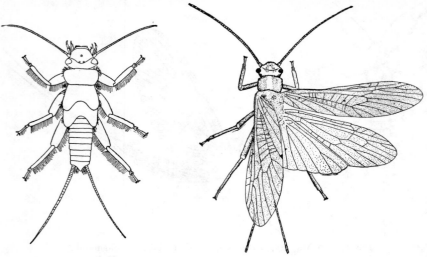

FIG. 55.—Stone fly, naiad and adult (Plecoptera).

**Corrodentia,** Burmeister, 1839 (psocids and book lice). These are very small insects. The wings, if present, are membranous, and the first pair is the larger. The antennae are long and slender and the cerci are wanting. The Corrodentia have mandibulate mouth parts and the metamorphosis is incomplete. Wardle placed them in the order Psocoptera.

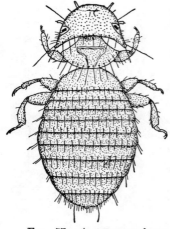

FIG. 56.—A winged psocid, *Cerastipsocus venosus* (Corrodentia). (*Redrawn after Comstock.*)

FIG. 57.—A common louse, *Menopon phaeostomum*, of the peafowl (Mallophaga). (*Redrawn after Herrick.*)

**Mallophaga,** Nitzsch, 1818 (bird lice). The members of this order are small insects with an acquired wingless condition. They have no metamorphosis, and the mouth parts are mandibulate. They are parasitic largely upon birds, but some attack mammals. Wardle places the Mallophaga in the order Anoplura. It is questionable whether biting and sucking insects should be placed in the same order.

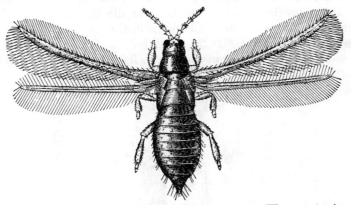

Fig. 58.—The pear thrips, *Taeniothrips inconsequens* (Thysanoptera).

**Embiidina,** Enderline, 1903. (Embidina, Hagen, 1861). These are small insects. The wings, if present, are four in number and about equal in size. They are folded upon the back when the insect is at rest. The cerci are two-segmented. The mouth parts are mandibulate. The metamorphosis is intermediate between incomplete and complete. They live in silken nests or galleries under stones and other objects and are very active insects. Wardle proposed a new name for this order, Embioptera, to correspond with the names of other winged orders. The Embiidina occur in warmer parts of the world. A few species have been taken in Florida, Texas, and California.

Fig. 59.—The hog louse, *Haematopinus suis* (Anoplura). (*After Metcalf and Flint.*)

**Thysanoptera,** Haliday, 1836 (thrips). These small insects may be winged or wingless. The wings, if present, are long and narrow and fringed with long hairs. They have haustellate mouth parts for rasping and sucking. The tarsi are usually two-jointed and bladderlike at the tip. The metamorphosis is incomplete. The majority of the species are plant feeders.   Injurious forms occur on onion, pear, bean, gladiolus, and many other crops.   A few species are predacious.

**Anoplura,** Leach, 1815 (lice).   These are small insects with an acquired wingless condition.   They have short antennae, short stout legs with a single claw apposing a spine on the tip of the tibia.   The species are parasitic upon mammals.   Wardle places the sucking lice, Anoplura, together with the Mallophaga in the order Anoplura.

**Heteroptera,** Latreille, 1825 (true bugs).   These are meduim to large insects.   The majority of the species have two pairs of wings.   The

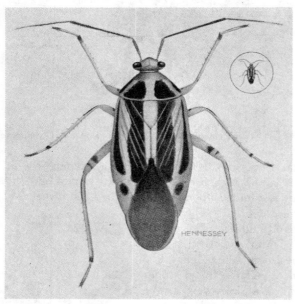

Fig. 60.—The four-lined plant bug, *Poecilocapsus lineatus* (Heteroptera).   (*After Dustan.*)

first pair is thickened at the base and membranous at the tips.   They have haustellate mouth parts and incomplete metamorphosis.   The proboscis or beak is borne on the front of the head.   They have a wide range of habits.   The majority are plant feeders.   Many are predacious. Some are aquatic and others are strictly terrestrial.   The order includes many serious pests.   The squash bug, the harlequin bug, and the chinch bug are common examples.   The bedbug is one of the best known species in this order.   There are also numerous beneficial insects in this order. Many of the species in the family Pentatomidae prey upon other insects.

**Homoptera,** Latreille, 1815 (cicadas, treehoppers, leafhoppers, aphids, scale insects, and others).   The majority of these species are small.   The winged members have two pairs of membranous wings.   Among the scale insects (Coccidae), the females are wingless and the males have but one pair of wings. The Homoptera have haustellate mouth parts and incomplete metamorphosis. The male Coccidae approach the holometabola in habits. All the species are plant feeders and the order includes many of the most injurious pests.

**Dermaptera,** Leach, 1815 (earwigs). These are moderately sized insects superficially resembling Coleoptera. The winged members of this order usually have two pairs of wings. The fore wings are leathery,

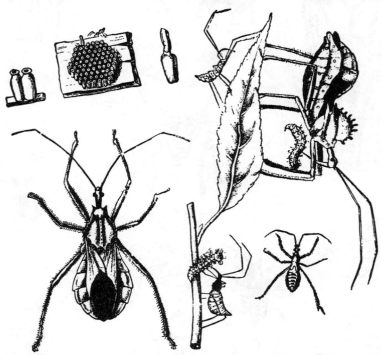

Fig. 61.—The wheel bug, *Arilus cristatus* (Heteroptera): (upper left) eggs natural size and enlarged; (left) adult; (right) nymphs attacking larvae. (*After Glover.*)

short, and without veins. The hind wings are often large with radiating veins. When the insect is at rest, the wings are folded lengthwise and crosswise. They have mandibulate mouth parts and incomplete metamorphosis, and the cerci are enlarged to form a pair of caudal forceps. The species are nocturnal in habit and are found in the ground or beneath bark. They are omnivorous feeders, sometimes eating insects and snails. Wardle includes the Hemimeridae in this order; most workers, however, place them in the Orthoptera.

**Coleoptera,** Linnaeus, 1758 (beetles). These are medium to large insects. When wings are present, there are usually two pairs; the first pair (*elytra*) is greatly thickened and meet along the back in a straight line; the second pair is membranous and, when the insect is at rest, is

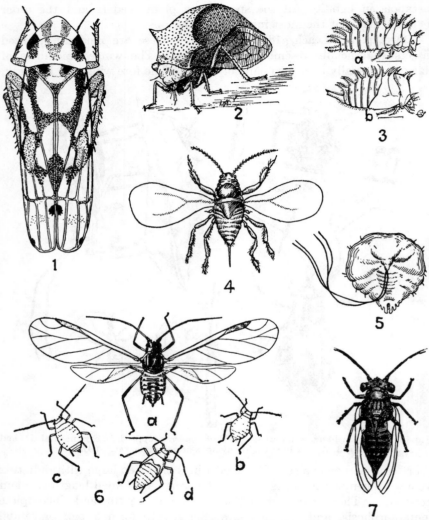

FIG. 62.—Some examples of common Homoptera; (1) grape leafhopper (Cicadellidae); (2) buffalo treehopper, adult (Membracidae); (3) buffalo treehopper nymphs; (4) San José scale, male; (5) San José scale, female removed from scale (Coccidae); (6) spinach aphid, adult and nymphal forms (Aphididae); (7) pear psylla (Psyllidae).

folded beneath the elytra. In a few species the second pair of wings is wanting, and the elytra are fused. The adults and the larvae have mandibulate mouth parts. The metamorphosis is complete. A few species are hypermetamorphic. This is a large order, including plant feeders, predacious forms, feeders on dried parts of animals, wool, etc. There are two suborders: Adephaga and Polyphaga. Many economic species are included in the families Chrysomelidae, Elateridae, Buprestidae, and Cerambycidae.

**Strepsiptera,** Kirby, 1813 (stylopids). These are minute insects which are parasitic upon other insects. Only the males have wings. The hind wings are large, but the fore wings are reduced to clublike appendages. The female is larviform and legless. The adult mouth parts are vestigial. All species are hypermetamorphic.

**Mecoptera,** Packard, 1886 (scorpion flies). The members of this order are moderate in size. If present, they have two pairs of wings

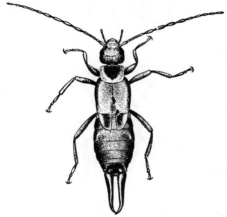

Fig. 63.—The European earwig, *Forficula auricularia* (Dermaptera). (*After Gibson.*)

which are about equal in size, with a considerable number of cross veins. The head is prolonged into a beak, at the end of which are located the mandibulate mouth parts. The antennae are long and slender. The metamorphosis is complete. The larvae resemble caterpillars, with three pairs of thoracic legs and with or without prolegs. Both larvae and adults are believed to feed largely upon dead or dying insects.

**Trichoptera,** Kirby, 1813 (caddis flies). These are moderately sized insects with two pairs of membranous wings which are more or less densely clothed with hairs. The metamorphosis is complete. The mouth parts of the adult are vestigial. The larvae have mandibulate mouth parts. They have well-developed thoracic legs and one pair of abdominal legs. The species are aquatic, some build cases of sticks, stones, grasses, and other materials. They feed upon microorganisms in the water. The pupae are exarate.

**Lepidoptera,** Linnaeus, 1758 (butterflies and moths). The majority of the species have membranous wings which are covered with dense

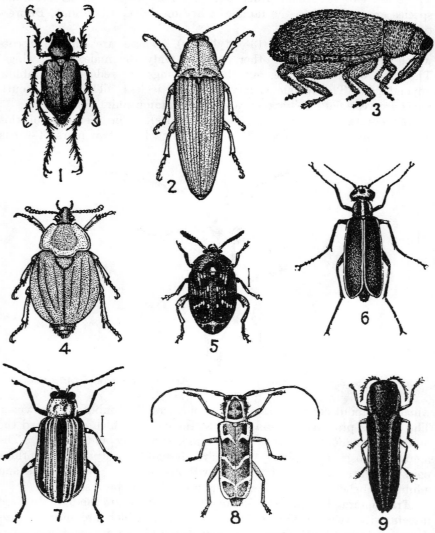

Fig. 64.—Some examples of common Coleoptera; (1) rose chafer (Scarabaeidae); (2) Elater (Elateridae); (3) apple flea-weevil (Curculionidae); (4) silphid (Silphidae); (5) bean weevil (Mylabridae); (6) margined blister beetle (Meloidae); (7) cucumber beetle (Chrysomelidae); (8) elm borer (Cerambycidae); (9) two-lined chestnut borer (Buprestidae).

overlapping scales. The females of the Psychidae and some of the Geometridae and Liparidae are wingless. The Lepidoptera have various types of wing-locking mechanisms: the *frenulum,* the *jugum,* etc.   The metamorphosis is complete.   The adults are haustellate except in a few cases.   The Micropterygidae are mandibulate.   In some, the mouth

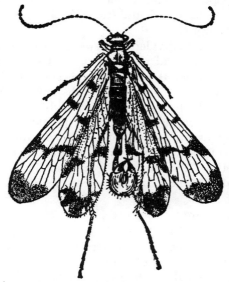

FIG. 65.—*Panorpa rufescens* (Mecoptera).

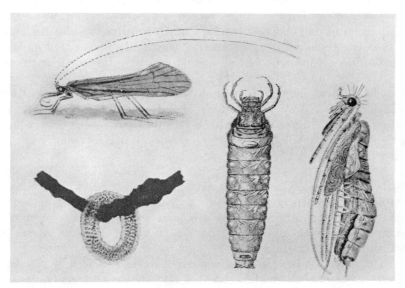

FIG. 66.   Life history of a typical caddis fly, eggs of *Phryganea,* larvae and pupae of *Halesus* and adult of *Leptocerus* (Trichoptera).

parts are vestigial. The adults feed upon nectar or not at all. The larvae have mandibulate mouth parts and are generally plant feeders. A few species are parasitic or predacious. The pupae are obtect except in the Micropterygidae. This order includes many serious pests, of which the codling moth, the oriental fruit moth, and the angoumois moth are outstanding examples.

This order is generally divided into two groups, the moths, which are nocturnal or night fliers, and the butterflies, which are diurnal or day fliers. This division is not as well marked as generally believed for there are no definite structural differences. Moths are generally said to have feathery antennae while the butterflies have clubbed or hooked antennae. It is a large order, comprising approximately 120,000 species. The larvae, often called caterpillars, are chiefly plant feeders and many rate as important pests of economic plants. The majority of the injurious species belong to the moths. The butterflies are usually considered harmless creatures collected by many for their beauty and interest. However, the caterpillars of a few, such as the cabbage butterfly and the celery or black swallow-tail, are injurious to plants.

**Diptera,** Linnaeus, 1758 (true flies). The members of this order are generally small insects. The majority have a single pair of wings. Wings are absent in most of the Pupipara. When wings are present, the second pair is represented by the balancers (*halteres*). Metamorphosis is complete. The adults have haustellate mouth parts and feed upon plant and animal juices. The larvae are parasitic, predacious, saprophagous, or feed upon living plants. Many injurious species are included, such as the apple maggot, the cherry maggot, the clover seed midge, and the botflies. Many beneficial insects are also included, such as the syrphus flies and the Tachinidae.

The Diptera include more than 80,000 species from different regions of the world. They range from small to minute species such as the biting midges, *Culicoides*, to the horse flies and robber flies which are generally large. The largest fly is a species of *Mydas* from South America which is about 4½ inches long.

The adults possess haustellate mouth parts. Many visit flowers, feeding upon pollen and nectar. The mosquitoes, black flies, horse flies, stable flies and others suck blood. Some feed on decaying material of plant or animal origin. Many transmit serious diseases of man and animal. The most notable is the Tse-tse fly of Tropical Africa which transmits sleeping sickness.

The larvae show a great diversity of habits. The majority feed on plants. Some, such as the fruit flies and the root maggots, cause considerable injury to crops. Others are predacious, parasitic or scatophagous. Thus, many are beneficial. Not a small number are aquatic.

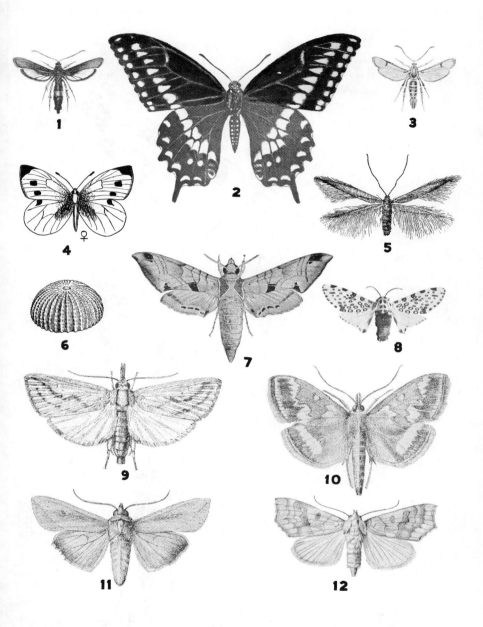

FIG. 67.—Examples of common Lepidoptera: (1) peach-tree borer, female; (2) swallowtail, butterfly, *Papilio;* (3) peach-tree borer, male; (4) imported cabbage butterfly, *Pieris rapae;* (5) larch case bearer, much enlarged; (6) egg of cutworm moth, much enlarged; (7) achemon sphinx moth; (8) Leopard moth, *Zeuzera pyrina;* (9) sugar cane borer, *Diatraea;* (10) European corn borer, *Pyrausta nubilalis;* (11) army worm, *Cirphis unipuncta;* (12) a noctuid.

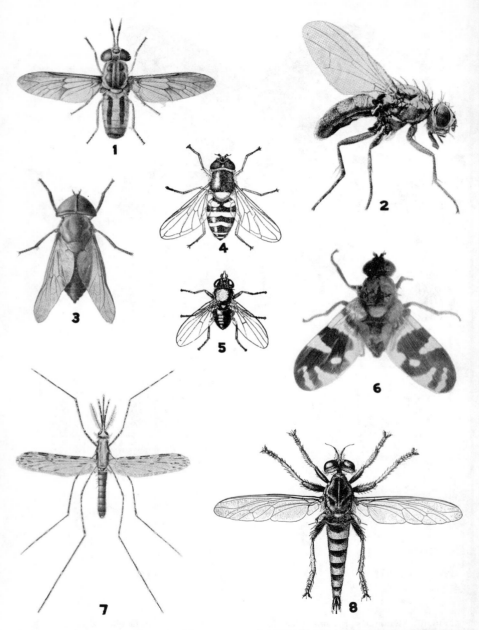

Fig. 68.—Examples of common Diptera: (1) deer fly, *Chrysops;* (2) cabbage root maggot fly, *Hylemyia;* (3) horsefly, *Tabanus;* (4) Syrphus or hover fly; (5) leaf-mining fly, *Agromyza;* (6) cherry fruit fly, *Rhagoletis;* (7) mosquito, *Anopheles;* (8) robber fly, *Promachus.*

The larvae show great differences in structure. Many of the common species are legless maggots. The aquatic forms show remarkable development as shown in figure 381.

**Siphonaptera,** Latreille, 1825 (fleas). These are small wingless insects which have the body strongly compressed laterally. They have haustellate mouth parts and complete metamorphosis. The adults are bloodsucking insects. The larvae are wormlike with mandibulate mouth parts and live upon decaying particles of animal and vegetable material.

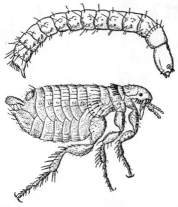

Fig. 69.—The cat and dog flea, *Ctenocephalus canis* (Siphonaptera); (above) larva; (below) adult. (*After Howard.*)

**Hymenoptera,** Linnaeus, 1758 (bees, wasps, ants, sawflies, and others). These insects are generally of medium size. Wings, if present, are four in number. The venation is generally reduced, and the longitudinal veins turn across the wing to take the position of cross veins. The mouth parts are mandibulate in the adult or exhibit a combination of mandibulate and haustellate types. The abdomen of the female is usually provided with a sting, piercer, or saw. The species have complete metamorphosis. The habits are varied. Many are social insects. Numerous species bore in the stems of plants or mine in leaves. Many are parasitic on other insects. The typical hymenopterous larva has a well-formed head, three thoracic segments, and nine to ten abdominal segments. In the sawflies, there are generally three pairs of thoracic legs and six to eight pairs of prolegs. In other Hymenoptera the larvae are legless. Evanescent thoracic appendages may appear in the early larval stages of parasitic forms. Hypermetamorphosis occurs in some of the parasitic forms. Polyembryony and parthenogenesis also occur. The pupae are exarate, and usually some sort of cocoon is formed.

Fig. 70.—A bee (Hymenoptera).

Fig. 71.—*Monobia quadridens*, a carpenter wasp (Hymenoptera).

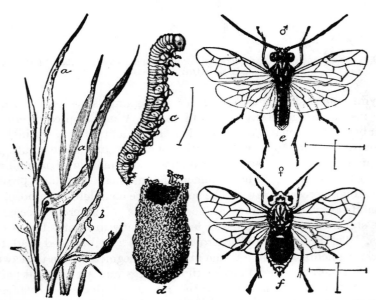

Fig. 72.—The grass sawfly (*Pachynematus extensicornis* Norton): (*a*) eggs on wheat blade; (*b*) young larva; (*c*) full-grown larva; (*d*) cocoon from which adult has emerged; (*e*) male; (*f*) female (Hymenoptera). (*After Riley and Marlatt, U. S. Department of Agriculture.*)

## A KEY FOR DETERMINING THE PRINCIPAL ORDERS OF INSECTS BASED ON ADULTS

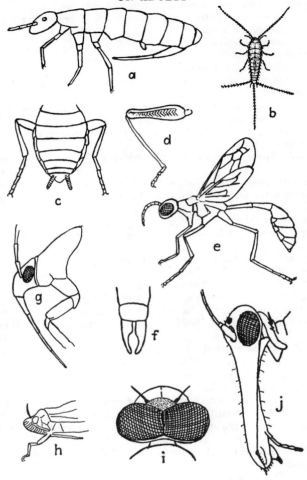

Wingless insects............................................................ **2**
Winged insects (The wing covers or elytra of beetles and earwigs are considered as wings.)................................................................ **28**
   **2.** Without a distinct head, often without legs, antennae, and eyes, living beneath scales or covered with a waxy secretion...................... *Homoptera*
   With a distinct head, legs, antennae, and eyes........................... **3**
   **3.** Abdomen consisting of not more than six segments, the fourth segment usually bearing a spring (*a*)................................. *Collembola*
   Abdomen consisting of more than six segments, the fourth segment without a spring................................................................ **4**
   **4.** Abdomen with two or three terminal bristlelike appendages (*b*)............ **5**
   Abdomen without terminal bristlelike appendages......................... **6**

**5.** Body clothed with scales, median terminal appendage present... *Thysanura*

Body not clothed with scales, median terminal appendage absent.. *Dicellura*

**6.** Mouth parts mandibulate, fitted for biting and chewing.................. **7**

Mouth parts haustellate, fitted for piercing, lapping, or sucking............ **22**

**7.** Head prolonged into a beak (*j*) ............................. *Mecoptera*

Head not prolonged into a beak...................................... **8**

**8.** Small louselike insects................................................ **9**

Usually larger insects, not louselike................................... **10**

**9.** Antennae with more than five segments, species that live in books or damp places.................................................... *Corrodentia*

Antennae with not more than. five segments, external parasites on birds and mammals, tarsi with strong claws for grasping................ *Mallophaga*

**10.** Cerci always present, usually well developed (*c*), roaches and crickets *Orthoptera*

Cerci minute or absent.................................................. **11**

**11.** Cerci always absent, antennae usually with no more than 11 segments......

....................................................... *Coleoptera*

Cerci present though often small, antennae with 15 or more segments....... **12**

**12.** Hind femora fitted for jumping (*c*), the Saltatoria, crickets, and grasshoppers

..................................................... *Orthoptera*

Hind femora not fitted for jumping..................................... **13**

**13.** Front legs enlarged adapted for grasping, prothorax long, the Raptoria, mantids.................................................... *Orthoptera*

Front legs not noticeably enlarged, prothorax not long.................... **14**

**14.** Body distinctly flattened or elongated................................. **15**

Body not distinctly flattened or elongated.............................. **16**

**15.** Body horizontally flattened, the mouth parts directed downward and backward between the anterior leg bases, roaches.................. *Orthoptera*

Body distinctly elongated, head not directed downward and backward, walkingsticks........................................... *Orthoptera*

**16.** Abdomen basally constricted (*e*)........................... *Hymenoptera*

Abdomen not noticeably constricted at the base........................ **17**

**17.** Tarsi two segmented, cerci unsegmented........................ *Zoraptera*

Tarsi three or four segmented........................................ **18**

**18.** Tarsi three segmented, basal segments of front tarsi greatly swollen, no species from Northeastern United States...................... *Embiidina*

Tarsi three or four segmented, basal segments of the front tarsi not greatly swollen........................................................ **19**

**19.** Body antlike in form but the base of the abdomen not constricted and the antennae not elbowed........................................ *Isoptera*

Body not antlike in form.............................................. **20**

**20.** Terminal abdominal segment provided with a pair of forceps (*f*).. *Dermaptera*

Terminal abdominal segment without a pair of forceps.................... **21**

**21.** Cerci present........................................... *Orthoptera*

Cerci absent............................................ *Coleoptera*

**22.** Tarsi with five segments............................................ **23**

Tarsi with less than five segments..................................... **25**

**23.** Body compressed laterally, hind legs adapted for jumping..... *Siphonaptera*

Body not compressed, hind legs not adapted for jumping.................. **24**

**24.** Antennae usually conspicuous, abdomen distinctly segmented, body bearing scales.................................................... *Lepidoptera*

Antennae usually inserted in grooves and inconspicuous, abdomen indistinctly segmented, body bearing hairs.................................. *Diptera*

**25.** Tarsi bladderlike and without claws....................... *Thysanoptera*

Tarsi not bladderlike, with distinct claws............................... **26**

**26.** Tarsi one segmented with single claws opposed by teeth on the tibiae, mouth parts inconspicuous, withdrawn within the head, parasites on mammals.... ............................................................... *Anoplura*

Tarsi with more than one segment or legs absent, mouth parts clearly visible. **27**

**27.** Cornicles absent, mouth parts arise from the front part of the head (*g*)...... ............................................................ *Heteroptera*

Cornicles present, mouth parts arise from the hind part of the head (*h*) aphids ............................................................ *Homoptera*

**28.** With one pair of wings.................................................. **29**

With two pairs of wings. (The elytra are considered as one pair of wings.).. **34**

**29.** Wings entirely membranous............................................. **30**

Wings in part horny or leathery, not membranous....................... **32**

**30.** Body with terminal filaments, wings with many cross veins..... *Ephemerida*

Body without terminal filaments, wings with few or no cross veins.......... **31**

**31.** Second pair of wings represented by a pair of knobs (halteres) which do not engage the front pair of wings.................................. *Diptera*

Second pair of wings represented by a pair of spines (hamuli) which engage the front pair of wings (male scale insects).................... *Homoptera*

**32.** Mouth parts mandibulate, fitted for biting and chewing.................. **33**

Mouth parts haustellate, fitted for piercing and sucking........ *Heteroptera*

**33.** Front wings horny, inner edges meeting in a straight line down the back..... ............................................................. *Coleoptera*

Front wings leathery or parchmentlike, usually with the inner edges more or less overlapping........................................... *Orthoptera*

**34.** Front and rear wings unlike in structure............................... **35**

Front and rear wings similar in structure but possibly differing in size and venation........................................................ **39**

**35.** Front wings thick at their bases, membranous and overlapping at their tips.. ............................................................. *Heteroptera*

Front wings the same texture throughout............................... **36**

**36.** Front wings stiff and horny without veins............................. **37**

Front wings leathery or parchmentlike with veins....................... **38**

**37.** Wing covers always short never covering the abdomen, abdomen provided with a pair of forceps (*b*)..................................... *Dermaptera*

Wing covers seldom short exposing much of the abdomen, abdomen without forceps......................................................... *Coleoptera*

**38.** Haustellate mouth parts, beak arising from the hind part of the head (*h*).... ............................................................... *Homoptera*

Mandibulate mouth parts................................... *Orthoptera*

**39.** Tarsi without claws, wings usually fringed with long hairs.... *Thysanoptera*

Tarsi with claws, wings seldom fringed with long hairs.................. **40**

**40.** Haustellate mouth parts.............................................. **41**

Mandibulate mouth parts, sometimes for lapping but never for piercing..... **42**

**41.** Wings usually covered with scales........................... *Lepidoptera*

Wings not scaly........................................... *Homoptera*

**42.** Wings with many longitudinal and cross veins......................... **43**

Wings with few longitudinal and cross veins........................... **47**

**43.** Abdomen with long segmented anal filaments (*b*)....................... **44**

Abdomen without long segmented anal filaments....................... **45**

**44.** Hind wings as large or larger than the fore wings.............. *Plecoptera*

Hind wing smaller than the fore wings or absent... ......... ..... *Ephemerida*

**45.** Antennae short and slender (i) .............................. *Odonata*
Antennae conspicuous................................................... **46**
**46.** Head prolonged into a beak with the mandibles at the tip........ *Mecoptera*
Head not prolonged into a beak............................ *Neuroptera*
**47.** Tarsi two or three segmented......................................... **48**
Tarsi four or five segmented........................................ **49**
**48.** Hind wings folded like a fan, tarsi three segmented............. *Orthoptera*
Hind wings not folded like a fan, tarsi two segmented......... *Corrodentia*
**49.** Fore wings slightly larger than the hind wings, mandibles well developed, wings membranous without hairs............................... *Hymenoptera*
Hind wings as large or larger than the fore wings, mandibles inconspicuous, wings sparsely clothed with hairs.......................... *Trichoptera*

## BIBLIOGRAPHY

**Classification of Orders.**

BALFOUR-BROWNE, F. (1920). Keys to the orders of insects. Columbia University Press, New York.

BEIER, M. (1938). Nachträge und Berichtigungen zuden einzelnen Insekten Ordunung. *In*, Kükenthal Handb-Zool. Berlin, Hefte 2, Lief 14.

BRUES, C. T., and MELANDER, A. L. (1932). Classification of insects. *Bull. Mus. Comp. Zool.* **63.** Cambridge, Mass. *Bibliographies.*

COMSTOCK, J. H. (1940). An introduction to entomology. 9th ed. Comstock Publishing Company, Inc., Ithaca, New York.

CRAMPTON, G. C. (1916). The orders and relationships of apterygotan insects. *Journ. N. Y. Ent. Soc.* **24** (4). *Bibliography.*

―――― (1928). The grouping of insect orders and their lines of descent. *The Entomologist* **61.**

―――― (1924). The phylogeny and classification of insects. *Journ. Ent. Zool. Claremont, Calif.* **16.**

FOLSOM, J. W. (1934). Entomology with special reference to its ecological aspects. Rev. by R. A. Wardle. The Blakiston Company, Philadelphia.

HANDLIRSCH, ANTON (1903). Zur Phylogenie der Hexapoden. *Sitzb. Akad. Wiss. Wien* **12** *Abth.* 1.

―――― (1925). Phylogenie. *In*, Schröder's Handbuch der Entomologie, Band 3, Gustav Fischer, Jena.

LANKESTER, E. R. (1904). The structure and classification of the Arthropoda. *Quart. Journ. Micros. Sci. Lond.* **47.**

LINNAEUS, CARL V. (1735). Systema Naturae. 1st. ed.

METCALF, Z. P., and C. L. (1928). A key to the principal orders and families of insects. 3d ed. Urbana, Ill.

MIRIAM, E. (1933). Les ordres des insectes. *Tabl. Anal. Fn. Russia, Leningrad,* 11.

PACKARD, A. S. (1886). A new arrangement of the orders of insects. *Amer. Nat.* **20** (808).

―――― (1903). Hints on the classification of the Arthropoda; the group a polyphyletic one. *Proc. Amer. Phil. Soc.* **42.**

PEARSE, A. S. (1936). Zoological names, a list of phyla, classes, and orders. *Sec. F. Amer. Assoc. Adv. Sci.*, Duke University Press, Durham, N. C.

SHARP, DAVID (1895). Insects. Cambridge Natural History, Ser. 1 and 2. The Macmillan Company, New York.

SHIPLEY, A. E. (1904). The orders of insects. *Zool. Anz.* **27.**

WARDLE, R. A. (1937). General entomology. The Blakiston Company, Philadelphia.

WEISS, H. B. (1926). James A. Turner and his "remarks on the Linnaean orders of insects." *Can. Ent.* **58** (12).

WILSON, H. F., and DONER, M. H. (1937). The historical development of insect classification. J. S. Swift & Co.

WOODWORTH, C. W. (1930). The arrangement of the major orders of insects. *Psyche* **37** (2).

**Outstanding References.**

ALDRICH, J. M. (1905). A Catalogue of North American Diptera. *Smiths. Misc. Coll.* **46** (1444). *Bibliography.*

BETTEN, C. H., *et al.* (1934). The caddis flies or Trichoptera of New York State. *N. Y. State Mus. Bull.* **292**. *Bibliography.*

BISCHOFF, H. (1927). Biologie der Hymenopteren. Verlag Julius Springer, Berlin.

BLATCHLEY, W. S. (1916). Rhynchophora or weevils of northeastern America. Nature Publishing Company, Indianapolis. *Bibliography.*

―――― (1920). Orthoptera of northeastern America. Nature Publishing Company, Indianapolis. *Bibliography.*

―――― (1926). Heteroptera or true bugs of eastern North America. Nature Publishing Company, Indianapolis. *Bibliography.*

――――, and LENG, C. W. (1910). An illustrated descriptive catalogue of the Coleoptera or beetles known to occur in Indiana. Nature Publishing Company, Indianapolis.

BOHART, R. M. (1941). A revision of the Strepsiptera with special reference to the species of North America. *Univ. Calif. Pub. in Ent.* **7** (6). *Bibliography.*

BRITTON, W. E. (1923). Insects of Connecticut. Part IV, The Hemiptera or sucking insects of Connecticut. *Geo. and Nat. Hist. Surv. Conn. Bull.* **34**.

CARPENTER, F. M. (1931). Revision of Nearctic Mecoptera. *Bull. Mus. Comp. Zool. Harvard* **72**.

―――― (1931). The biology of the Mecoptera. *Psyche* **38**.

CHOPARD, L. (1938). La biologie des Orthopterés. 20 Série à Encyclo. Entom. P. Lechevalier, Paris.

CLAASSEN, P. W. (1931). Plecoptera nymphs of America (north of Mexico). Charles C. Thomas, Publisher, Springfield, Ill.

CURRAN, C. H. (1934). The families and genera of North American Diptera. Pub. by C. H. Curran, New York.

DYAR, H. G. (1902). A list of North American Lepidoptera and key to the literature of this order of insects. *Bull. U. S. Nat. Mus.* **52**.

EWING, H. E. (1940). The Protura of North America. *Ann. Ent. Soc. Amer.* **33** (3). *Literature cited.*

FORBES, WM. T. M. (1920). The Lepidoptera of New York and neighboring states. *Cornell Univ. Agric. Exp. Sta. Mem.* **68**.

FOX, IRVING (1940). Fleas of Eastern United States. Collegiate Press, Inc., of Iowa State College, Ames, Iowa.

FRACKER, S. B. (1915). Classification of Lepidopterous larvae. *Ill. Biol. Monogr.* **2** (1).

FRISON, T. H. (1929). Fall and winter stoneflies or Plecoptera of Illinois. *Bull. Ill. Nat. Hist. Surv.* **18** (2).

―――― (1935). The stoneflies or Plecoptera of Illinois. *Bull. Ill. Nat. Hist. Surv.* **20** (4).

GARMAN, P. (1917). The Zygoptera or damsel-flies of Illinois. *Bull. Ill. State Nat. Hist. Lab.* **12**.

―――― (1927). The Odonata or dragonflies of Connecticut. Insects of Connecticut Part V. *Geo. and Nat. Hist. Surv. Conn. Bull.* **39**. *Bibliography.*

GAUSE, G. F. (1930). Studies on the ecology of the Orthoptera. *Ecology* **11** (2). *Literature cited.*

GURNEY, A. B. (1938). A synopsis of the order Zoraptera with notes on the biology of *Zorotypus hubbardi* Caudell. *Proc. Ent. Soc. Wash.* **40** (3).

———, and MUESEBECK, C. F. W. (1939). Nomenclatorial notes on Corrodentia, with descriptions of two new species of Archipsocus. *Wash. Acad. Sci.* **29** (11).

HINDS, W. E. (1902). Monograph of the Thysanoptera of North America. *Proc. U. S. Nat. Mus.* **26.**

HOLLAND, W. J. (1904). The moth book. Doubleday, Doran & Company, Inc., New York.

——— (1913). The butterfly book. Doubleday, Doran & Company, Inc., New York.

HORVATH, G., and PARSHLEY, H. M. (1927–1929). General catalogue of the Hemiptera, Fasc. 1 Membracidae. (Funkhouser, W. D.) Fasc. III. Pyrrocoriidae. (Sherman) *Bibliography.*

JACKSON, C. F. (1906). Key to the families and genera of the order of Thysanura. *Ohio Nat.* **6** (8).

METCALF, Z. P. (1936). General catalogue of the Hemiptera. Fasc. IV. Fulgoroidea Part 2 Cixiidae. Smith College.

MILLS, H. B. (1934). A monograph of the Collembola of Iowa. Collegiate Press, Inc., of Iowa State College, Ames, Iowa.

NEEDHAM, J. G., and CLAASSEN, P. W. (1925). A monograph of the Plecoptera or stoneflies of America, north of Mexico. Lafayette, Ind.

NEEDHAM, J. G., et al. (1929). A handbook of the dragon flies of North America. Charles C. Thomas, Publisher, Springfield, Ill.

——— (1935). The biology of mayflies. Comstock Publishing Company, Inc., Ithaca, New York.

PATCH, E. M. (1938). Food-plant catalogue of the aphids of the world, including the Phylloxeridae. *Me. Agric. Exp. Sta. Bull.* **393.** *Bibliography.*

ROSS, E. S. (1940). A revision of the Embioptera of North America. *Ann. Ent. Soc. Amer.* **33** (4).

TILLYARD, P. J. (1917). The biology of the dragonflies. Cambridge University Press, London.

TORRE-BUENO, J. R. DE LA (1939). A synopsis of the Hemiptera-Heteroptera of America north of Mexico. *Ent. Amer.* **19** (3 and 4). *Literature cited.*

——— (1941). A synopsis of the Hemiptera-Heteroptera of America north of Mexico. *Ent. Amer. n.s.* **21** (2). Part II families Corediae Alydidae, Corizidae, Neididae, Pyrrhocoridae, and Thaumastotheriidae.

VANDUZEE, E. P. (1917). Catalogue of the Hemiptera of America, north of Mexico. University of California Press, Berkeley. *Bibliography.*

VIERECK, H. L. (1916). Insects of Connecticut. Part III, The Hymenoptera or wasp-like insects of Connecticut. *Geo. and Nat. Hist. Surv. Conn. Bull.* **22.**

WADE, J. S. (1935). A contribution to a bibliography of the described immature stages of North American Coleoptera. Bur. Ent. Wash., D. C. E-358 mimeographed.

WALDEN, B. H. (1911). The insects of Connecticut. Part II, The Euplexoptera and Orthoptera of Connecticut. *Geo. and Nat. Hist. Surv. Conn. Bull.* **16.** *Bibliography.*

WATSON, J. R. (1923). Synopsis and catalogue of the Thysanoptera of North America. *Fla. Agric. Exp. Sta. Tech. Bull.* **168.**

WILLISTON, S. W. (1908). A manual of North American Diptera. 3d ed. New Haven, Conn. J. T. Hathaway.

# CHAPTER VI

## METAMORPHOSIS

The term *metamorphosis* is derived from the Greek words, *meta*, change, and *morphe*, form, designating a change in form. The plural is *metamorphoses*. The word has been used in many fields, as for example, metamorphosed rocks, Ovid's "Metamorphoses," etc. Among ento-

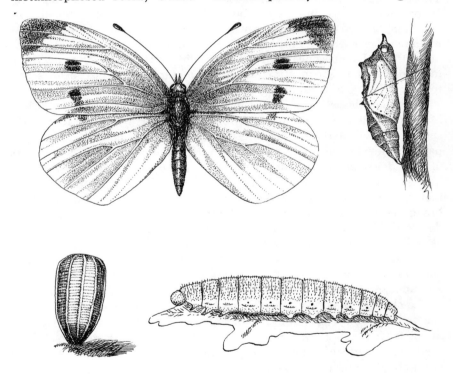

Fig. 73.—Metamorphosis of the common cabbage butterfly. Egg, larva, chrysalis and adult. *(After W. R. Walton.)*

mologists, there exists some difference of opinion as to the proper use of this term. Older workers confine it to the conspicuous changes from the pupa to the adult insect. A beautiful butterfly was once an ugly caterpillar, the busy bee had its origin in a footless grub, and the graceful dragonfly developed from a creature crawling upon the bottom of the pond. This is the popular conception of metamorphosis. Most entomologists, however, include all postembryonic changes in their idea of the metamorphosis of insects. Webster defines metamorphosis, in its application to zoology, as "a marked or more or less abrupt change in form, structure (and usually in habit and food) of an animal in its postembryonic development." In "A Glossary of Entomology," it is defined as the series of changes through which an insect passes in its growth from the egg through the larva and the pupa to the adult. Considering that larvae and pupae are stages of complete metamorphosis, this definition excludes insects with incomplete metamorphosis. Folsom (1906) states that it embraces all the conspicuous changes in form after birth. This is somewhat ambiguous because the birth of insects is often obscure. In parthenogenesis, insects are born after the eggs have hatched. In the Pupipara, larvae are full grown at the time of birth. The idea that metamorphosis refers only to postembryonic development no doubt arises through a misconception of the meaning of the term. Some writers interpret *meta* to mean after and hence metamorphosis to mean afterform, or change of form after the egg.

Metamorphosis is one of the most outstanding phenomena of insect life and, with the exception of a few orders and a comparatively few species, is well marked in all insects. In the higher animals, the larger part of development occurs before birth; in insects, the larger part of the development occurs after birth. The degree of development varies considerably with different insects. It is difficult to limit metamorphosis to any portion of the insect life or development. In a broad sense, it should embrace all the changes that occur during the development of an insect, including those that take place in the embryo as well as those that occur after the egg hatches. All life starts from a single cell and attains maturity through numerous changes. The changes that take place in the embryo, such as the development of legs, antennae, and segmentation, are equally as important and as remarkable as the changes that take place in the postembryonic development. The egg burster or special mouth hooks that serve to extricate the young insect from the egg are complete within the egg. As a matter of fact, the larva is perfect in the mature egg and often fully colored. After the adult stage is reached, there are further changes in some insects. Many insects, especially the Coleoptera, have distinct color changes. Insects also show outstanding sexual development after the adult stage is reached. In the Ephemerida, an additional moult occurs after the wings have fully expanded. May flies, of course, are not adults until the extra moult occurs.

Metamorphosis may be considered as a synonym of development because it involves the complete growth of an insect from the embryo to the adult. There are three basic principles underlying metamorphosis: growth, differentiation, and reproduction. Growth is conspicuously associated with the larva or nymph, differentiation is conspicuously associated with the pupa or transformation, and reproduction is asso-

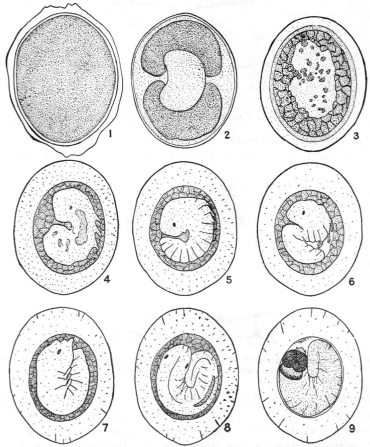

Fig. 74.—Embryonic development of the egg of the eye-spotted bud moth, *Spilonota ocellana*, showing development and change in body form. (1) Egg immediately after hatching; (2) egg showing migration of the yolk to the opposite ends; (3) yolk forming about the young embryo; (4) embryo 54 hours old showing leg buds; (5) embryo showing segmentation and leg buds pressed against the thorax; (6) horizontal revolution of the embryo; (7) embryo resting with legs between the thorax and abdomen; (8) embryo after the vertical revolution, head fully colored and yolk consumed.

ciated with the adult. These three basic principles are not, of course, clearly defined. For example, the germ cells, and later the gonads, are differentiated in the embryo but the genitalia are not completely formed until the adult. Likewise, the wings of insects appear as buds sometimes in the embryonic period but are not completely developed until the adult.

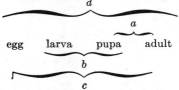

DIFFERENT OPINIONS CONCERNING THE LIMITS OF METAMORPHOSIS

a. Older and popular conception of metamorphosis.
b. General conception of most entomologists.
c. Conception of metamorphosis by recent workers.
d. The broad conception of metamorphosis.

**Change in Form.**—The change in form during the development of an insect is one of the outstanding characteristics of metamorphosis.    Form

FIG. 75.—The saddled prominent, *Heterocampa guttivitta;* (above) first instar larva showing nine pairs of horny appendages, (below) mature larva.

and structure, in turn, depend upon food and habits.    There are developmental periods in the life of an insect which are conspicuously different from other periods.    These are termed stages.    Thus an insect with complete metamorphosis has the egg stage, the larva stage, the pupa stage, and the adult stage.    These stages may in turn be divided into instars, marked, as we shall see, by moults.    During the larval stage, the conspicuous changes in form usually occur with each moult or at the time the insect passes into a new instar.    In the Lepidoptera, these changes are frequently so marked that it is difficult to recognize that two adjacent stages belong to the same species.    For example, the first instar of *Heterocampa guttivitta,* a feeder on oak, has a pair of large anterlike horns on the first thoracic segment and eight pairs of conspicuous horns on the abdominal segments.    In the second instar, all the horns are wanting except small vestiges of the first pair. The first, second, and third instars of the larvae of *Lithocolletis* are flat; the fourth and fifth instars are cylindrical. As a matter of fact, the first instars of most lepidopterous larvae are different from the subsequent instars. The most remarkable changes occur in insects that have extra larval stages. For example, the larval stages of the striped blister beetle, *Epicauta vittata*, differ remarkably from one another. The first instar is active and has long legs; the body is flattened and resembles the primitive insect (*campodeiform*). The second instar is flattened, has short legs, and resembles the larva of a carabid beetle (*carabiform*). The third instar is cylindrical and curved and resembles a white grub (*scarabaeiform*). The fourth instar resembles the pupa of an insect (*coarctate*).

Changes are often started in one instar and completed in a later one. The wings of plant bugs, Heteroptera, start their development in the first nymphal instar but do not become fully developed until the adult is reached. The wings of Diptera start their development in the larva but are not apparent until the adult is reached. Wing buds may even have their inception in the embryo.

Changes also take place during the interval of an instar. Considerable development takes place during the embryo. A larva assimilates food during each instar, but evident growth takes place only at the time the old larval skin is shed. In the adult, color develops and the insect matures.

**Change in Habits.**—The most conspicuous change in habit occurs when an insect transforms from the immature stage to the adult. Nearly all the larvae of the Lepidoptera are foliage feeders and have chewing mouth parts. The adults sip nectar or take no food at all; consequently, they have sucking mouth parts, or the mouth is vestigial or unfunctional. The immature stages of dragonflies, stone flies, and May flies live in the water and breathe by means of gills. The adults live in the air and breathe by means of tracheae. These abrupt changes in habits necessitate corresponding modifications in the structure of the insect.

A change of habit may occur within a single stage of the life history of an insect. The first, second, and third instars of a typical lithocolletid larva feed upon cell sap and produce a shallow mine. The fourth and fifth instars are tissue feeders and dig deeper into the parenchyma of the leaf. The first and second instars of the leaf-mining species of *Leucanthiza* and *Apophthesis* are sap feeders; the third and fourth instars are tissue feeders.

The larva of the ermine moth, *Yponomeuta malinellus*, mines at first in the leaf, then abandons the mine and feeds externally. The mixture of leaf-mining and external feeding habits is quite common among the microlepidoptera.

The first instars of the predacious *Epicauta* and *Mantispa* are very active; the legs are long and adapted for crawling and climbing. The later instars are more sedentary. By the time the larvae have reached the source of their food, locomotion is not so essential (see Fig. 83).

Individual species may also change their habits. The majority of the stinkbugs, Pentatomidae, are predacious. Frequently they turn phytophagous and attack fruits. "Cat-facing" is a common type of injury produced by certain species that attack peaches. The larvae of noctuids are herbivorous; nevertheless, they often turn cannibalistic, especially when confined with other larvae. Parasitic Hymenoptera generally feed upon nectar or sweetened liquids, but not a small number puncture their prey and drink the juices that exude.

**Change of Food.**—The change of food is closely correlated with the change of habits. The majority of the larvae of the Lepidoptera are plant feeders, but the adults sip nectar or take no food at all. The larvae of the majority of the Diptera, Trupaneidae, and Agromyzidae feed upon plant material but the adults feed upon liquids. The naiads of Odonata prey upon aquatic insects; the adults take food on the wing and attack terrestrial insects. The first instars of *Lithocolletis* are sap feeders; the later instars are cell eaters. The young of Hymenoptera are fed upon a variety of food. *Vespa* feed their larvae upon regurgitated food consisting of nectar and fruit juices. Later they feed them the softer parts of caterpillars, flies, and other insects. The adults prey upon flies and other insects. Young termites receive food of various kinds: wood, material regurgitated from the mouth or emitted from the rectum, cast skins, saliva, etc. At first, the nymphs receive only saliva; then they take stomodaeal food; finally they receive wood.

**Change of Size.**—Growth, in the sense of increase in size, seldom occurs in adults and rarely in the eggs of insects. The eggs of sawflies increase in size after they are laid. This is apparently due to the absorption of moisture from the plant. Growth, in the sense of development, occurs in all stages of insect life. The immature stage is the period of feeding and growing, and the largest part of an insect's life is spent in growing and developing. Insects that hibernate as eggs, larvae, or pupae, remain in these stages for a long time but these are periods of dormancy and inactivity. Normally insect growth is comparatively rapid. The majority of the species obtain their full size within a period of a year and many even within a few weeks. The rapidity of growth is often phenomenal. Trouvelat found that the caterpillar of *Telea polyphemus* increased in weight 4,140 times in 56 days. Scarabaeus eats its own weight in dung in 24 hours.

One of the characteristics of insects is the fact that there is no growth after the final moult and the adult is reached. Many insects take food in the adult stage.  Some have vestigial mouth parts and are incapable of taking food.  Adult insects that take food may increase considerably in size, but there is no growth.  Female mantids gorge themselves with food, and their abdomens increase tremendously in size.  Honey ants receive nectar from the workers and store it in their alimentary canals for the future food demands of the colony.  The abdomens of these repletes enlarge to such a degree that the sclerotized areas of the abdomen are scarcely visible and the membrane between the segments is stretched into a thin wall through which the golden honey is visible.  These ants become living honey casks.

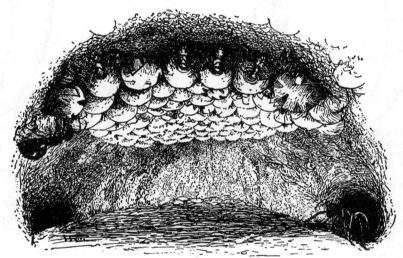

Fig. 76.—Repletes of *Myrmecocystus horti-deorum* hanging from the roof of the honey chamber.  These ants become living honey casks.  (*After McCook.*)

The abdomen of the female may become considerably enlarged by the development of eggs.  In most insects, the female can be recognized by the larger abdomen.  Extreme development occurs in the queen termite. An active queen in the Tropics may attain a length of seven to eight inches.  Her abdomen becomes greatly swollen by the development of eggs.  Such a queen is incapable of movement but lives with her mate in a royal chamber, eats food brought to her by the workers, and is attended by the workers, who clean her and carry the eggs away.

**Moulting.**—Growth or increase in size is not regular or gradual in insects as in many other animals.  The body of an insect is covered with a hard skin, known as the cuticula.  This is more prominent in nymphs, naiads, pupae, and adults.  Larval skins tend to be softer and less heavily sclerotized.  As a larva or nymph feeds and grows, its skin (cuticula) becomes tightly stretched.  At the same time a new skin is

formed beneath the old cuticula.  This new skin is soft.  A moulting
fluid, secreted by special hypodermal glands, is liberated between the

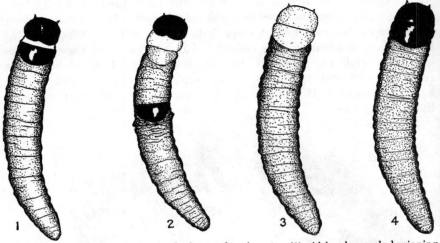

FIG. 77.—Moulting process in the larva of an insect.   (1) old head capsule beginning
to push off; (2) head capsule about half off, revealing the new head capsule beneath and the
thoracic shield pushing toward the posterior end of the body; (3) the larva has crawled
from the old larval skin, and the head capsule has popped off; at this time the new head
capsule and thoracic shield are usually uncolored; (4) the head capsule and thoracic shield
have assumed their normal color.

new and the old skins which dissolves the old cuticula and aids in separating them.   At the proper time, the skin splits, usually along the upper
side near the head, and the insect crawls out of its
old skin.   The head capsule of a caterpillar often
pops off separately as it is more heavily sclerotized
than the rest of the larval skin.   In the process of
moulting, the larva may appear to have two heads,
the old head capsule, in the process of moving off,
and the new head which at first is colorless.   After
the moult, the new skin soon hardens, the color
develops, and the insect is ready for another period
of feeding and growing.   This process is known as
moulting (molting), or *ecdysis*.   The old skin that is
thrown off is known as *exuviae*, the cast skin or

FIG.   78.—Silken
feeding case of the
eye-spotted bud moth,
*Spilonota ocellana*,
covered with fecula.

clothes.   The singular, *exuvia*, is never used in
entomology.   The period between moults is known
as a *stadium* (plural, *stadia*).   The form of the insect
between moults is an *instar* (plural, *instars*).

The moulting process is a delicate one, and the insect at this time is
often inactive, helpless, and very suceptible to injury.   Many insects

seek sheltered or secluded places in which to moult. The Lepidoptera often retreat to folded leaves or spin a few silken threads beneath which they find protection. The larva of the eye-spotted budmoth, *Spilonota ocellana*, moults within its feeding case, which serves as a retreat during the entire larval and pupal periods. The case is composed of silken threads tightly woven into a tube which is covered with pellets of frass. The construction of similar feeding cases is common among the Lepidoptera and is especially well developed in the leaf crumplers, *Acrobasidae*. The feeding cases of the *Acrobasidae* are trumpet shaped, one or more inches in length, and frequently considerably twisted. The apple bucculatrix, *Bucculatrix pomifoliella*, constructs special moulting cocoons on the leaves in which the larvae transform.

Fig. 79.—Fecular pellets discarded by lepidopterous larva feeding in captivity. Each bottle contained a larva of the same instar. The pellets are all the same size.

A remarkable feature associated with moulting is the fact that the lining of the fore and the hind portion of the alimentary canal, as well as that of the tracheae, is shed along with the cuticula. This is a difficult operation which compels the insect to cease activities for a short period of time.

As the time for moulting approaches, the insect ceases to feed and becomes inactive. In some insects, such as the poplar leaf roller, *Melalopha inclusa*, growth is somewhat uniform and all individuals hatching from the same batch of eggs moult about the same time. It is not unusual to see the larvae of this species congregated upon a leaf waiting for the moult to occur. This habit occurs in many gregarious insects. It is especially noticeable among sawflies.

Other interesting habits are associated with the moulting of insects. The consumption of food generally increases after each moult. A phytophagous species consumes larger areas of leaf surface following each moult. The particular instar can often be determined by the increased ravages on the plant. The sizes of the fecal pellets increase regularly with each successive moult. This is particularly noticeable when insects are reared in captivity. It has long been recognized that the sizes of the head capsules of the larvae of insects increase in a regular geometric

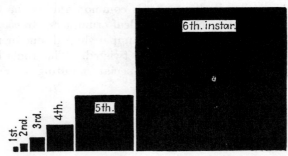

Fig. 80.—Relative amounts of foliage eaten in each larval instar by the larvae of *Cirphis unipuncta*. (*After Davis and Satterthwait.*)

progression. Dyar (1890) pointed out that the width of the heads of larvae is constant for a particular instar and species.

Measurements of Head Capsules of *Spilonota ocellan* Schiff

| Number of instar | Number of head capsules measured† | Average length,† millimeters | Average width,† millimeters | Dyar's* measurements, millimeters |
|---|---|---|---|---|
| 1 | 30 | 0.142 | 0.216 | 0.225 |
| 2 | 34 | 0.188 | 0.275 | 0.288 |
| 3 | 21 | 0.258 | 0.363 | 0.363 |
| 4 | 16 | 0.299 | 0.436 | 0.43 |
| 5 | 27 | 0.356 | 0.513 | 0.56 |
| 6 | 28 | 0.404 | 0.572 | 1.05 |
| 7 | 22 | 0.457 | 0.645 | |
| 8 | 15 | 0.510 | 0.733 | |
| 9 | 23 | 0.517 | 0.760 | |
| 10 | 16 | 0.629 | 0.975 | |

* Dyar (1890).
† Frost (1922). See also p. 116.

Little on this subject has been done with the nymphal and naiadal skins of insects, but these would probably show considerable uniformity in size in the different instars.

All the foregoing habits show a regular increase with the development of the insect. On the other hand, the duration of the stadium generally decreases as development increases; in other words, the rate of growth usually accelerates as the insect matures. It is well known that the last instar of some insects is distinctly shorter than the preceding ones. This is especially true of insects that hibernate in the immature stages. The fruit tree leaf roller, *Cacoecia argyrospila*, for example, lays its eggs on the bark of trees during June, but these do not hatch until the following May, a period of approximately 10 months. In contrast, the whole larval and pupal periods of this species are passed in less than two months.

**Number of Moults.**—The frequency of moulting varies considerably in different species of insects. In *Campodea* and *Japyx*, there is a single moult. The Acrididae have five moults. The Lepidoptera usually have four or five moults. The May fly, *Cloeon*, has 20 moults. The seventeen year cicada is said to have 25 to 30 moults. The number of moults varies according to the latitude and is affected by temperature, moisture, and food supply. Saunders and Dustan (1929) found that the eye-spotted budmoth, *Spilonota ocellana*, had seven moults in Canada. Frost (1922) recorded 10 moults for the budmoth in Pennsylvania. Since the latter figure was obtained from material reared in captivity, the larger number of moults may have been the result of undernourishment or other unfavorable conditions. In the case of the seventeen-year cicada, it is believed that the large number of moults is due to low temperatures and scarcity of food. The larvae of most insects, such as the clothes moth and the carpet beetle, *Anthrenus scrophulariae*, which feed on comparatively dry materials, may under starvation conditions continue to moult and become smaller in size with each successive moult. It is also well known that deficient or undesirable food produces smaller adults than normal.

**Types of Metamorphosis.**—We recognize two types of metamorphosis: internal and external. Internal metamorphosis takes place within the egg or other immature stages. These changes are not generally visible from the exterior of the insect. The wings of Diptera develop as buds in the larva and are completely obscured until the adult emerges. The most remarkable changes take place in the pupae of some insects. The tissues may be completely broken down. A process of degeneration takes place which is known as *histolysis*. Phagocytes feed upon the debris of the disintegrating tissues, and this material is diffused in the blood and serves as nourishment to build up new tissues. The prolegs of larvae and other structures are torn down. True legs, wings, and other structures may be formed.

On the other hand, it is customary to classify insects according to the types of metamorphosis, *i.e.*, upon the conspicuous external changes that occur during their development. Upon this basis we recognize five groups of insects: *Ametabola*, *Paurometabola*, *Bathmedometabola*,* *Holometabola*, and *Hypermetabola*.

* Hemimetabola of Comstock.

*Ametabola.*—Strictly speaking there are no insects without metamorphosis. Insects that have no distinct change in form between the egg and the adult, *i.e.*, those in which the immature forms resemble the adult, in which there is no indication of wings even in the embryo, and in which there is no change in food or habit, are said to have no metamorphosis. These insects are known as *Ametabola.* Thysanura, Dicellura, Collembola, Anoplura, and Mallophaga belong to this group. Some entomologists believe that Anoplura and Mallophaga have a slight change in their development and that they do not belong here. No term has been suggested for the immature forms of the Ametabola. For Paurometabola, we have nymph; for Bathmedometabola, we have naiad; for Holometabola and Hypermetabola, we have larva. The term gaead is therefore suggested for the young of Ametabola. The word is derived

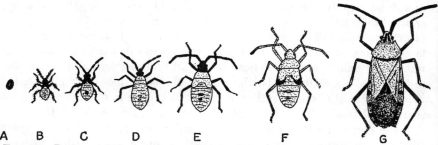

A       B       C       D       E           F               G

Fig. 81.—Paurometabolous growth. The development is gradual from egg to adult: (*A*) egg; (*B* to *F*) nymphal stages; (*G*) adult.

from the Greek *gaia*, earth, and *ad*, daughter of. The Thysanura, Dicellura, and Collembola are typically daughters of the earth.

*Paurometabola.*—In a large number of insects the immature stages, exclusive of the egg, comprise a number of active, feeding instars differing little from one another except in the gradual development of wings, genital organs, etc. These species are all terrestrial; at least, none has gills, there is no change in food or habit, and there is no distinct resting period before the adult stage. Insects with these characteristics are known as *Paurometabola.* They have three distinct life stages: egg, nymph, and adult. The grasshoppers and their relatives (Orthoptera), the termites (Isoptera) the, stinkbugs and their relatives (Heteroptera), the aphids, scale insects, and their relatives (Homoptera), the thrips (Thysanoptera), and the earwigs (Dermaptera) belong to this group. These insects are said to have gradual metamorphosis or direct metamorphosis.

*Bathmedometabola* (*Hemimetabola*).—Insects which have a gradual development similar to Paurometabola but in which there is a distinct change in food and habit and in which there is a more or less distinct resting period before the adult stage is reached are known as *Bathmedo-*

*metabola,* a word derived from the Greek, *bathmedo,* meaning gradual or step by step. It seems more descriptive than *Hemimetabola,* or half development, proposed by Comstock. The Bathomedometabola are aquatic in the immature stages and breathe by means of tracheal or blood gills. They have three distinct life stages: egg, naiad, and adult. The Odonata, Plecoptera, and the Ephemerida belong to this group and are said to have incomplete metamorphosis or direct metamorphosis. Both the Paurometabola and the Bathmedometabola have incomplete or direct metamorphosis, the difference depending upon the degree of development.

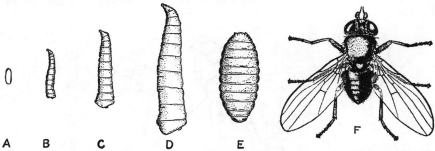

**A**    **B**     **C**      **D**      **E**       **F**

FIG. 82.—Holometabolous growth. Development is interrupted at *E.* (*A*) Egg; (*B* to *D*) larval stages; (*E*) puparium; (*F*) adult.

*Holometabola.*—In the orders Neuroptera, Mecoptera, Trichoptera, Lepidoptera, Diptera, Coleoptera, Siphonaptera, Strepsiptera, and Hymenoptera the immature forms are strikingly different from the adults and there is a distinct resting period, known as the pupa, that precedes the adult stage. When these insects hatch from eggs, they generally bear no resemblance to the adults. In their development, they pass

A SYNOPSIS OF THE HOLOMETABOLA

| Stage or phase | Egg | Larva | Pupa | Adult |
|---|---|---|---|---|
| Function | Beginning | Feeding Growing Transformation* | Reconstruction Transformation Adaptive | Reproduction |
| Characteristics | Inactive | Active | Inactive† Helpless | Active No growth |
| Additional names | Nit Pseudova | Grub Maggot Caterpillar "worm" | Chrysalis Puparium‡ Cocoon‡ | Imago Adult Insect |

\* Internal metamorphosis may start in the larva.
† Many pupae move when disturbed; mosquito pupae are active.
‡ Names for the coverings of certain pupae.

through four distinct life stages: egg, larva, pupa, and adult.    They are said to have complete or indirect metamorphosis.

*Hypermetabola.*—In a few insects there is a superdevelopment in which the larvae pass through four or more distinct instars.    This is really a form of complete metamorphosis (*Holometabola*).    Hypermetamorphosis, occurs conspicuously in four orders: Neuroptera, Coleoptera, Hymenoptera and Strepsiptera.    The more strikng examples are found in the life histories of *Mantispa,* a predator on the eggs of lycosid spiders; *Meloe,* a predator on grasshopper eggs and on certain solitary bees; *Stylopidae,* parasitic on bees and wasps; and *Platygaster,* a parasitoid on the hessian fly.

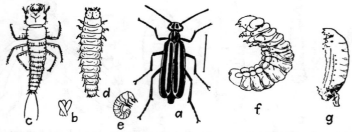

FIG. 83.—Hypermetabolous growth represented by the striped blister beetle (*Epicauta vittata*).    (*a*) Female beetle; (*b*) eggs; (*c*) triungulin larva; (*d*) second or caraboid stage; (*e*) and (*f*) scarabaeoid stage; (*g*) coarctate larva.

It is sometimes difficult to classify insects on the basis of metamorphosis, for there is considerable overlapping of habits in the different orders.

DEGREES OF DEVELOPMENT INDICATED BY TYPES OF METAMORPHOSIS

| Type of metamorphosis | Immature form | Resting period | Manner of transformation | Habit | Food |
|---|---|---|---|---|---|
| Ametabola............. | Gaead | None | None | No change | No change |
| Paurometabola......... | Nymph | None | In nymphal skin | No change | No change |
| Bathmedometabola (Hemimetabola). | Naiad | Slight | In naiadal skin | Distinct change | Slight change |
| Holometabola, Diptera.. | Larva | Complete, uninterrupted | In pupal and larval skins | Distinct change | Distinct change |
| Holometabola, orders other than Diptera. | Larva | Complete, uninterrupted | In pupal skin | Distinct change | Distinct change |
| Hypermetabola........ | Larva* | Complete, interrupted | In pupal skin | Distinct change | Distinct change |

* The larva has various forms: campodeiform, scarabaeiform, coarctate, etc.

*Mantispa*, for example, is hypermetabolous; however, the third larval instar is nymphlike, suggesting the Paurometabola. The Aleyrodidae and Coccidae are paurometabolous; nevertheless, the Aleyrodidae and the male Coccidae have a distinct resting period before the adult is reached, which suggest the Holometabola. A cocoon is even formed in the male Coccidae.

## BIBLIOGRAPHY

References may be found in any general text and in many isolated papers, a few of which are listed.

### I. General References.

BODENSTEIN, D. (1939). Investigations of the problem of metamorphosis. *Journ. Exp. Zool.* **82** (3). *Literature.*

CARPENTER, G. H. (1921). Insect transformation. Methuen & Co., Ltd., London.

DOWNES, J. A. (1936–1937). The metamorphosis of insects. *Proc. Trans. Soc. Lond. Ent. Nat. Hist. Soc.*

EVANS, A. C. (1932). Some aspects of the chemical changes during insect metamorphosis. *Journ. Exp. Biol. Lond.* **9**.

FOLSOM, J. W. (1934). Entomology with special reference to its ecological aspects 4th ed. rev. by R. A. Wardle. The Blakiston Company, Philadelphia.

HARRIES, F. H., and HENDERSON, C. F. (1938). Growth of insects with reference to progression factors for successive growth stages. *Ann. Ent. Soc. Amer.* **31** (4).

IMMS, A. D. (1937). Metamorphosis. *In*, Recent advances in entomology. The Blakiston Company, Philadelphia.

MALOEUF, N. S. R. (1938). The physiology of arthropodan development with special reference to the insects. *Bull. Soc. Ent. Egypte, Cairo* **21**.

MIALL, L. C. (1895). The transformations of insects. *Nat. Lond.* **53** (1364).

PÉREZ, C. (1899). Sur la metamorphose des insectes. *Bull. Sci. France et Belg.* **37**.

RICHARDS, A. G. (1937). Insect development analyzed by experimental methods; A review parts I and II. *Journ. N. Y. Ent. Soc.* **45** (2). *Literature cited.*

SCOTT, W. N. (1936). An experimental analysis of the factors governing the hour of emergence of adult insects from their pupae. *Trans. R. Ent. Soc. Lond.* **85** (13). *Literature.*

Symposium (1920). The life cycle in insects. *Ann. Ent. Soc. Amer.* **13** (2).

UVAROV, B. P. (1928). Insect nutrition and metabolism. *Trans. Ent. Soc. Lond.* **76** (2). *Literature summarized.*

WEBER, H. (1938). Physiological factors of metamorphosis. *In*, Grundriss der Insektenkunde. Gustav Fischer, Jena.

WIGGLESWORTH, V. B. (1934). The physiology of ecdysis in *Rhodnius prolixus* (Hemiptera) II. Factors controlling moulting and "metamorphosis." *Quart. Journ. Micros. Sci. Lond.* **77** (306).

XAMBEU, LE CAPITAINE (1899–1916). Moeurs et metamorphoses des insectes. Nearly 40 papers published in *Naturaliste Paris, Lyons Ann. Soc. Linn.*, and *Exchange Moulins.*

### II. References to Ecdysis.

BODENHEIMER, F. S. (1933). The progression factor in insect growth. *Quart. Rev. Biol.* **8** (1). *Literature.*

DIMMOCK, A. K. (1888). Variable number of moults of insects. *Psyche* **5**.

DYAR, H. G. (1890). The number of molts of lepidopterous larvae. *Psyche* **5**.

FRANKENBERG, G. V. (1939). Insektenhäute. *Microkosmos Stuttgart* **31**.

FROST, S. W. (1922). Ecdysis in *Tmetocera ocellana* Schiff. *Ann. Ent. Soc. Amer.* **15** (2).

GAINES, J. C., and CAMPBELL, F. L. (1935). Dyar's rule as related to the number of instars of the corn ear worm, *Heliothis obsoleta* (Fab.), collected in the field. *Ann. Ent. Soc. Amer.* **28** (4).

HARRIES, F. H., and HENDERSON, C. F. (1938). Growth of insects with reference to progression factors for successive growth stages. *Ann. Ent. Soc. Amer.* **31** (4).

HERIOT, A. D. (1936). Notes on the moulting of mites and insects. *Proc. Ent. Soc. Br. Columbia* **32**.

HESS, B. M. (1937). Moulting in some Lepidoptera. *Ent. News* **48** (6).

KNAB, F. (1909). The rôle of air in the ecdysis of insects. *Proc. Ent. Soc. Wash.* **11** (2).

―――― (1911). Ecdysis in Diptera. *Proc. Ent. Soc. Wash.* **13** (1).

MARLATT, C. L. (1890). The final molting of tenthredinid larvae. *Proc. Ent. Soc. Wash.* **2** (1).

METCALFE, M. E. (1932). On a suggested method for determining the number of larval instars. *Ann. Appl. Biol.* **19** (3). *References.*

MILES, MARY (1933). Observations on growth in larvae of *Plodia interpunctella* Hubn. *Ann. Appl. Biol.* **20** (2). *References.*

PACKARD, A. S. (1897). The number of molts in insects of different orders. *Psyche* **8**.

SHPET, G. (1935). Determining the instars by the variation-statistical method. (In Russian.) *Zashchita Rastenii Fasc.* **1**.

SLINGERLAND, M. V. (1893). Measurements of head capsules. A footnote in The bud moth. *Cornell Univ. Agric. Exp. Sta. Bull.* **50**.

SWAIN, R. B. (1938). On the number of molts in larvae of the fall webworm, *Hyphantria cunea* Drury. (Lep. Arctiidae.) *Can. Ent.* **70** (4).

SWEETMAN, H. L., and WHITTMORE, F. W. (1937). The number of molts of the fire brat (Lepismatidae, Thysanura). *Bull. Brooklyn Ent. Soc.* **32** (3).

TAYLOR, R. L. (1931). On "Dyar's rule" and its application to sawfly larvae. *Ann. Ent. Soc. Amer.* **24**.

WEBSTER, R. L. (1912). The number of moults of the pear slug, *Caliroa cerasi* Linne. *Journ. N. Y. Ent. Soc.* **20** (2). *Bibliography.*

WIGGLESWORTH, V. B. (1934). The physiology of ecdysis in *Rhodnius prolixus* (Hemiptera) II. Factors controlling moulting and "metamorphosis." *Quart. Journ Micros. Sci. Lond.* **77** (306).

WOODRUFF, L. C. (1939). Linear growth ratios for *Blattella germanica* L. *Journ. Exp. Zool.* **81** (2). *Literature.*

# CHAPTER VII

## IMMATURE INSECTS

Adult insects are better known and more frequently studied than the immature forms. Of the 640,000 described species of insects, it is doubtful if the immature forms of 10,000 are known. Entomology becomes complex when all the immature stages are taken into consideration. If we accept seven as the average number of instars in the majority of insects (egg, four immature instars, pupa, and adult), there are 4,480,000 forms for study.

The enticing feature of entomology is the magnitude of the field. No entomologist can hope to be thoroughly acquainted with the subject. It is a study in which something new can always be discovered. A typical life history illustrates the diversity of insect habits. Consider a small worm less than a quarter of an inch long that has hatched from a much smaller egg dropped without much ceremony upon a stone in a stream bed. This is the story of the caddis fly. It constructs a little house, selecting and fitting together small stones which are fastened firmly with a waterproof, gluelike cement manufactured within its own body. These shelters are carried about wherever the caddis worm goes in search of food. The food consists of diatoms and other small plants and animals that are obtained from the water. When the worm is full grown, it selects a stone and closes its case so that it can transform without interruption. After a period, during which many remarkable changes take place, the adult emerges with its long antennae and hairy wings. The development of the caddis fly is not more remarkable than that of many other species. The habits of no two species are identical, and each reveals some difference in food, structure, or habit.

**Eggs.**—Insects develop from eggs which are usually laid before the embryo is appreciably developed. Such insects are termed *oviparous*. In some cases the egg is retained within the body until it is hatched and the larva or the nymph is deposited. The larva may become full grown before it is deposited. These insects are called *viviparous*.* As a rule fertilization is necessary; however, there are many instances of partheno-

---

* Some writers use the term *ovoviviparous* to designate insects that produce eggs with well-developed shells which hatch within the body of the female. This term is unnecessary as all insects develop from eggs, whether they hatch within the body or are laid by the female in the normal manner.

genesis among insects. In ants, bees, and social wasps, the males (drones) are produced from unfertilized eggs, the workers and females (queens) are produced from fertilized eggs.

The egg is generally surrounded by a shell or *chorion*. Immediately beneath the chorion is a delicate lining, the *vitelline membrane*, enclosing the contents of the egg. The interior of the egg consists of the nutritive matter or yolk (*deutoplasm*), the nucleus, and its accompanying cytoplasm. There are two types of eggs: (1) those in which the germ band is arranged in a single layer of cells about and enclosing the yolk, as in *Musca*, and (2) those in which the germ cells are contained in the deutoplasm, as in the Collembola. The nucleus, the most vital part of the egg, contains chromatin from which the chromosomes are formed. The micropyle, a more or less conspicuous spot at one end of the egg, serves as an opening for the entrance of the male germ cell. The number and position of the micropylar openings vary considerably. As a rule there is a single opening but in the cotton moth there are several.

**Size of Egg.**—Insects eggs differ greatly in size but all are comparatively small. Eggs are generally smaller when they are produced in large numbers. *Corydalis cornuta*, a comparatively large insect, lays 2,000 or 3,000 eggs in a compact mass scarcely half an inch in diameter. On the other hand, some aphids lay single winter eggs which are nearly as large as the female. As a rule, insects tend to lay eggs according to their size. The smallest known eggs are those of the Collembola. The eggs of *Pterodontia flavipes*, a species of Cyrtidae, measure 0.15 by 0.18 millimeter. The eggs of the clover seed midge, *Dasyneura leguminicola*, are 0.30 by 0.075 millimeter. The eggs of the Tingidae are also minute. The other extreme is found in the eggs of the Citheroniidae. The egg of the polyphemus moth is 3 millimeters in diameter.

**Shape of Egg.**—Some insect eggs are flat and scalelike, as for example the eggs of the codling moth and the oriental fruit moth. Others are much thicker, such as those of polyphemus. The eggs of the swallowtail butterflies, *Papilio*, are nearly spherical. The eggs of the green June beetle, *Cotinis nitida*, and of many other Scarabaeidae are also spherical. The eggs of the imported cabbage worm *Ascia rapae*, the milkweed butterfly, *Danaus menippe*, and the violet tip, *Polygonia interrogationis*, are conical in shape and deeply ridged. Many eggs are elongate, as for example, the eggs of leafhoppers, tree hoppers, and tree crickets. The eggs of certain species of Drosophila have one or two appendages; others are without appendages. The eggs of *Nepa apiculata* have eight or more filaments radiating from the upper rim.

Pentatomid eggs are usually beset with a circle of spines around the upper edge. Reduviid eggs take various forms. They are generally cylindrical or elongate ovate. At one end there is always a definite cap

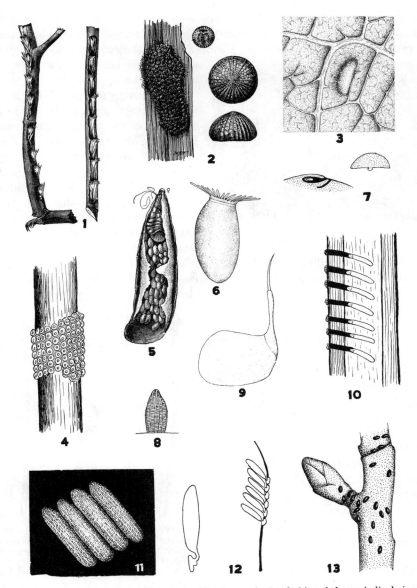

Fig. 84.—Eggs of various insects: (1) injury by egg-laying habits of the periodical cicada; (2) fall army worm; (3) egg of grape leafhopper inserted into leaf; (4) eggs of fall cankerworm; (5) ventral view of oyster shell scale showing female and eggs; (6) reduviid egg, *Melanolestes;* (7) plum curculio egg scar and egg inserted into fruit; (8) egg of imported cabbage butterfly; (9) egg mass of *Hydrophilus obtusatus;* (10) eggs of tree cricket, *Oecanthus nigricornis;* (11) eggs of *Pegomyia hyoscyami* on leaf; (12) eggs of ox warble attached to hair; (13) eggs of aphid on apple twig.

which is pushed off in the hatching process and which is often ornamented by raylike extensions. The poultry louse has a striking egg—white and covered with glasslike spines. The free end of the egg is furnished with a lid which bears at its apex a long lashlike appendage.

**Sculpturing of Eggs.**—The surfaces of insect eggs may be entirely smooth or variously sculptured. Eggs that are laid in wood, leaves, or in the ground are frequently without sculpturing. The eggs of Curculionidae and Scarabaeidae are perfectly smooth. On the other hand, many eggs are reticulated or strikingly marked. These reticulations are the imprints of

FIG. 85.—Eggs of a syrphus fly showing microscopic sculpturing.

the cells of the follicular epithelium. The eggs of Syrphidae are chalky white and microscopically sculptured. The leaf-mining *Pegomyia* usually have eggs that are well marked by hexagonal or polygonal areas. The eggs of Pieridae, Noctuidae, and certain other Lepidoptera are deeply ridged and strongly sculptured.

**Coverings of Eggs.**—As a rule, eggs laid during the summer have very thin chorions. The eggs of the codling moth, the oriental fruit moth, and in fact most Tortricidae are extremely thin and transparent. When first laid, they are so slightly visible that they can be clearly seen only by reflected light. The development of the embryo can be seen through this transparent egg shell. However, the eggs of the squash bug, the stinkbug, and other Heteroptera

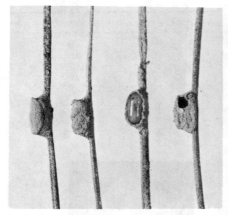

FIG. 86.—Eggs of Chlaenius in mud cells. One is cut to show egg. Another shows emergence hole of larva. (*Pennsylvania Department of Agriculture.*)

have hard shells although they are laid during the summer. The thick walls apparently resist desiccation and predacious enemies. The eggs of *Sturmia (Blepharipa) scutellata*, a tachinid that attacks the gypsy and brown-tail moths, have hard, thick shells. The eggs are laid upon the foliage and remain there until the host caterpillar engulfs the leaves and the eggs laid upon them. The eggs of the dobson fly, *Corydalis cornuta*, are surrounded by a thick white covering which no doubt protects them from prowling enemies. Eggs are often covered by materials such as fecula, regurgitated food, hair from the body of the female, etc. These coverings serve to obscure the eggs and render them less conspicuous.

Eggs that must withstand severe winter conditions, especially in open places, are protected by hard shells. An interesting comparison can be drawn between the eggs of two common leaf rollers that are closely related and that are injurious to the apple. The red-banded leaf roller, *Argyrotaenia velutinana*, hibernates as pupae and lays eggs only during the summer. In this species, the eggs are very thin and transparent. The fruit tree leaf roller, *Cacoecia argyrospila*, hibernates in the egg stage and the eggs have a thick varnishlike covering.

Eggs that are exposed during the winter are generally laid in compact masses. The apple tree tent caterpillar is an excellent example. On the other hand, the walkingstick drops its eggs to the ground one by one. They are protected by a thick chorion and by a covering of leaves and other debris. Aphids group their eggs loosely upon various woody plants. These eggs have very tough shells to withstand the winter.

Insect eggs are covered during the winter by various materials. The varnishlike covering of the eggs of the apple tree tent caterpillar has already been mentioned. Scale insects secrete a waxy covering. The spring egg sacs of the cottony maple scale *Pulvinaria vitis*, are unusually thick. The gypsy and brown-tail moths, and other species of Lepidoptera, incorporate considerable hair from their bodies in the coverings of their egg masses. The female bagworm deposits her eggs in the bag which serves as a home during the entire feeding period as a cocoon during the pupal period as a receptacle for the eggs, and as a nursery for the young.

Winter eggs that have no hard shells and no special covering are protected in various other ways. Some are inserted into the pith or wood of plants. Not a small number are placed in the ground. The short-horned grasshoppers and many of the June beetles deposit their eggs in the ground.

**Number of Eggs.**—We have previously mentioned the number of eggs laid by insects in connection with the reproductive capacity. The sheep tick and the true female of many aphids produce but a few eggs. On the other hand, the egg mass of the dobson fly may contain 3,000 eggs, and the parasitic *Pterodontia flavipes* (Cyrtidae) has been known to lay 3,977 eggs. The social insects lead the list. A termite may lay 1,000,000 eggs during her life.

**Modes of Laying Eggs.**—Insect eggs are generally laid in situations where the young, upon hatching, may readily find food. Species that

feed upon foliage usually lay their eggs upon leaves.  Aquatic insects
lay their eggs in or near the water.  Parasites generally lay their eggs
upon or within their host.  Some of the Syrphidae lay their eggs in
clusters of aphids or other soft-bodied insects.  The Mallophaga and
Anoplura lay their eggs upon the hair or feathers of their hosts.

There are of course exceptions.  Some insects lay their eggs upon
foliage or in the ground and the young larvae are compelled to seek their
hosts.  Some stylopids lay their young upon plants where they must
wait until certain solitary bees visit them.  The young grasp the legs of
these bees and are carried to nests where they find their hosts.  The eggs
of the walkingsticks hibernate beneath leaves or other debris upon the
ground.  With the approach of spring, the eggs hatch and the nymphs

Fig. 87.—Egg of mud dauber fastened to one of the spiders used as provision in its nest.

must find the leaves of their host plants.  Insects such as leafhoppers
and aphids, many of which feed upon herbaceous plants during the
summer, seek woody plants on which to lay their eggs when winter
approaches.

Numerous insect eggs are free or unattached.  The eggs of many
mosquitoes float upon the surface of the water.  The eggs of *Chironomus bathophilius* are dropped into the water and immediately begin
to sink slowly.  If they do not hatch before they reach a depth of 16 or
30 meters, they are crushed by the pressure of the water and never
hatch.  Some water scavenger beetles, Hydrophilidae, lay their eggs
in compact masses covered with a hard silklike secretion.  These often
float freely in the water.  A few species of Odonata drop their eggs in
the water.  The female flies back and forth over the water, sweeps
down at intervals to touch the surface with the tip of her abdomen, and
washes off an egg or two with each dip.  The Plectoptera and some
Ephemerida drop their eggs in masses into the water.

Eggs that are inserted into wood, leaves, fruit, and seeds are unattached but are held firmly in place by the medium in which they are deposited.   Many insects cut slits in the leaves and push their eggs within the tissues.   This is especially true of many leaf-mining insects of the orders Lepidoptera, Hymenoptera, Coleoptera, and Diptera.   The fruitflies, Trupaneidae, and the plum curculio insert their eggs in fruits.

FIG. 88.—Eggs of *Macremphytus varianus* inserted in the leaf of dogwood.

The tree crickets, tree hoppers, and leafhoppers lay their eggs within woody plants.   Many chalcids oviposit in seeds of various kinds.

The attachment of eggs may be superficial.   For instance, the egg mass of the May fly, *Heptagenia interpunctata*, has silklike expansions which entangle aquatic plants.   Other eggs are anchored firmly.   When an egg is laid, it is surrounded with a viscous substance which, upon drying, forms a cement that attaches the egg to the surface upon which it is laid.   The eggs of most insects are laid directly upon leaves, bark, stones, and similar objects.   A few eggs are attached by means of filaments.   The lacewing *Chrysopa* supports its eggs on delicate stalks of silk, which are so fragile that the eggs seem to float over the surface of the leaf. This is a remarkable development which prevents the young, upon hatching, from attacking the remaining eggs. The jug-builder,

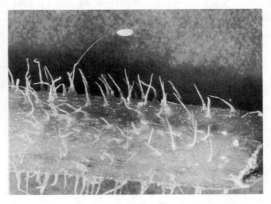

FIG. 89.—Egg of lace-wing fly, *Chrysopa*, attached by a slender filament to a flower stalk. Usually several eggs are laid in a group.

*Eumenes fraterna*, suspends its egg from the dome of its earthen cell which prevents it from being crushed by the paralyzed larvae placed there as food for the young wasp. *Chironomus meridionalis* suspends its small cluster of eggs from the surface film of the water by means of a disk of silklike material and a slender filament of the same kind.

FIG. 90.—Egg of the poly-phemus moth from which a para-site has emerged.

Many insects lay their eggs singly. This is probably the primitive method. The habit is well developed among insects such as Cecropia, Polyphemus, and Papilio, which feed upon foliage. Numerous leaf-eating beetles and sawflies exhibit the same habit. Social insects lay one egg in a cell. Parasitic insects usually deposit one egg at a time although several eggs may be laid in a caterpillar or several larvae may develop from a single egg (*polyembryony*).

The majority of insects lay their eggs in groups or masses. They may be loosely grouped as is the case in the squash bug and the aspar-agus beetle. On the other hand, they may be more closely grouped as is the case in the harlequin cabbage bug or the katydid. In many of the Tortricidae, the eggs overlap like shingles on a roof. Other insects lay their eggs in compact masses. There are many examples: most of the Hydrophilidae, certain May flies, apple tree tent caterpillars, and fall cankerworms. The laying of eggs in masses is sometimes associated with the gregarious habit later in life. This happens in the case of the apple tree tent caterpillar and the squash bug. On the other hand, many insects scatter when they hatch from their egg masses. This is particularly true of predacious insects that find it difficult to obtain food.

**Duration of Egg Stage.**—The egg stage usually lasts but a few days. Sometimes it is shorter or the eggs may hatch before they are laid, as is the case in the summer broods of aphids. Within a single species there may be considerable variation. The eggs of the oriental fruit moth, *Grapholitha molesta*, hatch in 3.5 to 6 days in warm weather, in 7 to 14 days in cool weather, and in 20 to 43 days in continued cool weather. Many insects hibernate in the egg stage, in which case the egg period may last five or six months. Although many species pass the winter as eggs, the embryos

FIG. 91.—Eggs of a Central American species of Orthoptera neatly fastened to a leaf.

may be completely formed in the fall and the larvae ready to emerge with the first favorable weather in spring. The eggs of *Bittacus* (Mecoptera) and the walkingstick are said to remain unhatched for two years.

**Special Adaptations for Egg Laying.**—Borers such as the raspberry cane borer, *Oberea bimaculata*, and the currant stem girdler, *Janus integer*, deposit their eggs in the stems of plants and girdle the plant above them. This causes the top of the plant to wilt and prevents its growth from crushing the egg. The plum curculio inserts its egg in a small hole in the fruit. Then it cuts a crescent-shaped slit which extends obliquely under the egg so that the egg is left in a flap of flesh and is not crushed by the growth of the fruit. In spite of this habit, the eggs seldom develop in the fruits of apple, although they develop normally in plum, cherry, and peach.

Insects that oviposit in wood, leaves, and similar tissues have sharp ovipositors which are used to cut slits or cavities into these materials. The sawflies are especially well developed in this respect. The ovipositor is a complex structure with at least two movable blades which are often serrated. Species that lay their eggs in wood have longer ovipositors. Leaf-mining Agromyzidae, which insert their eggs within the tissues of the leaves, have highly specialized, sharp, stiff ovipositors. In contrast, the leaf-mining species that lay their eggs on the surface of leaves, such as the Anthomyiidae, have generalized ovipositors which are membranous and incapable of piercing leaves. Both types of ovipositors are found among the leaf-mining Lepidoptera. The primitive Micropterygidae lay their eggs on the surface of the leaves and exhibit the extreme lack of ability to insert eggs within the tissues. The Eriocrania and Incurvaridae have horny piercing ovipositors with powerful muscles with which they easily insert eggs within the tissues of leaves.

**Devices That Trip Egg Laying.**—What are the stimuli that cause the female to lay her eggs in the right place and at the proper time? Our knowledge of this subject is still elementary. Many factors such as touch, odor, sight, temperature, and humidity may contribute to these impulses; they seldom operate alone. Richardson (1925) groups these stimuli under two headings: internal and external factors. The internal factors relate to fertility and internal periodicities. In the cotton boll weevil, the female will not oviposit unless she is fertilized. On the other hand, the female cecropia moth oviposits freely whether or not she is fertilized. In social insects and in many parasitic Hymenoptera, males are produced from unfertilized eggs. The pressure of developing eggs within the body of the female may play a part, in some cases, in determining when the eggs are laid.

In the case of temperature and moisture, there is probably an optimum for each species of insect; variations from these affect the rapidity of oviposition and the number of eggs laid. Certain insects oviposit more freely at low temperatures. The housefly prefers high temperatures. An Indian mosquito lays eggs most freely at 95°F. Females usually oviposit more freely at higher humidities. *Glossina* does best at the saturation point. On the other hand, *Habrobracon juglandis* is said to be little affected by degrees of moisture.

It is well known that light affects insects differently in respect to oviposition. Cockroaches oviposit at night. The codling moth and oriental fruit moth start ovipositing at dusk and continue during the early evening, activities depending also upon temperatures. Light stimulates the housefly but apparently has no effect upon the egg-laying responses of *Drosophila melanogaster*. V. E. Shelford, working with *Spogostylum anale*, a parasite of the larva of a tiger beetle, found that the sight of the hole or the entrance to the tiger beetle burrow caused the female to release her eggs.

The tactile response seems to guide many females in releasing their eggs. The corn ear worm, *Heliothis obsoleta*, prefers to oviposit on corn that has rough, hairy stalks or leaf surfaces. Roaches usually oviposit in cracks and narrow places. The angoumois moth does not require grain upon which to deposit her eggs but will lay them in any available crevice. Coarse muslin is used to obtain the eggs of this moth in rearing *Trichogramma* parasites. The female pushes her eggs through the openings in the fabric. Social insects are no doubt stimulated to lay their eggs when the ovipositor touches the cell in which the egg is laid. Some dragonflies hover over the water and drop eggs each time the ovipositor touches the water.

Odor may not feature in oviposition responses so frequently as might be suspected. This occurs more often in insects that oviposit on their food plants and insects in which the young and the adults have the same food plant. Flesh flies may oviposit in response to o d o r s. R i c h a r d s o n, McIndoo, and others suggest that taste and contact with chemical substances may function more frequently than smell. Certain salt solutions are conducive to egg laying in mosquitoes; while corresponding acids are repellent.

**Transportation of Eggs.**—Not a small number of insects rely upon other species to carry their eggs or young. This h a b i t is developed among many different orders. The majority of the parasitic and predacious insects lay their eggs or deposit their young upon their hosts. The Tachinidae oviposit upon many caterpillars and upon certain beetles. The Anoplura and Mallophaga deposit their eggs upon the fur or feathers of their hosts. The females of some insects carry their own eggs. The

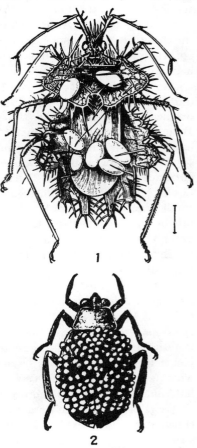

Fig. 92.—Male Heteroptera that carry eggs on their backs: (1) *Phyllomorpha laciniata*, Spain (*after Sharp*); (2) Western water bug, *Abedus* (*After Kellogg*).

hydrophilid beetles of the subfamily Sphaeridiinae carry the eggs attached to their hind legs. Certain May flies may carry two egg capsules adhering to the posterior end of the body until an opportunity is found to drop them into the water. Roaches often carry their oötheca in a similar manner. The females of the giant water bugs, *Belostoma, Serphes,* and *Abedus,* deposit their eggs on the backs of the males. A European coreid, *Phyllomorpha laciniata*, has somewhat the same habit. The sides of the body of the male are directed upward and provided with numerous spines which form a pocket for the reception of the eggs.

The most interesting cases are those insects that impose upon other species. The water boatman, *Ramphocorixa acuminata*, attaches its eggs to the body of a crayfish. The botfly of man, *Dermatobia hominis*, uses the mosquito to transport its eggs to man. As far as known, this is the only species of botfly that does not oviposit on its host. This botfly has no direct interest in man or other animals that might serve as hosts. On the contrary, this botfly visits marshy places where mosquitoes are emerging. It seizes a mosquito and deposits 10 or 12 eggs on the abdomen and legs of the mosquito, after which it releases its hold. When the mosquito visits man, the warmth of his body causes the botfly eggs to hatch and the young maggots dig into the flesh of the victim.

A European beetle, *Clythra quadrimaculata*, resorts to strategy to gain entrance to ants' nests. The female deposits her eggs on the foliage of birch or other trees. These are covered with excrement and resemble small bracts of the plant. The ants pick these up, mistaking them for bits of vegetable refuse, and take them into their nests. When the eggs hatch the larvae live as guests (*inquilines*) in the ants' nest.

**Extrication of Young from Eggs.**—There are many devices for liberating the young from eggs. This is an important problem when the insects are confined in eggs with extremely hard shells. In the mandibulate species, the larvae can chew their way to freedom. The mouth hooks of Diptera are efficient tools to pick away at the interior of the eggshells. Leaf-mining species not only have the problem of exit from the egg but many species must dig immediately into the leaf. The first-instar larvae of *Pegomyia* have sawlike mouth hooks for this purpose, which are replaced by the typical mouth hooks in the later instars. The hatching of the egg is more difficult in haustellate insects. Many of the eggs of the Heteroptera have caps intended to break off and release the young. The neat holes made by emerging Pentatomidae indicate that there is a line of weakness about the top of the eggs. The embryos of many insects possess special organs, known as the egg burster, egg tooth, hatching spine, or *ruptor ovi*, which are used to cut through the eggs. The Colorado potato beetle has three pairs of hatching spines. The embryo of the flea has a knifelike egg opener on the head.

**Nymph.**—A nymph is one of the immature instars of a paurometabolous insect, *i.e.*, the young of a terrestrial insect with a gradual metamorphosis. The term nymph is obtained from the Greek word meaning bride or maiden. In mythology, a nymph is one of the inferior deities of nature, represented by a beautiful maiden, who inhabited the mountains, forests, and water. The immature stages of Orthoptera, Dermaptera, Isoptera, Heteroptera, Homoptera, and Thysanoptera (exclusive of eggs) are known as nymphs. There is a gradual growth of the body, wings, and genital appendages from the time the insect hatches until the adult is reached. In Thysanoptera, there is no indication of wing buds until the second or third instar. In Corrodentia, the nymphs lack wing buds even in species that develop wings. The nymph generally has no resting stage before the adult is reached. In Thysanoptera

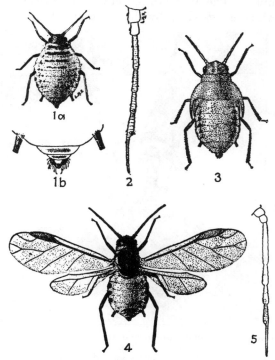

Fig. 93.—Development of the corn-root aphid, *Anuraphis maidi-radicis;* (1*a*) and (3) are nymphal forms; (1*b*) the posterior end of a nymph showing the cornicles; (4) winged adult; (5) antenna of adult; (2) antenna of nymph.

and the male Aleyrodidae and Coccidae, there is what appears to be a pupa. In the male Coccidae, even a cocoon is formed. Nymphs have certain things in common: the wings develop on the exterior of the body (Thysanoptera excluded) and the species are terrestrial (at least none have closed spiracles or breathe by means of gills). The nymphs of Notonectidae, Corixidae, Belostomidae, Nepidae, and other smaller families of Heteroptera are semiaquatic; they descend beneath the water and remain there for a considerable period of time, but they are air breathers. Nymphs are chiefly plant feeders, *phytophagous*. A few

Thysanoptera, Orthoptera, and Heteroptera are predacious. Nymphs may be mandibulate or haustellate. The Orthoptera, Corrodentia, Isoptera, and Dermaptera are mandibulate. The Thysanoptera, Heteroptera, and Homoptera are haustellate. Both nymphs and adults are injurious to crops. Conspicuous examples are the thrips, leafhoppers, grasshoppers, lace bugs, aphids, and scale insects.

**Wax and Scent Glands.**—Exudations of various kinds are produced by most nymphs. The Isoptera secrete a substance through pores in the cuticula which is licked up by the workers. Roaches produce a dis-

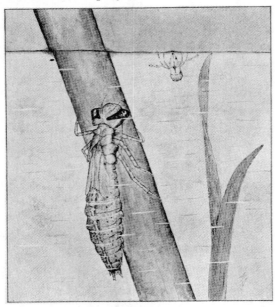

Fig. 94.—A typical naiad, the immature form of the dragonfly, *Aeschna constricta.* (*After Needham.*)

agreeable odor which results from an oily liquid secreted by special scent glands. The stinkbugs (Pentatomidae) and the bedbug (Cimicidae) have scent glands on the dorsal aspect of the abdomen. Certain earwigs (Dermaptera) possess stink glands which open through tubercles situated on each side near the hind margin of the second and third visible abdominal segments. It is said that a foul-smelling fluid can be shot a distance of three or four inches. The whiteflies (Aleyrodidae), aphids (Aphididae), the scale insects (Coccidae), and the phylloxera (Phylloxeridae) produce an abundance of wax which exudes through special pores in the body wall.

**Naiad.**—A naiad is one of the immature instars of bathmedometabolous insects, *i.e.*, the young of an aquatic insect with incomplete metamorphosis. In mythology, a naiad is one of the nymphs believed to live in, and give life and perpetuation to, lakes, rivers, springs, and fountains. The immature stages of Plecoptera, Ephemerida, and Odonata (exclusive of eggs) are naiads. All naiads are aquatic;* they have closed spiracles, breathe by means of gills, and are mandibulate. The adults are aerial and are mandibulate or have vestigial mouth parts. Most of the naiads are predacious. The naiads of Ephemerida are believed to be herbaceous. None of the naiads is known to be injurious to crops but many serve as food for fish and other animals.

Naiads are generally quite uniform in appearance. They are campodeiform; the legs are long, the body is flattened, and they are very active. In the Plecoptera and the Ephemerida, the naiads have con-

Fig. 95.—A typical larva, the caterpillar of *Polia adjuncta*.

spicuous caudal filaments, varying from two to three in number. In the damsel flies, Zygoptera, the caudal appendages are modified to form tracheal gills. Tracheal gills are located on various parts of the body. In Plecoptera, they are usually located on the underside of the thorax, although some species have gills on the head or on the abdomen. In Ephemerida, the gills are located on the abdomen. In the dragonflies, the rectum is modified to form a tracheal gill chamber. Water is alternately taken in and forced out. This mechanism not only brings oxygen to the gills but serves as a means of locomotion. In the damsel flies, there are three platelike gills at the posterior end of the abdomen.

**Larva.**—A larva in the limited sense is one of the immature instars of holometabolous or hypermetabolous insects, *i.e.*, the young of insects with complete or hypermetamorphosis. The term larva is derived from the Latin word for mask, having reference to the ancient belief that the adult form was masked or obscured in the larva. The term larva is

* Except a few exotic species.

used in the broad sense for the immature stages of all insects with the exception of eggs and pupae.

**Common Names of Larvae.**—The larvae of Coleoptera and some Hymenoptera are known as grubs. The larvae of Lepidoptera are called caterpillars. The caterpillars of the family of Eucleidae are known as slug caterpillars. The larvae of the more specialized Diptera are called maggots. No satisfactory common name has been suggested for the larvae of sawflies, which resemble in a superficial way the larvae of Lepidoptera. In the orders that are not so well known, such as Neuroptera, Siphonaptera, Trichoptera, and Strepsiptera, common names are not so numerous.

Fig. 96.—A campodeiform type of larva, immature of the alder fly, *Sialis infumata* (Neuroptera).

Many common names have been given to the larvae of certain groups or species of insects. Some are suggested by the habits of the larvae: loopers, inchworms or measuring worms for Geometridae, bagworms or basketworms for Psychidae, wigglers or wrigglers for mosquito larvae, ant lions for Myrmeleonidae, aphis lions for Chrysopidae, and hellgrammites for the larvae of the dobson flies. Other groups of insects are known by their larval habits, such as leaf miners, leaf rollers, bud-moths, and casebearers. Still other groups are named from their structure, such as slugs for certain sawfly larvae, slug caterpillars for the larvae of Eucleidae, wireworms for Elateridae, water pennies for the larvae of Psephenus, and white grubs for the larvae of the June beetles. From the food of the larvae we establish names such as plum curculio, apple and cherry maggots, green fruitworms, etc. It is unfortunate that the term *worm* has been used so frequently in entomology; it should be reserved for the phylum Annelida. The term *slug* should properly be applied to certain Mollusca.

**Types of Larvae.**—Larvae present certain similarities that are striking. Those that live under similar environmental conditions, feed upon the same type of food, and have similar habits tend to develop similar form and structure. This is known as convergent development. Many interesting comparisons can be drawn from the larvae of insects. Two systems of nomenclature have been adopted, one using the suffix -*form* as in carabiform, the other using the suffix -*oid* as in caraboid.

*Campodeiform or Thysanuriform Larva.*—The characteristics of a campodeiform larva are flattened body, long legs, cerci or caudal filaments usually present. These forms are usually active. This is the primitive larval form and is best represented in the Thysanura. It is found in some insects with complete and in others with incomplete metamorphosis. The larvae of most of the Neuroptera and Trichoptera, many of the Coleoptera, Dytiscidae, Carabidae, Staphylinidae, and the naiads of Plecoptera, Ephemerida, and Odonata are campodeiform.

FIG. 97.—Abbott's sawfly, *Neodiprion pinetum*, a typical eruciform larva.

*Carabiform Larva.*—This a modified form of the campodeiform larva in which the body is flattened but the legs are shorter. Generally there are no caudal filaments. The majority of the chrysomelid beetles and many others Coleoptera (Lampyridae, Carabidae, Melyridae) exhibit

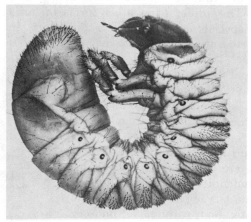

FIG. 98.—A typical scarabaeiform larva, a white grub, *Phyllophaga rugosa* Melsh. (*Pennsylvania Department of Agriculture.*)

this type. The second instar of the Meloidae is designated as carabiform. In the Coccinellidae, the legs are longer, the larvae are more active, and these species approach the campodeiform type.

*Eruciform Larva.*—The eruciform larva is cylindrical, the thoracic legs and prolegs are present, and the head is well formed. The larvae are not so active as the campodeiform larvae. This type is well illustrated in the Lepidoptera, Tenthredinidae, and Mecoptera.

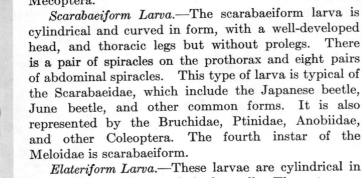

*Scarabaeiform Larva.*—The scarabaeiform larva is cylindrical and curved in form, with a well-developed head, and thoracic legs but without prolegs. There is a pair of spiracles on the prothorax and eight pairs of abdominal spiracles. This type of larva is typical of the Scarabaeidae, which include the Japanese beetle, June beetle, and other common forms. It is also represented by the Bruchidae, Ptinidae, Anobiidae, and other Coleoptera. The fourth instar of the Meloidae is scarabaeiform.

*Elateriform Larva.*—These larvae are cylindrical in form with a thick tough body wall. The setae are much reduced; the legs are usually present but short. They resemble the vermiform and carabiform larvae. This type is found in several families such as the Elateridae, Tenebrionidae, Alleculidae, Ptilodactylidae,

Fig. 99.— Larva of a click beetle. Popularly known as a w i r e w o r m . (*Pennsylvania Department of Agriculture.*)

and Eurypogonidae.

*Platyform Larva.*—Many larvae represented in different orders are short, broad, and extremely flat. They differ from other forms of larvae and deserve a special name. The legs are short, inconspicuous, or absent. Syrphid larvae of the genera *Microdon* and *Xanthogramma* are typical examples. The larvae of some slug caterpillars (*Euclea*) and those of the water pennies, Psephenidae, and the hister beetles, Histeridae, also belong here. The larvae of *Blepharocera* and some *Stratiomyiidae* approach this type.

*Vermiform Larvae.*—Larvae that are more or less wormlike are termed *vermiform.* This designation is indefinite but is usually considered to include larvae that are cylindrical in form, elongate, and without locomotory appendages. Most of the larvae of the Diptera, especially those of the higher or muscoid type, belong here. The larvae of many other orders also represent this type. This is especially true of the larvae of wood-

Fig. 100.—A platyform larva, the tortoise beetle, *Metriona*, bearing an accumulation of dung and cast skins.

boring beetles and sawflies and flea beetles of the genera *Systena* and *Epitrix.* The larvae of fleas and many parasitic Hymenoptera also belong here. Even the Australian plant-boring cricket, *Cylindrodes campbellii*, approaches this form.

*Limaciform Larva.*—These larvae have the form of a slug such as the larvae of certain sawflies. They are known also as molluscoid or limacoid larvae.

*Onisciform Larva.*—This term, meaning shaped like a wood louse, *Oniscus*, is applied to the unusual shaped larvae of the lycaenids and similar larvae.

*Coarctate Larva.*—The fifth instar of the meloid larva is known as a *coarctate* larva. It is intermediate in form between a larva and a pupa. It has the appearance of a larva with the fourth larval skin adhering. Although it is inactive during this period and thus resembles a pupa, it becomes active after the hibernating period and assumes a typical larval form again. The term coarctate is more frequently applied to a pupa

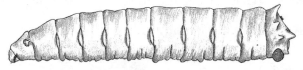

Fig. 101.—A vermiform larva, the beet-leaf miner, *Hyleymia betarum* (Diptera).

that is enclosed within the hardened larval skin, as is the case in most Diptera. These are generally called puparia.

*Special Names for Larvae.*—The larvae of insects with hypermetamorphosis often have special names. The first instar of the larvae of Meloidae, Strepsiptera, and Mantispidae are called *triungulins*. They receive this name because the feet have three claws. The fifth larval instar of the Meloidae, as mentioned above, is called a coarctate larva or a pseudopupa. The first instar of *Platygaster*, a parasite of the hessian fly, resembles a crustacean more than an insect and is called a *naupliiform* larva. The first instar of *Perilampus*, a secondary parasite of the fall webworm, is called a *planidium*, meaning a diminutive wanderer.

**Larvae of Lepidoptera.**—The Lepidoptera present three distinct larval types: (1) larvae with five pairs of prolegs, (2) larvae with three or four pairs of prolegs, (3) larvae without prolegs. (The latter group really belong to the vermiform type.)

The crochets or hooks with which the prolegs of Lepidoptera are armed offer useful characters for classification. Their arrangement varies from a single to a triple band, or the crochets may be arranged in a circle or a part of a circle. Occasionally the prolegs are reduced to mere stubs and the crochets appear to be located on the abdominal wall.

The body of the caterpillar is usually clothed with hairs or setae, which are sometimes long and dense but often inconspicuous. In either case, they have a definite arrangement, which is useful in classifying the larvae. Scattered among these setae are the nettling hairs of stinging caterpillars such as the io moth and the saddleback. When hairs are numerous, they are grouped upon tubercles, called *verrucae*.

FIG. 102.—Arrangement of the hooks (crochets) on the prolegs of caterpillars. (1) *Thyridopteryx ephemeraeformis*, uniordinal; (2) *Pyralis farinalis*, biordinal; (3) *Psorosina hammondi*, triordinal.

The head of the caterpillar is always well formed except in a few leaf-mining species. The epicranial suture is usually present, and the antennae are minute and three segmented. The ocelli vary from one to six pairs on each side of the head. In the larger caterpillars, there are

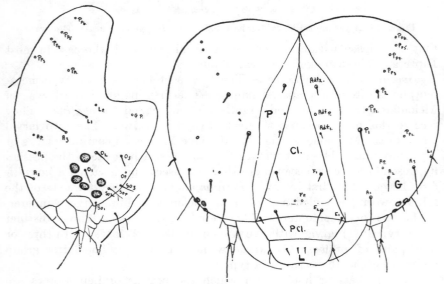

FIG. 103.—Head capsule of a lepidopterous larva; (*Cl*) clypeus sometimes designated as front; (*P cl*) preclypeus; (*P*) parietal or adfrontal; (*G*) gena.

usually six pairs of ocelli. (In Micropterygidae, Gracilariidae, and Nepticulidae, there is only one ocellus and in Heliodinidae there are two ocelli on each side.)

The spiracles have a definite position on the larvae of Lepidoptera. There is a spiracle on each side of the prothorax (correctly speaking on the membrane between the pro- and mesothorax). There are eight pairs of spiracles on the abdomen, occurring on segments one to eight.

**Sawfly Larvae.**—The larvae of sawflies (Tenthredinidae) are caterpillarlike and somewhat uniform in structure. The wood-boring and leaf-mining forms show considerable reductions in many respects and depart from the general type. In the typical sawfly larva, the head is well developed, there is only one ocellus on each side of the head, and the antennae are always present but minute. The epicranial suture is present in all the external feeders but is absent in the wood-boring forms. There are seven or eight pairs of larvapods (prolegs) except in the leaf-

Fig. 104.—Young sawfly larvae feeding on the edges of holes cut in the leaf and assuming characteristic "*S*" attitudes.

Fig. 105.—A mature sawfly larva, *Macremphytus* sp., assuming a characteristic curled attitude.

mining Fenusinae and Scolioneurinae. All sawfly larvae have thoracic legs except the Oryssidae. There is a pair of spiracles on the mesothorax and occasionally a pair on the metathorax. There are usually eight pairs of abdominal spiracles. (In Acordulescerinae, there are only five pairs of abdominal spiracles.)

**Mecoptera Larvae.**—The larvae of Mecoptera are caterpillarlike in form with three pairs of true legs. Prolegs are generally well developed. *Panorpa* has eight pairs of prolegs, *Bittacus* has from nine to ten pairs. In the mecopterous larvae, the head is large, the antennae are short and thick, and the eyes are composed of many facets. The body is beset with many conspicuous spines, which are stronger toward the posterior end of the body.

**Dipterous Larvae.**—The larvae of Diptera vary considerably in structure and form. They often take odd shapes with numerous lobes at the posterior end, with filamentous expansions of the body wall or with groups of bristles or setae. In some aquatic species, gills are present. In the more specialized Diptera, two pairs of spiracles are the rule, one pair at the anterior end and one pair at the posterior end. Many of the more primitive Diptera have a larger number of spiracles. There are two general groups of larvae: (1) the primitive type, in which the head capsule is more or less complete and mandibles are conspicuous, and (2) the specialized type, in which the head is absent and mouth hooks replace the mandibles.

*Larvae of the Primitive Diptera, Orthorrhapha.*—The Orthorrhapha are the flies that escape from their pupa by means of a T-shaped slit. The adults usually have antennae with more than three segments. The Dixidae, Culicidae, Tipulidae, and Chironomidae are typical examples. The larvae vary a great deal in form and structure and are too com-

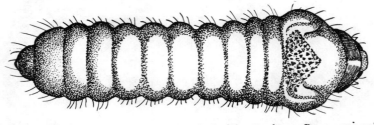

Fig. 106.—A typical wood-boring larva, the cloaked knotty horn, *Desmocerus palliatus.*
*(After Craighead.)*

plicated to permit discussion here. The Dixidae, Culicidae, and Chironomidae are peculiar in that they have active pupae.

*Larvae of the Specialized Diptera, Cyclorrhapha.*—These larvae show more uniformity in structure than the Orthorrhapha. The larvae are usually cylindrical in form, legless, and often taper toward the anterior end. They pupate in their last larval skin, which is known as a *puparium.* The antennae of the adults have three segments. The Syrphidae, Muscidae, Tachinidae are typical examples. Not all the larvae of the Cyclorrhapha are vermiform, for the genera *Microdon* and *Xanthogramma* are platyform.

**Larvae of Coleoptera.**—The larvae of the Coleoptera are varied as to form. The vestiture is often reduced, especially in the leaf-mining and wood-boring forms, but is profusely developed in the Dermestidae and takes the form of barbed spines in some of the Coccinellidae and Chrysomelidae. The head is always well developed although sometimes telescoped into the thorax, especially in the leaf-mining and boring species, or is hidden beneath the greatly expanded prothorax. The thoracic legs are usually well developed although they are absent in some of the wood-boring species such as Micromalthus, Buprestidae, and Cerambycidae and are reduced or missing in some of the leaf-mining forms. The thoracic spiracles are generally located on the mesothorax. The Dytiscidae, Gyrinidae, and Staphylinidae are conspicuously campodeiform. The larvae of the Tenebrionidae, Alleculidae, Elateridae, and

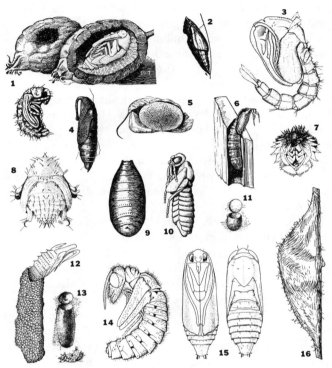

Fig. 107.—Pupae and cocoons of various insects: (1) pupa of apple curculio in fruit; (2) chrysalis of alfalfa caterpillar, *Eurymus eurytheme;* (3) mosquito pupa; (4) pupa of tomato worm; (5) cocoon of clover leaf weevil, *Hypera punctata;* (6) empty pupal skin of *Zeuzera* protruding from burrow; (7) pupa of tortoise beetle, *Metriona bicolor;* (8) pupa of strawberry root worm, *Paria canella;* (9) puparium of housefly, *Musca domestica;* (10) pupa of cherry-hawthorn leaf miner, *Profenusa canadensis;* (11) cocoon of lacewing fly, *Chrysopa occulata;* (12) empty pupal skin of peach tree borer projecting from cocoon; (13) cocoon of a parasite, *Apanteles glomeratus;* (14) pupa of *Sialis infumata;* (15) pupa of clover-seed caterpillar, *Laspeyresia interstinctana;* (16) cocoon of Cecropia moth.

others are elateriform. The larvae of the water pennies, Psephenidae, and others are platyform. The larvae of the Lucanidae, Scarabaeidae, Ptinidae, Bruchidae, and other families are scarabaeiform.

The larvae of the Coleoptera have a wide range of habits. Some are terrestrial, others are subterranean, and still others are aquatic. Some feed upon vegetable material, others upon animal material, and still others upon decayed material. The phytophagous species feed upon the surface of the plant, mine in leaves, or bore in seeds, stems, or roots. The zoophagous species prey upon other insects. The scatophagous species feed upon carrion.

**Pupa.**—The term *pupa*, derived from the Latin word meaning baby or child, was proposed by Linnaeus on account of its resemblance to a papoose or baby bound in garments. The term was first used in connection with the chrysalis of Lepidoptera. The pupa is defined as the resting or inactive period of all holometabolous insects, the intermediate stage between the larva and the adult. The insect is usually delicate and helpless at this time. It is the period during which the most conspicuous changes take place.

The rapidity and nature of transformation vary considerably in different insects. In the Paurometabola there is usually no resting period and transformation is gradual from the time the insect hatches until the adult is reached. In the Bathmedometabola, there is usually a superficial resting period before the adult is reached and the development is not so gradual as in the preceding insects. In the Holometabola, the change from larva to adult is so great that a distinct resting period or pupa is essential.

The term pupa is strictly applicable to the transformation period of holometabolous and hypermetabolous insects. The term *prepupa* refers to the last larval instar of some insects which retain the larval form and mobility but cease to feed. This condition exists in many orders of insects, notably the Diptera, Hymenoptera and Coleoptera.

**Outward Appearance of the Pupa.**—The pupa generally has a shrunken mummylike appearance and is usually shorter in length than the larva. The mouth parts, antennae, legs, genital appendages, and wings appear as buds upon the outside of the pupa (obscured in the more specialized Diptera and male Strepsiptera, because the insects are enclosed in the last larval skin).

The pupae of insects can be classified with reference to the degrees of freedom of the appendages. If the appendages are closely appressed to the body, it is said to be an *obtect pupa*. This is the common type in the Lepidoptera, in many of the Coleoptera, and in more primitive Diptera. If. the appendages are not closely appressed to the body but are free, it is said to be an *exarate pupa*. The Neuroptera, Trichoptera, most of the Coleoptera, and a few of the Lepidoptera (Tischeriidae) have exarate pupae. If the appendages are not visible at all and are obscured by the last larval skin, it is said to be a *coarctate pupa*. This type is found in the more specialized Diptera (Cyclorrhapha) and in certain Coccidae and Stylopidae.

The pupae of insects may also be classified on the basis of degrees of movement. The pupae of mosquitoes and certain midges are exceedingly active.    They are aquatic and are able to swim by means of movements of the caudal end of the body.    The pupae of some Neuroptera (*Chrysopa, Hemerobius,* and *Raphidia*) become active and crawl about before transforming to the adult stage.    A slight degree of movement can be detected in any pupa when it is disturbed.    The movements of some pupae are very evident at the time of emergence.    Many pupae, formed in wood, beneath bark, and in the ground, free themselves from their resting places before the adults emerge.    Pupae often have rows of strong spines along the posterior edge of the abdominal segments which aid them in working their way from the ground.    Movements of the body may cause the pupa to progress in a forward direction. Thus a pupa may work its way from a depth of several inches to the surface of the ground.    The pupae of the Sphingidae work their way from the ground and the pupae of the Cossidae work their way from cells in ᶜsolid wood.    The codling moth, oriental fruit moth, hessian fly, and many other insects have similar habits.

Pupae may be further classified as to the nature and amount of covering.    If the pupa is not protected by a cocoon or definite covering, it is termed naked. Many Hymenoptera, Coleoptera, Lepidoptera, and Diptera have naked pupae, which are often found in the ground, in solid wood, or beneath rubbish.    In the Lepidoptera, all butterflies have naked pupae which are generally called *chrysalids* or *chrysalides* (the singular being *chrysalid* and *chrysalis.*)

Fig. 108.—Coccinellid pupae suspended by their posterior ends which are attached to the cast larval skins.

The naked pupa often possess a *cremaster,* which is especially well developed in the chrysalids.    Before pupating, the larvae of butterflies spin a disk of silk into which the cremaster is implanted to hold the chrysalis after the larval skin is shed.    Some larvae spin a girdle of silk to help support the chrysalis.

**Internal Changes in the Pupa.**—The most remarkable changes that take place during the pupal stage are not visible from the outside.    During pupation, the insect undergoes many changes that are essential to adapt it for a new environment or new food habits.    The prolegs of the caterpillar are lost, the legs of Diptera are acquired.    The gills of larvae may be replaced by spiracles.    Mandibles may be lost and haustellate mouth parts acquired.    Wings are acquired in the majority of forms.

Genitalia are developed in all forms. These changes may have their inception in the larva or even in the embryo, but their completion takes place in the pupa.

These changes often require drastic modifications. There is a degeneration or dissolution of the tissues known as *histolysis*. The leucocytes or phagocytes feed upon the disintegrating tissues and debris and return, to the blood, products that serve to nourish the new tissues. After the old organs have been broken down, the new tissues are built up. This process is known as *histogenesis*. At certain stages in development, the contents of the pupa may be largely semiliquid. The extent of these changes varies greatly in different insects. In the Coleoptera, Lepidoptera, Hymenoptera, and nematocerous insects, the mid-intestine

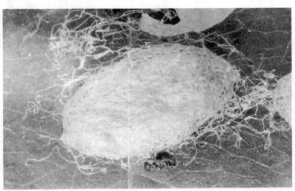

FIG. 109.—A silken cocoon of one of the sawflies which is spun of secretions from the cephalic silk glands.

and some of the other larval organs are greatly modified. In *Musca*, all the organs break down and are reformed, except the central nervous system, the heart, and the reproductive organs.

**Protective Coverings of Pupae.**—Many provisions are made to protect the pupa during the period when the insect is helpless and decidedly vulnerable. Insects frequently retreat to protected places before transforming. Many insects pupate in the ground, within curled leaves, or beneath bark. If the natural protection is sufficient, no further covering is necessary. Other insects produce various protective coverings. Silk is the most common. Silken cocoons are spun by the larvae of the majority of the moths, by Neuroptera, Trichoptera, a few Coleoptera, some Diptera, Siphonaptera, and Hymenoptera.

The cocoon-making habit is most highly developed in the Lepidoptera. The silkworm, *Bombyx mori*, produces one of the finest cocoons, each composed of a single continuous thread of silk often 1,000 feet in length. The cocoons are compact and of even texture. Our American silkworm, *Samia cecropia*, spins a double-walled cocoon. A layer of tough paperlike silk forms the exterior of the cocoon. The inner capsule, containing the pupa, is spun of very tough silk, and the interior is covered with a secretion from the larva which gives it a hard shiny surface. Between these two layers is an air space formed of loose strands of silk.

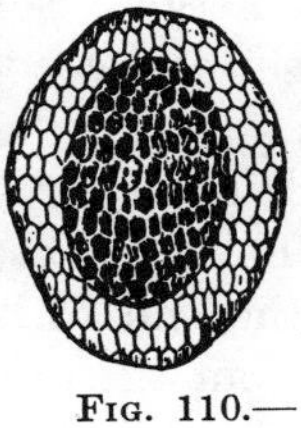

The promethea moth, *Callosamia promethea*, incorporates a leaf in its cocoon and spins silk about the petiole of the leaf as well as the twig to which it is attached so that the cocoon will not drop when the rest of the leaves fall from the tree. To the uninitiated, the cocoon has the appearance of a dead leaf hanging upon the tree.

The cocoon of *Climacia* (Neuroptera), one of the most beautiful pieces of workmanship in nature, is formed of hard tough silk and is covered with a delicate netlike veil. It resembles the cocoon of *Hypera* (Coleoptera) but is more fragile and delicate.

Fig. 110.—Cocoon of *Climacia*, Neuroptera; a delicate lacelike structure spun of secretions from the malpighian tubules. (*After Comstock.*)

**Some Unusual Pupae.**—The pupae of the Micropterygidae and the Trichoptera have conspicuous cruciate jaws. Many of the Sphingidae have pitcherlike pupae. The maxillae are unusually long and are not appressed to the body but are looped like the handle of a jug. The pupa of the catalpa sphinx does not possess such a handle. Many naked pupae resemble seeds. The small black pupae of *Comedo* and other parasitic Hymenoptera are seedlike. The puparia of many Diptera, especially the Agromyzidae, also resemble seeds. Nymphlike pupae are found among some insects. The pupa of *Mantispa* bursts from its cocoon and becomes active. During this period it resembles a miniature mantid. Later it changes to the adult form. The pupae of *Feniseca* (Lepidoptera) is mollusclike, resembling a spiral shell. Some have likened this pupa to a monkey's face.

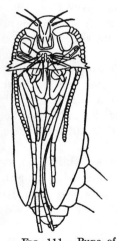

Fig. 111.—Pupa of *Eriocrania auricyanea* (Micropterygidae) with cruciate mandibles.

**Eclosion.**—The emergence of an insect from its pupa is known as *eclosion*. The phenomenon is sometimes improperly called hatching, a term that should be reserved for the emergence of the immature insect from the egg. Insects have developed numerous ways of extricating themselves from their pupae and from their cocoons. Chewing insects may employ their jaws to gnaw their way to freedom. Sucking insects must meet the problem in other ways. Some secrete a liquid which softens the silk at one end of the cocoon. Two of our large species of moths, *Samia cecropia* and *Callosamia promethea*, construct a valvelike structure at one end of the cocoon that separates easily when the adult is ready to emerge. Some Lepidoptera possess a series of spines which are used to slit the cocoon. *Megalopyga* constructs a hinged partition that serves as a trap door through which the adult emerges. The cyclorrhaphous Diptera have a bladderlike swelling on the head, known as the *ptilinum*. This is inflated and pushed against the end of the puparium, which causes the end to break off and release the fly. Adults may emerge directly from their cocoons leaving their larval and pupal skins in the cocoons, or the pupae may wiggle their way from the cocoons before the adults emerge.

Most naturalists have watched a moth emerge from a cocoon. Other insects emerge in somewhat the same way. At first the wings appear dwarfed, soft, and moist. In a few minutes the blood begins to flow through the veins of the wings, causing them to expand. In about 20 minutes, the wings attain their full size but they are still limp and moist. The insect begins to move them slowly. After about two hours, they assume their full color and firmness. As a matter of fact the legs, antennae, and other parts of the body go through considerable changes during the process of eclosion. In many insects, the full color may not be attained for many days; in certain cases, not for several months.

Dragonfly, stone fly and May fly naiads crawl from the water before the adults emerge. Empty naiadal skins are left along the streams or ponds, attached to stones or grasses, to tell the tale.

There are some deviations from the normal method of eclosion. After the May fly has emerged from its naiadal shell, it is still covered by a thin skin, which gives it a milky appearance. An additional moult follows before the insect is mature. The term *subimago* is applied to the form of the May fly when it first emerges from its naiadal skin and before the final moult.

In the net-winged midges, *Blepharocera*, the wings are fully formed and folded in the pupa. This permits the fly to take immediate flight as soon as it emerges from its pupa. This is essential because the fly emerges from swift water where there is nothing for it to rest upon and no opportunity for it to tarry to dry its wings.

A few insects never emerge from their cocoons. This occurs only among the females. The female bagworm is fertilized in the bag in which she lays her eggs and then dies. Female scale insects never leave their scales. Eggs or young are produced beneath the scales.

**Fecula of Insects.**—Fecula is the accepted term for the waste material or excrement voided by insects. Frass, often used for the waste material deposited by leaf-mining insects, is inappropriate for it is the German for food, meat, or pasture. The chips or particles cast aside by wood borers are also known as frass and are frequently mixed with fecula. One should not confuse the exudations from the mouths of leafhoppers, grasshoppers, and certain flies with fecula. The honeydew produced by aphids and other insects is fecula enriched by certain food constituents in the alimentary canals of these insects. The froth of spittle insects, although it issues from the anus, is not fecula but a secretion from special glands.

The finding of "insect droppings," fecula, beneath a tree or plant may clearly betray the presence of an insect. To the uninitiated, these pellets may appear

FIG. 112.—A mixture of frass and fecula resulting from the feeding of a wood-boring beetle larva, *Monochamus*. (*After Craighead.*)

FIG. 113.—Quince, showing the abundant frass pushed from the burrows by the larvae of the oriental fruit moth.

as particles of tobacco carelessly dropped by a smoker. "Frass" protruding from an apple is a definite indication that the fruit has been attacked by the larva of the codling moth. The entomologist knows that a peach can be infested by an oriental fruit moth larva and show no conspicuous external injury, but the presence of a few minute fecular pellets at the stem end of the fruit betrays the culprit. This criterion is frequently employed to determine whether fruits are infested, without cutting them.

Fecula is produced conspicuously by immature insects. Adults generally feed little or none and naturally void little or no waste material.

Social insects sometimes feed considerably and produce much fecal matter. *Polistes* have occasionally been observed to rest, night after night, in small clusters in protected places not far from their paper nests. Under such circumstances, the surface upon which they rest becomes stained with their fecal discharges. The common house spider frequently selects a site where it remains for several weeks retreating to this place to consume its victims. Accumulations of limelike fecula are deposited beneath these roosts.

Entomologists have paid little attention to the fecula of insects. A better knowledge of this subject would be exceedingly valuable to the economic entomologist, because it is often necessary to have some means of identifying a pest after the insect has disappeared or left the scene of injury. The characteristic form, texture, size, and color of the fecula can often be used to identify the species just as the type of feeding or the cast skins have frequently been used for this purpose.

Fig. 114.—A portion of the mine of *Phytomyza nigricornis* on the leaf of peach. Frass laid down in central row of dots.

The size of the voided pellet is determined by the species concerned and by the stage of the development of that insect. It has been noted previously that the size of the pellets increases proportionally with each successive moult of the insect. When insects are reared in captivity, the change of the size of the fecular pellets often indicates a moult more plainly than the small head capsule or shriveled skin which the larva casts off.

The color of the fecula depends chiefly on the character of the food upon which the insect feeds. The pellets of wood-boring larvae are usually pale brown in color, dry, and woody in texture. The fecula of foliage feeders is generally green in color; however, the excrement of many insects, such as the larvae of the Chrysomelidae, is pitchy black. The pellets of some of the tropical caterpillars are said to be brilliantly colored. The fecula of zoophagous insects is usually black although there are outstanding exceptions. The discarded pellets of *Comedo koebelei* are pale brown and contrast conspicuously with the black pupae adjacent to which they are always heaped following the final, larval moult.

The texture of the pellets voided by insects is also determined largely by the nature of the food eaten by these insects. Foliage feeders tend to produce pellets that are soft when first formed. All pellets become hard after drying. Juritz (1920) gives an analysis of the fecula of the caterpillar of *Antheraea cytherea* which feeds on *Acacia cyclopis;* in comparison, he gives an analysis of the leaves of the host.

| | Analysis of the fecula of *Antheraea cytherea,* per cent | Analysis of the leaves of *Acacia cyclopis,* per cent |
| --- | --- | --- |
| Water | 11.10 | 71.96 |
| Ash | 11.06 | 1.39 |
| Nitrogen | 2.07 | 1.07 |
| Potash | 2.87 | 0.25 |
| Lime | 1.86 | 0.13 |
| Phosphoric oxide | 0.87 | 0.12 |

Since the percentages compare somewhat favorably with those of horse and cow manure, the abundance of the insect pellets suggested their use as fertilizer.

The fecula of many of the Chrysomelidae is soft and has no definite form. The frass of leaf-mining insects is also soft and, as a rule, has no definite form but may be laid down in characteristic manners in the mine. A few examples will illustrate the great diversity of habits in connection with the disposal of this material. The hop hornbeam leaf miner, *Lithocolletis ostryarella,* smears its fecula over the floor or lower surface of the mine. The larvae of the Nepticulidae usually lay their fecula in a central black line. In *Phyllocnistis populiella,* the frass line is most striking and follows the numerous convolutions of the long tortuous mine which is otherwise white or pale in color. The elm leaf miner, *Agromyza aristata* (*ulmi* Frost), forms two lines of black dots, and the peach leaf miner, *Phytomyza nigricornis* lays its fecula down in a single row of conspicuous dots. The fecula of *Lyonetia speculella* is pushed from holes cut in the mine and adheres to the lower surface of the leaf as minute black pellets. By reference to Chap. XVI, one will see that leaf-mining insects have many unique methods of disposing of their waste materials.

Fig. 115.—Fecula of a leaf-mining species, *Lyonetia speculella,* pushed from the lower side of a leaf, like miniature links of sausage.

The pellets of most free-feeding caterpillars and sawfly larvae have definite shape and texture. In these insects, the moisture is extracted from the food in the alimentary canal and the resulting pellets are comparatively dry and highly compressed. The shape of the pellet is often determined by the internal structure of the rectum, which is impressed upon the fecula.

A few examples will illustrate the variety and consistency of the form and structure of the fecula pellets of insects. Snyder has figured the pellets of various tropical termites, including *Cryptotermes thompsonae*, *Termopsis angusticollis*, *Neotermes castaneus*, and *Kilotermes*. In general, the pellets are smooth and compact, slightly longer than wide, truncate at both ends, and feebly ridged. The pellets of *Cryptotermes thompsonae* are 0.85 millimeter long and 0.54 millimeter wide.

Fig. 116.—Fecula of the leopard moth, *Zeuzera pyrina*, pushed from the end of the burrow.

The squash vine borer, *Melittia satyriniformis*, pushes most of its soft yellowish or greenish fecula through holes cut in the squash vine stem and it collects upon the ground under the plant, plainly indicating that the plant is infested.

The fecula of the leopard moth caterpillar is pushed from its burrows as dry woody pellets which are held more or less together by means of silken threads. The individual pellets are cylindrical in form and about 2 millimeters in diameter and 3 millimeters long.

The grape and apple twig pruner, *Hypermallus villosus*, pushes an abundance of borings from its burrow. The individual pellets are dry, light brown in color, 0.05 millimeter in diameter, and 0.5 millimeter long. They adhere to one another, end to end, to form rods about 2 millimeters long.

The pellets of the Abbott's sawfly, *Neodiprion pinetum*, are quadrangular in shape about 2.5 millimeters long, 1 millimeter wide, and 0.5 millimeters thick. The individual pieces cut from the pine needles are arranged crosswise in each pellet.

The pellets of caterpillars are usually circular or hexagonal in cross section with markings impressed by the structure of the rectum. The pellets of the Eucleidae, as far as known, are cup-shaped and resemble the fruits of raspberries.

The voided pellets of the walkingsticks are elongate, irregular in shape, from 6 to 7 millimeters long, and 1 millimeter in diameter.

The method of the disposal of the fecula may also help to identify the species concerned. Some insects retain their fecula until the final larval moult, when it is voided. This habit is common in the social and parasitic species which feed upon highly concentrated food, live within hosts, or within cells or chambers that must be kept clean. Such modifications of the usual method of fecal discharge occurs in the hymenopterous families Proctotrypidae, Ichneumonidae, Formicidae, Vespidae,

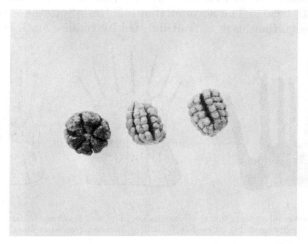

FIG. 117.—Fecula of the larva of the polyphemus moth. The six-sided pellets are typical of most Lepidoptera.

Apidae, and Cynipidae; in the dipterous family Hippoboscidae; in the neuropterous families Myrmeleonidae, Osmylidae, Sisyridae; and in the Strepsiptera.

Fecula may be retained in burrows, in leaf mines, in rolled leaves, by means of silken threads, or in other ways. On the other hand, it may fall to the ground, be washed away in the case of aquatic insects, or be disposed of in numerous other manners.

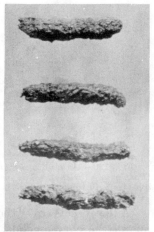

The larvae of the oriental fruit moth and many free-feeding caterpillars possess an anal comb which is located at the posterior end of the larva near the anal opening. As the fecula flows out of the posterior end of the alimentary canal, the anal comb is bent downward until it touches the pellet, which is tossed with a sudden snap away from the body. Casebearers, leaf miners, and borers seldom possess an anal comb as it would be a useless structure for larvae, confined as they

FIG. 118.—Fecular pellets of the walking stick.

are. Incidentally the anal comb is a valuable character for distinguishing the larvae of the codling moth, *Carpocapsa pomonella*, which has no comb, from the larvae of the oriental fruit moth, *Grapholitha molesta*, which has a distinct anal comb.

The codling moth larva pushes its fecula from the burrow in the fruit by means of its head. The accumulation of "frass" at the exit hole is a familiar recognition mark of fruit infested by codling moth larvae.

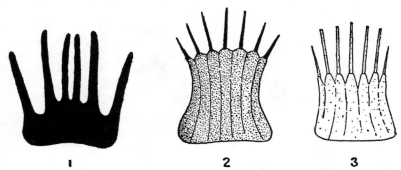

Fig. 119.—Anal combs (dung forks) of various Lepidoptera: (1) *Grapholitha molesta* Busck; (2) *Sparganothis idaeusalis* Walk; (3) *Cacoecia rosaceana* Harris.

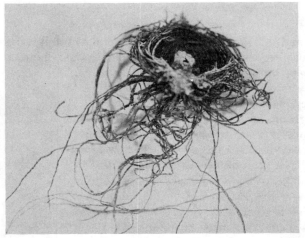

Fig. 120.—A minute nest made of fecular material by the larva of a tropical beetle (*Porphyraspis tristis*). The nest is scarcely a quarter of an inch in diameter. The larva lives beneath this nest which is closely pressed to the leaf.

A few insects cover their eggs with fecula. It is said that *Comptosomes*, one of the chrysomelids, has a groove near the extremity of the body. When the egg is extruded, the female holds it in this hollow by means of the hind legs and envelops it with a covering of excrement. Many other chrysomelids cover their eggs with a viscous material which issues from the anus but is produced by glands associated with the alimentary canal.

The larvae of the argus potato bettle and the tortoise beetles cover themselves with fecula. These larvae possess a "dung fork," *fecifork*, which is used to place the excrement and cast skins in a pack over the back. This accumulation of debris is attached to the fecifork and the larva can, at will, elevate it and tilt the mass backwards. The fecifork is analogous to the anal comb of Lepidoptera.

The larva of *Porphyraspis tristis*, a common insect of Bahia which lives on the cocoa palm, covers itself with a dense coat of fecular fibers, each many times the length of the body and elaborately molded, so as to form a round nest beneath which the larva lives and finally pupates.

The most interesting habits are found in species of Mecoptera, namely *Bittacus pilicornis*, *punctiger*, and *stigmaterus*. Dirt and sand are swallowed by the larvae; after each moult the body is sprayed with a mixture of sand and a glutinous material which is emitted through the anus. The particles adhere to the back and sides of the larvae.

Many species may be recognized because of the habit of constructing silken feeding tubes to which an abundant supply of fecula is added. The eye-spotted budmoth, *Spilonota ocellana*, makes a short slightly curved case which is covered with black fecula. Species of the genus *Acrobasis* make larger, cornucopia-shaped cases which are covered with dense black excrement. The apple leaf crumpler, *Mineola indigenella* is one of our common species.

## BIBLIOGRAPHY

References are given to the immature forms of only the more important orders of insects. A bibliography of keys for the identification of immature insects is being prepared by W. P. Hayes. Part I, Diptera, and Part II, Odonata, have already appeared.

### I. General References.

BALDUF, W. V. (1939). The bionomics of entomophagous insects. John S. Swift & Co., St. Louis.

COMSTOCK, J. H. (1918). Nymphs, naiads and larvae. *Ann. Ent. Soc. Amer.* **11** (2).

HAYES, W. P. (1932). Method of procedure in insect life-history investigations. *Can. Ent.* **54** (4).

———— (1939). The present status of the classification of immature insects. *Trans. Ill. Acad. Sci.* **24**.

HYSLOP, J. A. (1938). Giving meaning to the terms, brood and generation. *Journ. Econ. Ent.* **31** (5).

PETERSON, ALVAH (1939). Keys to the orders of immature stages of North American insects. *Ann. Ent. Soc. Amer.* **32** (2).

———— (1945). Some insect infants. *Sci. Monthly* **40** (6).

RICHARDS, A. G. (1937). Insect development analysed by experimental methods; a review, Part II larval and pupal stages. *Journ. N. Y. Ent. Soc.* **45** (2). *Literature*.

Symposium (1920). The life cycle in insects. *Ann. Ent. Soc. Amer.* **13** (2).

### II. References to Moulting in Insects. See bibliography, Chap. VI.

**III. Some Outstanding References to the Eggs of Insects.**

COMSTOCK, J. H. (1940). An introduction to entomology. Comstock Publishing Company, Inc., Ithaca, New York.

FOLSOM, J. W. (1934). Entomology with special reference to its ecological aspects. Revised by R. A. Wardle. The Blakiston Company, Philadelphia.

FULTON, B. B. (1915). The tree crickets of New York; Life history and bionomics. *Geneva Exp. Sta. Tech. Bull.* **42.** *Bibliography.*

HEIDEMANN, O. (1911). Some remarks on the eggs of North American species of Hemiptera-Heteroptera. *Proc. Ent. Soc. Wash.* **13** (3).

HOLLAND, W. J. (1907). The butterfly book. Doubleday, Doran & Company, Inc., New York.

METCALF, C. L., and FLINT, W. P. (1939). Destructive and useful insects. 2d ed. McGraw-Hill Book Company, Inc., New York.

PETERSON, A. (undated). Some studies on the eggs of important apple plant lice. *N. J. Agric. Exp. Sta. Bull.* **332.**

READIO, P. A. (1926). Studies of the eggs of some Reduviidae (Heteroptera). *Univ. Kan. Sci. Bull.* **16** (4).

RICHMOND, E. A. (1920). Studies on the biology of the aquatic Hydrophilidae. *Amer. Mus. Nat. Hist.* **42** (1). *Literature.*

**IV. References to the Immature Stages of Mecoptera.**

BALDUF, W. V. (1939). The bionomics of entomophagous insects. John S. Swift & Co., St. Louis. *Literature cited.*

CARPENTER, F. M. (1931). The biology of the Mecoptera. *Psyche* **38** (1).

CRAMPTON, G. C. (1917). A phylogenetic study of the larval and adult head in Neuroptera, Mecoptera, Diptera and Trichoptera. *Ann. Ent. Soc. Amer.* **10** (4).

SETTY, L. R. (1940). Biology and morphology of some North American Bittacidae (Order Mecoptera). *Amer. Midl. Nat.* **23** (2). *References.*

―――― (1941). Descriptions of the larvae of *Bittacus apicalis* and a key to bittacid larvae (Mecoptera). *Journ. Kans. Ent. Soc.* **14** (2).

**V. References to Immature Stages of Neuroptera.**

CRAMPTON, G. C. (1917). A phylogenetic study of the larval and adult head in Neuroptera, Mecoptera, Diptera and Trichoptera. *Ann. Ent. Soc. Amer.* **10** (4).

DAVIS, K. C. (1903). Sialididae of North and South America. *In,* Aquatic insects of New York State. Part 7. *N. Y. State Mus. Bull.* **68** (18).

SMITH, R. C. (1922). The biology of the Chrysopidae. *Cornell Mem.* **58.** *Bibliography.*

―――― (1923). Life histories and stages of some hemerobiids and allied species (Neuroptera). *Ann. Ent. Soc. Amer.* **16** (2).

―――― (1926). The trash-carrying habits of certain lace-wing larvae. *Sci. Monthly* **23.**

VANDUZEE, E. P. (1935). Some recent neuropteroid papers. *Pan-Pacific. Ent.* **11** (1).

WHEELER, W. M. (1930). Demons of the dust. W. W. Norton & Company, Inc., New York.

**VI. References to the Immature Stages of Trichoptera.**

BETTEN, C., *et al.* (1934). The caddis flies or Trichoptera of New York State. *N. Y. State Mus. Bull.* **292.**

CLARKE, C. H. (1891). Caddis-worms of Stony Brook. *Psyche* **6.**

CRAMPTON, G. C. (1917). A phylogenetic study of the larval and adult head in Neuroptera, Mecoptera, Diptera and Trichoptera. *Ann. Ent. Soc. Amer.* **10** (4).

DODDS, G. S., and HISAW, F. L. (1925). Adaptation of caddisfly larvae to swift streams. *In*, Ecological studies on aquatic insects. III. *Ecology* **6** (2). *Literature cited*.

KRAFKA, J. (1915). Key to the families of trichopterous larvae. *Can. Ent.* **47** (7).

—— (1923). Morphology of the head of trichopterous larvae as a basis for the revision of the family relationships. *Journ. N. Y. Ent. Soc.* **31** (1). *Bibliography*.

LLOYD, J. T. (1915). Notes on the immature stages of some New York Trichoptera. *Journ. N. Y. Ent. Soc.* **23** (4).

—— (1921). Biology of North American caddis-fly larvae. *Cincinnati Bull.* **21.** Lloyd Library. Ent. ser. No. 1.

LUTZ, F. E. (1930). Caddice fly larvae as masons and builders. *Nat. Hist.* **30.**

NEEDHAM, J. G., and BETTEN, C. H. (1901). Aquatic insects of the Adirondacks. *N. Y. State Mus. Bull.* **47.**

NOYES, A. A. (1914). The biology of the net-spinning Trichoptera of Cascadilla Creek. *Ann. Ent. Soc. Amer.* **7** (4).

SLEIGHT, C. E. (1913). Relations of Trichoptera to their environment. *Journ. N. Y. Ent. Soc.* **21** (1).

**VII. References to the Immature Stages of Ephemerida.**

DODDS, G. S., and HISAW, F. L. (1924). Adaptations of mayfly nymphs to swift streams. *In*, Ecological studies of aquatic insects I. *Ecology* **5** (2). *Literature cited*.

MORGAN, A. H. (1913). A contribution to the biology of may-flies. *Ann. Ent. Soc. Amer.* **6** (3).

MURPHY, H. E. (1922). Notes on the biology of some of our North American species of mayflies. Lloyd Library 22. Ent. ser. No. 2. *Bibliography*.

NEEDHAM, J. G., *et al.* (1935). The biology of the mayflies. Comstock Publishing Co. Inc., Ithaca, New York.

**VIII. References to the Immature Stages of Odonata.**

BYERS, C. F. (1927). Notes on some American dragonfly nymphs (Odonata, Anisoptera). *Journ. N. Y. Ent. Soc.* **35** (1).

HAYES, W. P. (1941). A bibliography of keys for the identification of immature insects, Part II Odonata. *Ent. News* **52** (3–4).

HOWE, R. N. (1917–1920). Manual of the Odonata of New England. Parts I–VI and Supplement, *Mem. Thoreau Mus. Nat. Hist.* **11.**

NEEDHAM, J. G. (1903). The life histories of Odonata suborder Zygoptera. *In*, Aquatic insects of New York State. Part 3. *N. Y. State Mus. Bull.* **68** (18).

——, and FISHER, E. (1936). The nymphs of North American libelluline dragonflies (Odonata). *Trans. Amer. Ent. Soc.* **62.**

——, and HART, C. A. (1903). The dragon-flies (Odonata) of Illinois with descriptions of the immature stages, Part I. Petaluridae, Aeschnidae and Gomphidae. *Bull. Ill. State Nat. Hist. Lab.* **6** (1). *Literature*.

——, and HEYWOOD, H. B. (1929). A handbook of the dragonflies of North America. Charles C. Thomas, Publisher, Springfield, Ill.

**IX. References to the Immature Stages of Plecoptera.**

CLAUSSEN, P. W. (1931). Plecoptera of America, north of Mexico. Thomas Say Foundation.

FRISON, T. H. (1929). Fall and winter stoneflies, or Plecoptera, of Illinois. *Bull. Ill. Nat. Hist. Survey* **18** (2).

—— (1935). The stoneflies or Plecoptera, of Illinois. *Bull. Ill. Nat. Hist. Survey* **20** (4).

## X. References to the Immature Stages of Thysanoptera.

SHULL, A. F. (1914).    Biology of the Thysanoptera.    Parts 1 and 2.    *Amer. Nat.*
    **48** (567–568).    *Bibliography.*

## XI. References to the Immature Stages of Coleoptera.

ANDERSON, W. H. (1936).    Comparative study of the labium of coleopterous larvae.
    *Smiths. Misc. Coll.* **95** (13).

BALDUF, W. V. (1935).    The bionomics of entomophagous Coleoptera.    John S.
    Swift & Co., St. Louis.    *Literature cited.*

BARRETT, R. E. (1930).    A study of the immature forms of some Curculionidae
    (Coleoptera).    *Univ. Calif. Pub. Ent.* **5**.    *Bibliography.*

BEESON, C. F. C., and BHANTIA, B. M. (1939).    On the biology of the Cerambycidae
    (Coleoptera).    *Indian For. Rec.* **5** (1).

BLAIR, K. G. (1934).    Beetle larvae.    *Proc. Trans. Soc. Lond. Ent. Nat. Hist. Soc.*
    1933–1934.

BÖVING, A. G., and CRAIGHEAD, F. C. (1932).    Illustrated synopsis of the principal
    larval forms of the order Coleoptera.    *Bull. Brooklyn Ent. Soc.* **27** (1).

CHAPMAN, R. N. (1920).    The life cycle of Coleoptera.    *Ann. Ent. Soc. Amer.* **13** (2).

HATCH, M. H. (1925).    Habitats of Coleoptera.    *Journ. N. Y. Ent. Soc.* **33** (4).
    *Bibliography.*

PERRIS, M. E. (1877).    Larves des Coléoptères.    Deyrolle Naturaliste, Paris.

PEYLRIMHOFF, P. DE (1933).    Les larves des Coléoptères.    *Ann. Soc. Ent. Fr.* **102**.

PIERCE, W. D. (1907).    On the biologies of the Rhynchophora of North America.
    *Nebraska State Bd. Agr.*    *Bibliography.*

SCHAUPP, F. G. (1878–1879).    List of the described coleopterous larvae of the U. S.
    *Bull. Brooklyn Ent. Soc.* **2** (2, 3, 4).

SHELFORD, V. E. (1908).    The life histories and larval habits of tiger beetles.    *Journ.
    Linn. Soc. Zool.* **30**.

SMITH, J. B. (1889).    Notes on cerambycid larvae.    *Ent. Amer.* **5** (8).

STICKNEY, F. S. (1923).    The head capsule of Coleoptera.    *Ill. Biol. Monogr.* **8** (1).

WADE, J. S. (1935).    A contribution to a bibliography of the described immature
    stages of North American Coleoptera.    Mimeographed.    Washington, D. C.,
    E-358.

WEISS, H. B. (1922).    A summary of the food habits of North American Coleoptera.
    *Can. Ent.* **56**.

WILLIAMS, J. W. (1938).    The comparative morphology of the mouth parts of the
    order Coleoptera.    *Journ. N. Y. Ent. Soc.* **46** (3).

## XII. References to the Immature Stages of Lepidoptera.

CAPPS, H. W. (1939).    Keys for the identification of some lepidopterous larvae fre-
    quently intercepted in quarantine.    Mimeographed.    *U. S. Dept. Agr. Bur.
    Ent. & Plant Quarantine E-475.*

DAVENPORT, D., and DETHIER, V. G. (1937).    Bibliography of the described life
    histories of the Rhopalocera of America north of Mexico.    *Ent. Amer.* **17** (4).

DE GRYSE, J. J. (1916).    The hypermetamorphism of the lepidopterous sapfeeders.
    *Proc. Ent. Soc. Wash.* **18** (3).    *Bibliography.*

DYAR, H. G. (1894–1904).    Ten papers on lepidopterous larvae published in *Journ.,
    N. Y. Ent. Soc., Amer. Nat., Proc. Ent. Soc. Wash.,* and *Ann. N. Y. Acad. Sci.*

EDWARDS, H. (1889).    A bibliographical catalogue of described transformations of
    North American Lepidoptera.    *Bull. U. S. Nat. Mus* **35**.

ELIOT, J. M., and SOULE, C. G. (1902).    Caterpillars and their moths.    D. Appleton-
    Century Company, Inc., New York.

FORBES, WM. T. M. (1907–1910).   New England caterpillars.   *Journ. N. Y. Ent. Soc.* **15** (1) and **18** (3).
——— (1909).   On certain Pieris caterpillars.   *Psyche* **16.**
——— (1910).   The aquatic caterpillars of Lake Quinsigamond.   *Psyche* **17** (6).
——— (1910–1914).   Structural studies of caterpillars.   *Ann. Ent. Soc. Amer.* **3** (2), **4** (3), and **7** (2).
——— (1916).   On certain caterpillar homologies.   *Journ. N. Y. Ent. Soc.* **24** (2).
FRACKER, S. B. (1915).   The classification of lepidopterous larvae.   *Ill. Biol. Monogr.* **2** (1).   *Bibliography.*
——— (1920).   The life cycle of the Lepidoptera.   *Ann. Ent. Soc. Amer.* **13** (2).
HEINRICH, CARL (1916).   On the taxonomic value of some larval characters in the Lepidoptera.   *Proc. Ent. Soc. Wash.* **18** (3).   *Bibliography.*
HILTON, W. A. (1902).   The body sense hairs of lepidopterous larvae.   *Amer. Nat.* **36** (427).   *Bibliography.*
LLOYD, J. T. (1914).   Lepidopterous larvae from rapid streams.   *Journ. N. Y. Ent. Soc.* **22** (2).
McDUNNOUGH, J. (1935).   Notes on early stages of certain Canadian Microlepidoptera.   *Can. Ent.* **67** (4).
MOSHER, EDNA (1916).   A classification of the Lepidoptera based on character of pupae.   *Ill. Bull.* **12** (2).   *Bibliography.*
PACKARD, A. S. (1895–1914).   Monograph of the bombycine moths of North America, including their transformations and origin of larval markings and armature. Part I, Notodontidae, Parts II and III, Ceratocampidae.   *Mem. Nat. Acad. Sci.* **7–9, 12.**
POULTON, E. B. (1891).   The external morphology of the lepidopterous pupa.   *Trans. Linn. Soc. Zool.* ser. **2** (5).

XIII. References to the Immature Stages of Diptera.
BHATIA, M. L. (1939).   Biology, morphology and anatomy of the aphidophagous syrphid larvae.   *Parasitology* **31** (1).   *References.*
FROST, S. W. (1924).   A study of the leaf-mining Diptera of North America.   *Cornell Univ. Agric. Exp. Sta. Mem.* **78.**   *Bibliography.*
GREEN, C. T. (1921).   An illustrated synopsis of the puparia of 100 muscoid flies (Diptera).   *Proc. U. S. Nat. Mus.* **60** (11).
HAYES, W. P. (1938–1939).   A bibliography of keys for the identification of immature insects.   Part I, Diptera.   *Ent. News* **49** (9), **50** (1).
HEISS, E. M. (1940).   A classification of the larvae and puparia of the Syrphidae of Illinois.   *Univ. Ill. Bull.* **36.**
JOHANNSEN, O. A. (1935–1937).   Aquatic Diptera.   Parts I, II, III, IV, V.   *Cornell Mem.* **164, 177, 205,** and **210.**   *References.*
MALLOCH, J. R. (1917).   A preliminary classification of Diptera exclusive of Pupipera, based upon larval and pupal characters with keys to imagines in certain families.   *Bull. Ill. State Nat. Hist. Lab.* **12** (3).   *References.*
METCALF, C. L. (1920).   The life cycle in the Diptera.   *Ann. Ent. Soc. Amer.* **13** (2).
PETERSON, ALVAH (1916).   The head capsule and mouth parts of Diptera.   *Ill. Biol. Monogr.* **3** (2).   *Bibliography.*

XIV. References to the Immature Stages of Homoptera and Heteroptera.
BALL, E. D. (1920).   The life cycle in Hemiptera.   *Ann. Ent. Soc. Amer.* **13** (2).
BUTLER, E. A. (1923).   A biology of the British Hemiptera-Heteroptera.   London.
COMSTOCK, J. H. (1916).   Reports on scale insects.   *Cornell Univ. Agric. Exp. Sta. Bull.* **372.**
——— (1918).   Nymphs, naiads and larvae.   *Ann. Ent. Soc. Amer.* **2.**

DIETZ, H. F., and MORRISON, H. (1916).   The Coccidae of Indiana.   Office State Entomologist, Indianapolis, Indiana.

DOZIER, H. L. (1926).   The Fulgoridae or plant-hoppers of Mississippi.   *Miss. Agr. Exp. Sta. Tech. Bull.* **14.**

ESSENBERG, C. (1915).   The habits and natural history of the back-swimmers, Notonectidae.   *Journ. Animal Behavior* **6** (5).   *Literature.*

FUNKHOUSER, W. D. (1917).   Biology of the Membracidae of the Cayuga Lake Basin. *Cornell Univ. Agric. Exp. Sta. Mem.* **11.**   *Bibliography.*

HART, C. A. (1919).   The Pentatomidae of Illinois with keys to nearctic genera. *Bull. Ill. Nat. Hist. Surv.* **13** (7).

HERRICK, G. W. (1911).   Some scale insects of Mississippi.   *Miss. State Coll. Agric. Tech. Bull.* **2.**

HOTTES, F. C., and FRISON, T. H. (1931).   The plant lice or Aphididae of Illinois. *Bull. Ill. Nat. Hist. Surv.* **19** (3).

HUNGERFORD, H. B. (1919).   The biology and ecology of aquatic and semi-aquatic Hemiptera.   *Univ. Kan. Sci. Bull.* **11.**

METCALF, Z. P. (1923).   Fulgoridae of eastern North America.   *Journ. Elisha Mitchell Sci. Soc.* **38.**

OSBORN, H. (1928).   The leafhoppers of Ohio.   *Ohio State Univ. Bull.* **32.**

——— (1938).   The Fulgoridae of Ohio.   *Ohio Biol. Survey Bull.* **35, 6** (6). *References.*

——— and DRAKE, C. J. (1916).   The Tingitoides of Ohio.   *Ohio State Univ. Bull.* **20.**

READIO, P. A. (1928).   Biology of the Reduviidae of North America, north of Mexico. *Univ. Kan. Sci. Bull.* **17.**

## XV. References to the Immature Stages of Hymenoptera.

ALSTERLUND, J. F. (1937).   The larva of the *Chalcodermus collaris* Horn with key to related species.   *Proc. Ent. Soc. Wash.* **39** (8).

BISCHOFF, H. (1927).   Biologie der Hymenopteren.   Eine Naturgeschichte der Hautflüger.   Verlag Julius Springer, Berlin.

COCKERELL, T. D. A. (1920).   The life cycle in the Hymenoptera.   *Ann. Ent. Soc. Amer.* **13** (2).

DYAR, H. G. (1895).   The larvae of the North American sawflies.   *Can. Ent.* **27** (12).

——— (1900).   On the larvae of *Atomacera* and some other sawflies.   *Journ. N. Y. Ent. Soc.* **7** (1).

HILL, C. C., and PINCKNEY, J. S. (1940).   Keys to the parasites of the hessian fly based on remains left in the host puparium.   *U. S. Dept. Agric. Tech. Bull.* **715.**

HOPPING, G. R., and LEACH, H. B. (1936–1937).   Sawfly biologies No. 1 and 2. *Can. Ent.* **68** (4) and **69** (10).

MARLATT, C. L. (1891).   The final molting of tenthredinid larvae.   *Proc. Ent. Soc. Wash.* **2** (1).

MIDDLETON, WM. (1917).   Notes on the larvae of some Cephidae.   *Proc. Ent. Soc. Wash.* **19** (1–4).

——— (1922).   Sawflies injurious to roses.   *U. S. Dept. Agric. Farmers' Bull.* **1259.**

——— (1922).   Sawflies of the subfamily Cladiinae.   *Proc. U. S. Nat. Mus.* **60** (1).

——— (1922).   Descriptions of some North American sawfly larvae.   *Proc. U. S. Nat. Mus.* **61** (21).

MILES, H. W. (1935).   Biological studies of certain species of *Caliroa, Costa* and *Endelomyia* Ashmead (Hymenoptera, Symphyta).   *Ann. Appl. Biol.* **22** (1). *References.*

——— (1936).   On the biology of certain species of *Holcocneme* Kon.   (Hymenoptera-Symphyta).   *Ann. Appl. Biol.* **23** (4).   *References.*

——— (1939).   On the biology and habits of British sawflies.   *Trans. Lancs. Nat. Un.*

PACKARD, A. S. (1897). Notes on the transformation of the higher Hymenoptera, Parts I to III. *Journ. N. Y. Ent. Soc.* **5** (2–3).

PARKER, H. L., and THOMPSON, W. R. (1925). Notes on the larvae of the Chalcidoidea. *Ann. Ent. Soc. Amer.* **18** (3).

ROHWER, S. A. (1922). North American sawflies of the subfamily Cladiinae with notes on habits and descriptions of larvae by Wm. Middleton. *Proc. U. S. Nat. Mus.* **60** (1).

WHEELER, W. M. (1910). Ants their structure and behavior. Cambridge University Press, London.

YUASA, HACHIRO (1922). A classification of the larvae of the Tenthredinoidea. *Ill. Nat. Hist. Survey Monograph* **7** (4). *Bibliography*.

**XVI. References to the Fecula of Insects.**

CLARKSON, F. (1884). The origin and limitation of the term scarabaeus. *In*, The dung pellet-makers. *Can. Ent.* **16** (1).

CRUMB, S. E. (1956). The larvae of the Phalaenidae. *U. S. Dept. Agric. Tech. Bul.* **1135**.

ECKSTEIN, K. (1939). Das Bohremehl der Anobien. *Forstarchiv* **15** (13–14).

———— (1939). The frass and the debris in galleries of *Myelophilus piniperda* L. *Arb. Physiol. Angew. Ent. Berl.* **6** (1).

EDGE, E. R. (1934). Faecal pellets of some marine invertebrates. *Amer. Midl. Nat.* **15** (1). *Bibliography*.

FROST, S. W. (1919). The function of the anal comb in certain lepidopterous larvae. *Journ. Econ. Ent.* **12** (19).

———— (1928). Insect scatology. *Ann. Ent. Soc. Amer.* **21** (1). *Literature cited*.

FUKAYA, S. (1938). On the mechanism of excretion in the last larval instar of *Acanthostoma insidiator* Smith (Ichneum). *Oyo-Dobuts Zasshi, Tokyo* **10** (1).

HODSON, A. C. and BROOKS, M. A. (1956). The frass of certain defoliators of forest trees in the north central United States and Canada. *Can. Ent.* **88**.

JURITZ, C. F. (1920). Analysis of the dropping of the caterpillar (*Antheraea cytherea*). *Chem. News* **121** (3157).

KNAB, F. (1915). Dung bearing weevil larvae. *Proc. Ent. Soc. Wash.* **17** (4).

MORRIS, R. F. (1942). Use of frass in the identification of forest insect damage. *Can. Ent.* **74** (9).

PETERSON, A. (1951, 1956). The larvae of insects. Parts I and II. Edwards Brothers, Ann Arbor, Mich.

POLL, M. (1935). La methode des injections colorées appliquée à étude de l'excretion chez les insectes. *C. R. Congr. Nat. Sci.*

SCHWARTZ, E. A. (1901). "Sawdust" of wood borers. *Proc. Ent. Soc. Wash.* **4** (4).

STOREY, H. H., and NICHOLS, R. F. W. (1937). Defaecations by a jassid species. *Proc. R. Ent. Soc. Lond.* **12** (10–12). *References*.

WEISS, H. B. and BOYD, W. M. (1950, 1952). Insect feculae. *Journ. N. Y. Ent. Soc.* **58** and **60**.

## CHAPTER VIII

## INSECT MORPHOLOGY

The mouth is associated with feeding, biting, and the transmission of diseases; the legs are associated with walking, jumping, and feeding; the wings are associated with flight, stridulation, and other functions; therefore, it is essential to be familiar with the morphology of insects in order to understand their habits. The three sciences, anatomy, morphology, and physiology, are closely related but each covers a definite field. *Anatomy* is the study the structure and relation of the various

parts of organisms, usually by dissection. *Morphology* is the science of form of living organisms. *Physiology* is the study of the vital functions performed by the organs and structures of living organisms. The present chapter deals primarily with the morphology of insects.

The body of an adult insect is usually divided into three well-marked major divisions: head, thorax, and abdomen. As a matter of fact these divisions constitute one of the characteristics of insects. The head, with rare exceptions, is distinctly differ-

FIG. 121.—Head of praying mantis. The two compound eyes, the three simple eyes, the mandibles, and the antennae are conspicuous morphological features.

entiated. In some highly specialized species, such as the scale insects, the division between head and thorax is obscure. The division between the thorax and abdomen is usually distinct; however, in certain insects, it may be obscure. In the cicada, the division between thorax and abdomen is complicated because the organs of sound production, which morphologically belong to the thorax, extend backward and overlap several of the ventral abdominal segments. In the grasshopper, the first ventral abdominal segment is dovetailed into the sternum of the metathorax. In many insects, such as the stink bugs (Pentatomidae), the treehoppers (Membracidae), and certain of the Orthoptera, the pronotum is considerably produced and covers a large

158

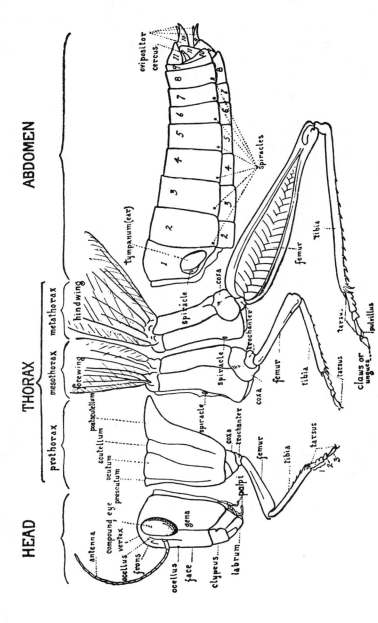

FIG. 122.—Outline of the body of a grasshopper as seen from the side. Dissected to show the three body regions and the parts commonly referred to in books and bulletins. (*After Metcalf and Flint.*)

par⌄ of the dorsal aspect of the abdomen. In the pigmy locusts (Acrydiinae), the pronotum covers the entire dorsal aspect of the abdomen and sometimes extends beyond its extremity. In the Membracidae, the pronotum extends not only backward but upward and sidewise to form grotesque little creatures sometimes known as brownie bugs or tree hoppers. In the Hymenoptera, the division between the thorax and abdomen seems to be distinct, especially in the thread-waisted wasps,

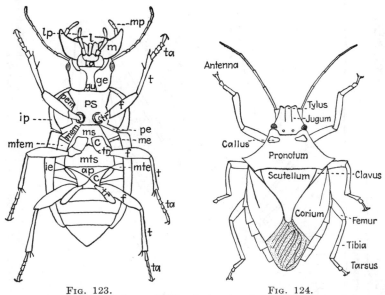

FIG. 123.    FIG. 124.

FIG. 123.—Under surface of a carabid beetle showing the principal parts: (*m*) mandible; (*l*) labrum; (*la*) labium; (*ge*) gena; (*gu*) gula; (*ps*) prosternum; (*ms*) mesosternum; (*mts*) metasternum; (*ap*) antecoxal piece; (*c*) coxa; (*tr*) trachanter; (*f*) femur; (*t*) tibia; (*ta*) tarsus; (*pe*) prosternal epimeron; (*me*) mesosternal epimeron; (*mte*) metasternal epimeron; (*pem*) prosternal episternum; (*mem*) mesosternal episternum; (*mtem*) metasternal episternum; (*ip*) inflexed pronotum; (*ie*) inflexed elytron; (*lp*) labial palpus; (*mp*) maxillary palpus. (*After Smith.*)

FIG. 124.—Dorsal view of a pentatomid Heteroptera, showing the principal taxonomic characters.

Sphecinae, but as a matter of fact the division is frequently not clear. The first abdominal segment is often so closely associated with the thorax as to appear as a part of the thorax rather than the abdomen. In the ants, the first two abdominal segments may be completely separated from the rest of the abdomen to form one or more scales or nodes. In most insects, the thorax can be recognized as the three leg-bearing segments.

In immature insects, the division between the head and thorax is usually distinct but the division between the thorax and abdomen may be obscure.

**Exoskeleton.**—The framework that supports the softer parts of the body of an insect is a hard, tough body wall known as the *exoskeleton*. It consists of three distinct layers: the *cuticula*, the *hypodermis*, and the *basement membrane*. The hypodermis* is a single layer of comparatively large cells, with large nuclei. It is the only living portion of the body wall. The hypodermis produces a horny covering known as the cuticula. The cuticula is not cellular but composed of dead material. It contains chitin, pigments, and various substances that give it color, toughness, and hardness. Beneath the hypodermis there is a thin basement membrane which binds the lower ends of the hypodermal cells together. In most larvae, a large part of the body wall is soft and unsclerotized; the

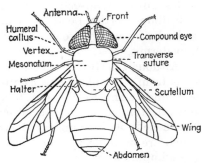

Fig. 125.—Tabanus species, dorsal view, showing the principal structures.

head alone is usually hard and is called a head capsule. Most nymphs, naiads, pupae, and adults have exoskeletons that are heavily sclerotized. To permit movement between the segments, the conjunctivae or portions of the body wall between these segments are less heavily sclerotized.

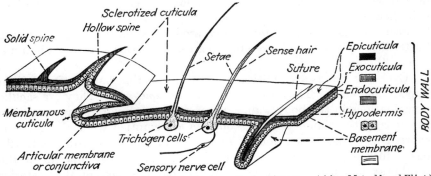

Fig. 126.—Diagrammatic section of the body wall of an insect. (*After Metcalf and Flint.*)

Each body segment is composed of several separate horny areas known as *sclerites*. The lines separating the sclerites are known as *sutures*. Thus the labrum is separated from the clypeus of the head by a suture.

**Structures Associated with the Body Wall.**—Many structures are derived from or closely associated with the body wall of insects. Hairs

---

* The epidermis of some writers.

are formed from enlarged hypodermal cells (*trichogens*).   Spines are produced by the thickening and attenuation of the cuticula.   Scales are produced from hypodermal cells.   As a matter of fact, scales are modified hairs.   We are sometimes prone to think that only Lepidoptera have scales, but many insects have scales.   Those found on other insects, such as bedbugs, certain beetles, the larvae of fleas, and Cyrtidae, are not true scales because they are developed from the cuticula rather than from the hypodermis.

Although the body wall is a protective covering, it also receives impressions from without and transmits impulses from within.   The receptors of sound, sight, taste, and touch are located in the body wall.   In general, thin areas of the cuticula serve to receive outside impulses. Nerve cells and complicated structures, such as chordotonal organs, which are more deeply removed from the surface of the body, transmit these impulses to the nervous system.

Many of the glands of insects are located in the body wall.   Moulting fluid glands are common to practically all insects.*   They are formed from hypodermal cells.   These unicellular glands secrete a fluid which is liberated between the new and the old cuticulae at the time of moulting and which aids in separating these two layers.

Nettling hairs, common in certain caterpillars, have poison glands at their bases.   These are modified hypodermal cells which are connected with the lumina of the nettling hairs.   When the tip of the hair is broken off, the venom is liberated.

Scent scales, known as *androconia*, are located on the wings of certain butterflies.   These glands secrete an agreeable fluid that attracts individuals of the opposite sex.

Eversible glands are multicellular structures that have their origin in hypodermal cells.   They are known also as *osmeteria*.   They are saclike invaginations in the body wall.   These pockets fill with secretions from the cells of the glands and the contents are emptied when the gland is everted or turned inside out.   The larva of the swallowtail butterfly has a forked gland which protrudes from the anterior margin of the prothorax just behind the head.   Similar glands appear on the posterior end of the larva of the puss moth, *Cerura*, on various parts of certain noctuid, sawfly, and other larvae.

The wax glands of honeybees and scale insects are multicellular glands that develop from hypodermal cells.   They open through pores in the cuticula.   The stink glands of the Pentatomidae and the froth glands of spittle insects likewise have their origin in the hypodermal cells.

---

* It is believed that insects living in the soil, in burrows, or in cells do not possess definite glands but the moulting fluid is secreted by the entire hypodermal layer.

We should note that the digestive glands, associated with the alimentary canal, the reproductive glands, associated with the genital organs, and the silk glands, are not of hypodermal origin.

**Head.**—The head bears the mouth parts, antennae, and eyes. Adults may have two kinds of eyes: compound and simple. Compound eyes are situated one on each side of the head and are composed of many facets or *ommatidia*. Immature insects seldom possess compound eyes. Simple eyes occur in many immature insects and are generally present in adults. The typical arrangement in the adults is in the form of a triangle on the front of the head. Sometimes the middle ocellus is absent.

The movable parts of the head consist of the antennae and the mouth parts. Insects have but one pair of jointed antennae, which vary considerably in form, and numerous names have been proposed to describe them. The antennae of adults are usually conspicuous and often long. In a few insects, such as the dragonfly, the antennae are minute. The antennae of immature insects are often small, reduced, or absent.

It is interesting to note that other appendages may look like and even function as antennae. The front legs of the tailless scorpion, *Tarantula marginemaculata*, are two or three times as long as the other legs and are modified to function as feelers. Functional antennae are, of course, absent in the Arachnida. The fleshy filaments of the swallowtail butterfly larva, *Laertias philenor*, seem to serve as tactile organs. The front legs of the Central American longicorn, *Acrocinus longimanus*, are almost as long as the antennae but as far as known have no sensory function.

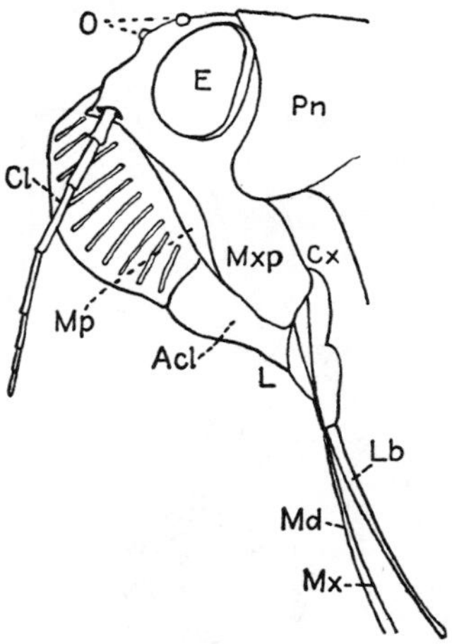

FIG. 127.—Head of a cicada showing typical haustellate mouth parts: (*Acl*) ante clypeus; (*Cl*) clypeus; (*E*) compound eye; (*Cx*) coxa; (*L*) labrum; (*Lb*) labium; (*Mp*) mandibular plate; (*Mxp*) maxillary plate; (*Md*) mandible; (*Mx*) maxilla; (*Pn*) pronotum; (*O*) ocelli.

The mouth parts of insects are remarkable structures. They vary tremendously in form and function. Insects are divided on the basis of the mouth parts into two groups: chewing insects, *mandibulate*, and sucking insects, *haustellate*. Although the two types are quite different in appearance, their parts are homologous. The mandibulate type is the more primitive and shows the more complete development. There are three pairs of appendages that more or less oppose each other. One pair of these operates in a vertical plane when the head is held in the normal position.* These are known as the *labrum*, or upper lip, and the *labium*, or lower lip. The labrum is not truly a part of the mouth but a sclerite of the head. The labium, although a single appendage, is formed by the fusion of two appendages. Two other pairs of mouth parts operate in a horizontal plane. The upper pair are the mandibles, which are usually horny and possess teeth and grinding surfaces. Below the mandibles are the *maxillae*. Each maxilla has a segmented appendage, the maxillary palpus, which aids in manipulating the food and in directing it toward the mouth. Other less conspicuous parts are the *epipharynx*, located on the lower side of the labrum, and the *hypopharynx*, a tonguelike organ borne on the floor of the mouth cavity.

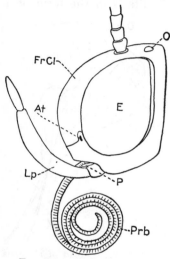

The haustellate type of mouth is often considerably reduced or highly modified. In the Lepidoptera, the proboscis is composed almost entirely of two maxillae which fit together to form a tube. In the resting position, the maxillae are coiled like a watch spring and held close to the lower side of the head often concealed by the abundance of scales. The labial palpi are frequently much enlarged, scaled, and form conspicuous features of the head. In the

FIG. 128.—Haustellate mouth parts of a butterfly: (*At*) anterior tentorial pit; (*E*) compound eye; (*FrCl*) front and clypeus; (*Lp*) labial palpi; (*O*) ocellus; (*P*) pilifer (rudimentary mandible); (*Prb*) proboscis (maxillae).

Heteroptera, the labium is well developed and serves as a sheath in which four stylets are enclosed, namely, two mandibles and two maxillae. The labrum is usually short and inconspicuous. The mouth parts vary considerably in the Diptera. They are often developed for piercing and sucking. In other species, they are developed for lapping and sucking. In the more typical forms, the mouth parts consist of six bristlelike organs enclosed in a sheath, and a pair of jointed palpi. In the muscoid Diptera, the labium and the maxillary palpi are the most conspicuous organs. The mouth parts of the Hymenoptera are formed primarily for chewing. In the more specialized members, they are fitted for chewing and lapping.

The mouth parts of immature insects take many forms and tend to show more variation than those of the adults. The greatest departure probably exists in the muscoid Diptera. They have an internal pharyngeal skeleton to which the mouth hooks are attached. The nymphs of Isoptera and the larvae of Trichoptera, Mecoptera, Lepidoptera, and

* Normal position as in the grasshopper.

Hymenoptera are mandibulate. Although the larvae of the Neuroptera are mandibulate, they have sickle-shaped jaws with openings at the tips through which the juices of their victims are sucked. The nymphs of the Orthoptera and the naiads of the Odonata, Ephemerida, and the Plecoptera are mandibulate. The labium of the dragonfly naiad is elongate, geniculate, and is normally folded so that the mandibles are concealed. It is a powerful weapon of offense. During the process of metamorphosis, the Lepidoptera have a complete change of mouth types; the caterpillar is mandibulate and the adult is haustellate.

**Thorax.**—The thorax is divided into three regions termed the *prothorax, mesothorax,* and *metathorax.* Theoretically each region is divided into four parts represented dorsally by four sclerites: *prescutum, scutum, scutellum,* and *postscutellum.* On the prothorax of the grasshopper, these sclerites are rather plainly indicated. In other insects, they often overlap one another or are turned inward so that it is difficult to trace the limits of these divisions. The structure of the sides of the thorax (*pleurae*) and the venter of the thorax (*sternum*) is usually complicated and not easily observed. Two pairs of thoracic spiracles are usually distinguished; the mesothoracic and the metathoracic spiracles. It is believed that the pair of prothoracic spiracles have migrated forward to form the silk glands. As a matter of fact the spiracles are located not directly on the segments but rather on membranes between the segments. In the larvae of insects, there is usually only one pair of thoracic spiracles. The thorax of the adult also bears the locomotory appendages: legs and wings.

**Legs.**—The segments of the thorax are most readily distinguished by the fact that each bears a pair of legs. Thoracic legs are always present in adults although the first pair may be considerably reduced. In the Nymphalidae, they are so small and useless that this group has been called the four-legged butterflies. Thoracic legs may be absent in immature insects.

Each leg consists of the following parts: *coxa, trochanter, femur, tibia,* and *tarsus.* The coxa is the proximal segment and the one by which the leg is articulated to the thorax. In several orders each coxa is composed of two, more or less distinct, parallel parts. The trochanter is usually a short segment. In some Hymenoptera it consists of two parts. The femur and tibia are usually long. The tarsus consists of a series of segments varying from one to six. The first tarsal segment is sometimes more elongate than the others. The pollen basket of some of the bees is located on the first tarsal segment. The terminal segment of the foot of the adult, nymph, and larva usually bears two claws. In the Lepismatidae three claws are present. The third claw of each tarsus of the triungulin larva of the meloid beetle is probably a spine. In a few insects, such as the Thysanoptera, the claws are minute or absent and the foot terminates in a bladderlike arolium. The Coccidae, Pediculidae, and certain Mallophaga have but a single claw on each distal tarsal segment.

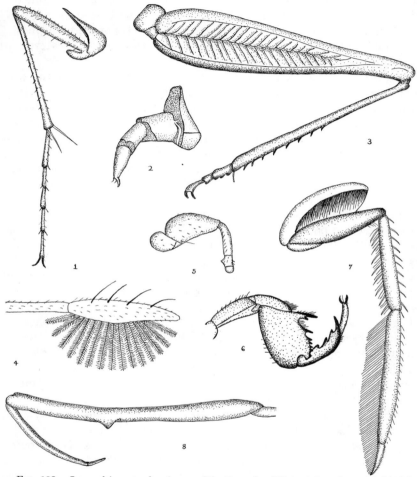

Fig. 129.—Legs of insects showing modifications for different functions: (1) hind leg of *Cicindela sexguttata*, cursorial; (2) thoracic leg of caterpillar, ambulatorial; (3) hind leg of grasshopper, saltatorial; (4) middle tarsus of *Rhagovelia obesa*, natatorial; (5) leg of *Anthothrips nigricornis*, arolial; (6) prothoracic leg of Cicada nymph, fossorial; (7) hind leg of *Corixa*, natatorial; (8) prothoracic leg of *Ranatra fusca*, raptatorial.

The primary function of legs is locomotion, but they are built for different types of movement. *Ambulatorial* legs are long or short and intended for walking. *Cursorial* legs are long and adapted for running.

Legs adapted for leaping are known as *saltatorial* legs, examples of which are numerous among the grasshoppers, fleas, and flea beetles. In the grasshoppers and the flea beetles, the femora are enlarged; in the fleas, the coxae are enlarged. Legs developed for swimming are known as *natatorial* legs, which are well developed in the back swimmers and the water boatmen. Claws are absent and the legs are armed with long hairs or bristles which increase the area of the leg to serve as an oar or paddle. Grasping or *raptorial* legs are found among many different insects. The anterior pair of legs is usually modified for this purpose. The femur and tibia are armed with sharp spines which oppose each other when closed. The mantids, mantispids, nepas, and phymatids have legs of this sort. In the panorpids, the hind legs are adapted for grasping. All of these species are predacious and feed upon other insects. Insects that dig have *fossorial* legs. The femur and tibia of the mole cricket are expanded, armed with spurs, and adapted for digging. Scarab beetles and certain wasps also have fossorial legs; in fact, one large group of wasps is known as the fossorial wasps. Anoplura and Mallophaga have legs modified to cling to feathers and hair. In Anoplura, the foot has a single strong claw which folds against the end of the tibia to form an excellent grasping organ. This is known as a *scansorial* leg. Many insects employ their legs in the production of sound. Such legs might be called *stridulatorial*. Certain Locustidae have a series of pegs on the femur of the hind leg which are rubbed against the edge of the wing covers to produce sound. Legs are used for other miscellaneous purposes. The pollen baskets and wax pincers of bees are located on the legs. Legs are frequently used in the manipulation of wax or silk. Even the ears of some species are located on the legs.

It is necessary to differentiate between the prolegs of larvae and the true or thoracic legs. Prolegs occur only on immature insects, their primary function being to cling. They are lost before the adult stage is reached.

**Wings.**—Wings, with the exception of those of the subimago of the May fly, occur only on mature insects. Although most insects have wings, the Thysanura, Collembola, Dicellura, Anoplura, Mallophaga, and Siphonaptera are entirely wingless. Other orders may have wingless representatives. There are typically two pairs of wings. The first pair is borne by the mesothorax, the second pair by the metathorax. Certain fossil roaches show expansions of the prothorax which indicate that insects may at one time have had three pairs of wings.

**Modifications of Wings.**—The primitive condition was probably membranous wings such as those now found in the Hymenoptera and Diptera. From this type, wings have developed various textures: hairy wings (Trichoptera), scaly wings (Lepidoptera), leathery wing covers (Orthoptera), horny wing covers (Coleoptera), half leathery wings (Heteroptera), etc. The leathery wing covers of the Orthoptera are called *tegmina*. The horny wing covers of Coleoptera are called *elytra*. The half membranous and leathery wing covers of Heteroptera are called *hemelytra*.

In the Diptera, the second pair of wings is replaced by the balancers or *halteres*. In the Strepsiptera, the first pair of wings is replaced by the *pseudohalteres*. In the male Coccidae, the second pair of wings is replaced by two spines, one on each side of the thorax. They hook into the first pair of wings and give them rigidity. Although the Coleoptera usually

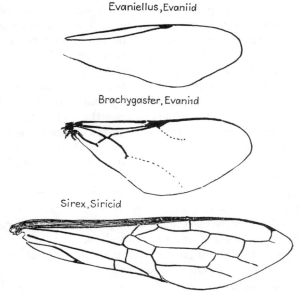

Evaniellus,Evaniid

Brachygaster, Evaniid

Sirex,Siricid

Fig. 130.—A series of front wings of Hymenoptera showing different degrees of venational development. Lack of wing veins often indicate poor fliers.

have two pairs of wings, the second pair is lacking in some of the Tenebrionidae, Carabidae, and other beetles, and the elytra are firmly sealed down the center of the back.

**Loss of Wings.**—The Thysanura, Collembola, and Dicellura are completely wingless and it is believed that, during the course of evolution, they never had wings. They are therefore said to be primitively wingless. The Anoplura, Mallophaga, and Siphonaptera are likewise wingless but it is believed that they, at one time, had wings which have been lost through disuse.

Wings are sometimes lacking in the individuals of other orders. It is usually the female that is wingless. All female scale insects are wingless.    The females of the families Psychidae, and some of the Geometridae and Liparidae, are wingless.    Rarely is the male wingless.    The fig insect *Blastophaga* has an unusual life history.    The wingless male remains in the cavity of the fig and waits for the female to visit the fruit.

**Development of Wings.**—Wings develop as buds starting, in some insects, early in the immature stages.    Wings are generally formed within pads or buds on the exterior of the body of insects with complete metamorphosis.    The development of wings in the higher Diptera is internal and is not visible from the outside.    The wings usually develop externally in insects with gradual or incomplete metamorphosis.    There are exceptions.    In Corrodentia for example, the development is internal and there is no indication of wings until the adult stage is reached.    In the Thysanoptera, the wing buds do not appear on the surface until the third instar.

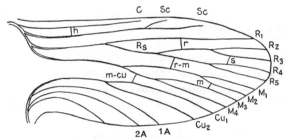

Fig. 131.—Course of veins in a hypothetical wing: (*C*) costa; (*Sc*) subcosta; (*R*) radius; (*M*) media; (*Cu*) cubitus; (*A*) anal; (*h*) humeral cross vein; (*r*) radial cross vein; (*r-m*) radio-medial cross vein; (*m*) medial cross vein; (*m-cu*) medio-cubital cross vein; (*s*) sectorial cross vein.

**Wing Venation.**—The wing is an expansion of the body wall and consequently has two surfaces.    In the early embryonic development these two surfaces are separate but during subsequent development they become joined together.    Intervening channels or spaces are left which carry nerves, blood, and temporary tracheae and determine the courses of the future veins of the wings.    In insects with incomplete metamorphosis, the tracheae usually take their places in the wings before the veins are formed.    Thus a study of the tracheation in the wing buds of these insects serves as a key to the homology of the wing veins of the adult.    In insects with complete metamorphosis, the veins may be formed in advance of the permanent tracheae.    A study of the veins of insects is known as wing venation, neuration, or tracheation.    There is a definite arrangement of veins in the wings of each species.    J. H. Comstock and J. G. Needham were the first to recognize and study the relationships among the veins of different orders.

**Areas of the Wing.**—The veins and the areas between the veins take their names largely from the margins of the wing. The wing has three margins: costal or frontal margin, apical or distal margin, and anal or basal margin. A hypothetical wing (Fig. 131) shows the various veins and cross veins that occur in the wings of insects.

There are two general types of veins: the longitudinal veins that run lengthwise of the wing and the cross veins that run across it. Fre-

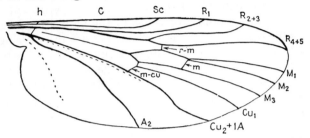

Fig. 132.—Wing of *Anisopus*, Diptera: (*C*) costa; (*Sc*) subcosta; (*R*) radius; (*M*) media; (*Cu*) cubitus; (*A*) anal; (*h*) humeral cross vein; (*r-m*) radio-medial cross vein; (*m*) medial cross vein; (*m-cu*) medio-cubital cross vein.

quently, as in the Hymenoptera and Corrodentia, the longitudinal veins turn across the wing and have the appearance of cross veins. In the Neuroptera, Odonata, Ephemeridae, and Plecoptera, numerous additional longitudinal and cross veins occur. These are known as accessory

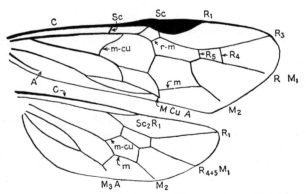

Fig. 133.—Wing of a tenthredinid, *Pteronus*, Hymenoptera: (*C*) costa; (*Sc*) subcosta; (*R*) radius; (*M*) media; (*Cu*) cubitus; (*A*) anal; (*r-m*) radio-medial cross vein; (*m*) medial cross vein; (*m-cu*) medio-cubital cross vein.

veins and take their names from the principal veins. The number of veins may thus be increased. On the other hand, the number of veins may be reduced. In certain Hymenoptera, especially the cynipids, the wing is almost veinless. Veins may be lost when two coalesce or they may be lost by actually atrophying.

**Methods of Joining Wings.**—Insects with two pairs of wings have various devices for joining the front and the hind pair. In the butterflies, there is a seta or group of setae, the *frenulum*, at the base of the costal margin of the hind wing which engage a fold on the basal margin of the front wing. Most moths have on the anal margin of the fore wing a lobe or projection, the *jugum*, that fits into a fold on the costal margin of the hind wing. Many bees and wasps have a series of recurved spines on the costal margin of the hind wing which fit into a fold on the anal margin of the front wing. These are called the *hamuli*.

**Abdomen.**—The abdomen, the third region of the body, is divided into a series of somewhat similar rings or segments. Each is composed of two parts: the tergum above and the sternum below. They are united by lateral folds, the *conjunctiva*. These folds are weakly sclerotized and permit expansion of the body for breathing. The number of segments in the abdomen of the adult varies from 3 to 10 or 11.

**Appendages of the Abdomen.**—In the embryo, each segment of the abdomen, except the telson, bears a pair of appendages. Many of these disappear before the egg hatches. Certain larvae have prolegs on the abdominal segments. Adults are usually without abdominal appendages except on the eighth and ninth segments, which bear the reproductive appendages, *genitalia*. The eleventh segment in some species bears the cerci and a caudal filament. Spiracles normally occur on the first to the eighth segments.

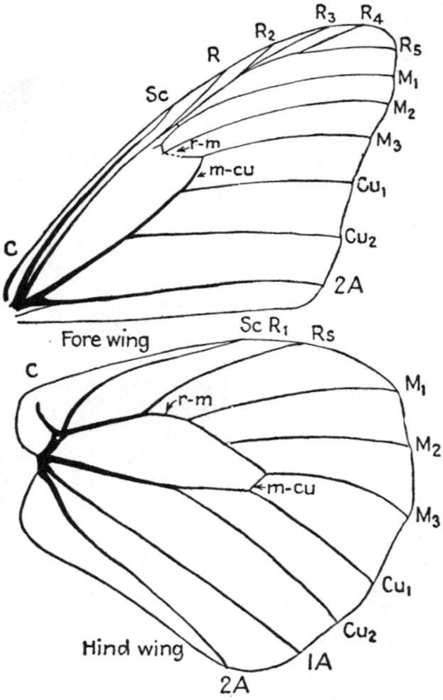

Fig. 134.—Wing venation of monarch butterfly: (*C*) costa; (*Sc*) subcosta; (*R*) radius; (*M*) media; (*Cu*) cubitus; (*A*) anal; (*h*) humeral cross vein; (*r*) radial cross vein; (*r-m*) radio-medial cross vein; (*m*) medial cross vein; (*m-cu*) medio-cubital cross vein.

**Other Structures on the Abdomen.**—The light organs of the fireflies (Lampyridae) are located on the under side of the fourth to the sixth abdominal segments and are exposed when the wings are raised. They occur close to the hypodermal cells and are profusely supplied with tracheae. Little is known concerning the light mechanism of the fireflies. Two highly complex organic compounds, *luciferase* and *luciferin*, are produced in these organs. One is an oxidizing agent, the other a reducing agent.   Light is produced by rapid oxidation.   Very little heat is produced and there is very little loss of energy.   The efficiency is rated between 92 and 100 per cent.   In comparison, the efficiency of an ordinary oil lamp is 1 per cent.

The sound-producing organs of the cicada are located on the dorsal surface of the basal abdominal segment.   Within the first abdominal

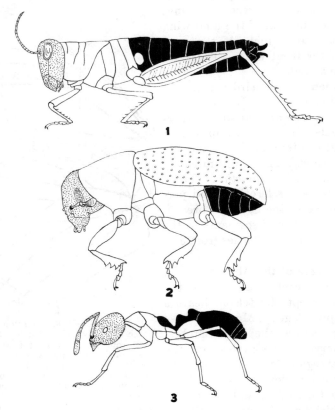

FIG. 135.—Principal body regions of insects: (1) grasshopper; (2) bark beetle; (3) ant; head is stippled, thorax is white and the abdomen is black.

segment there are two large cavities or chambers.   The cuticula in this area is thin and forms a cover or *operculum*.   Within the chamber are three membranes: the timbal, the folded membrane, and the mirror. Powerful muscles attached to the timbal membrane set it into vibration and produce the characteristic drumming of the cicada.   The chamber acts as a resonator.   The mirror acts as a sounding board to deflect the sound waves upward.

**Internal Anatomy.**—There are seven principal systems that make up the internal anatomy of an insect: digestive, circulatory, nervous, respiratory, muscular, excretory, and reproductive. Each presents certain features that are peculiar to insects.

*Digestive System.*—The alimentary canal follows in general the plan of other higher animals with mouth, pharynx, esophagus, crop, gizzard, mid intestine, and hind intestine. These organs vary considerably according to the type of food eaten. The muscles of the pharynx are especially well developed in the haustellate insects. The crop or food reservoir may be absent in some insects. In the roach it is especially large. Insects that feed upon hard substances have strongly sclerotized gizzards (*proventriculi*) with powerful muscles. The mid intestine is long in herbivorous insects but short in parasitic, predacious insects and those that feed on highly concentrated food such as nectar.

In general, there are three main divisions to the alimentary canal: the fore gut, *stomodaeum*, the mid gut, *mesenteron* and the hind gut, *proctodaeum*. To understand the alimentary canal we must remember that the fore and hind guts are invaginations of the body wall and show similar structures, which, however, have received different names. The cuticula of the body wall becomes the *intima* of the alimentary canal, the hypodermis becomes the *epithelium*. The mid intestine is developed from other tissues and unlike the fore and the hind guts has no intima. The epithelial cells are enlarged and secrete digestive fluids. These changes would be expected as this is the digestive portion of the alimentary canal. Since the stomodaeum and the proctodaeum are invaginations of the body wall, the intimae of these portions are shed each time the insect moults. In some insects, there is no connection between the mid intestine and the hind intestine until the insect pupates; then a connection is made and, for the first time, the insect voids its waste material.

*Circulatory System.*—The blood of insects does not circulate through their bodies in veins and arteries as is the case in higher animals. On the contrary, it flows through the open spaces in the insect's body, bathing all tissues. The heart is a comparatively simple organ. It is a tube closed at the posterior end and open at the anterior end. This tube is provided with pairs of valves which open inward and which have paired muscles that cause the heart to contract. The muscles start to contract at the posterior end of the heart and the impulse is carried forward. As the muscles contract, the valves close and the blood is forced towards the head. The blood flows through the aorta, a constricted portion of the anterior end of the heart. It first bathes the brain and then flows to other parts of the body. After the muscles of the heart contract, they relax and the valves open and admit a new supply of blood.

*Nervous System.*—The nervous system, typical of all arthropods, is ventrally located. In brief, it consists of an esophageal ganglion in the head above the alimentary canal and a subesophageal ganglion below the alimentary canal. These are connected by commissures, one on each side of the alimentary canal. From the esophageal ganglion, nerves lead to the clypeus, mouth parts, antennae, eyes, and other parts of the head. Along the ventral wall of the insect, arranged segmentally in pairs, are the ganglia which form the sympathetic nervous system. These are connected by two fused commissures from the subesophageal ganglion. In adults, there is a tendency towards the cephalization of the nervous system. In some insects, there are only three ventral ganglia: one located in the head, one in the thorax, and one in the abdomen. The cephalization may continue to such a degree that there is only one large fused ganglion in the thorax. This condition is found in many Diptera.

*Respiratory System.*—The respiratory system of insects is unique. Air is taken to all parts of the body by means of air tubes, known as *tracheae* and *tracheoles*. The openings to the exterior are known as *spiracles*. Each spiracle opens into a trachea which is an invagination of the body wall and shows similar structure. The inner lining of the trachea is the *intima*. This in turn is furnished with a spiral thickening known as the *taenidium*. When the insect moults, the intima is shed along with the cuticula. The tracheae divide into small tubes, which in turn divide into smaller tubes that ramify all parts of the body. These small tubes are called *tracheoles*. They differ in structure from the tracheae for they are formed from individual cells and there is no intima. The linings of the tracheoles therefore are not shed when the insect moults. As a matter of fact, there is no connection between the trachea and the tracheoles until the first moult occurs.

The immature forms of aquatic insects, such as the dragonflies, stone flies, and the May flies, have tracheal gills. These are thin expansions of the body wall profusely supplied with tracheoles. These permit an interchange of gases between the tracheoles and the water outside. The rectal gills of the Odonata are forms of tracheal gills. The blood of certain Chironomidae, known as bloodworms, contains hemoglobin, which extracts oxygen from the water.

## BIBLIOGRAPHY

Abbott, C. E. (1936). The physiology of insect senses. *Ent. Amer.* **16** (4). Bibliography.

Berlese, A. (1809). Gli insetti. Società Editrice Libraria, Milano.

Comstock, J. H., and Kellogg, V. L. (1912). The elements of insect anatomy. Comstock Publishing Company, Inc., Ithaca, New York.

Ferris, G. T. (1934). Setae. *Can. Ent.* **66** (7).

FRAENKEL, G., and RUNDALL, K. M. (1940). A study of the physical and chemical properties of the insect cuticula. *Proc. Roy. Soc. Lond. Biol. Sci.* **129**.

HAYES, W. P. (1930). Contributions to morphology of insects. *Ann. Ent. Soc. Amer.* **23** (3).

HENNEGUY, L. F. (1904). Les insectes. Masson et Cie, Paris.

HOSKINS, W. M., and CRAIG, R. (1935). Recent progress in insect physiology. *Physiol. Rev. Baltimore* **15**.

JANET, CHARLES (1909). Sur la morphologie de l'insectes. Limoges Ducourtieux et Gout.

JOHANNSEN, O. A. and BUTT, F. H. (1941). Embryology of insects and myriapods. McGraw-Hill Book Company, Inc., New York.

MACGILLIVRAY, A. D. (1923). External insect anatomy; a guide to the study of insect anatomy and an introduction to systematic entomology. Scarab Co., Urbana, Ill.

PETERSON, ALVAH (1916). The head capsule and mouth parts of Diptera. *Ill. Monogr.* **3** (2).

PHILLIPS, E. F. (1930). Contributions to physiology of insects. *Ann. Ent. Soc. Amer.* **23** (3).

RICHARDS, A. G. (1951). The integument of arthropods. Univ. Minn. Press.

ROEDER, K. D. (1953). Insect physiology. John Wiley & Sons, Inc., New York.

SNODGRASS, R. E. (1935). Principles of insect morphology. McGraw-Hill Book Company, Inc., New York. Bibliography.

———— (1952). A textbook of arthropod anatomy. Comstock Publishing Company. Ithaca, New York.

STICKNEY, F. S. (1923). The head capsule of Coleoptera. *Ill. Biol. Monogr.* **8** (1).

WEBER, HERMANN (1938). Grundriss der Insektenkunde. Gustav Fischer, Jena.

WIGGLESWORTH, V. B. (1940). The principles of insect physiology. E. P. Dutton & Company, Inc., New York.

WILLIAMS, J. W. (1938). The comparative morphology of mouth parts of the order Coleoptera. *Journ. N. Y. Ent. Soc.* **46** (3).

Consult also general texts on entomology and references to immature insects in Chap. VII.

# CHAPTER IX

## COLOR

Color is a remarkable phenomenon. It is one of the most conspicuous features of the world, yet it exists only as a manifestation of light. Insects are frequently colorful and take their hues from the range of the spectrum. The larger moths and butterflies are frequently sought by collectors because of their beauty but some of the microlepidoptera are equally as handsome. The species of *Lithocolletis* are indeed jewels,

FIG. 136.—A common Central American grasshopper, *Tropidacris latreillei.* One of the many brilliantly marked insects. The forewings are mottled with greenish yellow and brown; the hind wings are brick red with numerous cross bars of black; the legs are green with brick-red tarsi.

set with scales of gold and silver and etched with delicate markings. Many of our Orthoptera, Heteroptera, Homoptera, Odonata, and other insects compare in beauty with the Lepidoptera. Color is most profusely displayed in the tropical species.

Although many insects are colorful, others are gray, brown, black, or inconspicuous. The immature of aquatic, subterranean, and subcutaneous insects are usually white or pale in color. Adults, immediately after emerging, and insects, just after moulting, lack color. The eggs of insects are seldom conspicuously marked.

Color is important from the standpoint of protection and, in some cases, possibly as recognition marks. It is well known that insects are eaten by birds and other animals and that protective coloration is ineffective in numerous instances. However, it is an advantage in many cases. Color often forms identification marks for the taxonomist. Many of the Ottidae and Trupaneidae can be recognized by the color patterns of the wings.

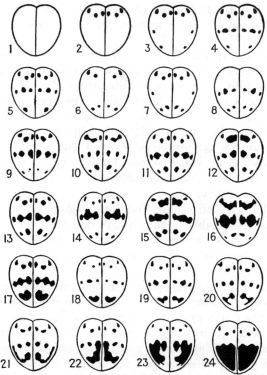

Fig. 137.—Variations in the color patterns of the elytra of the Mexican bean beetle, *Epilachna varivestris*. Figures 2 to 8, 14, 17, 19 and 23 represent beetles from Mexico. Figure 16 is a beetle from Alabama. The remaining are from Ohio. (*After Landis and Mason.*)

Light, by absorption or reflection, is the source of all color; upon this basis, insect colors are classified as pigmental and structural.

**Pigmental Coloration.**—The ordinary red, yellow, and blue colors of insects, which fade after death, are pigmental in nature and generally occur in the cuticula and the hypodermis. The *melanins* are usually located in the cuticula; the other pigments are located in the hypodermal cells. Thus the black stripes of the Colorado potato beetle occur in a different layer of the body wall from that of the yellow ground color, like a slip applied to a piece of pottery or a lace dress over a colored slip.

The pigments found in insects are obtained from the food they consume. The green color of caterpillars is due largely to the ingestion of chlorophyll. The green pigment of the wings of *Pieris* is a chromoproteid.

The yellows and reds of insects are due largely to *lyochromes* (flavines) and *lipochromes* (carotene and xanthophyll). These are derived from the chlorophyll of plants. It is interesting to note that *Perillus bioculatus*, a predator of the Colorado potato bettle, obtains its yellow pigment from the potato bettle, which in turn derives its color from the potato leaves.

Red colors have various sources. *Hemoerythrin* is responsible for the red color of bloodworms, Chironomidae. Insects that feed upon the blood of higher animals become red because of the ingested hemoglobin which shows through the thin body wall of the insect. The pigment in this case is located in the alimentary canal, not in the body wall of the insect. Carminic acid is the red pigment of the cochineal insect. *Anthocyanin*, a substance which produces blue, red, and purple colors in flowers, fruit, leaves, and stems, is the source of reds, purples, and possibly blues, in many insects. The vermilion color of the aphid *Tritogenaphis* has been isolated as anthocyanin.

Little is known concerning the blue pigments of insects. *Carotinalbumins* are responsible for the pink, purple, and green hues of Orthoptera. Blue pigments, which are somewhat rare in insects, may be produced by anthocyanin. The violets, blues, and greens are generally structural colors.

The black and brown pigments of insects are nitrogenous substances known as *melanins*. They are by-products of metabolism and have their source in *tyrosine* (monoxyphenylalanine) and *dopa* (dioxyphenylalanine) which are formed in the blood of insects. They are probably derived from amino acids which are constituents of animal and vegetable proteins.

Purine pigments are somewhat uncommon in insects. The chalky white color of the pierid butterflies is due to uric acid. The reds and yellows of *Pieris, Eurymus*, and *Papilio* are derivatives of uric acid. Purine pigments occur also in the hypodermis of Neuroptera, Cicadidae, Syrphidae, wasps, hornets, and doubtless other insects.

**Structural Coloration.**—The iridescent blues, greens, and violets of insects are permanent and do not fade in the sunlight. They are produced by the refraction or diffraction of light by the surface of the insect's body. The sheen of the hyaline wings of Hymenoptera, Diptera, and certain other insects is due to the interference of the light waves reflected from the two surfaces of the wings much as light is reflected by the film of a soap bubble.

The brilliant changeable colors of the wings of many butterflies, especially the tropical morphos, are due to the diffraction of light by fine closely parallel striae on the scales of the wings. The scales of the wings of the morphos have approximately 1,400 striae per millimeter. The light is broken by these fine ridges into its component parts, much as a prism forms a rainbow.

The metallic sheen of many insects is due to the refraction of light in the same manner that metals, such as gold, silver, or copper, give forth their characteristic reflections. The surfaces in these cases are extremely opaque and throw back a large proportion of the light that strikes them. The reflected light therefore is very brilliant. The distance to which light can penetrate these surfaces is a small fraction of the wave length. The depth to which lights of different wave lengths penetrate varies, and

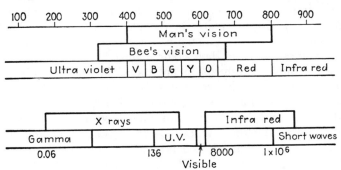

Fig. 138.—Color vision. Upper figure, a comparison of man's and bee's vision, wave lengths given in millimicrons. Lower figure, radiant energy in the vicinity of the visible spectrum, wave lengths in angstroms.

some colors are transmitted more freely than others. Thus the reflected portion of the light differs from the original source and the characteristic reflections are produced. The reflected light is approximately complementary to the transmitted light. The greenish sheen of certain tiger beetles, Cicindelidae, and wood-boring buprestid beetles is of this nature.

The silvery white color of certain insects is due to the reflection of light from air-filled pockets such as those located in, or between, overlapping scales or air-filled tracheae. Diving bettles and other insects that descend below the surface of the water reflect light from the film of air incorporated in the hairs of the body.

**Invisible Emanations.**—Insects, flowers, and other objects that seem colorful to man, may appear quite different to insects or other animals. Experiments have been conducted with bees and other insects which show that many—perhaps all—insects see ultraviolet rays but do not see red rays. Thus they perceive much that is beyond the range of man at the short end of the spectrum but less than he sees at its long end.

These facts clearly illustrate that animals may perceive colors quite differently from man. A portulaca flower that looks red, yellow, or pink to a man appears ultraviolet to the insect. Insects see their own kind through eyes different from those of man. The delicate green of the female luna moth is so strongly tinged with ultraviolet that it appears the same to both man and insect with the exception of the reddish edge on the anterior margins of the wings which appears black to the insect. The male luna moth, however, lacks the ultraviolet and therefore appears dark gray or black to the insect. The yellow tiger swallowtail butterfly, *Papilio glaucus turnus*, with its contrasting black lines appears dingy to the insect except for a row of spots on the hind wing.

The wings of certain insects, especially butterflies, emit a form of energy, the nature of which is not exactly understood. Clark(1932) and others have shown that, when the wings of butterflies are placed in contact with a photographic plate, in complete darkness, they affect the plate in such a way that an image of the wings is produced. The best images were obtained from dark-colored butterflies and from the upper sides of the wings. Fresh specimens gave the most brilliant images. However, a specimen of *Basilarchia arthemis astyanax*, thirty years old, gave a faint image after a week's exposure; other specimens forty and fifty years old gave very faint images after long exposures. Pieces of quartz 0.2 millimeters in thickness placed between the object and the photographic plate cut the emanations completely even after an exposure of 12 days. Pieces of glass approximately 0.2 millimeters in thickness also cut these emanations completely. Cellophane, after exposures of 72 hours, cut the emanations only slightly. It should be noted that light of short wave lengths will pass through quartz and glass as readily as it will pass through cellophane. In the experiments, quartz and glass completely obliterated the images, but faint images were produced through cellophane. Gases, on the other hand, penetrate cellophane but not quartz or glass. It is therefore believed that photographic plates in contact with the wings of butterflies were affected not by light but by a slow decomposition of the pigments of the scales and possibly by sulphur which is released during this process.

**Color Changes.**—The change of color in insects may be of four kinds. Somewhat rapid changes in color follow a moult. This is due to the development and accumulation of pigments in the hypodermis or cuticula. After moulting the insect is usually pale in color; in a comparatively short time, the new color is formed which may differ from that of the preceding instar. One must look to enzymes such as *tryosinase*, *trypsin*, or *cellulase* to find the regulators of pigment formation.

A more gradual change in color occurs as an insect matures. Frequently it does not acquire its mature coloration for a considerable period of time after it emerges; sometimes, not until the following year. A species of leaf beetle, known in literature as *Lina lapponica,* is pale when it emerges but soon becomes yellow with black markings. After several hours the yellow changes to red. The term *teneral* is used to indicate the condition of an adult insect when it is not entirely hardened or fully colored. Green grasshoppers occasionally become pink toward the end of the summer. The Calverts describe color changes in the Odonata. When the male of the Costa Rican *Erythrodiplax funerea*

FIG. 139.—Two stages in the development of the larva of the Western parsley caterpillar, *Papilio zelicaon,* showing change in color pattern. (*After Duncan and Pickwell.*)

emerges, the wings are a pale yellowish brown except the extreme tips which are edged with darker brown. As it becomes older, the color darkens until it is blackish brown. Similar color changes take place in certain flowers. The flowers of *Trillium grandiflorum* are white when they first open but later turn rose color. These changes are due largely to the oxidation or reduction of the pigments, the effect of acids, alkalies, etc.

Some insects change their color in response to external stimuli, such as temperature, humidity, or light. This is the result of the alteration of the arrangement of the pigments and is comparable to the remarkable color changes that occur in the chameleon and frog. The movements of the pigments of the eyes of certain insects for day and for night vision are forms of color change induced by light intensity. Considerable variation in colors has been noted in the Mantidae and Phasmidae. These insects are generally dark at night and pale by day.

Certain color changes also take place after the death of the insect. The eyes of many insects, especially the Tabanidae, lose their brilliant color after death. The golden tortoise beetle, *Metriona bicolor*, loses its golden sheen after death. Comstock remarks, "The brightness of the colors is said to depend upon the emotions of the insect. What a beautiful way to express one's feelings, to be able to glow like melted gold when one is happy!" The alteration of color after death is due to changes in the nature of the pigments. The brilliancy may fade or the source of the pigments may be lost after the cessation of metabolism.

**Seasonal Coloration.**—Overwintering adults are often darker in color than the summer generations of the same species. The hibernating elm leaf beetles, for example, are dark greenish in color and the newly emerged beetles are pale brown. Adults that emerge during wet weather are frequently darker in color than adults that emerge during dry weather. *Ceratomyza dorsalis* and *C. femoralis* were described as two distinct species. However, it is believed that *dorsalis* is a pale variety of *femoralis*, and that the differences are associated with the relative amount of moisture at the time of transformation. In like manner *Acidia heraclei* has been described as a darker form of *A. fratria*.

The most prominent seasonal color variations are known as *polymorphism*.* This subject has been carefully studied in the Lepidoptera. Many species have two or more color forms. These color forms are well marked in the imported cabbage worm, *Ascia rapae*, the painted lady, *Cynthia cardui*, the red admiral, *C. atlanta*, the swallowtails, *Papilio philenor*, *P. glaucus turnus*, and others. Butterflies with a single brood also show polymorphism. The spangled fritillary, *Argynnis cybele*, emerges over a long period from May to October. Specimens that emerge early in the season are smaller and duller in color than those that emerge late in the season.

**Varietal Coloration.**—Many species show considerable variation in coloration, which makes identification difficult. It is well recognized that color in certain groups of insects is not reliable as a means of identification. For example, the leaf roller, *Argyrotoxa semipurpurana*, ranges from specimens that are pale yellow to specimens strongly suffused with

---

* The condition of having several color forms in the adult might better be called *polychromorphism.*

brown. Some individuals have whitish markings, others have orange bands. Such diversity of color markings is very common among the Lepidoptera.

**Sexual Coloration.**—Sexual color dimorphism is most prevalent among the Lepidoptera. The imported cabbage worm, *Ascia rapae*, is a common example. The adults are white or cream color. The female has two conspicuous black spots on the anterior margin of the upper side of each front wing; the male has but one spot on each of the front wings. In many of the larger moths, as for example, *Automeris io* and *Callosamia promethea*, the two sexes are often remarkably different.

Sexual color differences are well developed in the Membracidae. The female locust tree hopper, *Thelia bimaculata*, for example, is uniformly brown in color; the male is much darker, almost black, with a conspicuous broad yellow line on each side of the body.

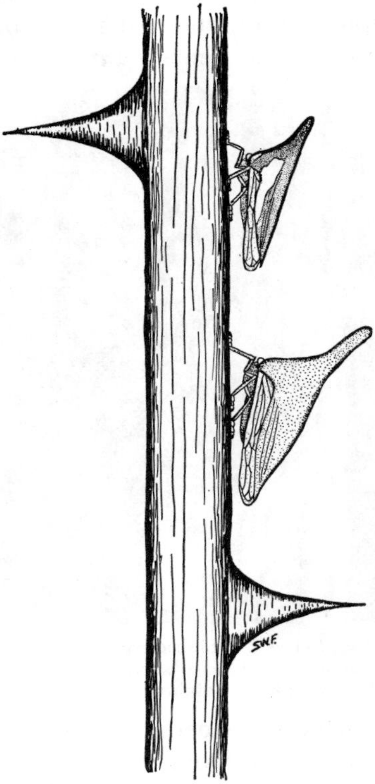

The Odonata often show conspicuous differences in sexual color markings. The wings of the male of *Calopteryx maculata* are velvety black, and those of the female are smoky in color with a distinct white stigmatal spot on the tip of each wing.

Sexual color differences are not common in the Coleoptera although there are some examples. The male of *Hoplia trifasciata* is grayish in color and the female is reddish.

Color dimorphism is not conspicuous in the Hymenoptera but it is perhaps best represented in the Tenthredinidae.

FIG. 140.—Sexual color differences in the locust treehopper, *Thelia bimaculata*. Below, the female is brown in color. Above, the male is darker in color with distinct yellow marks on the sides of the body.

In the cherry-hawthorn sawfly, *Profenusa canadensis*, the abdomen of the female is metallic black and the prothorax is rufous; the abdomen of the male has a broad whitish dorsal band and the thorax is whitish in color.

The Orthoptera show some differences in sexual coloration. Wood describes the Tasmanian grasshopper, *Acripeza reticulata*, in which the color markings of the two sexes are quite different. The second pair of wings is even absent in the female.

The sexes of Diptera more frequently show differences in form than in color. The male *Dolichopodidae*, for example, are characterized by the elongate, ornate fore tarsi. The eyes of many small Diptera are contiguous; those of the females are more widely separated. In the Dioptidae, the eyes of the males are conspicuously stalked and those of the female are only slightly so. The two sexes of the Indian species, *Achias longividens*, show distinct differences of color. Some of the Tipulidae of the tribe Ctenophorini also show sexual dimorphism. Species of the genus *Tanyptera* are often brilliantly colored with black and

reddish yellow simulating Hymenoptera. In *T. fumipennis* the legs and abdomen of the male are black and the legs and the abdomen of the female are reddish yellow.

**Protective Coloration.**—Protective coloration in insects differs from that of other animals. The wings are generally the most conspicuous part, especially when in flight; they are frequently folded out of sight when the insect is at rest, and they have two surfaces both of which may be brightly colored. Insects are also small in size. They can hide more readily than larger animals and their

Fig. 141.—The pine sawyer, *Monochamus*, at rest on the bark of a tree. An excellent case of an insect matching its background.

movements are, on the one hand, often rapid and, on the other hand, very deliberate. Counter-shading, a prominent feature of larger animals, is seldom a factor in insects. Protective coloration depends upon the form and color of the insect, the background, illumination, and sometimes upon motion or cessation of motion.

**Insects That Resemble Their Background.**—The simplest form of protective coloration is the matching of the background. A catocala moth, resting upon the bark of a tree, fades from view because it resembles its background. The moths of this group have front wings that vary in color from white to gray and brown. The hind wings are generally brilliantly colored with red, orange, and black. The somber-colored front wings cover the showy hind wings when the moths are at rest. Many grasshoppers also have showy hind wings which are obscured by the wing covers when the insects are at rest. Species that live upon the beaches often have pale-colored wing covers which resemble the white sand of the beach.

Transparent wings or wings with transparent areas reveal the background and serve the same purpose as though the insect were colored like the background. A Central American butterfly, *Callitaera menander*, has wings that are almost entirely transparent, a slight iridescent color being visible when the insect is viewed from an angle. Aside from a small spot on each of the hind wings there are no markings. The background shows through the wings almost as clearly as through a pane of glass.

Fig. 142.—*Callitaera menander*, a Central American butterfly with wings so transparent that the background is evident through them.

Protective coloration can be divided into aggressive and defensive coloration. On the one hand, insects may be hidden or obscured by their surroundings so that they can prey upon other insects. The crab spider, *Misumena vatia*, in its yellow form, and the greenish ambush bug, *Phymata erosa*, are protectively colored when resting in the flower heads of goldenrod where they await their victims. On the other hand, insects may be hidden or obscured so that they can avoid their enemies. The locust borer, *Cyllene robiniae*, with its waxy yellow bands and dusting of pollen visits the goldenrod for food but apparently finds protection from its enemies.

**Insects Resembling Other Objects.**—The adult of *Stenoma algidella* is white in color with faint gray markings. When at rest upon a leaf, it resembles a bird dropping.

Fig. 143.—A moth, *Stenoma algidella* Walk., at rest on an apple leaf resembles a bird dropping.

Some of the noctuid moths resemble fungus growths. This is particularly true of *Euthisanotia grata*. The resting attitude of this moth with thoracic tufts raised and palpi projecting helps to complete the resemblance. The front wings are creamy white with brownish-purple markings resembling the colors of certain fungi.

The *Heliconias* of Florida present an interesting problem of resting attitude and protective coloration. In the late afternoon, according to

Jones, they drift one by one to the vicinity of their sleeping quarters, lighting on their roost or darting up again for a short flight.  At sunset, the air about the roost is filled with butterflies, which gradually decrease in numbers as they settle down for the night.  When at rest, they hang

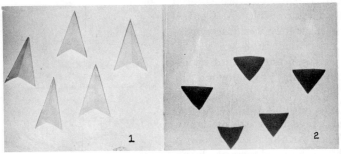

FIG. 144.—Paper models to represent the effect produced by moths casting their shadows upon the trunk of a tree: (1) light directed from below casts no shadows; (2) light directed from above casts triangular shadows which, in nature, give the effect of short stout spines on the tree.

with their wings drooping downward.  The pale stripes on the wings merge with dry roots and grasses on which they often roost.

When in Panama, the writer observed certain small moths* at rest in large numbers upon the trunk of a tree, all with their heads upward.

FIG. 145.—Larva of *Schizura unicornis* feeding upon the edge of a hickory leaf.

The sun striking them at a sharp angle produced short triangular shadows that gave the appearance of stout spines on the trunk of the tree.  The moths themselves, because of their color, were quite invisible and were discovered by accident.  When an attempt was made to detect the nature of the spines, a cloud of moths darted out from the tree but soon returned to their original positions.

The larvae of *Schizura* vary greatly in appearance but resemble the color of various hosts upon which they feed.  *Nerice bidenta* which feeds on elm is entirely green in color.  The body segments of the larva are deeply cut and, when feeding upon the edge of the leaf, it resembles very closely the serrated edge of the elm leaf.  Another species that feeds on hickory is green with brown spots which resemble dead portions of the leaf.  Still another species that feeds on black walnut is dark brown in color and resembles a dead portion of a black walnut leaf.

* J. F. G. Clarke determined that this species belongs to a new genus near Ocalaria.

The dead-leaf butterfly, *Kallima inachis*, of India, is a classical example of protective coloration. The upper sides of the wings are strikingly marked with blue and shades of brown. When at rest, these surfaces are folded together and the lower surfaces resemble a dead leaf. A dark line through the center of the wings even simulates a leaf vein. It is said that these butterflies rest among dead leaves, thus enhancing the protection.

The larva of Harris's sphinx, *Lapara bombycoides*, feeds in a group of pine needles and is well protected because the body is marked with longitudinal green and white stripes that simulate a bunch of pine needles (see Fig. 287).

Fig. 146.—The dead-leaf butterfly, *Kallima inachis*, is a classical example of protective coloration: (1) wings expanded revealing the upper surfaces which in nature are brilliant orange, brown, and blue in color; (2) wings folded revealing the under surfaces which resemble a dead leaf.

The strange elongate walkingsticks are green or brown in color and are difficult to see because of their color, form, and slow movement.

Geometrid larvae often imitate twigs by holding themselves rigid. Their color frequently resembles very closely the bark of the tree upon which they rest.

The rose mantid of India, *Gongylus gongyloides*, is a remarkable example of protective coloration. The thorax is long, slender, green in color, and has the appearance of the stem or petiole of a flower. The second and third pairs of legs have green leaflike expansions which permit the insect to hide very well in the foliage. The lower surface of the insect is rose or lavender in color and simulates a flower so perfectly that insects actually fly straight into its clutches. This is at the same time alluring and aggressive coloration.

**Vanishing Coloration.**—Many insects are conspicuous during flight; when they come to rest, they fold their brilliantly colored wings beneath the fore wings and become inconspicuous. The hind wings of the bella moth, *Utetheisa bella*, are conspicuously red with black markings; the front wings are cream colored and less conspicuous. This moth has the habit of dropping suddenly into the grass and folding its wings, which makes it difficult to follow the insect to its resting place.

The author has observed groups of morphos feeding upon rotten seeds of almedro and other nuts in the jungles of Panama. When feeding, they fold the wings over the body, and the iridescent blue color of the upper

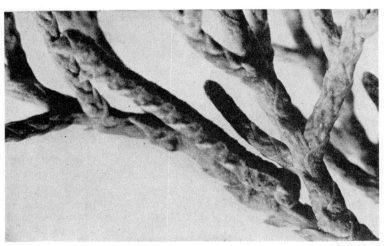

Fig. 147.—A measuring-worm (Geometridae) resting upon a cypress branch. The markings of the caterpillar resemble the scales of the cypress. (*After Duncan and Pickwell.*)

surfaces of the wings is hidden. The lower surfaces of the wings are gray or brown in color and it is almost impossible to locate these butterflies when they are resting upon the ground. If disturbed, they quickly take flight and exhibit a gorgeous flash of color as they drift lazily away. It is not uncommon to find 20 or 30 of these butterflies feeding beneath a single tree, yet they may easily be passed by.

Tiger beetles and grasshoppers depend upon a sudden halt and a few side steps to throw the enemy off the trail. The cessation of motion is often combined with the folding of brilliant wings, which obscures the insect.

**Warning Coloration.**—Many insects display hideous colors or markings that may frighten their enemies as, for example, the two large eye spots on the thorax of the eyed elater or the two spots on the larva of *Papilio troilus* or *turnus*. Many butterflies and moths have similar spots that serve the same purpose. One of the most conspicuous species is the owl butterfly, *Caligo*, of tropical America. The upper surfaces of the wings are chocolate brown in color with a slight iridescence in certain lights. The lower surfaces of the wings are largely mottled gray with a large black eye spot on each of the hind wings surrounded by a conspicuous yellow edge. Hingston describes a two-headed butterfly in which the tips of the wings form what appears to be a head. The fine tails on the posterior margin of the hind wings appear as antenna and legs, and dark spots form the eyes. These marks are supposed to frighten their enemies. Some of the spots and markings on the wings of butterflies are said to direct the attack by birds in such a way that they avoid striking the more vital part of the insect's body.

**Mimicry.**—Mimicry is a form of protective coloration in which an insect resembles another species that is obnoxious or distasteful and thus obtains the protection of the species it mimics. The protection takes many forms: stings, hairs, bristles, unpleasant odor or taste, warning coloration, etc. How this habit developed is beyond the scope of this discussion but we wish to point out some of the interesting cases of mimicry. Entomologists have, perhaps, gone too far at times in assuming that because one insect resembles another, it is protected by the second species. We must remember that insects and other animals do not see alike and that insects that seem to be similarly colored to man

Fig. 148.—The elater beetle, *Alaus oculatus*, showing two eye spots supposed to frighten other insects and animals. (*Champlain and Kirk.*)

may appear quite different to other animals. Conversely, two or more insects that appear different to man, may look the same to other animals. In like manner, odors or sounds produced by certain insects, which may contribute to protective resemblance, may not be perceived by other insects and animals in the same manner as they are by man.

The conditions of mimicry, which were formulated largely by Wallace, are as follows: the mimic and the species it mimics must occupy the same area and generally the same station; the mimic lacks the protective device of the species it mimics; the mimic is usually less numerous than the species it mimics; the imitation is usually external and visible except in those cases where internal conditions may affect external appearance or internal structures may contribute to protection.   Form, structure, and motion may contribute to protective resemblance as much as color.

A familiar example of mimicry is the viceroy, *Basilarchia archippus*, which imitates the monarch butterfly, *Danaus menippe*.   The monarch is unmolested by birds because it is unpalatable.   The viceroy shares this protection because it is mistaken for the monarch.   There is, however, no evidence that the viceroy profits at present by its color pattern although it may have done so in the past.

One of the most remarkable cases of protective coloration is found in the African *Papilio dardanus*.   The female of this species has three distinct color varieties which mimic, respectively, three entirely different species of *Danaus*.   Strangely none of the females has color markings like those of the males.

A single distasteful species may have a large number of imitators. Marshall and Poulton cite six beetles of the genus *Lycus* that are imitated by almost 40 species of other genera, families, and orders.

Many cases of mimicry occur among the Diptera.   The syrphus fly, *Eristalis tenax*, mimics the honeybee.   Another syrphus fly, *Spilomyia vittata*, mimics the yellow jacket.   A syrphus fly, *Volucella evecta*, and a species of robber fly both mimic bumblebees.

Numerous inquilines in the nests of ants mimic the ants they visit. There are Coleoptera, Orthoptera, and Heteroptera that mimic ants.

Many larvae and pupae perhaps mimic one another.   No one has attempted to study these relationships.   Larvae and pupae are not generally highly colored and do not attract so much attention as the adults.   However, many are unpalatable and no doubt have mimics. The larva of the pipe-vine moth, *Papilio philenor*, might be mistaken for the larva of the mourning-cloak butterfly, *Hamadryas antiopa*.   Both are dark in color, armored with spinelike protuberances, and have conspicuous yellow or reddish spots.   The larva of the pipe-vine moth is protected by an eversible gland on the dorsum just back of the head which emits a disagreeable odor.   The larva of the mourning-cloak butterfly is without this protection but may share the protection of the other species.

Alluring and aggressive coloration has been discussed in connection with insects resembling their background and various objects.

**Photogenic Insects.**—One of the most interesting and the most complicated forms of light emitted by insects and other low forms of animal life is known as cold light.   Early writers called it *phosphorescence* but it is now well known that phosphorus is not involved in the light produced by these animals.   *Luminescence* has been suggested for the light emanating from cold bodies in contrast to *incandescence*, the light produced by substances heated to the glowing point. *Photogenic*—light generating—describes the nature of this phenomenon better than luminescent.

Light is produced by comparatively few insects although many may glow because of the presence of photogenic bacteria within them. Certain springtails, beetles, and rarely flies and moths are photogenic.

Springtails of the genera *Lipura, Neanura, Onychiurus,* and *Achorutes* are photogenic. *Lipura* is exotic; the remaining genera have representatives in this country. Some of these species give forth a continuous greenish light and the whole body glows except the antennae and legs; other species flash only when disturbed.

The larva of a New Zealand mycetophilid,* *Bolitophila,* glows by a light produced in the four modified Malpighian tubules. Fulton (1939) describes the habits of a luminescent mycetophilid from North America. This is a web-spinning species which apparently belongs to the genus *Platura.* The light is produced in the integument.

A European moth, *Arctia caja,* is said to give forth a pungent secretion which glows. A luminous substance exudes from the tubercles of other moth larvae, notably *Zygaena.*

The best known photogenic species belong to the Coleoptera, namely the fireflies, Lampyridae, and the elater beetles, Elateridae. Both larvae and adults glow; even the eggs and pupae may, at times, give forth light.

One of the most brilliant species is the tropical American "cucujo," an elater of the genus *Pyrophorus.* The beetles are powerful fliers and dart swiftly through the jungles. They inhabit the higher shrubs and trees and are sometimes difficult to collect. In this species, two greenish lights glow on the prothorax and one reddish light glows on the underside of the first abdominal segment. They generally fly in a straight line and flash at irregular intervals. The Calverts described the larva of the "cucujo"† which they observed in Costa Rica. In daylight, these larvae are pale yellow in color and present no remarkable difference from other larvae of this family. At night, they are very striking creatures. The head is wholly illuminated, and two spots, on each of the abdominal segments from three to eleven, glow brilliantly. All the spots do not glow at the same time but the larvae can be coaxed "to turn on their lights" by touching or disturbing them.

The fireflies or lightning bugs are the most familiar light producers and have been observed by people of many parts of the world. Unfortunately, the common names of these insects are inappropriate for they are neither flies nor bugs, but these names are so well impressed upon the literature and upon the popular mind that even the entomologist will

---

* A tipulid of early authors.

† This, according to H. S. Barber, is apparently a species of Phengodes.

hesitate to change them.   The larvae and the females of the wingless
species are known as glowworms.   These conspicuous insects have deeply
impressed musicians and writers.   Japanese art frequently features
them.   As to our own works, one thinks first of the ever popular idyl
"The Glow Worm" by Lincke.   "Firefly" is typical of many of the
verses that have been written to these insects.

FIREFLY

A little light is going by,
Is going up to see the sky,
    A little light with wings.
I never could have thought of it,
To have a little bug all lit
    And make it go on wings.
                —ELIZABETH MADOX ROBERTS.

There are scarcely half a dozen common kinds of fireflies that are
observed in the Eastern United States.   *Photinus marginellus* frequently

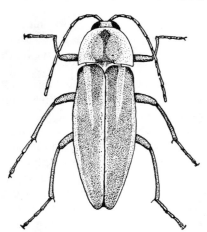

amuses young and old during the
hours of dusk in July and August.
The species is hardly more than a
quarter of an inch long and dusky
brown in color; the thorax is rosaceous
with yellow margins and a central
black spot; the margins of the elytra
are yellowish.   This species inhabits
low shrubs.   The females remain
close to the ground, often lurking
in the grass waiting for the flash of
the male which eventually brings them
together.   The male has a stronger
flash than the female and as a keen
observer once remarked, "the males
use fifty-watt lamps while the females
use twenty-five-watt lamps."   They
start flashing at dusk, usually about

FIG. 149.—*Photuris pennsylvanica,*
one of the large and more common fireflies of Eastern United States.

eight in the evening, and cease when the sun has set fully and darkness
has completely settled, which is about ten o'clock.   Stragglers may continue to flash far into the night but the conspicuous display of lights is
over early in the evening.   It is said that the European *Lampyris
noctiluca* become inactive after midnight.

*Photinus pyralis* is marked much like *P. marginellus* but is slightly
larger in size.   It has, as will be noted later, a different manner of flashing.   In certain areas, this species replaced *P. marginellus.*

A much larger species, *Photuris pennsylvanica*, is also common in the Eastern United States. It is about three-quarters of an inch long and generally brown in color; the prothorax is broadly bordered with yellow, and the disk is red with a narrow median dark stripe; the elytra are margined with yellow and have a central yellowish line. This species flies higher than the two species mentioned previously and has a much stronger flash.

Fireflies are not so common in Western United States, where the females are generally wingless. The pink glowworm, *Microphotus angustus*, occurs only in dry grass on the foothills. The light is bright and attracts much attention but these species do not make a spectacular showing like our eastern species.

Each species of firefly has a characteristic method of flashing distinguished by intensity, duration, number, and intervals between flashes and flight levels. *Photinus pyralis* generally fly close to the ground and prefer recently cultivated fields. They fly in an undulating manner but always flash as they ascend; following the flash, they drop down. The beetles thus always appear to be ascending. They look like brilliant sparks shooting from the ground. As the evening advances, they gradually ascend higher but never attain a great height. This species flashes with a yellow light which is less brilliant but longer than those of *Photinus scintillans* and *P. marginatus*. They start flashing at sunset and cease about nine o'clock.

The flash of *Photinus marginellus* is yellow but much sharper and quicker than that of the preceding species. These beetles are often seen in the afternoon long before sunset. They inhabit shrubs and trees and reach their highest altitude about nine or ten o'clock in the evening. The flash of *Photinus scintillans* is quite similar to that of *P. marginellus*. It consists of a single sharp yellow flash. *P. consanguineus* differs in that the males and the females give two short flashes during each period of flashing.

*Photuris pennsylvanica* occupies the tree tops and flashes with a greenish-blue or pale blue light. The males flash three, four, or five times in rapid succession; the females flash one, two, or three times each period. The females remain in the grass and attract the males with their lights.

Synchronous flashing has been repeatedly observed and described by numerous naturalists. The phenomenon is particularly spectacular in the Tropics. In Siam, members of a species come forth at dusk and seek a certain kind of tree. At first, they flash at random but later they flash in unison. These displays may occur every evening for many months irrespective of weather. Some factor, still little known, stimulates these insects to flash simultaneously. Temperature, pressure, and various other conditions have been cited as possible explanations of this habit. Displays of this sort are very uncommon or are probably imaginary in the temperate climate. Some writers believe that synchronous flashing is accidental and compare it with the ticking of clocks in unison, when a large number are grouped in a room.

**Structure of the Light Organs of Fireflies.**—The photogenic organs of the males of *Photinus* are located on the hind margin of the fourth and the whole of the succeeding ventral abdominal segments. The light organs of both sexes of *Photuris* occupy the whole of the fifth and following segments. The light organs of the female of *Pyractomena* are located on the sides of the abdomen.

The light organs of *Photinus marginellus* have been described by Townsend and others. They are located on the ventral aspect of the

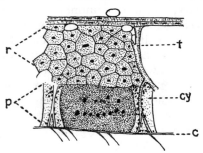

Fig. 150.—Transverse section of a portion of the photogenic organ of the firefly, *Photinus marginellus:* (*r*) reflecting layer; (*p*) luminescent or light-producing area; (*t*) trachea; (*cy*) cylinder; (*c*) cuticula. (*After Townsend.*)

abdominal segments. Two distinct portions can be differentiated: the dorsal mass of comparatively large, nucleated cells, which is the reflecting layer, and the ventral mass of larger cells, which is the photogenic layer. The reflecting layer is chalky white owing largely to the presence of urate crystals. The photogenic layer is yellowish and luminous. The photogenic layer is well supplied with tracheae and nerves arising from the last two abdominal ganglia. It is divided into numerous areas by the large tracheae which come down from above. These tracheae are surrounded by cylinders which are well supplied with tracheoles that anastomose profusely. It is here that an enzyme, *luciferase*, and a substance, *luciferin*, react to produce light. Luciferin is oxidized in the presence of luciferase under the control of the nervous system. The flow of air through the tracheae and the tracheoles of course determines the brilliance of the flash. The raising of the wing covers during flight helps to display the light to the best advantage. The light thus produced is most unusual. The spectrum is perfectly continuous without any trace of lines either bright or dark. No infrared and no ultraviolet light is produced and no rays are formed capable of penetrating opaque objects. It is cold light and its efficiency is estimated to be between 92 and 100 per cent. In comparison, illuminating gas light has only 3 per cent efficiency, an incandescent lamp about 10 per cent efficiency, and sunlight 25 per cent efficiency.

Although the display of fireflies on warm summer evenings is commonplace, the larvae are seldom seen except by those who seek them. The larvae of *Photuris pennsylvanica* are quite common and are frequently seen in the late summer, crawling upon the surface of the ground. They are predacious or carnivorous and of course are not injurious to plants. This modest species displays only two luminous spots. On the other hand, the larvae of *Phengodes laticollis* are rarely seen but exceedingly attractive. There are ten conspicuous luminous spots along each side of the larva and a narrow band across the top of each segment.

The writer has observed the larvae of a species of *Pyractomena* which are aquatic in habits. They crawl upon grasses completely submerged and sometimes walk in an inverted position on the underside of the surface film.

## BIBLIOGRAPHY

An extensive list of references is given by Folsom (1934). More recent papers and some not listed by these authors are appended here.

**I. General References.**

BACK, J. R. (1941). Studies on the firefly. III. Spectrometric data on thirteen Jamaican species. *Proc. Rochester Acad. Sci.* **8** (1) *Literature cited.*

BECKER, C. (1937). Die Farbstoff der Insekten. *Ent. Rdsch. Stuttgart* **54**.

BRUNNER, W. C. VON (1897). Observations on the coloration of insects. Translated by E. J. Bles.

CARPENTER, G. D. H. (1936). Insect coloration and natural science. *Nat. Lond.* **138** (3484).

CLARK, A. H. (1932). Emanations from butterfly wings. *In,* The butterflies of the District of Columbia and vicinity. *Bull. U. S. Nat. Mus.* **157**.

COCKERELL, T. D. A. (1890). Evolution of metallic colors in insects. *Ent. News* **1** (1).

DAVIS, WM. T. (1927). The rearing of pink katy-dids. *Journ. N. Y. Ent. Soc.* **35** (2).

DOBZHANSKY, TH. (1933). Geographical variation in lady beetles. *Amer. Nat.* **67** (709). *Literature cited.*

FOLSOM, J. W. (1934). Entomology with special reference to its ecological aspects. Revised by W. R. Wardle, 4th ed. The Blakiston Company, Philadelphia.

FORBES, WM. T. M. (1938–1939). Iridescence. *Ent. News* **49** (10), **50** (2, 3, and 4).

Fox, D. L. (1936). Structure and chemical aspects of animal coloration. *Amer. Nat.* **70** (730). *Literature cited.*

GORTNER, R. A. (1911). The origin of the pigment and the color pattern in the elytra of the Colorado potato beetle (*Leptinotarsa decemlineata* Say). *Amer. Nat.* **45** (540). *Literature cited.*

HANCOCK, J. L. (1916). Pink katy-dids and the inheritance of pink coloration. *Ent. News* **27** (2).

HARVEY, E. N. (1940). Living light. Princeton University Press, Princeton, N. J.

HORN, G. H. (1892). Variations in color-markings in Coleoptera. *Ent. News* **3** (2).

IMMS, A. D. (1937). Coloration. *In*, Recent advances in entomology. The Blakiston Company, Philadelphia.

KALAMUS, HANS. (1941). Physiology and ecology of cuticular color in insects. *Nature Lond.* **148** (3754).

KNAB, FREDERICK (1907). Color varieties of Locustidae. *Science* **26** (670).

——— (1909). Nuptial colors in the Chrysomelidae. *Proc. Ent. Soc. Wash.* **11** (3).

KNIGHT, H. H. (1924). On the nature of the color patterns in Heteroptera with data on the effects produced by temperature and humidity. *Ann. Ent. Soc. Amer.* **17** (3).

LANDIS, B. J., and MASON, H. C. (1938). Variant elytral markings of *Epilachna varivestis* Muls. (Coleoptera: Coccinellidae). *Ent. News* **49** (7).

LONGLEY, W. H. (1917). Studies upon the biological significance of animal coloration. *Amer. Nat.* **5** (605).

MANI, M. S. (1935). The nature and origin of insect colors. *Curr. Sci. Bangalore* **4**.

MCATEE, W. L. (1934). Protective resemblances in insects, experiment and theory. *Science n. sr.* **79** (2051). *Literature.*

MOSER, J. (1936). Ueber den metallglanz der Käfer. *Ent. Jb. Leipzig.*

PROCHNOW, O. (1929). Die Farbung der Insekten. *In*, Schroeder's Handbuch der Entomologie, Band Ill. Gustav Fischer, Jena. *Literature.*

REINIG, W. F. (1937). Melanismus, Albinismus und Rufinismus. Ein Beträg zum Problem der Entstehung und Bedeutung tierischer Farbungen. Georg Thieme Verlag, Leipzig.

SLATER, J. W. (1886). On the origin of colours in insects. *Proc. Ent. Soc. Lond.*

WHEELER, W. M. (1907). Pink insect mutants. *Amer. Nat.* **41**.

## II. Mimicry and Protective Coloration.

CARPENTER, G. D. H. (1937). The needs of the mimetic theory. *Science* **86** (2224).

——— (1938). Mimicry in relation to other forms of protective coloration. *Proc. Bournemouth Nat. Sci. Soc.* **30**.

———, and FORD, E. B. (1933). Mimicry. Methuen & Co., Ltd., London.

CARRICK, R. (1936). Experiments to test the efficiency of protective adaptations in insects. *Trans. R. Ent. Soc. Lond.* **85** (4).

COTT, H. B. (1940). Adaptive coloration in animals. Methuen & Co., Ltd:, London.

FOLSOM, J. W. (1934). Entomology with special reference to its ecological aspects. Revised by R. A. Wardle, 4th ed. The Blakiston Company, Philadelphia.

JONES, F. M. (1932). Insect coloration and the relative acceptability of insects to birds. *Trans. Ent. Soc. Lond.* **80** (2).

MCATEE, W. L. (1932). Effectiveness in nature of the so-called protective adaptations in the animal kingdom, chiefly as illustrated by the food habits of Nearctic birds. *Smiths. Misc. Coll.* **85** (7).

MYERS, J. G. (1937). Mimetic and other associations between neotropical insects and spiders. *Proc. R. Ent. Soc. Lond.* **12** (1–2).

POULTON, E. B. (1902). Five years' observations and experiments (1896–1901) on the binomics of South African insects chiefly directed to the investigation of mimicry and warning colors. *Trans. Ent. Soc. Lond.* **3**.

——— (1921). Recent advances in the knowledge of insect mimicry. *Trans. Oxford Univ. Jr. Sci. Cl.* **3** (5).

THAYER, A. H. (1909). Concealing-coloration in the animal kingdom. The Macmillan Company, New York.

**III. Photogenesis.**—Harvey's recent book "Living Light," summarizes numerous references on the subjects of bioluminescence, light measurements, light efficiency, and luminescence in Coleoptera and other animals. A few papers not included by Harvey are added here.

ALLARD, H. A. (1917). Synchronism and synchronic rhythm in the behavior of certain creatures. *Amer. Nat.* **51** (607).

———— (1935). Synchronous flashing of fireflies. *Science* **82.**

FULTON, B. B. (1939). Lochetic luminous dipterous larvae. *Elisha Mitchell Sci. Soc.* **55.**

HARVEY, E. N. (1940). Living light. Princeton University Press, Princeton, N. J.

HESS, W. N. (1917). Origin and development of the photogenic organs of *Photuris pennsylvanicus* De Geer (Col.). *Ent. News* **28** (7).

KNAB, FREDERICK (1905). Observations on Lampyridae. *Can. Ent.* **37** (7).

LECONTE, J. L. (1880). On lightning bugs. *Proc. Amer. Assoc. Adv. Sci.* **29.**

LUND, E. J. (1911). On the structure, physiology and use of photogenic organs, with special reference to the Lampyridae. *Journ. Exp. Zool.* **11** (4).

McDERMOTT, F. A. (1911). Photogenic organs of certain Lampyridae. *Amer. Nat.* **45.**

———— (1910–1912). Five papers on photogenic organs of American Lampyridae. *Can. Ent.* **42, 43, 44.** Also *Amer. Nat.* **45** (1911) (533).

MANN, B. P. (1875). Notes on luminous larvae of Elateridae. *Psyche* **1** (16).

MAST, S. O. (1912). Behavior of fire-flies (*Photinus pyralis*)? with special reference to the problem of orientation. *Journ. Animal Behavior* **2** (4). *Literature.*

SCHWARZ, E. A. (1901). Luminosity of lampyrid larvae. *Proc. Ent. Soc. Wash.* **4.**

TOWNSEND, A. B. (1904). The histology of the light organs of *Photinus marginellus.* *Amer. Nat.* **38** (446). *Bibliography.*

WILLIAMS, F. X. (1916). Photogenetic organs and embryology of lampyrids. *Journ. Morphol.* **28** (1). *Bibliography.*

———— (1917). Notes on the life-history of some North American Lampyridae. *Journ. N. Y. Ent. Soc.* **25** (1). *Bibliography.*

# CHAPTER X

## SONIFICATION

The adults of many insects, and the larvae and pupae of a few, produce sounds by rubbing one part of the body against another, by vibrating the wings, or by forcing air through the spiracles. Man is handicapped in studying insect sounds because his ears perceive sound waves only between 30 and 30,000 vibrations per second. Insects often produce sounds that are beyond this range. Some people can detect a wider range of vibrations than others but at best man's range is limited. With age, the sensitiveness of the human ear becomes less acute.

FIG. 151.—A click beetle, Elateridae, showing the spine on the underside of the prothorax which fits into a groove on the underside of the mesothorax. The movement of this spine over the sharp edges of the groove causes the beetle to jump and produces a faint but audible click. (*Champlain and Kirk.*)

**Purpose of Insect Sounds.**—Many of the sounds produced by insects, such as those caused by the gnawing of wood-boring larvae, the click of the elater beetles, or the buzz of flying insects, are probably incidental. They correspond to noises such as the swish of silk clothing, the sound of the woodchopper, or the whine of a bullet. Such noises are produced without purpose and are the result of work or motion. The bee, the fly, and the roach are continually cleaning their antennae and legs by passing them through their mouths. If we could detect the slight vibrations produced by such activities, we would interpret them as insects sounds; however, they are faint and entirely incidental. To assume that the human mouth is intended to hold tacks or pins would be folly, for it serves many purposes: eating, breathing, talking, smiling, etc. The sounds produced by the cicadas and Orthoptera, which have well-developed sound-producing organs and complicated auditory organs, are apparently produced for some definite purpose. Attempts have been made to classify sounds thus: those which frighten enemies, call social insects together, protect or defend the hive, or attract or charm the opposite sex.

It cannot be doubted that insects often frighten man by their ferocious sounds. Even the harmless May beetle disturbs some. How much fear insects instill in other species is questionable. The sand cricket, *Stenopelmatus longispina*, "throws a great bluff," as the expression goes. It faces its foe and boldly defies the aggressor by rubbing its hind legs vigorously against the sides of the abdomen and producing a distinct rasping noise. On the other hand, the common cricket and the tree cricket are readily intimidated and quickly cease singing when disturbed.

Many of the cerambycids and the curculionids produce faint noises when handled. The cicada often produces a characteristic murmur when squeezed gently. These sounds though faint to the human ear may cause birds or other animals to drop them.

Many insects sing apparently to attract the other sex. It is known that the male mosquito is tuned to the vibrations of the wings of the female. The female snowy tree cricket responds in a peculiar manner to the singing of the male. As we shall later see, the female is attracted by odors liberated during the singing activities of the male.

**Seasons of Insect Songs.**—It is a moot question whether the bird, the bug, or the bullfrog opens the chorus of the world's "spring song." The ornithologist, on his early morning walks, listens with intent for the first note of the purple finch or the call of the phoebe. The zoologist turns a sensitive ear for the early pipings of the spring peeper or the trill of the toad. The entomologist is cognizant of the chirps of the common black cricket, *Gryllus assimilis*, which may be heard very early in the spring. To a true naturalist, all will blend to recall vivid recollections of bygone experiences. Perhaps he can see the cricket elevating his wings as he stridulates, the spring peeper puffing out its throat with each shrill note, or hear the bluebird's soft, melodious warble.

Insect songs are most conspicuous toward the middle of the summer or during the fall. This is due to the fact that insects seldom sing until they are mature and that the outstanding noise makers are the Orthoptera, which do not usually mature until the middle of the season. The late summer and fall are the seasons that reverberate with insect songs. The long-horned grasshoppers, the crickets, and the katydids are some of the common songsters. During mild autumns, the snowy tree crickets may sing until the first of November but they are usually hushed by early frosts that come in October.

The piping of the little spring peeper, *Hyla crucifer*, is often mistaken by the uninitiated for the song of an insect. Another small frog, *Acris gryllus*, has been so named because its song resembles that of a cricket.

As a rule, the male insect does the singing. Thus, Xenarchus, approximately 500 B.C., wrote "Happy the cicadas' lives, for they have voiceless wives." However, the females of mosquitoes and other flies and bees produce noises by the vibration of their wings. Rudimentary stridulatory organs also occur in the female of *Stenobothrus*.

Insects are usually soloists. When they sing in choruses, it is generally accidental. Much has been written about their singing in unison, but duets or quartets are probably coincidental. One insect may stimulate another to sing. The mass effect of a swarm of bees humming in clover, a horde of seventeen-year cicadas pouring forth their shrill songs, or the continuous chirps of numerous tree crickets is undeniable, but there is no coordination and no attempt on the part of the insects to join their neighbors.

Fig. 152.—The Carolina locust, *Dissosteira carolina*, is one of the many short-horned grasshoppers that produce a crackling sound when flying.

**Sounds Produced by the Vibration of Wings.**—The sounds produced by the rapid vibration of wings or halteres are the results of normal activities of insects and apparently have no purpose. Comstock has termed this "the music of flight." The sound of the bees is the most familiar and has been caught and beautifully reproduced by Nicolas Rimsky-Korsakov, in his delightful and popular "Flight of the Bumblebee," Die Hummel-Schertzo. The high-pitched violin gives a realistic representation of the flight of the queen, and the lower pitched violins depict the workers that swarm with her.

The honeybee produces different tones which apparently result from degrees of activity; the contented hum of the worker as it collects nectar and pollen, the enraged note of the belligerent bee, and the characteristic hum of the swarm. The notes produced by the wings of insects depend upon the number of vibrations. The vigorous bee produces $A$ of 435 vibrations; the tired bee hums $E$ of 326 vibrations per second. The housefly vibrates its wings 21,120 times per minute to produce the note $F$ of the middle octave.

Other insects reveal similar differences in their songs. The mud d a u b e r, *Sceliphron coementarium,* produces a faint hum as it gathers a load of mud at the edge of the pond. While it is applying mud to the nest, the pitch is raised to a rasping sound that is audible for a considerable distance.

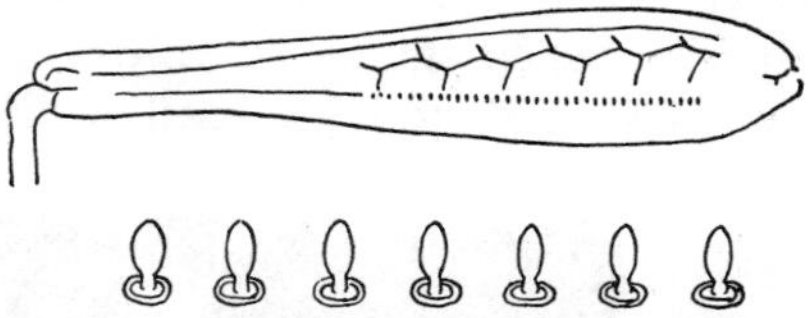

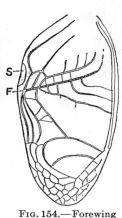

Fig. 153.—The hind femur of *Stenobothrus* showing the pegs which are rubbed against the forewings to produce a sound. Below, pegs enlarged. (*After Comstock.*)

The Scarabaeidae, Orthoptera, Heteroptera, Diptera, and Hymenoptera commonly produce music of flight. The loud buzz of the May beetle instills fear among those unfamiliar with insects. Many of the Heteroptera are powerful fliers and produce a loud noise when on the wing. The high-pitched notes of the mosquito and housefly are well known to all.

**Stridulation.**—The majority of the insect choristers stridulate; that is, they rub one portion of the body against another. The wings and wing covers of insects, because of their structure, give the strongest vibrations. They are operated in three different ways. The Carolina locust and some of the other short-horned grasshoppers (Locustidae) produce a cracking sound by rubbing the edges of the two wings together, while in flight. The upper surface of the front margin of the hind wing is rubbed against the thickened veins on the lower surface of the fore wing, producing a loud harsh sound.

Fig. 154.—Forewing of the male cricket, *Gryllus,* showing the scraper (*s*) on the inner margin of the wing and the file (*f*) on one of the larger cross veins.

*Stenobothrus* (Locustidae) has a series of pegs, the file, on the inner surface of each hind femur. When stridulating, the insect raises both hind legs and rasps the femora against the outside surface of the front wings.

The long-horned grasshoppers (Tettigoniidae), including the katydids and the crickets (Gryllidae), have a file on one of the larger veins at the base of the wing cover which is rubbed by a scraper, a thickened area at the edge of the other wing. In the katydid, the file is developed only on the left wing, thus making them left-handed singers. Certain crickets use only the file on the right wing cover and are therefore right-handed singers.

The chirping of the tree crickets is a highly complicated procedure. The male differs in appearance from the female: his wings are broader and lie flat on the body; the wings in the female are narrower and bend about the sides of the body. When stridulating, the male raises its wings nearly perpendicular to the body. The scraper on the left wing cover is rasped against the file on the right wing cover. When the wings are raised, a gland is exposed on the upper surface of the metathorax and

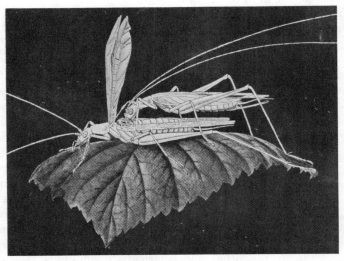

Fig. 155.—Female snowy tree cricket, *Oecanthus niveus*, feeding at the metanotal gland of the male. (*After Fulton.*)

emits an odor which attracts the female; as she climbs upon the back of the male to imbibe this liquid, coition takes place. The female has no auditory organs and does not hear the earnest chirps of the male. The mating process therefore depends upon chemotropic responses.

The rate of chirping in crickets depends upon the temperature. During warm weather, the chirp is rapid and high pitched but during cold periods it slows down and becomes a rattle. Dolbear worked out a method of determining the temperature by means of the number of chirps per minute. For the tree cricket, *Oecanthus niveus*, the formula is as follows, where $T$ = temperature Fahrenheit, $N$ = the number of chirps per minute.

$$T = 50 + \frac{N - 92}{4.7}$$

For the house cricket the formula is

$$T = 50 + \frac{N - 40}{4}$$

and for the katydid, *Cyrtophyllus perspicalis*, the formula is

$$T = 60 + \frac{N - 19}{3}$$

The tree cricket chirps during both day and night but is more plainly heard at night because other sounds are not so loud. One can scarcely believe that so much noise could come from such a small insect. From a distance of a few feet, the sound pierces the eardrums. Tree crickets are incessant chirpers and have been known to chirp 2,640 times without stopping and probably would continue much longer if not interrupted.

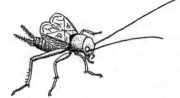

Fig. 156.—The male of the common cricket, *Gryllus assimilis*, with wings raised in the singing attitude.

It is said that the piercing note of the European *Brachytrypes megacephalus* can be heard for a distance of a mile.

The larvae of some insects produce faint sounds. The most interesting example is the larva of the horned passalus, *Passalus cornutus*. The coxae of the mesothoracic legs are provided each with a row of fine ridges. The metathoracic legs are minute and modified to form rasping organs which are rubbed against the sides of the coxae of the mesothoracic legs to produce a noise. The third pair of legs is so small that the larva appears to have only two pairs of legs. The larvae of this species live in rotten wood and it is believed that they keep together by communicating with one another.

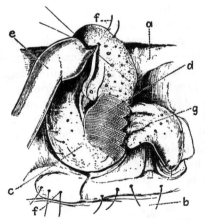

Fig. 157.—Stridulating organs of the larva of *Passalus*. (*a*, *b*) portions of metathorax; (*c*) coxa of second leg; (*d*) file; (*e*) basal part of femur of middle leg; (*f*) hairs; (*g*) diminutive third leg modified as a scraper. (*After Sharp.*)

Stridulating organs are found on nearly any part of the body. In the Nitidulidae and Erotylidae, they are located on the front of the head. Certain tenebrionid beetles have striations on the underside of the head. Mouth parts are frequently used in producing sounds. *Corixa americana* rubs its front legs against the proboscis. *Buenoa limno-*

*castoris* makes a clicking sound by rubbing the femora and tibiae against the base of the proboscis. In the Sphingidae, the palpi are sometimes rubbed against the proboscis. The larvae of the coprides stridulate by rubbing a roughened area of the stipes of the maxillae against the under surfaces of the mandibles.

One of the bostrichids is said to rub its front legs against a projection at the posterior angle of the prothorax. The Elateridae make a sound by means of a ventral prothoracic spine which normally rests in a groove on the ventral surface of the mesothorax. The connection between the pro- and mesothorax is more flexible than in most insects. As the prothorax is flexed upward, the ventral spine slips over a sharp edge on the anterior margin of the mesothorax and produces a clicking noise. Some of the ants have roughened areas on the node or scale joining the thorax and the abdomen. *Siagona*, an exotic carabid, has a file on the underside of the prothorax against which the anterior femora are rasped.

The wings or wing covers, most commonly used in the production of sound, afford considerable area as a resonator or sounding board. The Orthoptera produce sounds by vibrating their wings in various ways. A few Lepidoptera also make sounds in the same way. An Indian species of Agaristidae, *Aegocera tripartitia*, has structures on the anterior pair of wings which are rubbed against spines on the legs. A South American nymphalid makes a clicking sound when it settles down. The noise is produced by a small appendage bearing two hooks at the extreme base of each fore wing which is rasped against a process on the thorax.

Sounds are produced by rubbing the legs against the proboscis, the head, the sides of the thorax or abdomen, the wings, or against each other. The abdomen is also used in the production of sound. In the cicada, the sound organs occupy a large part of the basal segment of the abdomen. In the Heteroceridae there is a slightly elevated, curved line on each side of the base of the abdomen which is set in vibration by roughened areas on the femora. Certain Scarabaeidae have ridges on the middle of the pygidium which are rubbed against the extremities of the wing covers. An Old World species, *Pelobius*, has a series of ridges on the under surfaces of the wing covers which are rubbed by the tip of the upturned abdomen. Forbes (1941) believes that the male genetalia of *Tamphana marmorea*, a Central American species, of Lepidoptera bear stridulatory organs.

**Spiracular Sound Organs.**—Certain insects are said to force air from the spiracles and in so doing produce a faint noise. In the Diptera, there is a series of leaflike folds in the intima of the trachea which are set in vibration by the rush of air through the spiracles. The queen bee, the blowfly (*Calliphora*), and the May beetle produce sounds in this manner.

**Drummers.**—Drumming may be produced in different ways. The deathwatch beetles, *Anobium striatum* and *A. tesselatum,* produce faint ticking sounds by bumping their heads against the sides of their burrows. The soldiers of some termites strike their heads against the ground. Clicking sounds of a different sort are produced by woodboring larvae such as the Buprestidae and Cerambycidae, as they gnaw within solid

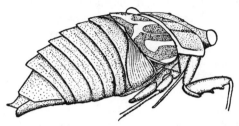

Fig. 158.—The common cicada, *Tibicen linnei*, with wings and legs removed to show the position of the sound-producing organs.

wood. Even the sounds produced by the minute powder-post beetles can be heard a considerable distance.

The cicadas are the classical drummers of the insect world and are perhaps the best known of the sound producers. The two-year cicada has broods emerging every year so that its shrill note is familiar to all. The doleful murmurs of the seventeen-year cicada come less frequently. A Central American species, more than twice as large as the common cicada of North America, has an exceedingly shrill song. Virgil remarked concerning the shrill calls of the cicadas, "They burst the very shrubs with their noise."

The sound organs of the cicada are the most complicated found anywhere in the animal kingdom. Although drummers, they do not beat their drums but operate them by means of powerful muscles. On the underside of the third

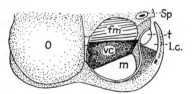

Fig. 159.—Underside of the metathorax of *Cicada plebeia*, showing the musical apparatus: (*o*) operculum; (*t*) timbal membrane; (*m*) mirror; (*fm*) folded membrane; (*vc*) ventral cavity; (*lc*) lateral cavity; (*sp*) spiracle. (*After Carlet.*)

thoracic segment of the male, there are two large plates, *opercula*, which cover the sound organs. Within the body are two large cavities each containing three membranes: the tymbal, folded membrane, and mirror. The tymbal is set into motion by muscles which are attached to it. The air cavity acts as a resonator and the other membranes amplify the sound. The air of the resonator chamber communicates with the exterior through the metathoracic spiracle. Without such an opening to the exterior, the membranes would not vibrate.

**Organs of Hearing.**—Since certain male insects produce sounds, it is natural to suppose that the females have ears to hear them. This, however, is not exactly the case. Some females have auditory organs but other insects, such as the snowy tree cricket, have no ears and do not hear the persistent notes of the male. However, auditory organs occur in many insects and on many parts of the body. The body wall itself is comparatively thin and may serve to receive numerous impressions, provided mechanisms exist to transfer these impressions to the nervous system. The hairs of certain caterpillars, especially *Vanessa antiopa*, the antennae of mosquitoes, ants, and possibly butterflies and moths, and the cerci of *Gryllus* and *Periplaneta* respond to sound vibrations. The cerci of certain insects pick up vibrations that are inaudible to man.

The more highly developed ears of insects consist of a thinned portion of the body wall, known as the *tympanum*, for the reception of sound, and complicated structures, chordotonal organs, consisting of rods, nerves, and nerve endings, which transmit the vibrations to the nervous system. Special sound receptors may be located on different parts of the body. In *Plea*, *Corixa*, *Nepa*, and *Naucoris*, the tympana are located on the thorax. Tympana are well developed on the metathorax of many Lepidoptera. In the short-horned grasshoppers, they are located on each side of the first abdominal segment and are large and conspicuous. The katydids, crickets, and termites have sound-perceiving organs on the tibiae of the front legs. In the katydid, the tympana are oval and quite conspicuous; in the cricket and termite, they are sunk within the leg and only two small slits appear on the surface. The air is equalized by special ducts which lead through the center of the leg and open on the dorsal aspect of the thorax.

**Common Insect Musicians.**—The crickets, katydids, and the grasshoppers (Orthoptera) and the cicadas (Homoptera) are the conspicuous choristers of the insect world. The grasshoppers generally sing by day and usually striadulate during flight. The crickets sing by day or night but do not move about when they chirp. The cicadas* sing during the day. As a matter of fact, they cease their noise even during cool rainy periods. Scudder remarks that the short-horned grasshoppers shuffle, rustle, or crackle; the crickets shrill and creak; and the long-horned grasshoppers scratch and scrape. A good entomologist soon becomes acquainted with the songs of many of these insects and can recognize them as an ornithologist knows his bird songs.

---

* Cicada is the common name for the seventeen-year cicada and its allies. The term locust, improperly used for these species, should be reserved for the short-horned grasshoppers.

The majority of the short-horned gassshoppers (Locustidae) make no characteristics sound. For example, our common injurious species, the red-legged grasshopper, *Melanoplus femur-rubrum*, the differential grasshopper, *M. differentialis* and the lesser migratory grasshopper, *M. atlanis*, are "voiceless." On the other hand, some of the band-winged species are illustrious musicians. The cracker locust, *Circotettix verruculatus*, is a common species throughout the Northern States. They are blackish brown in color; the basal halves of the wings are sulphur yellow, the distal halves are smoky brown each with a wide brown, transverse band. These grasshoppers are generally found at high elevations, along bare ridges where they harmonize with the gray of the rocks and lichens.

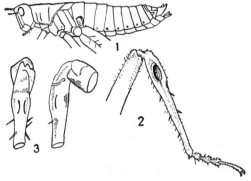

FIG. 160.—Ears of some insects: (1) ear of grasshopper on side of first abdominal segment; (2) ear of katydid on the tibia of the foreleg; (3) tibia of locustid showing openings to covered tympana, front view and side view. (*After Schwabe.*)

The males produce a loud rattling sound which Scudder says can be heard for a quarter of a mile.

The clouded locust, *Encoptolophus sordidus*, is another common musician of the Eastern United States. It is abundant during autumn in meadows and pastures, and attracts attention by the crackling sound produced by the males during flight. Blatchley describes its note as a "harsh droning a buzzing sound."

The Carolina grasshopper, *Dissosteira carolina*, is perhaps the best known of the short-horned grasshoppers although it does not stridulate loudly. The general color of this grasshopper is gray or brown depending upon its habitat. The pitchy black wings with the contrasting yellow outer margins surprise the beginner when the insect takes flight and exposes its gorgeous markings.

The coral-winged locust, *Hippisus apiculatus* (*tuberculatus*), is one of the larger of the beautiful band-winged grasshoppers. The general color is brown; the basal portion of the wings is coral red (rarely yellow): the colored area is bordered distally with a dark band. This species is found in dry bushy pastures, on hillsides, or on light uncultivated soils. The male produces a crackling sound as it flies.

Many of the long-horned grasshoppers, Tettigoniidae, produce distinct sounds. They make soft but audible notes that have been described as *zeep, zeep, zeep* or *zip, zip, zip*. Their songs are dwarfed by the louder notes of other insects and can be heard only short distances. Many of these species are predominantly green in color and have exceedingly long antennae. The species most commonly seen or heard is the angular winged or false katydid, *Microcentrum retinerve*. It is frequently attracted to artificial lights and is often mistaken for the true katydid. The long-horned grasshoppers produce high pitched notes which are often difficult for some people to perceive.

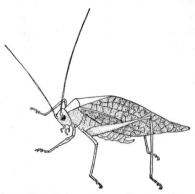

FIG. 161.—The angular-winged or false katydid, *Microcentrum retinerve*.

The true katydid, *Pterophylla camellifolia*, is heard more frequently than it is seen. It inhabits the tree tops and is difficult to obtain. It is green in color but the wings are shorter and more rounded than those of the angular-winged katydid and there is a dark brown triangle on the back where the bases of the two wing covers overlap. The eyes of the true katydid have spots like pupils. Its song is too familiar to describe.

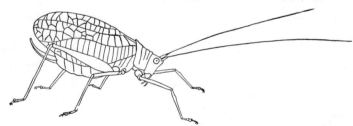

FIG. 162.—The true katydid, *Pterophylla camellifolia*.

The chirp of the cricket is perhaps the most common insect note. The black field cricket, *Gryllus assimilis*, is the species most frequently heard and its incessant chirping is one of the predominant sounds of the late summer and fall. This cricket is frequently found beneath stones and rubbish. "Chirp" is an excellent description of the cricket's song for it comes in broken, vibratory notes. We have already learned that the number of chirps per minute is determined largely by the temperature.

The house cricket, *Gryllus domesticus,* was introduced from Europe and is the cricket that Charles Dickens immortalized in "The Cricket on the Hearth." The chirp is much like that of the common field cricket but is more rapid. It is probable that in many instances the so-called "house crickets" are species of field crickets.

The tree crickets form a group quite different in structure and habits from the other crickets. There are many different species belonging to the genus *Oecanthus.* Although they are generally pale green in color, a few are conspicuously marked with black. The wing covers of the female are somewhat tightly held about the sides of the body, but those of the male are broad and flat. Their songs are varied. Sometimes they produce short high-pitched notes repeated regularly 100 or more times a minute. At other times, they make soft purring notes about 30 per minute. The unmated male apparently chirps more vigorously than the mated male, which purrs the so-called contented note. Burroughs described this noise as "rhythmic beat"; Thoreau called it "slumbrous breathing"; and Hawthorn called it "audible stillness." Tree crickets generally become very numerous toward the end of summer; but, so elusive are they, it is

FIG. 163.—The male snowy tree cricket, *Oecanthus niveus,* with wings raised in the singing attitude.

especially difficult to locate the source of the sound.

The remaining conspicuous singers are the cicadas, which belong to a different order, the Homoptera. There are a great many species ranging in size from half an inch to three inches in length. All have the same general appearance, with four glassy wings which are held over the body like a roof. The seventeen-year cicada, or "locust," has probably been seen and heard by all in this country where it is a native. The injury

FIG. 164.—Four species of Cicadas from Central America showing the great range in size.

done by egg laying, perhaps, attracts more attention than the insects themselves. The shells or cast skins left by the adults when they crawl forth from the ground have been observed and collected by most children. These cicadas produce a doleful murmur known as the "Pharaoh" note. When disturbed by birds or other animals, they produce a more shrill note. The two-year cicada or harvest fly produces a much higher pitched, shrill sound that is frequently associated with hot dry weather and prognosticators predict that fall is not far away. The common Central American species is much larger than any of our North American species and has a much shriller call. This is probably the species that Darwin speaks of in his description of insects heard near the shores of Bahia.

## BIBLIOGRAPHY

A bibliography of papers on sonification is given by Frost (1936). Some of the more important and the more recent references are cited here.

ABBOTT, C. E. (1936). The physiology of insect senses. *Ent. Amer.* 16 (4).

ACKLAND, MYRON (1929). Animals in orchestration. *Am. Mus. Nat. Hist.* 29.

ALLARD, H. A. (1937). Some observations on the behavior of the periodical cicada. *Megacicada septendecim* L. *Amer. Nat.* 71 (737).

BAIER, L. J. (1930). Contributions to the physiology of stridulation and hearing of insects. *Zool. Jahrb. Abt. Allgem. Zool. u. Physiol. Tiere* 47 (2).

BUNSEL, RÉNE-GUY (1954). Collugue sur L'acoustiques des Orthoptères. Institut National de la Recherche Agronomique, Paris.

EDWARDS, H. (1889). Notes on noises made by Lepidoptera. *Insect Life* 2 (1).

FEDERLEY, H. (1905). Sound produced by lepidopterous larvae. *Journ. N. Y. Ent. Soc.* 13 (3).

FORBES, WM. T. M. (1941). Does he stridulate? (Lepidoptera; Eupterotidae). *Ent. News.* 52.

FRINGS, M. (1954). Bibliographie sur L'acoustique des insectes. *Annales des Epiphyties. (Special fascicle.)*

FRINGS, MABLE AND FRINGS, H. (1960). Sound production and sound reception by insects. Penn. State Univ. Press. *Bibliography.*

FROST, S. W. (1936). Insect musicians. *In*, Ancient artizans. The Van Press, Boston. *Bibliography.*

GOTTFRIED, I. (1936). Notes on the song of the periodical cicada. *Proc. Ent. Soc. Wash.* 38 (7).

HANCOCK, J. L. (1905). The habits of the striped meadow cricket (*Oecanthus fasciatus* Fitch). *Amer. Nat.* 39 (457).

ISLEY, F. B. (1936). Flight-stridulation in American acridians (Orthoptera, Acrididae). *Ent. News* 47 (8).

LUTZ, F. E. (1924). Insect sounds. *Amer. Mus. Nat. Hist.* 50 (6).

———— (1938). The insect glee club at the microphone. *Nat. Hist.* 42 (5).

McCLURE, H. E. (1933). The click beetle's click (Coleop.: Elateridae). *Ent. News* 44 (6).

McINDOO, N. E. (1928). Communications among insects. *Rept. Smiths. Inst. Wash.* 81.

McKEOWN, K. C. (1937). Insect musicians. *Aust. Mus. Mag. Sydney* 6.

MEYERS, J. G. (1929). Insect singers, a study of natural history of the cicadas. George Routledge & Sons, Ltd., London. *Bibliography.*

PACKARD, A. S. (1904). Sound produced by a Japanese saturnian caterpillar. *Journ. N. Y. Ent. Soc.* 12 (2).

PIERCE, G. W. (1948). The songs of insects. Harvard Univ. Press.

PROCHNOW, O. (1928). Die Organe zur Hautäusserung. *In*, Schroder's Handbuch der Entomologie. Gustav Fischer, Jena, Band I.

RAU, PHIL. (1940). Auditory perception in insects, with special reference to the cockroach. *Quart. Rev. Biol.* **15** (2).

RIDLEY, H. N. (1908). The crackling moth of Singapore. *Journ. Straits Asiat. Soc.* **19.**

SCUDDER, S. H. (1868). The songs of the grasshoppers. *Amer. Nat.* **2** (3).

―――― (1875). The note of the katydid. *Psyche* **1** (16).

―――― (1876). The chirp of the mole-cricket. *Amer. Nat.* **10** (2).

―――― (1892). The songs of our grasshoppers and crickets. 23rd *An. Rept. Ent. Soc. Ont.*

SNODGRASS, R. E. (1925). Insect musicians, their music and their instruments. *Smiths. Rept.* **2775.**

SWINTON, A. H. (1889). Stridulation in *Vanessa antiopa. Insect Life* **1** (10).

TURNER, C. H. (1910–1917). Literature on the behavior of spiders and insects other than ants. *Journ. Animal Behavior* **1–6.**

# CHAPTER XI

## INSECT BEHAVIOR

Insect behavior is considered under three more or less closely related subjects: tropisms, instincts, and intelligence. Insects, like other organisms, are continually subjected to various environmental influences to which they directly or indirectly respond. These stimuli are of chemical or physical nature and fall in various categories. The reactions to these stimuli are known as *tropisms* (*tropotaxes*).* Since tropisms may be positive or negative in nature, the insects are said to be positively or negatively tropic. Tropisms seldom operate individually. The response to touch, *thigmotropism*, may, for example, involve *rheotropism*, the response to water currents, *anemotropism*, the response to air currents, or *geotropism*, the response to gravitation. The study is further complicated by the fact that the reactions of insects may result from hidden or internal stimuli as well as the more obvious external stimuli.

Fig. 165. —Scent scale or andro-conium from the wing of the imported cabbage but-terfly, *Ascia* (*Pieris*) *rapae*.

**Chemotropism.**—The reaction of insects to chemical stimuli is known as *chemotropism*. These stimuli often aid insects in locating places to oviposit. The bluebottle fly lays its eggs on fresh meat, the bacon beetle lays its eggs on dried meat, the carrion beetle lays its eggs on decayed meat. All are apparently directed by the odors of these substances.

Odors may assist insects in locating their food. *Drosophila* are attracted to over-ripe fruits. Under experimental conditions, *Drosophila ampelophila* is positively chemotropic to amyl alcohol, ethyl alcohol, acetic acid, lactic acid, and other products that occur in fermenting fruit. Bees and butterflies may be attracted to flowers by odor as well as by color. Many experiments have been conducted, during recent years, to show that insects are definitely attracted to various chemicals. The Japanese beetle responds freely to geraniol, eugenol, and other monatomic alcohols. The oriental fruit moth is strongly attracted by linaloöl, oleic acid, safrol, and many other

---

* Some writers define *tropism* as the *movement* in response to a stimulus and *tropotaxis* as the response to the stimulus. A neutral condition may also exist.

compounds. Insects are not always attracted by materials that are agreeable to man. The oriental fruit moth, for example, is rather freely attracted to formalin.

Chemical stimuli may also aid insects in finding their mates. The scent scales, *androconia*, of the wings of certain male butterflies and the scent glands on the thorax of tree crickets attract the females. Many female moths have glands associated with the genital organs that attract the males. On the other hand, various offensive fluids are emitted by insects, such as the secretions of the poplar leaf beetle, *Lina tremulae*, and other Coleoptera, which render them unpalatable to numerous animals. Certain carabids secrete a pungent fluid from a pair of anal

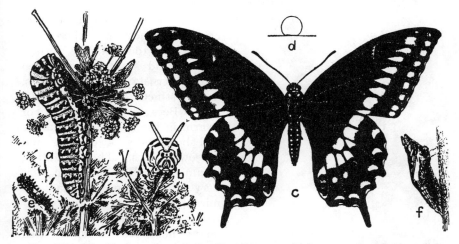

Fig. 166.—The black swallowtail, *Papilio polyxenes.* (*e*) Immature larva; (*a*) and (*b*) mature larvae, the latter showing the scent gland or osmeterium extended from the neck of the larva; (*c*) adult ♂ ; (*d*) egg; (*f*) chrysalis. (*After Chittenden, U. S. Department of Agriculture.*)

glands. The caterpillars of the papilio butterflies, when disturbed, put forth a forked organ just back of the head, which liberates a strong odor disagreeable to man.

Negative chemotropic reactions enable insects to avoid injurious substances. Insects will not deliberately walk into a cyanide bottle. Chemicals have been used in a practical way to repel certain insects. For many years Bordeaux mixture has been recognized as a repellent for leafhoppers. Sulphides of thallium, aluminum, and thiuram have been used to prevent feeding by Japanese beetles. Cedar and camphor are well-known repellents for the clothes moth.

**Thigmotropism (Stereotropism).**—The withdrawal from or contact with certain substances is known as *thigmotropism*. It is commonplace to state that an insect generally moves when touched. However, when insects are in death feints, they resist all touch stimuli. Most insects like freedom; as a rule they do not thrive in close quarters. A dragonfly may scour the country for a mile or more.

*Aphis helianthi* align themselves in a more or less definite row on the stems of *Helianthus*. If one of the aphids is touched, it will move and cause all the others to shift their positions. The movement may proceed up or down the stem like a wave. Many insects also require considerable freedom at the time of pupation. The chrysalids of butterflies are always found in the open; at the most a girdle of silk is used as a support.

Other insects thrive best under crowded conditions where one individual touches the other. Aphids and scale insects often form incrustations on plants. Such instincts lead to gregariousness. Bees cluster in the winter or swarm in the summer. Roaches and bedbugs seek crevices and narrow places. The Aradidae, commonly called flat bugs, are very much compressed and admirably adapted to live under bark. Leaf miners and borers adapt themselves to very limited quarters.

Fig. 167.—Bees clustering on a limb. An excellent example of thigmotropism. (*Pennsylvania Department of Agriculture.*)

**Hydrotropism.**—*Hydrotropism* might be defined as the lure of the water. In the spring toads are attracted to ponds and streams where they lay their eggs. The remainder of the year they are seldom seen near the water. Experiments show that they have unusual ability in locating water, and often wander a considerable distance to find it. Insects are likewise attracted by water. *Haliplus* and *Hydroporus* are strongly hydrotropic. If they are removed from the shore, they quickly orient themselves and head for the water.

Many wood-boring beetles are negatively hydrotropic; they prefer dry wood. Termites and ants seek dry places to construct their nests. Ants store their food only in dry places. Ant lions can construct their pits only in comparatively dry areas. All of these forms exhibit negative hydrotropism.

Atmospheric moisture plays an important part in the habits of many insects. Such reactions might be called *humidotropism*. Most insects oviposit at high humidities. *Glossina* oviposits most frequently at the saturation point. *Habrobracon juglandis* is said to be little affected by degrees of moisture. Grain insects, the larvae of the clothes moth, and similar insects have a relatively low water content and may exist in an environment of low humidity.

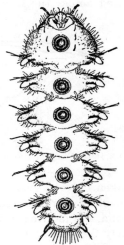

**Rheotropism.**—The ability of organisms to adjust themselves to the currents of water is known as *rheotropism*. Trout orient themselves parallel to the currents of the stream with their heads upstream. Various theories have been propounded for this adjustment but pressure certainly plays an important part. Thus rheotropism coincides in part with thigmotropism. Many insects adjust themselves to the currents in streams. Some caddis worms live only in sluggish streams; others select swift waters. As a rule, the species that live in swift water build cases of stones which act as sinkers. They may inhabit grasses where there is less danger of being washed away. Other insects such as the Simulidae and the Blepharoceridae select the swiftest part of the stream and attach themselves by means of suckers. Needham and Christenson studied the adjustment of insects to swift streams, from which we glean the following observations.

Fig. 168.—Larva of *Bibiocephala*, showing the sucking disks on the ventral surface by which it attaches itself to a stone in swift-flowing water.

ADJUSTMENT OF INSECTS TO SWIFT WATER (FROM NEEDHAM AND CHRISTENSON)

| Exposure | Species |
|---|---|
| In swift water | Black fly larvae, Simulidae |
|  | Net-wing midge larvae, *Bibiocephala* |
| In swift water protected by tubes | Midge larvae, *Chironomus* and *Tanytarsus* |
|  | Caddis worms, *Brachycentrus* |
| On sheltered side of obstruction | Caddis worms, *Glossosoma* and *Hydropsyche* (often in crevices), snipe fly larvae, *Atherix* (often in crevices) |
| Underside of stone or obstruction | May fly naiads, *Iron* and *Rhithrogena* |
| On bottom in slack water | May fly naiads, *Baetis* and *Leptophlebia* |
| In trash or sheltered places | May fly naiads, *Ephemerella grandis* |
|  | Stone fly naiads, *Acroneuria* and *Pteronarcys* |

**Anemotropism.**—Various insects orient themselves with respect to the direction of the wind or air currents. Midges often dance in swarms and seek sheltered places such as the protection of a hill. Dragonflies and other predacious species often seize this opportunity to obtain food upon the wing. Some recent observations showed that leafhoppers were lured across the Delaware River by means of powerful lights but were

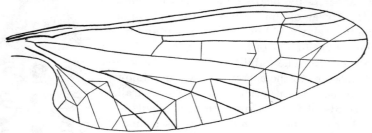

FIG. 169.—Wing of *Bibiocephala*, showing the creases upon which it is folded within the puparium. (*After Comstock.*)

taken only on evenings when the wind carried them toward the lights. In other words, these insects were not attracted to the light when the wind was blowing in the opposite direction. On the other hand, insects may fly against the prevailing air currents. Wheeler observed swarms of *Bibio albipennis* with all the individuals headed directly toward the

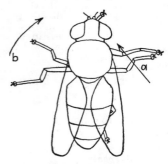

FIG. 170.—*Eristalis*, illustrating how the legs on one side move forward while those on the other move backward when the light is directed toward the posterio-lateral surface of the eye. (*a*) Direction of the light; (*b*) direction of the movement of the insect. (*After S. O. Mast.*)

gentle wind that was blowing. When the wind shifted, the insects at once changed their position to face the wind.

**Phototropism.**—Light and color play an important part in the reactions of insects. If the source of light is the sun, the reaction is called *heliotropism*. In considering light, one must take into account the direction, quality, intensity, and duration. It is a well-known fact that many insects are attracted to light, the response to which is often greater than life itself. Light of short wave lengths, such as blue and violet, is usually more attractive to insects than other colors. Black infuriates bees but white seems to have no effect upon them. It is said that enraged bees will congregate upon and sting the black spots on Holstein cattle but will not attack the white spots. Beekeepers know that white clothing is preferable to black, when handling bees. The larvae of clothes moths are partial to red and attack red flannels in preference to materials of other colors. On the other hand, white is repellent to the adults of the oriental fruit moth. As these moths oviposit reluctantly upon white surfaces, heavy applications of lime and talc have been recommended as a control for them.

Although moths are attracted to artificial light, they shun sunlight Butterflies on the contrary are attracted to sunlight but are repelled by artificial light. Butterflies are tuned to high intensities of light and moths to low intensities, so that the bright light attracts the butterflies and feeble light attracts the moths.

Roaches scamper for darkness when a light is turned on. Termites, ants, and many beetles take cover when a stone is overturned, thus exposing them. Termites build tubes in which to travel and are never seen in the open except at the time of the nuptial flight.

Social insects disregard heliotropic responses when the instinct to gather honey and pollen is operating, for they go from darkness to light in leaving the nest and from light to darkness in returning to it.

Insects generally orient themselves with their heads directly toward or directly away from the light source. The direction of the light is therefore an important factor. Loeb states that the moth is not attracted by the light but is oriented by it and, in constantly adjusting its head to the light, is drawn into it. On cool mornings, grasshoppers often turn their bodies perpendicular to the sun's rays for another reason, *i.e.*, to receive the greatest advantage of the heat rays.

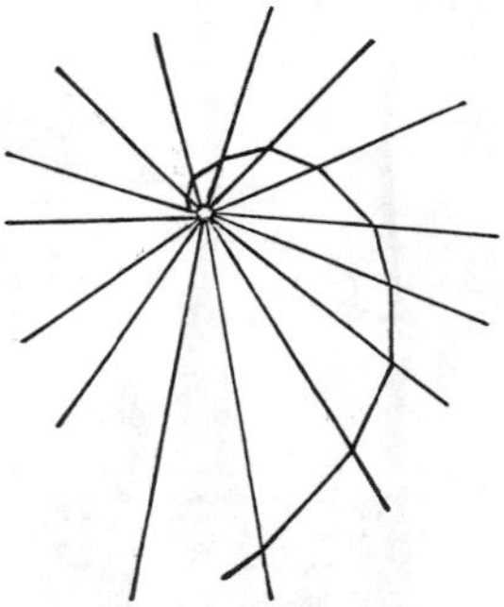

Fig. 171.—Diagram showing how a moth, attracted by a near source of light, turns regularly toward the beam until it eventually approaches the light. (*After Budden-brock.*)

Experiments have shown that all lights of the same color do not attract insects proportionally. Much depends upon the intensity. The intensity in turn depends upon the character of the glass transmitting the light and the intensity of the source of the light.

**Thermotropism.**—The reaction of insects to heat is often difficult to separate from other tropisms. Parasites of warm-blooded animals may be attracted by heat as well as by odor. The mosquito that attacks a frog is certainly guided to its host by other stimuli. The eggs of the human botfly, which are carried to man by the mosquito, are said to hatch in response to the warmth of the human body. Heat is essential for the laying and hatching of most eggs. An Indian mosquito oviposits freely at 95°F. The carabid, *Geopinus*, is active on the sand dunes all night but disappears before daybreak, apparently responding to the increase in temperature rather than the increase in light. Bees, wasps, and ants cease to work as dusk approaches, after which they rest. These responses may result from light deficiency as well as temperature deficiency.

**Geotropism.**—The response to gravity is seldom uniquely associated with insects. A negative stimulus must direct insects to work their way upward to the surface of the ground when they emerge.    *Lyctus* has the remarkable habit, when emerging, of boring outward regardless of impediments that might be in the way.    In experiments, they have been known to bore through several layers of lead.

Most borers, such as the raspberry cane borer, *Oberea bimaculata*, the currant stem girdler, *Janus integer*, and the bronzed birch borer, *Agrilus*

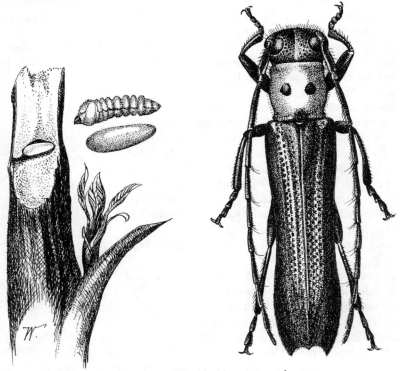

FIG. 172.—The blackberry cane borer *Oberea bimaculata* showing egg, larva and adult. *(After W. R. Walton.)*

*anxius*, burrow downward.    On the other hand, the red-necked raspberry cane borer, *Agrilus ruficollis*, burrows upward in a spiral, eventually girdling the cane and causing a gall.

The pupae of many borers, as for example, the locust borer, *Cyllene robiniae*, and the leopard moth, *Zeuzera pyrina*, are formed with their heads directed downward.    On the contrary, the pupa of the carpenter worm, *Prionoxystus robiniae*, is formed with its head directed upward.

Insects generally rest with their heads upright but, when certain crambid moths alight on stalks of grass, they quickly place their bodies parallel with the stalks, their wings are tightly wrapped about their bodies, and the heads and palpi project downward. Some of the Noctuidae rest with their heads downward. The cotton leaf worm, *Alabama argillacea*, invariably rests with its head downward. The larva of Harris's sphinx, *Lapara bombycoides*, has a way of hanging with its head downward in a group of pine needles. The larvae of many aquatic Diptera, notably *Culex*, normally rest with their heads downward. The chrysalids of many butterflies hang head downward. Stinkbugs, when feeding on various prey, often swing their burden free and hang in an inverted position. This attitude results partly from the ease of handling the victim and partly to prevent the captured prey from obtaining a foothold and more easily freeing itself.

Instinct.—It is difficult to define instinct; the term is vague and often used without much meaning. Darwin made no attempt to define instinct as he made no attempt to define the species. Lamarck defined instinct as "inherited habit." This involves the inheritance of acquired characteristics which, by some entomologists, is not believed to occur in insects. Folsom (1906) defines instinct, as distinguished from intelligence, as "the attainment of adaptive ends without prevision and without experience." A caterpillar for example feeds in preparation for, but without knowledge of, the wonderful transformation that is to take place in the pupa. A wasp, emerging from its nest, probably never saw its mother and was certainly never instructed in the art of building a nest; nevertheless, it constructs a nest identical to the one from which it came and the same as the kind made by all individuals of the same species. It provides its nest with the correct kind of food and lays an egg in the proper manner, without instruction or experience.

An instinct is the response of an organism to a single or to several stimuli. Instinctive actions consist of coordinated reflexes. They are complex chains of tropisms. Some stimuli are received from without; others are received from within the insect. Instincts are not inherited but the potentiality to produce instinctive reactions is inherited. Instincts are the property of protoplasm; and the ability to produce secretions or to respond to outward stimuli are embodied in this life-giving substance.

It is easier to cite the characteristics of instincts than it is to define them. Instincts are executed without prevision or without experience or training. They are therefore not associated with reasoning. They are complex reflexes in response to numerous chemical or physical stimuli. They proceed in a regular sequence characterized by rhythm and repetition. They are generally inflexible but submit feebly to modification.

The reactions of insects can seldom be attributed to a single reflex. Folsom and Wardle (1934) have shown that the egg-laying instinct of *Drosophila melanogaster* is not the result of a single stimulus but follows a chain of responses incurred by stimuli such as odor, moisture, taste, and touch. Simple reflexes thus combine to form an instinctive act.

In general, the responses of insects are inflexible. All mud daubers of the same species inhabit the same type of location, provision their nests with the same kind of food, and employ the same type of architecture. The inflexibility or stubbornness in insects is also illustrated by the habits of *Pieris brassicae* in their annual migrations in India which are described by Hingston. In March, a few individuals were seen flying up the sides of the Himalayas; by the end of April, thousands per minute were flying in one direction and headed for the snow-covered summit of

Fig. 173.—Mud at the edge of a pool, showing the tiny footprints made by the wasps and depressions resulting from loads of mud carried away in the form of pellets by mud daubers.

the mountain where they undoubtedly perished. One wonders what force drives these butterflies to their destruction but the fact remains that they are unable to forestall this wholesale catastrophe. Migrating locusts know no barriers. If a river chances to be in their path, they will attempt to cross only to fall in the water and be drifted to the shores by the waves.

Many cases can be cited of the inability of insects to adjust themselves to unusual conditions. A certain wasp always carries its prey, a grasshopper, by the antennae; when the antennae of the grasshopper are removed, the wasp is unable to contrive a way to carry it. Generally a larva cannot adjust itself to new food. As a rule, the eggs of insects are deposited on the food intended for the larva. If by chance the caterpillar is transferred to another plant, it will starve rather than adopt the new food.

On the other hand, insects are capable of slight modifications in their habits. The mud dauber may fasten its nest to an iron beam or to a wooden joist; in one case, a female attached its nest to the mohair lining of an automobile. Individual mud dauber wasps show diversity of type of architecture. The mud is generally smoothed on the nest in an even coat. Occasionally, the individual pellets of mud are left clinging to the exterior of the nest and are not smoothed down. The most remarkable ability to cope with an unusual situation was exhibited by a mud dauber, *Sceliphron coementarium*, that built a nest inside of the writer's car. The

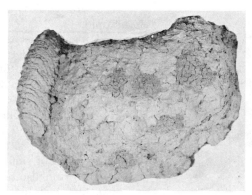

Fig. 174.—Nest of the mud dauber, *Sceliphron coementarium*, showing the usual method of smoothly applying the mud.

car had been out of the garage several hours each day and no doubt the wasp, on account of these interruptions, had some difficulty in finding its nest. The mud cell was first noticed on June 26 at 1 P. M., when it was about completed but not provisioned or sealed. During the remainder of June 26, the car was out of the garage and was not returned until 7 A. M. on June 27. From this time until the afternoon of June 29, the car was in the garage and little attention was paid to the nest. As a matter of fact it was quite forgotten by man but not by wasp, for, on the afternoon of June 29 when the car was taken out again, it was noticed that the cell had been provisioned and sealed. The following morning it was discovered that a second cell had been started. The car was backed about 10 feet out of the garage but the wasp failed to locate its nest. At 11 A. M. the car was shifted to the right side of the two-car garage rather than the left, the original position. By 1 P. M. the nest was completed but not sealed. On July 1 no additional change was evident and apparently the nest was left open to be provisioned. An extended auto trip interrupted the observations, for the car was out of the garage from July 2 to July 5. It was supposed that the wasp had surely abandoned the nest but the writer was surprised on July 9 to find that the second cell, upon which work had of necessity ceased on July 2, had been provisioned and sealed.

The repetition of responses or persistence is characteristic of lower animals. A paramecium that is placed in a solution unfavorable for its development will continue to give avoiding reactions until it dies. It is a matter of commonplace observation that a bee or a fly, in response to heliotropic and geotropic stimuli, consumes many hours in unsuccessful attempts to escape although the window may be open at the bottom.

FIG. 175.—Nest of the mud dauber, *Sceliphron coementarium*, showing a variation in architecture. Individual pellets of mud are left on the surface of the cells.

When left to their judgment, insects are helpless and surprisingly stupid.

Hingston discusses the persistence of a spider. Immediately after a spider had finished making a spoke of its orb web, he cut it. The spider replaced the spoke. He cut the spoke and it was replaced again. Thus he repeated the experiment 25 times before the spider tired of this little game.

Hingston also relates an experiment in which he put a small piece of camphor in the nest of a potter wasp, *Eumenes*, after it had completed its mud cell. In spite of the fact that the young could never mature under such circumstances, the wasp continued to provision the nest with caterpillars, lay its egg, and seal the cell in the usual fashion. Insects have, so to speak, little upon their "minds," but what they accomplish they do completely and perfectly.

Instincts frequently oppose one another. This is best illustrated by an example taken from the birds. The migratory instinct frequently overcomes the maternal instinct. Darwin states that the swifts and swallows often leave the young in the nest to starve, in order to follow the instinct to migrate southward.

The rhythm of instinct is well illustrated by the potter wasp, *Eumenes fraterna*. The nesting habits proceed in four distinct steps: (1) the building of the mud cell, (2) provisioning the cell with caterpillars, (3) suspending the egg from the interior of the ceiling of the cell, and (4) closing and sealing the cell. If the operations of the wasp are interrupted by cutting a small hole in the side of the jug at the time the wasp is provisioning the nest, *i.e.*, during step 2, she will not mend or repair the injury but will continue to provision the cell, lay her egg, and close the cell. She does not go back to step 1 but continues with steps 3 and 4.

Thus the nest will be completed with the hole or injury in the side. If, on the other hand, a hole is cut at the rim at the time she is finishing the nest, she will respond and repair the injury.

Fabre relates a similar procedure in the case of the wall bee, *Chalicodoma muraria.* There are two sets of responses in connection with its nesting habit: (1) the bee goes in the nest head first, disgorges the contents of the crop, and backs out of the nest, (2) the bee goes in the nest tail first and brushes off the pollen. Fabre interrupted the bee as it

Fig. 176.—*Eumenes fraterna*, a solitary wasp, and its small jug-like nests. (*Champlain and Kirk.*)

was about to go in tail first, after it had already disgorged the food from its mouth. As the second act was prevented, she started the performance all over again and went in head first, although it had nothing to disgorge. This was repeated many times before the wasp gave up.

From Fabre, we also take the interesting account of the habits of a species of *Sphex* that nests in the ground. There are three principal steps in its nesting habits: (1) the wasp drags a cricket to the edge of its burrow, (2) she goes within the nest to make a preliminary inspection, and (3) she returns to obtain the cricket and take it into the nest. Fabre interrupted the wasp between steps 1 and 2, *i.e.*, when the wasp was in the burrow making the preliminary inspection. When the wasp came out and found the cricket removed from the edge of the burrow, she returned to step 1, dragged the cricket to the edge of the burrow again, and then went into the nest for the inspection. Fabre repeated this experiment 40 times before the wasp tired.

The writer watched some mud daubers that had gained entrance to an attic through a broken window pane. A spider's web stretched across the window impeded the flight of the wasps as they entered the window carrying pellets of mud for their nests. They immediately dropped their pellets and flew away for more material, never attempting to pick them up and save themselves considerable effort. The window sill

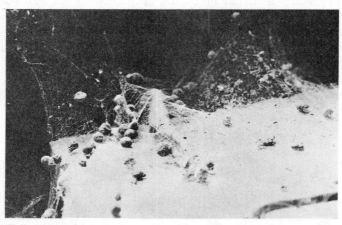

FIG. 177.—Pellets of mud left upon a window sill by mud daubers, which were caught in a spider web, on their way to construct nests in an attic.

eventually became strewed with mud pellets. These examples illustrate that the stimuli or forces that call forth instincts must be strong and that they proceed in a regular sequence. If interrupted, the insect cannot go back and do something that does not fall in the sequence of events. It can only do one thing at a time. It can go back to the beginning and start all over again or it can proceed with the subsequent steps in the sequence even though the results may be futile.

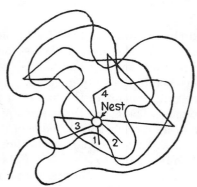

FIG. 178.—Orientation flight or location study made by the wasp, *Astata bicolor*. Four flights were made in four different directions before the wasp took off on its last tortuous flight. (*After the Peckhams.*)

**Intelligence.**—No satisfactory definition has been proposed for intelligence. The opinions of modern educators differ on this subject. Webster gives eight distinct definitions, all implying the faculty to understand, the capacity to know or apprehend, the acuteness of acquiring understanding or knowledge, and the accumulation of knowledge or general information. Psychologists exclude the latter from their conception of intelligence. Under the circumstances, it is somewhat difficult to discuss the subject as far as insects are concerned.

Instinct is no doubt dominant in insects. At times, however, it seems difficult to account for their habit on the basis of instinct alone. The trial flights of *Chlorion* described by the Peckhams, the attempts of wasps to enlarge their burrows to accommodate large insects and the recognition of one ant by another lead us to believe that there might be a trace of intelligence in insects. On first thought these acts seem like intelligence, but a closer observation and a more fundamental knowledge of the habits of these insects indicate that these astounding actions may after all be tropic responses. It has been proved that the recognition of one ant by another of the same colony is due to a characteristic odor. When an ant is washed and placed back in its own colony, it receives hostile treatment by the others.

It seems phenomenal that a spider, in constructing its orb web, can measure distances accurately, draw parallel lines, and calculate angles. A closer examination shows, however, that these are purely mechanical responses. In making the beautiful geometric orb web, the spider probably uses its feet to measure off the distance between the spirals so that succeeding sections of the spiral are equally spaced. If the strands are equally spaced, they are naturally parallel. If they are parallel, it follows by the simple geometric theorem that the angles are equal for "parallel lines cut by a diagonal line form equal opposite angles."

Even though we agree that insects seem to memorize certain acts or images, we must admit that they cannot recall them at will. They are therefore incapable of reasoning or understanding.

## BIBLIOGRAPHY

The references concerning the responses of insects to various stimuli are exceedingly numerous. A few of the more important papers dealing with their responses to light, moisture, temperature, and chemical stimuli are cited here. Uvarov (1931) summarizes the more important papers dealing with the relation of insects and climate.

### I. Responses of Insects to Light and Electrical Waves.

ABBOTT, C. E. (1936). The vision of insects; color perception. *In*, The physiology of insect senses. *Ent. Amer.* **16** (4). *Bibliography.*

BERTHOLF, L. M. (1931). Reactions of the honeybee to light. *Journ. Agr. Res,* **42** (7). *Literature cited.*

BUDDENBROCK, W. VON, *et al.* (1933). Light-compass orientation. *Zeit. verlag. Physiol.* **15.**

CLARK, L. B. (1933). Modifications of circus movements in insects. *Journ. Exp. Zool.* **66** (2). *Literature cited.*

COLLINS, D. L., and NIXON, M. W. (1930). Responses to light of the bud moth and leaf roller. *Geneva Agr. Exp. Sta. Bull.* **583.**

DIRKS, C. O. (1937). Biological studies of Maine moths by light trap methods. *Me. Agric. Esp. Sta. Bull.* **389.** *Literature cited.*

DOOLEY, W. L. (1932). The factors involved in stimulation by intermittent light. *Journ. Exp. Zool.* **62.** *Literature.*

————, and WIERDA, J. L. (1929). Relative sensitivity to light of different parts of the compound eye in *Eristalis tenax*. *Journ. Exp. Zool.* **52** (2). *Literature.*

DRIGGERS, B. F., and PEPPER, B. B. (1932). Effect of artificial illumination on the oriental fruit moth under orchard conditions. *Journ. Econ. Ent.* **25** (2).

HEADLEE, T. J. (1933). The effects of radio waves on internal temperatures of certain insects. *Journ. Econ. Ent.* **26** (2). *Literature.*

————, and JOBBINS, D. M. (1938). Progress to date on studies of radio waves and related forms of energy for insect control. *Journ. Econ. Ent.* **31** (5).

HEDRICK, U. P. (1932). The responses of orchard insects to light. *Geneva Agr. Exp. Sta. Rept.* **51.**

HERMES, W. B., and ELLSWORTH, J. K. (1934). Field tests of the efficacy of colored lights on trapping insect pests. *Journ. Econ. Ent.* **27** (5). *Bibliography.*

HESS, C. VON (1917). New experiments on the light reactions of plants and animals. *Journ. Animal Behavior* **7** (1).

HOLMES, S. J., and McGRAW, K. W. (1913). Some experiments on the method of orientation to light. *Journ. Animal Behavior* **3** (5). *References.*

HUTCHINGS, R. E. (1940). Insect activity at a light trap during various periods of the night. *Journ. Econ. Ent.* **33** (4).

KELSHEIMER, E. G. (1932). Leafhopper response to colored lights. *Ohio Journ. Sci.* **32** (2).

LAWSON, P. B. (1929). Leafhoppers and the trap light. *Journ. Kan. Ent. Soc.* **2** (2).

———— (1930). Another season's trap-lighting of leafhoppers. *Journ. Kan. Ent. Soc.* **3** (2).

LOVELL, J. H. (1910). The color senses of the honeybee; can bees distinguish colors? *Amer. Nat.* **44** (527). *Literature.*

LUTZ, F. E. (1924). A study of ultra violet light in relation to flower visiting habits of insects. *Ann. N. Y. Acad. Sci.* **29.**

———— (1933). Color vision of insects. *Nat. Hist.* **33.**

MAST, S. O. (1917). The relation between spectral color and stimulation in the lower organisms. *Journ. Exp. Zool.* **22** (13). *Bibliography.*

———— (1923). Photic orientation in insects with special reference to the dronefly, *Eristalis* spp. *Journ. Exp. Zool.* **38** (1). *Bibliography.*

———— (1926). Photic orientation in insects. *Amer. Nat.* **60** (670). *Literature.*

MINNICH, D. E. (1919). The photic reactions of the honeybee, *Apis mellifera* L. *Journ. Exp. Zool.* **29** (3). *Bibliography.*

PACKARD, A. S. (1903). Color-preference in insects. *Journ. N. Y. Ent. S.* **11** (3).

PARROTT, P. J., and COLLINS, D. L. (1935). Some further observations on the influence of artificial light upon codling moth infestations. *Journ. Econ. Ent.* **28** (1).

PEARSE, A. S. (1911). The influence of different color environments on the behavior of certain arthropods. *Journ. Animal Behavior* **1** (2). *Bibliography.*

PETERSON, ALVAH, and HAEUSSLER, G. J. (1928). Responses of the oriental peach moth and codling moth to colored lights. *Ann. Amer. Ent. Soc.* **21** (3). *References.*

STEARNS, L. A. (1932). Effects of artificial illumination on the oriental fruit moth under orchard conditions. *Journ. Econ. Ent.* **25** (2). *Literature cited.*

WEISS, H. B. (1927). Insects captured in the lookout stations in New Jersey. *Bur. Sta. Insp. N. J. Dept. Agric. Bull.* **106.**

————, *et al.* (1941). Notes on the reactions of certain insects to different wavelengths of light. *Journ. N. Y. Ent. Soc.* **49** (1).

## II. Responses of Insects to Moisture and Temperature.

ABBOTT, C. E. (1936). Responses of insects to temperature and humidity. *In,* Physiology of insect senses. *Ent. Amer.* **16** (4).

CHAPMAN, R. N. (1928). Temperature as an ecological factor in animals. *Amer. Nat.* **62** (681). *Bibliography.*

—— (1928). The quantitative analysis of environmental factors. *Ecology* **9** (2). *Literature cited.*

FLEMION, F., and HARTZÉLL, A. (1936). Effect of low temperature in shortening the hibernation period of insects in the egg stage. *Boyce Thompson Inst.* **8** (2). *Literature cited.*

FLINT, W. P. (1936). The effect of winter temperatures of 1935–1936 on some common Illinois insects. *Trans. Ill. Acad. Sci.* **29.**

GATES, B. N. (1914). The temperature of the bee colony. *U. S. Dept. Agric. Bull.* **96.**

GLENN, P. A. (1922). Codling moth investigations of the state entomologist's office. *Ill. Bull.* **14** (7).

HEADLEE, T. J. (1917). Some facts relative to the influence of atmospheric humidity on insect metabolism. *Journ. Econ. Ent.* **10** (1).

—— (1928). Some data relative to the relationship of temperature to codling moth activity. *Journ. N. Y. Ent. Soc.* **36** (2).

HINE, J. S. (1908). Some observations concerning the effects of freezing on insect larvae. *Ohio Nat.* **5** (4).

HOPKINS, A. D. (1938). Bioclimatics. A science of life and climate relations. *U. S. Dept. Agric. Misc. Pub.* **280.**

IMMS, A. D. (1932). Temperature and humidity in relation to problems of insect control. *Ann. Appl. Biol.* **19** (2). *References.*

KOZHANTSHIKOV, I. W. (1938). Physiological conditions of cold-hardiness in insects. *Bull. Ent. Res.* **29** (4). *References.*

MACGILL, E. I. (1931). The relation between temperature and humidity and the life cycle. *In,* The biology of Thysanoptera with reference to the cotton plant. *Ann. Appl. Biol.* **18** (4). *References.*

—— (1937). The relation between variations in temperature and life cycle. *In,* The biology of the Thysanoptera with reference to the cotton plant, VIII. *Ann. Appl. Biol.* **24** (1). *References.*

MAIL, G. A. (1930). Winter soil temperatures and their relation to subterranean insect survival. *Journ. Agr. Research* **41** (8). *Literature cited.*

MARCHLAND, W. (1930). Thermotropism in insects. *Ent. News* **31** (6). *Literature cited.*

MAYER, A. G. (1918). Toxic effects due to high temperature. *Dept. Marine Biol.* **12.**

MELLANBY, K. (1932). The influence of atmospheric humidity on the thermal death point of a number of insects. *Journ. Exp. Biol. Edinburgh* **9.**

—— (1936). Humidity and insect metabolism. *Nature Lond.* **138** (3481).

PAYNE, N. M. (1926). The effects of environmental temperatures upon insect freezing points. *Ecology* **7** (1). *Literature cited.*

—— (1926). Freezing and survival of insects at low temperatures. *Quart. Rev. Biol.* **1** (2). *Literature.*

—— (1927). Two factors of heat energy involved in insect cold hardiness. *Ecology* **8** (2). *Literature cited.*

—— (1930). Some effects of low temperature on internal structure and function of animals. *Ecology* **11** (3). *Literature cited.*

PEAIRS, L. M. (1914). The relation of temperature to insect development. *Journ. Econ. Ent.* **7** (2).

—— (1927). Some phases of the relations of temperature to development of insects. *W. Va. Agric. Exp. Sta. Bull.* **208.** *Bibliography.*

PIERCE, W. D. (1916). A new interpretation of the relationships of temperature and humidity to insect development. *Journ. Agric. Res.* **5** (25).

ROBINSON, WM. (1927). Water-binding capacity of colloids a definite factor in winter hardiness of insects. *Journ. Econ. Ent.* **20** (1).

—— (1928). Determination of the natural undercooling and freezing points in insects. *Journ. Agric. Res.* **37** (12).

SACHAROV, N. L. (1930). Studies in cold resistance of insects. *Ecology* (3). *Literature cited.*

SALT, R. W. (1936). Studies on the freezing process in insects. *Tech. Bull. Minn. Agric. Exp. Sta.* **116**. *Literature cited.*

SANDERSON, E. D. (1908). The relation of temperature to hibernation of insects. *Journ. Econ. Ent.* **1** (1).

—— (1908). The influence of minimum temperature in limiting the northern distribution of insects. *Journ. Econ. Ent.* **1** (4).

——, and PEAIRS, L. M. (1913). The relation of temperature to insect life. *N. H Agric. Exp. Sta. Tech. Bull.* **7**.

SHELFORD, V. E. (1927). An experimental investigation of the relations of the codling moth to weather and climate. *Ill. Bull.* **15** (5 and 6). *Bibliography.*

—— (1932). An experimental and observational study of the chinch bug in relation to climate and weather. *Ill. Bull.* **19** (6).

TOWNSEND, C. H. T. (1924). An analysis of insect environments and responses. *Ecology* **5** (1).

UVAROV, B. P. (1931). Insects and climate. *Trans. Ent. Soc. Lond.* **97**. *Bibliography.*

—— (1932). The influence of temperature on the life history of insects. *Trans. Ent. Soc. Lond.* **80** (2). *Literature.*

## III. Responses of Insects to Chemical Stimuli.

ABBOTT, C. E. (1936). The physiology of insect senses. *Ent. Amer.* **16** (4).

BOBB, M. L., *et al.* (1939). Baits and bait traps in codling moth control. *Bull. Va. Agric. Exp. Sta.* **320**. *References.*

BRUES, C. T. (1920). The selection of food-plants by insects. *Amer. Nat.* **54**.

DETHIER, V. G. (1941). Chemical factors determining the choice of food plants by Papilio larvae. *Amer. Nat.* **75** (756). *Literature cited.*

EYER, J. R., *et al.* (1937). Analysis of attrahent factors in fermenting baits used for codling moth. *Journ. Econ. Ent.* **30** (5). *Literature cited.*

FROST, S. W. (1933). Baits for oriental fruit moths. *Pa. State College Bull.* **301**. *Bibliography.*

—— (1936). A summary of insects attracted to liquid baits. *Ent. News* **47** (3).

——, and DIETRICH, H. (1929). Coleoptera taken from bait traps. *Ann. Ent. Soc. Amer.* **22**.

IMMS, A. D. (1931). Sense organs and reflex behavior. *In,* Recent advances in entomology. The Blakiston Company, Philadelphia.

—— and HUSAIN, M. A. (1920). Field experiments on the chemotropic responses of insects. *Ann. Appl. Biol.* **6** (4). *Bibliography.*

McINDOO, N. E. (1926). An insect olfactometer. *Journ. Econ. Ent.* **19** (3).

—— (1928). Responses of insects to smell and taste and their value in control. *Journ. Econ. Ent.* **21** (6). *Literature cited.*

—— (1929). Tropisms and sense organs of Lepidoptera. *Smiths. Misc. Coll.* **81** (10). *Literature cited.*

—— (1931). Tropisms and sense organs of Coleoptera. *Smiths. Misc. Coll.* **82** (18). *Literature cited.*

——— (1938).   The senses of insects compared to those of higher animals.  *Proc. Ent. Soc. Wash* **40** (2).  *Bibliography.*

MARSHALL, J. (1935).   The location of olfactory receptors in insects.  *Trans. R. Ent. Soc. Lond.* **83.**

MINNICH, D. E. (1929).   The chemical senses of insects.  *Quart. Rev. Biol.* **4** (1). *Bibliography.*

MORGAN, A. C., and CRUMB, S. E. (1928).   Notes on the chemotropic responses of certain insects.  *Journ. Econ. Ent.* **21** (6).

RICHMOND, E. A. (1927).   Olfactory response of the Japanese beetle (*Popillia japonica* Newm.).  *Proc. Ent. Soc. Wash.* **29** (2).   *Bibliography.*

TRÄGÅRDH, IVAR (1913).   On the chemotropism of insects.  *Bull. Ent. Res.* **4.**

WEISS, H. B. (1913).   Odour preference of insects.  *Can. Ent.* **45.**

**IV. Responses of Insects to Air Currents.**

See Bibliography, Chap. II, part III, Wind and air currents as factors in the distribution of insects.

**V. References to Instinct and Intelligence.**

BALFOUR-BROWN, F. (1936).   The evolution of the social life in insects.  *Trans. Soc. Br. Ent.* **3.**

BEAUMONT, J. DE (1935).   L'instinct et l'intelligence chez les insectes.  *Bull. Soc. vaud. Sci. Nat. Lausanne* **58.**

BOUVIER, E. L. (1922).   The phychic life of insects.  Trans. by L. O. Howard.  T. Fisher Unwin, London.

CROWELL, M. F. (1929).   A discussion of human and insect societies.  *Psyche* **36.** *Literature.*

DAVIS, W. T. (1897).   Intelligence shown by caterpillars in placing their cocoons. *Journ. N. Y. Ent. Soc.* **5** (1).

EMERSON, A. E. (1939).   Social coordination and superorganism.  *Amer. Midl. Nat.* **21.** *References.*

FOLSOM, J. W. (1934).   Entomology with special reference to its ecological aspects. Revised by Wardle.  4th ed.  The Blakiston Company, Philadelphia.

FOREL, A. (1928).   The social world of the ant compared with that of man.  G. P. Putnam's Sons, New York.

FRANCON, J. (1939).   The mind of the bees.  Trans. by H. Eltringham.  Methuen & Co., Ltd., London.

HINGSTON, R. W. G. (1928).   Problems of instinct and intelligence.  Edward Arnold & Co., London.

——— (1929).   Instinct and intelligence.  The Macmillan Company, New York.

HOLMES, S. J. (1911).   The evolution of animal intelligence.  Henry Holt and Company, Inc., New York.

JENNINGS, H. S. (1906).   The behavior of the lower organisms.  Columbia University Press, New York.

McINDOO, N. E. (1938).   The senses of insects compared to those of higher animals. *Proc. Ent. Soc. Wash.* **40.**

REEDINGER, E. (1935).   Die Triebhandungen der Kerfe.  *Ent. Rdsch. Stuttgart* **53.**

SNODGRASS, R. E. (1928).   The mind of an insect.  *Smiths. Inst.*  1927.

WHEELER, W. M. (1928).   Foibles of insects and man.  Alfred A. Knopf, New York.

——— (1939).   On instincts.  *In,* Essays in philosophical biology.  Harvard University Press, Cambridge.

# CHAPTER XII

## INSECT ASSOCIATIONS, INCLUDING SOCIAL INSECTS

Continuing the discussion of insect behavior, we turn our attention to insect associations. The subject suggests two general headings: associations of individuals of the same species and associations of individuals of different species. These associations may be passive or active

FIG. 179.—May flies, *Hexagenia bilineata*, congregated on a building, following emergence. (*After Needham.*)

and the individuals may be assembled during a part or during the entire life of the insects.

**Passive Insect Associations.**—Insects are frequently collected in tremendous numbers by waves, tides, or air currents. Such aggregations are usually detrimental to the species but may serve as food for other animals. The pupae of the Koo-tsabe fly, *Ephydra hians*, of the

230

Mono Lake regions of California and Nevada, are often drifted in great windrows along the shores of the lake. The Indians collected them and used them as food.

The wind frequently drives May flies, at the time of their annual emergence, against buildings and other structures. Much has been written by E. P. Felt and others on the insects of the higher atmosphere. They are frequently carried upward by convectional air currents and even to remarkable heights by cyclonic storms. Wasps and mosquitoes have been taken at an altitude of 6,000 feet. Insects at high altitudes are not generally numerous but during migration or swarming large numbers may be found in the air. The moth of cotton leaf worm, *Alabama argillacea*, is frequently aided by the fall winds in its flight from the Gulf States to the northern part of the United States or Canada. The gypsy moth larvae are often carried considerable distances by the wind. Collins states that the wind is an important factor in the distribution of this species. In certain experiments 346 larvae were taken from a sticky surface 1,644 square feet in area. They had been blown a distance of from ⅛ to 13½ miles.

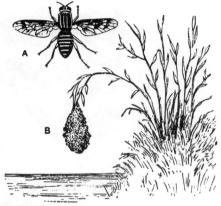

Fig. 180.—*Atherix ibis*, a European species of Rhagionidae: (*A*) adult; (*B*) mass of dead flies overhanging water. (*After Sharp.*)

**Active Associations.**—Feeding, mating, sleeping, swarming, migration, estivation, or hibernation may lead insects to gather in small or in extensive groups. Those that live in societies or in communities, but are not truly social, are said to be gregarious.

Insects that hatch from a batch or group of eggs may remain together during a large part of their lives. This habit is more common among phytophagous insects. The tent caterpillar, squash bug, and grape leaf skeletonizer can obtain food without moving far from the location where the eggs are deposited. The tent caterpillars have the advantage of a tent or nest to which they can retreat during cool, rainy days and at night. Predacious insects, such as many of the Pentatomidae, would find it difficult to obtain food if they remained together.

Japanese beetles, leafhoppers, grasshoppers, and other insects often hatch or emerge in considerable numbers at the same season of the year. They feed gregariously and, as the source of food is depleted, move to new areas. Thus grasshoppers, chinch bugs, and armyworms frequently migrate in large numbers. Ants, termites, midges, and numerous other insects emerge in great swarms and take to the air for nuptial flights. Hundreds of sap-feeding beetles, *Glischrochilus*, may gather upon sap oozing from a wound on a tree. Drosophilidae gather in large numbers to feed and oviposit on decaying fruit.

Upon the basis of food, many smaller associations are often formed. Several individuals may occupy the same feeding area. This often happens when several eggs are laid upon the same leaf. Leaf-mining *Pegomyia* lay their eggs in groups of two to ten upon the undersides of the leaves. When they hatch, all the larvae enter the leaf and occupy the

FIG. 181.—Japanese beetles clustered upon fruit. An aggregation resulting from the simultaneous emergence of adults.

same mine. It is not uncommon to find ten or more larvae of this species feeding side by side. Twenty-five larvae of the black gum leaf miner, *Antispila nyssaefoliella*, have been found in a single leaf. Many parasitic larvae may also be found in a single host and may be the progeny of a single egg, the result of polyembryony.

**Estivating or Hibernating Aggregations.**—Insects that gather to estivate, pupate, or hibernate are impelled by various forces, one of which is the selection of a protected place. The ladybird beetle, *Ceratomegilla maculata*, is known to congregate by the thousands in the fall to pass the winter beneath bark or in other sheltered places. It is natural for insects to gather in this manner since the food supply usually becomes depleted toward the end of the summer and the insects need protection from extreme winter conditions. Hungerford describes a most unusual case of hibernation.    He found a large number of back swimmers clustered in a small crevice in thick ice.

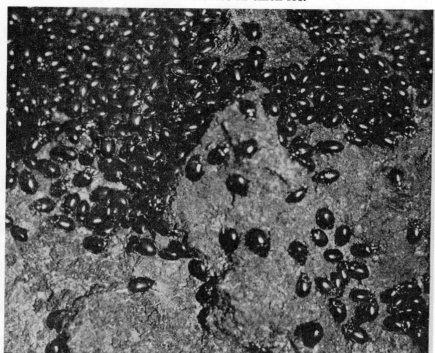

FIG. 182.—Convergent ladybird beetles, *Hippodamia convergens* var. *ambigua*, assembled in large numbers to hibernate.    (*After Duncan and Pickwell.*)

FIG. 183.—A sphinx caterpillar bearing the cocoons of *Apanteles glomeratus*.    (*Champlain and Kirk.*)

The nature of the congregating habit may vary in the same genus *Apanteles glomeratus*, a common parasite on the tomato worm, forms its cocoons in a loose mass upon the back of the caterpillar. *A. congregatus*, a closely related species that attacks the larvae of the imported cabbage worm, leaves the host and spins its cocoons in a compact mass upon the foliage or some near-by object.

**Protective Aggregations.**—Insects probably secure protection from parasites and predacious enemies and from adverse weather conditions by their congregating habits. This is especially true of hibernating

insects. Swarms of whirligig beetles no doubt gain some protection by their numbers as well as their activity. The yellow-necked apple worms and the red-humped apple worms are protected by congregating in large masses. The former have a characteristic manner of raising the anterior and posterior ends of their bodies as a warning threat. Psocids often rest in large groups on the bark of trees. Protective coloration also plays a part.

**Migrating Aggregations.**—Insects often migrate in large groups. The painted lady, *Cynthia cardui*, and the milkweed butterfly, *Danaus menippe*, make annual flights southward, in swarms frequently so great that they cover the branches of trees when they roost for the night.

The larvae of a certain species of *Sciara* travel over the ground in snakelike masses consisting of hundreds of individuals. As they advance, they leave a glistening trail of silken threads

Fig. 184.—The gregarious feeding habit of the red-humped oak worm, *Symmerista albifrons.*

upon the ground.

The procession moth, *Cnethocampa (Bombyx) processionea*, is a common feeder upon oaks in the Old World. Our fall webworm, *Hyphantria cunea*, is a close relative and has habits somewhat similar; however, its caterpillars do not live in a nest and they do not march in a procession.

The habits of the procession moth are remarkable. The caterpillars build silken nests in which they rest during the daytime. At night they come forth and wander in search for food. One caterpillar takes the lead and is closely followed by the others, the head of the one behind usually touching the tail of the one in front. They start out in single file but soon form double and triple lines; eventually, the procession may be seven or eight lines in width and include hundreds of individuals. The caterpillars can neither see nor smell and are guided by two strong

FIG. 185.—Larvae resting between moults; (a) dogwood sawfly, *Macremphytus varianus;* (b) io moth, *Automeris io.*

FIG. 186.—The gregarious habit exemplified by the alder-blight or maple-leaf aphid, *Prociphilus tessellata.*    (a) On maple; (b) on alder.    The alternate host compells migration in the fall.

impulses: to move and to follow a leader.   It is supposed that each one spins a silken thread as it moves forward and the succeeding caterpillars follow these threads. (The habit of spinning silken threads is common among other caterpillars and may serve the same purpose.) J. H. Fabre, the famous French entomologist, studied these caterpillars in his laboratory and discovered that the leader could be guided to follow the tail end of the column, in which case the caterpillars marched in a circle. He watched them march in this formation for seven days. By accident one of them got out of line and all the others followed this leader. Generally, the caterpillars continue to travel in a circle until they collapse from fatigue. Although these insects are said to be dumb, they have senses that warn them of approaching storms. Like the apple tree tent caterpillars, they return to their nests when unfavorable weather threatens.

The gregarious habit of the foraging or legionary ants is unique. Species chiefly of the genera *Anomma*, *Eciton*, and *Dorylus* occur in the warmer parts of Africa, Asia, or America. Most of the species have no permanent nests but wander about searching for food. They are encumbered in a number of ways. The workers are blind and depend upon their keen sense of touch and upon their olfactory organs to follow the trail and to work together. They must drag their queen and carry their eggs, larvae, and pupae with them as they go, resting only for brief periods to feed or to make temporary nests. Thousands of these ants march in dense columns, sometimes two or three yards wide. They move rapidly over the ground and remind one of a stream of molten material. If they encounter an obstacle, they surmount it or detour it but they continue to march on. It is almost impossible to brush the ants aside and separate the column. If the column is separated, the ants pile up at the front end of the new column but soon they succeed in bridging the gap and join the original column. As the column advances, the ants drive out all animal life. They work systematically, examining every hole and crevice until they have secured all the desirable food. Their presence is often detected by the commotion produced when the roaches, grasshoppers, and other insects attempt to avoid this mighty army. The ants attack insects and other small animals, feed upon the more desirable parts, and cast aside the legs and wings. A retinue of alert parasites and predators accompanies this great army. Ant birds follow the column and pick up the insect remains.

Thomas Belt and other naturalists have described the visits by these ants to human dwellings. They swarm over the floors and the walls, driving out roaches, ants, spiders, rats, and mice. In a sense, they are welcome because they thoroughly cleanse the house of vermin. Their stay in the house makes life unpleasant for the rightful owners and one is fortunate if he can remain within the narrow limits of a bed or chair. Man usually retreats and waits until the ants retire.

These legionary ants are exceedingly ferocious; some are known to attack animals including lizards and pythons. They are fatal to larger animals especially those in captivity or animals that are sluggish or weakened.

Armyworms, chinch bugs, and grasshoppers, as mentioned previously, often move to new grazing fields. The migratory locust received its name because it makes extended journeys each year.

The migrations may be of daily occurrence; for example, June beetles feed on roots of grasses during the day but at night move to various fruit and nut trees to feed upon their foliage. The Japanese beetles feed on the tops of trees during the daytime but descend to the bases of the trees at night.

**Swarming Aggregations.**—Groups of insects in motion are often referred to as swarms. Thus many insects may swarm. Williston described a swarm of Mycetophilidae that danced in the air in incredible numbers and produced a noise like a distant waterfall. Sharp relates that he has seen great swarms of mosquitoes in New Zealand measuring three-quarters of a mile long, 20 feet high, and 18 inches thick. He also describes great swarms of *Chlorops* of many millions of individuals that came into houses in England and France.

The true swarming habit, however, is associated with mating. It relieves the congestion in the nests of social insects and affords a means of distribution. We commonly think of the honeybee, ant, and termite.

**Sleeping Aggregations.**—It is well known that insects cease their labors to rest and often to sleep. Schwarz found about 60 bees, *Priononyx atratum*, asleep on a single shrub with others near by in smaller numbers. Bradley collected 490 sleeping wasps from a small area. The writer has seen small groups of *Polistes* congregate night after night in the same situation close to their paper nest. The mourning-cloak and the heliconias have frequently been observed in sleeping attitudes. They quickly refrain from work when the sun goes under and resume activity as soon as it becomes bright again. The milkweed butterflies congregate to sleep during their annual southern migrations.

**Dissociation.**—Closely associated with the habit of aggregation is the habit of dissociation. Certain species are impelled by various forces to scatter. The females of wolf spiders and scorpions carry their young on their backs; as soon as the young begin to feed, they leave their parents and scatter. Predacious insects can seldom afford to remain in groups for they compete keenly with one another for food. The predacious Pentatomidae, Reduviidae, and Phymatidae hatch from comparatively large groups of eggs; soon after hatching, the nymphs scatter and search for food, which is often scarce. On the other hand, the phytophagous Heteroptera, such as the harlequin bug, *Murgantia histrionica*, or the squash bug, *Anasa tristis*, commonly feed in groups.

Dissociation may also be caused by external forces such as heavy rains or attacks by parasites. Caterpillars have frequently been observed crawling back upon trees from which they have been washed by storms. As the time for transformation approaches, some species tend to congregate and others to scatter. The larvae of the eastern tent caterpillar, which feed gregariously nearly all summer, disperse widely when seeking places to transform.

Fig. 187.—Nest of the carton ant, *Azteca trigona*, Panama.

**Social Aggregations.**—Insects of a given species that live together in organized groups or colonies, in which there is division of labor and in which the workers provide food for the young, are known as social insects. They possess certain characteristics in common: a comparatively large population, cooperation, division of labor, generally progressive provisioning, parental care, trophallaxis, swarming in some cases, and the construction of more or less elaborate nests. The social habit occurs conspicuously among insects with complete metamorphosis: ants, honeybees, bumblebees, and the paper wasps, *Vespa* and *Polistes*. However, the social habit occurs also among a few insects with gradual metamorphosis, namely the Isoptera and Dermaptera. The Embiidina and a few of the Scarabaeidae care for their young and are said to be incipiently social or subsocial.

Social insects naturally fall into two groups. Species that have permanent or perennial colonies, and in which swarming occurs, include the ants, termites, honeybees, and stingless bees. Species which have annual colonies and in which swarming does not occur, at least in the temperate climate,* include the bumblebees and the paper wasps, *Vespa* and *Polistes*. It is probable that the honeybee and the bumblebee were originally tropical species and that both were perennial and swarmed to produce new colonies. Bumblebees have adjusted themselves to the temperate climate by adopting the annual method of starting new colonies in the spring. The honeybees have adapted themselves to the temperate

* All representatives of these groups are perennial and swarm in the Tropics.

climate by the habit of clustering in the winter, a method by which the temperature of the hive is sufficiently raised to protect the queen.

Social insects frequently build up large populations. The size varies from about 50 individuals in *Polistes* to several millions in some of the tropical species of termites. A populous colony of bumblebees may consist of 300 or 400 individuals. The nest of the bald-faced hornet, *Vespula maculata,* may contain 3,000 individuals. A strong colony of honeybees, *Apis mellifera,* has from 35,000 to 50,000 individuals.

Fig. 188.—Nest of *Polistes annularis* and its builder.    A single comb is constructed without an envelope.

The nests of certain African and Australian termites extend eight or ten feet above the ground and many feet below the ground. The population of such a nest may exceed several millions. A Brazilian ant, *Atta sexdens,* may have a population of 600,000. The mound-building ants, when located in a desirable area, often build 1,600 nests in a community. The total population may reach hundreds of millions of inhabitants.

In organizations of this sort there is need for cooperation and division of labor. These two go hand in hand. Cooperation follows as a result of the division of labor. There is no effort on the part of the insects to cooperate. The queen does not rule the colony, as the term is generally used. Bees solve their problems without legislative, judicial, or

Fig. 189.—Nest of the bald-faced hornet, *Vespula maculata*, one of our large social wasps.

Fig. 190.—An unusual location for the comb of the honey bee, *Apis mellifera*. (*Pennsylvania Department of Agriculture*.)

executive bodies.   Virgil recognized some of the characteristics of bees
and aptly expressed them in Book IV of the "Georgics."

> The bees have common cities of their own,
> And common sons: beneath one law they live,
> And with one common stock their traffic drive.
> Each has a certain home, a several stall.
> All is the state's, the state provides all.

*   *   *   *   *

> Their toll is common.   Common is their sleep:
> They shake their wings when morn begins to peep;
> Rush through the city gates without delay:
> Nor ends the work, but with declining day.
> They work for the common good—die for their queens and homes.
> They are efficient and work apparently without fatigue.
> They exhibit anger, pain, panic and playfulness.

**Castes.**—One of the characteristics of social insects is the caste
system upon which division of labor rests.   In general, there are three
castes: the females or queens, the workers, and the males (drones in
bees).   As a rule only one mature queen exists in a colony at a time,
although new queens may be developed for emergencies and to lead
forth new colonies at the time of swarming.   Among the termites,
complimental queens often exist.   The queen's duty is to lay eggs and
build up the population of the colony.   The queens of bumblebees
and the paper wasps also start the new colonies in the spring.   The
workers carry on the duties of the colony, forage for food, store food, and
feed and care for the young as well as for the queen.   The function
of the male is to fertilize the queen.   After the swarming season the
males (except in the termites) are driven from the colony and killed.
Soldiers are found in the colonies of termites and ants, whose duty is
generally to drive off enemies.

The caste system is most highly developed in the termites.   Each
caste includes both male and female individuals.   In most species there
are four castes: (1) the reproductive caste, consisting of males and
females, in which wings become fully developed, are used for swarming,
and are then shed, (2) the reproductive caste, consisting of males and
females, in which the wing buds remain short, (3) the worker caste, and
(4) the soldier caste.   In certain species of termites there are individuals
that have an elongated projection on their heads.   These are known as
*nasuti* and are a special soldier caste.   The association of the male and
female is permanent.   In the Tropics, the king and queen live together
in the royal chamber and there is repeated coition.   Both sexes are
represented in the worker caste and occasionally workers are capable

of laying eggs.   In certain species of termites, the worker caste is absent and the duties are taken over by the third reproductive caste.

The paper wasps, Vespinae and Polistinae, have three castes: females, workers, and males.   In the parasitic species, which of course are not social, there are no workers.   Swarming does not occur among the species of the temperate climate, and the new colonies are founded by overwintering females.

Ants, like the bees and wasps, have three castes: females or queens, males, and workers.   The males and females are usually winged and the workers are wingless.   In a few species the female is wingless.   The

Fig. 191.—Castes of termites: (1) queen; (2) worker; (3) soldier.   (*After Baerg.*)

workers are modified females.   In the parasitic species there are no workers, as might be expected.   In many of the species, certain of the workers have large heads and strong mandibles.   These are soldiers. In the honey· ants, the repletes become reservoirs for the storage of honey.   Polymorphism is common among the ants.   All three castes may have large forms or dwarf forms.   The workers have been called, workers major and workers minor.   Males and females may be *ergatoid;* that is, they may resemble the workers in having no wings.

Honeybees, bumblebees, and stingless bees have the usual three castes: female, worker, and male.   The queen bumblebee differs from the queen honeybee in that she has pollen baskets on her hind legs.   She must collect pollen when she is founding a colony in spring.   In the parasitic bumblebees, *Psithyrus*, the females do not possess organs for collecting and carrying pollen and there is no worker caste.   The workers among honeybees are imperfect females.

**Nests of Social Insects.**—There is a great deal of difference in the structure and location of the nests of social insects.   Most of them, especially the bees and the wasps, build small hexagonal cells of wax, paper, or other material in which the brood is reared and food is stored. The ants and the termites, on the contrary, do not construct cells but fashion chambers for their broods.

Fig. 192.—Nest of a bumblebee, opened to show developing brood.

Honeybees and bumblebees construct nests of wax.   The combs of the honeybees are more regular than those of bumblebees.   The honeybees originally nested in hollow trees but they have become highly domesticated.   Wax and propolis are used solely in the construction of their nests.   Bumblebees generally build in the ground and use considerable moss and grass in the construction of their nests.

The stingless bees of the Tropics make nests of clay, using wax and resin for brood and storage cells.   Their nests are usually made in the ground or at the base of old tree stumps.   A few species nest in the open.

Yellow jackets, wasps, and hornets make their nests of paper.   The bald-faced hornet builds its nest in the open, often upon the limb of a tree.   Several combs are constructed one below the other and are covered by an envelope of tough, gray paper.   The large European hornet builds its nest in hollow trees or within buildings.   The yellow jacket usually builds its nest in the ground.   The paper of its nest is poor in texture, generally made from rotten wood, and is often brown in color.   *Polistes* constructs a single horizontal comb of gray paper which is suspended by a pedicle and is not protected by an envelope.

Termites build their nests in the ground or in wood. Small colonies are often found under stones. Certain tropical species build their nests above ground. They construct no permanent cells for their brood. In the tropical species, the king and queen live in a royal cell but in the North American species no permanent royal cell is formed.

The ants show the greatest variety in the construction of their nests. They do not construct permanent brood cells but place their young in chambers. The eggs, larvae, and pupae are kept in separate places. They frequently move their brood according to changes in temperature or moisture. Many species, such as the cornfield ant, *Lasius niger americanus*, build their nests in the ground. The mound-building ant, *Formica exsectoides*, constructs the largest of our ant hills. Many colonies often inhabit the same area. Individual nests may measure one meter in height and two meters across.

Fig. 193.—The small paper nest of *Protopolybia sedula* on the underside of a palm leaf, Panama.

The slave-making ants, *Formica sanguinea* and *Polyergus lucidus*, also build nests in the ground, but the queen is unable to raise a colony. These species depend upon their raiding habits to bring in the larvae and pupae of other species, which are subjugated to raise the brood.

The black carpenter ant, *Camponotus herculeanus pennsylvanicus*, builds its nest in dead trees, excavating complicated series of chambers. This large black species is often the source of abundant borings and frass pushed from the interior of the tree.

The most extraordinary nests are those made by certain ants of the genera *Oecophylla*, *Camponotus*, and *Polyrhachis* that occur in tropical parts of the world. They construct nests of leaves and silk, making a sacklike structure in which the brood is raised. The ants cannot produce silk, nor can many adult insects. However, the larvae spin silk which is normally used in the construction of cocoons. It issues from a small opening at the base of the labium. The ants manipulate these larvae in such a way that a nest can be formed. Some of these ants seize the edge of a leaf with their jaws and pull it toward the edge of a second leaf.   At the same time, others appear, each with an ant larva in its jaws.   The ants work these larvae back and forth over the edges of the leaves.   Material from the silk glands adheres to the leaf when the mouth parts of the larvae are pressed against it.   As the larvae are moved toward the opposite leaf, silken threads are spun which are attached to it.   This is continued until the leaves are properly joined. The silk is fine and delicate and many strands must be spun to form the entire nest.   Vast numbers of larvae are required to complete the job.

The wandering or legionary ants build no permanent nests but utilize any convenient spot they may chance to find during their wanderings.

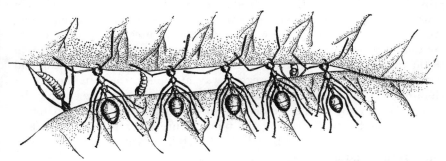

Fig. 194.—A brigade of *Oecophylla smaragdina* workers drawing the edges of leaves together while others bind them with silk spun by the larvae.   (*After Doflein.*)

A single genus *Eciton* is represented in the Southern United States.   They do not form large armies like the species in the Tropics although the colonies of some species may contain thousands of individuals.

**Progressive Provisioning.**—Progressive provisioning is one of the factors that demands a cohort of workers when large colonies are involved. Most of the social insects feed their young from day to day.   It is interesting to note that this is not the habit of the stingless bees, which are very closely related to the honeybees and bumblebees.   Stingless bees, on the contrary, provide sufficient food, at the time the egg is laid, for the complete development of the larva (*mass provisioning*).   Progressive feeding occurs in most of the social insects, and in a few solitary wasps as, for example, *Bembex* and *Ceramius*.   *Odynerus tropicalis*, a Congoese species, feeds its young one or two caterpillars a day.   This is rather unusual as the majority of the potter wasps practice mass provisioning. In progressive provisioning, the larvae require considerable attention for a long period of time, thus, only a few young can be raised by a solitary female.   When workers cooperate in taking over the duties of the nest, larger colonies can be produced.   The type of food is often varied.   The newly hatched termites receive only saliva, later they get stomodaeal food, finally they receive wood, the staple food of termites.

The larvae of the bald-faced hornet is first fed regurgitated food consisting of nectar and fruit juices; later it receives parts of masticated insects. Honeybees and bumblebees are fed upon nectar and pollen.

The food of ant larvae varies with different species. The following comprise a partial list of foods of the common species: nectar, sap, honeydew, fruit juices, fungi, leaves, other plant material, and parts of insects and small animals. Like termites and honeybees, the young of ants are fed from day to day, often receiving different types of food as the larvae grow.

**Parental Care.** Parental care is embodied largely in progressive provisioning. This method of feeding contrasts with mass provisioning where daily attention is not necessary. In the social insects, the young are generally cleaned by the workers and dead insects and other debris are removed by them from the nests. The workers pay special attention to the queens. In the ants and termites, the workers not only clean and feed the queens but carry the eggs away and place them in the proper chamber. The workers of the honeybees cool the hive in the summer by fanning their wings and protect the queen in the winter by clustering about her and actually raising the temperature several degrees by their activities. Ants move the brood about to take advantage of changes in temperature and moisture.

Parental care has been described in connection with insects other than the typical social species. The Dermaptera guard their eggs until they hatch. The horned passalus, *Passalus cornutus*, guards its young and assists them in the construction of the cocoons. The tumble beetles, *Copris*, *Canthon*, and *Phanaeus*, also exhibit parental care.

FIG. 195.—A case of trophallaxis. An ant attending a staphylinid beetle, *Atemeles*, which gives a few drops of agreeable substance in return for the attention. (*After Wasmann.*)

**Trophallaxis.**—The exchange of nourishment between one insect and another is known as *trophallaxis*. It is a common practice for ants to feed one another from mouth to mouth. When the workers return to the nest, they offer food to those that remain at home. As a matter of fact, these workers solicit food from those that forage in the field and in return offer regurgitated food. The exchange of food may also take place between the young and the adult. The young of both termites and ants give forth a substance that is eagerly sought by the workers. The young of the paper wasps, *Vespa* and *Polistes*, have similar habits. The workers are stimulated to feed the brood in response to the material which they receive from the young. The adults of a few solitary wasps feed upon a liquid that the larva emits from its mouth. Trophallaxis, however, does not occur among the social bees.

The exchange of food may take place between individuals of different species. Many staphylinid beetles have this habit. *Atemeles emarginatus*, for example, is tolerated by certain ants and is fed as a member of the colony. In return the ant imbibes of a fluid which is secreted by the beetle from glandular hairs on the sides of the abdomen.

**Swarming.**—Swarming occurs in many social insects as a means of alleviating congestion in the colony and as a means of distribution.

Fig. 196.—A winter cluster of honey bees.  (*E. J. Anderson.*)

It generally takes place among social insects that are perennial in habit and that produce colonies of large proportions. Thus we find swarming in the honeybees, termites, and ants. The bumblebees do not swarm in the temperate climate; however, they do swarm in South America. Tropical species of *Polybia* are also known to swarm. Among honeybees, the old queen leaves the hive accompanied by most of the workers. Among stingless bees, a new queen leads forth the workers to another site to start a new colony. The immediate stimulus that produces swarming, according to Kellogg, is positive phototropism. Other forms of swarming have been noted in connection with associations of insects for feeding, migrating, and mating.

**Clustering.**—Clustering, a remarkable habit that occurs in the true sense only in the honeybee, is the way a tropical species adapts itself to the temperate climate. These insects do not hibernate in the true sense. If the weather becomes too cold during the winter, the workers consume food and become active, thus generating enough heat to protect the queen. As the temperature of the air in the hive falls, the bees become less active until a temperature of 14°C. is reached; then the workers imbibe freely of their winter stores, cluster about the queen, and by vibrating their wings produce enough heat to protect her. On some occasions, temperatures of 30° to 35°C. are produced, even when the temperature outside the colony is 0°C. Since these activities naturally drain the stores of the bees after a long winter, the honey supply may become depleted and the colony much weakened.

HABITS OF SOCIAL INSECTS

| Habit | Termite | Ant | Honey-bee | Bumble-bee | Stingless bee | Paper wasps, *Vespa* and *Polistes* |
|---|---|---|---|---|---|---|
| Position of nest | In wood or ground | In leaves, wood, or ground | In tree, etc. | In ground | Exposed or in ground | Exposed or in ground |
| Material of nest | Wood, soil | Wood, earth, leaves | Wax | Wax, moss, grass | Mud, resin, wax, etc. | Paper |
| Nest started by | ♀ and workers | ♀; or by ♀ and workers | ♀ and workers | ♀ | ♀ and workers | ♀ |
| Nature of brood | Perennial | Perennial | Perennial | Annual* | Perennial | Annual* |
| Food of brood | Regurgitated food, insects, wood | Vegetable, insects, juices, etc. | Pollen, nectar | Pollen, nectar | Pollen, nectar | Regurgitated food,† nectar |
| Swarm | Yes | Some species | Yes | No* | Yes | No* |
| Method of hibernation | Brood | Brood | Brood | ♀ | No | ♀ |
| Type of feeding | Progressive provisioning | Progressive provisioning | Progressive provisioning | Progressive provisioning | Mass provisioning | Progressive provisioning |

* Except in the Tropics.
† *Polistes*, pollen; *Vespa*, parts of insects.

**Protective Devices of Social Insects.**—The sting is the most formidable weapon possessed by most social insects. Nearly every one has had some experience with the stings of bees or wasps. Certain ants also have well-developed stings. The fire ant, *Solenopsis geminata*, which is found only in tropical America, is pugnacious and can sting severely. *Formica rufa* and its allies can spray poison to a distance of 20 or 30 centimeters. The stingless bees of the Tropics are without functional stings but bite viciously. Certain of the wasps, such as *Polistes*, are not easily provoked but are nevertheless powerful stingers.

Jaws are frequently used as a method of defense. The stingless bees come forth in great numbers when they are disturbed and swarm over one's body, settling in the hair, eyes, ears, and nostrils. They bite viciously. The soldier ants and termites use their powerful jaws to ward off enemies. The nasuti of the termites use the secretions of their pointed heads for much the same purpose.

The aggressiveness of most social insects adds to their ferocious nature. The *Polistes* of North America and the *Polybia* of the Tropics are not inclined to attack. The author has carried whole colonies of *Polybia* wasps through the woods for miles without receiving a sting. They swarm forth and congregate on the exterior of the nest but seldom fly at or attack the intruder.

Fig. 197.—Nest of *Polybia*. on the underside of a palm leaf, Panama. The colony is comparatively small and the wasps are docile.

Guards are used efficiently by social insects to protect the nests. Honeybees, bumblebees, termites, ants, and wasps delegate a few workers to stand guard at the entrance of the nest. It is a commonplace observation that a colony of bald-faced hornets can easily be aroused by a few movements. The stingless bees are said to post a large number of guards at the entrance of the nest.

Stingless bees have other unique methods of protection. Many species close the nest at night by sealing the funnellike entrance with a sticky material. Some species have numerous exits through the sides of the nest through which they can make a rapid escape when alarmed.

The nests of social insects are often placed in protected locations, such as the ground or hollow trees used by some species. Species that nest in the open frequently enclose their nests in several layers of paper. Certain species of *Polybia*, common in the Tropics, construct the exterior of the nest of mud or very hard paper.

**Stingless Bees.**—The habits of the stingless bees, *Melipona* and *Trigona*, of the Tropics depart so much from those of other social insects that it seems desirable to include a few notes on these species. There are about 250 species; all are small in size and seldom exceed one-quarter of an inch in length.    As the name implies, they have no functional sting but have the unpleasant habit, when disturbed, of swarming from the nest and climbing over one's body.

The nests vary considerably in size.    The smaller colonies are built in the open and consist of only 200 or 300 individuals.    The larger colonies are built in trees or in the ground and often exceed 80,000 individuals.    The nests are made of gum, resin, plant fuzz, feces, clay, wax, and other materials.    As a matter of fact, they are a veritable refuse

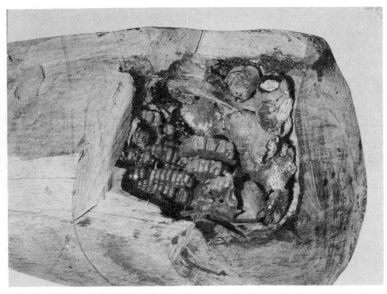

Fig. 198.—A colony of stingless bees.    Log opened to show the brood chamber.    The combs are very irregular, dark in color, and composed of wax, clay, and miscellaneous materials.    (*U. S. Department of Agriculture, Bureau of Entomology and Plant Quarantine.*)

pile.    The nest is divided into two parts, one for the brood and one for the storage of pollen, nectar, and other materials.    The brood combs are placed one above the other.    There is only one layer of cells in each comb and these open upward.    The wax has a very dark appearance and is quite different from the light-colored comb of the honeybee. Unlike other social insects, the colony practices mass provisioning. Sufficient food is placed in the cells for the entire development of each bee.    Even the queens are reared in this manner although they receive a greater supply of food.    The pupae are formed in papery cocoons which are at first enclosed in wax.    As the time approaches for the bees to emerge, the workers remove the wax from their chambers; this aids the bees in making their escape and at the same time conserves the wax for use in the construction of new combs.    (This form of economy occurs in the bumblebees also.)  The exterior of the nest is covered with a mixture of clay, wax, resin, and other materials which make a hard covering.  The opening to the nest is usually made in the form of a spout or funnel.  A few species close these funnels at night to keep out intruders.

**Other Social Insects.**—Many insects that are not truly social live in communities or show social tendencies.  The horned passalus, *Passalus cornutus*, work together for the common good.  The larvae congregate in small numbers in damp rotten wood.  They produce a faint noise by

Fig. 199.—A female earwig brooding over her nest of eggs.  (*After Fulton.*)

rubbing a pair of aborted legs against a series of ridges on the coxae of the middle legs.  This is apparently a means of keeping the colony together.  When the time for pupation arrives, the adults assist the larvae in making their cocoons and, during pupation, stand guard until the new adults emerge.

The ambrosia beetles Ipidae and Platypodidae also exhibit social tendencies.  They guard their young, provide them with fungus food (ambrosia) from day to day, and remove the excrement from their burrows.  Nevertheless, there is no division of labor and no caste system.

The earwigs, Dermaptera, and the tumble beetles, *Copris, Canthon,* and *Phanaeus,* are known to guard their eggs until they hatch and thus exhibit parental care.

Many of the so-called solitary insects form loose colonies, but the parents seldom practice progressive provisioning.  The burrowing bees, Andrenidae, cooperate to form a community.  They make nests in the ground and frequently 100 or more individuals may live in the same colony. The community consists of a series of tunnels which have a common entrance at the surface of the ground. Each tunnel is the home of a single bee. Cells are constructed along the sides of the tunnel and are provided with a mixture of pollen and nectar. In each an egg is laid. (Although mass provisioning is the rule in the Andrenidae, *Allodape* feeds a group of young from day to day.) The individual females take care of their own nests. The main entrance to the community is guarded by a sentinel that rests with its head plugging the entrance. The sentinel allows the rightful owner to pass but drives intruders away. In the genus *Halictus*, the female shows further social habits because she survives the development of her progeny.

**Extrapopulation of the Nests of Social Insects.**—This introduces the subject of the interrelation of different species of insects, which is known as *symbiosis*, abundantly illustrated among the social insects. Of the many definitions of symbiosis, we select one with a broad interpretation, namely, the living together of two or more species in more or less intimate association. About 1,200 species of small crustacea and arthropods inhabit the nests of social insects. Many factors attract these species: they find refuge from predacious and parasitic enemies, the temperature of the colony is frequently higher than that outside, and there is usually a surplus of food. The species found in ants' nests are known as *myrmecophiles;* those found in termite nests are known as *termitophiles.* Upon the same basis, we might suggest the term *bombophiles* for the inquilines of the nests of bumblebees, and *vespophiles**\** for the inquilines of the nests of the vespids. The extra inhabitants of the nests of social insects are known as captives, guests, intruders, scavengers, parasites, and predators.

*Captives.*—Insects that are enslaved by social insects are known as captives. The ants have many of them. The shiny Amazon ant, *Polyergus lucidus*, and the red slave-making ant, *Formica sanguinea*, are common species of the temperate climate that make raids on the other species of ants and bring back larvae and pupae which they raise to take over the duties of the nest. The queens of the raiding ants are not able to raise broods themselves.

Many aphids and coccids are attended by ants. It is common to see ants stroking the cornicles of aphids to induce the flow of honey-dew from their anal openings which they so eagerly seek. In some instances, the ants build shelters of mud which protect the enslaved insects. Some ants actually transport the aphids in the fall from the roots of corn to deeper locations where they can survive the winter. In the spring, they return the aphids to the roots of the corn plant.

\* An unavoidable combination of Latin stem and Greek suffix.

*Guests.*—Guests of social insects may be commensals, *i.e.*, table companions or messmates that live in intimate relation with the social insects without essentially benefiting or injuring them.   On the other hand, guests may take the form of mutualism in which both individuals profit by the association.   Many insects are welcomed or at least amicably treated by the inhabitants of the social colony.   Some feed upon

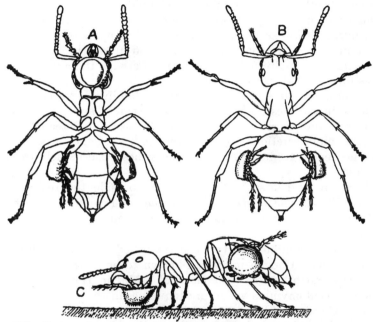

Fig. 200.—*Lasius mixtus*, worker ant carrying three unwelcomed guests, *Antennophores pubescens*, in the normal position; (*A*) ventral; (*B*) dorsal; and (*C*) lateral views.   (*After Janet.*)

debris or waste material in the nest; others feed upon the surplus of food found there.

The larvae of certain *Volucella* feed upon excrement and dead bees found in the nests of bumblebees and wasps.   Their presence in no way interferes with the activities of the colony.   Experiments have been performed to show that they do not touch the larvae of active bees but quickly pounce upon a bee that has been injured or disabled.

*Hetaerias brunneipennis* and other Histeridae enter the nests of ants as scavengers.   Two Lepidoptera, *Vitula edmandsii* and *Aphomia sociella* are found in the nests of bumblebees and are apparently scavengers.   A number of species of Syrphidae live in the nests of social insects, feed upon debris found there, and are unmolested by the individuals of the colony.

The guest ant, *Leptothorax emersoni*, lives only in association with the species *Myrmica brevinodis*.   The *Myrmica* builds its nest in boggy soil or in clumps of moss. The two species occupy the same nest but the broods are brought up separately. The guest ant, *Leptothorax*, does not leave the nest to forage but licks the surfaces of the bodies of the *Myrmica* to obtain the secretion that covers them. By caressing the *Myrmica*, they cause them to regurgitate food upon which they also feed. What benefit do the *Myrmica* obtain from this association?

Many beetles of the families Pselaphidae and Staphylinidae are welcome visitors in ants' nests. Their bodies bear tufts of hairs upon which materials are secreted that are eagerly sought by the ants. The beetles in return accept regurgitated food from the ants. *Atemeles emerginatus* is actually fed by the ants. Perhaps the ants mistake this species for one of their own kind.

Numerous crickets belonging to the subfamily Myrmecophilinae live as guests in the nests of ants. They resemble ants quite closely in appearance and thereby gain entrance to nests, undetected, and feed upon the oily secretions covering the bodies of the ants. The ants derive no benefit from their guests unless these visitors help to keep the nest clean.

There are of course numerous guests outside the realm of social insects. The larvae of *Drosophila sigmoides* live in the froth of the spittle insect and obtain shelter as well as food but contribute nothing to their hosts.

Numerous inquilines have been described in the abodes of leaf-mining insects. Certain mites are known to enter the mines and feed upon frass and other decayed material. Martin Hering describes the larvae of Cecidomyiidae which are inquilines in the mines of *Napomyza xylostei*. Cecidomyiid larvae occur as guests in galls. Many inquilines have been found in the pine-cone willow gall. Some Cynipidae lay their eggs in the galls made by other species. They feed upon the galls produced by their host but do not interfere with the rightful owner. Some of the Gelichiidae have similar habits.

There are also many cases of commensalism where the relationship exists between insects and plants. These will be discussed in Chap. XV.

*Intruders, Thieves.*—Many insects prey upon the supplies of social insects or attack their brood. They are not tolerated by the individuals of the colony and are frequently driven from one part of the nest to another or driven from the hive. A thysanuran, *Atelura formicaria*, steals food from the mouths of ants when two ants are in the act of exchanging food. A small ant known as the thief, *Solenopsis molesta*, makes its burrow in the walls of the nest of a larger species of ant. It emerges and preys upon the larvae and pupae of the larger species. A fly, *Metopina pachycondylae*, steals food from a species of ponerine ant that occurs in Texas. The larva of this species clings to the neck of the young ant by means of its suckerlike posterior end. It encircles the host like a collar and in this position takes food from the mouth of the ant larva. Many species of mites also cling to ants and take regurgitated food from them. These mites often adjust themselves in a symmetrical position on the body of the ant and thus relieve the load upon the ant.

*Predacious and Parasitic Enemies of Social Insects.*—There are innumerable species that prey upon social insects. Like guests and intruders, they find many conditions favorable for existence in the nests of these insects. Stylopids attack many bees and wasps. The bee moth attacks the wax of social bees and makes silken galleries in the comb. The social groups themselves contribute many parasitic species. There are parasitic Bombidae, Formicidae, Vespidae, and *Polistes.* Wheeler has summarized the parasites that attack ants in his classical work.

**Phoresy.**—*Phoresy* is another form of symbiosis in which one insect is carried on the body of another, usually a larger insect, upon which however it does not feed. The borborid fly, *Limosina sacra,* travels habitually upon the backs of dung beetles. The triungulin larvae of certain Meloidae are carried on the bodies of social Hymenoptera to their nests where they eventually consume the brood. Certain parasites of the family Scelionidae take their position on the backs of female grasshoppers and are carried about until the grasshoppers lay their eggs, when the parasites oviposit in the eggs of their hosts. The larvae of the Japanese chalcid, *Schizaspidia tenuicornis,* attach themselves to the legs of ants and are carried to the nest, where they leave the ant and attack the ant larvae and pupae. The most interesting example is the human botfly. The female attaches her eggs to the body and legs of mosquitoes. When the mosquito visits man to feed, the eggs hatch and the maggots attack their host.

### BIBLIOGRAPHY

I. Insect Aggregations, Nonsocial Groups.

ALLEE, W. C. (1937). Animal aggregations. *Quart. Rev. Biol.* **2** (3).

BARBER, G. W. (1938). The concentration of *Heliothis obsoleta* moths at food. (Lepidoptera; Noctuidea.) *Ent. News* **49** (9).

BRADLEY, J. C. (1908). A case of gregarious sleeping habits among aculeate Hymenoptera. *Ann. Amer. Ent. Soc.* **1** (2).

BRUES, C. T. (1926). Remarkable abundance of a cistelid beetle with observations on other aggregations of insects. *Amer. Nat.* **60** (761). *Literature.*

FRAENKEL, G. S., and D. L. DUNCAN (1940). The orientation of animals. Oxford University Press, New York.

GURNEY, W. B. (1938). Grasshopper swarms. *Agric. Gaz. N.S. Wales* **49** (8).

HAWKES, O. A. (1936). On the massing of the ladybird, *Hippodamia convergens* (Coleoptera) in the Yosemite Valley. *Proc. Zool. Soc. Lond.* **2**.

HESSE, RICHARD, et al. (1937). Ecological animal geography. John Wiley & Sons, Inc., New York.

JONES, F. M. (1930). The sleeping Heliconias of Florida. *Nat. Hist.* **30**. *Bibliography.*

KELLOGG, V. L. (1904). Gregarious hibernation. *Proc. Ent. Soc. Lond.* **23, 24.**

KNAB, FREDERICK (1908). Swarming of a reduviid. *Proc. Ent. Soc. Wash.* **10** (1–2).

LINSLEY, E. G. (1919). Observations on swarming of Melanophila. *Pan-Pac. Ent.* **9.**

LOESSER, J. A. (1940). Animal behavior, impulses, intelligence and instinct.

MCNEILL, F. A. (1937). Notes on the gregarious resting habits of Danaine butterfly, *Danaus melisa humata* W. S. *Proc. R. Ent. Soc. Lond.* **12.**

MARLATT, C. L. (1890). Swarming of *Lycanena comyntas* Godt. *Proc. Ent. Soc. Wash.* **1** (4).

MORLAND, D. M. T. (1930). On the causes of swarming in the honeybee (*Apis mellifera* L.). An examination of the brood food theory. *Ann. Appl. Biol.* **17** (1). *Bibliography.*

MUKEYI, D. (1935). Gregarious Collembola. *Curr. Sci. Bangalore* **1.**

O'BYRNE, H. J. (1937). Gregarious caterpillars. *Proc. Mo. Acad. Sci.* **3** (4).

PARK, O. (1930). Studies in the ecology of forest Coleoptera with observations on certain phases of hibernation and aggregations. *Ann. Amer. Ent. Soc.* **23** (1).

PEARSE, A. S. (1939). Animal ecology. 2d ed. McGraw-Hill Book Company, Inc., New York.

RAU, PHIL (1939). Notes on the behavior of certain social caterpillars (Lepid.: Notodontidae, Arctiidae.) *Ent. News* **50** (3).

SHACKLEFORD, M. W. (1929). Animal communities of an Illinois prairie. *Ecology* **10** (1). *Literature cited.*

SHANNON, H. J. (1917). Monarchs resting in Autumn migrations of butterflies. *Amer. Mus. Journ.* **17.**

SHELFORD, V. E. (1937). Animal communities in temperate America. University of Chicago Press, Chicago. *Geo. Soc. Chicago Bull.* **5.**

SHERMAN, F. (1938). Massing of convergent ladybeetle at summits of mountains in Southeastern United States. *Journ. Econ. Ent.* **31** (2).

SMYTH, E. G. (1934). The gregarious habit of beetles. *Journ. Kan. Ent. Soc.* **7** (3–4).

THRONE, A. L. (1935). An unusual occurrence of the convergent lady beetle. *Ecology* **16** (1).

VAN DYKE, E. C. (1919). A few observations on the tendency of insects to collect on ridges and mountain snowfields. *Ent. News* **30** (9).

WICKHAM, H. F. (1915). Swarm of monarch butterflies. *Ent. News* **26.**

WORTH, C. B. (1937). Colonial habits of certain caterpillars (Lepidoptera: Notodontidae). *Ent. News* **48** (8).

See also sleep and hibernation, Chap. XXIII.

## II. Social Insects.

ALLEE, W. C. (1938). The social life among animals. W. W. Norton & Co., Inc., New York.

BEESON, C. F. C. (1938). Carpenter bees. *Indian For.* **64** (12).

BODENHEIMER, F. S. (1937). Population problems of social insects. *Biol. Rev.* **12.**

BROMLEY, S. W. (1931). Hornet habits. *Journ. N. Y. Ent. Soc.* **39** (2).

CREIGHTON, W. S. (1950). The ants of North America. *Bull. Mus. Comp. Zool.* Cambridge, Mass.

DAVIS, W. T. (1914). The fungus-growing ants on Long Island, N. Y. *Journ. N. Y. Ent. Soc.* **22** (1).

DUNCAN, C. D. (1939). A contribution to the ecology of North American vespine wasps. *Stanford Univ. Ser. Biol. Sci.* **8** (1). Stanford University Press, Stanford University, Calif.

——— (1940). The biology of North American Vespine wasps. Stanford University Press, Stanford University, Calif.

EMERSON, E. A. (1939). Populations in social insects. *Ecol. Monogr.* **9** (3). *Literature cited.*

ENTEMAN, W. M. (1902). Some observations on the behavior of social wasps. *Pop. Sci. Mo.* **61.**

FABRE, J. H. C. (1912). Social life in the insect world. D. Appleton-Century Company, Inc., New York.

FOREL, A. (1929). The social world of the ants. Boni & Liveright, New York.

FRISCH, K. VON (1950). Bees, their vision, chemical senses and language. Cornell Univ. Press.

———— (1953). The dancing bees. Methuen & Company, Ltd., London.

FROST, S. W. (1936). The paper makers. *In,* Ancient artizans. Van Press, Boston. *Bibliography.*

IMMS, A. D. (1931). Social behavior in insects. Methuen & Co., Ltd. London. Lincoln Mac Veagh, Dial Press, Inc., New York.

KOFOID, C. A., *et al.* (1934). Termites and termite control. University of California Press, Berkeley.

LUBBOCK, SIR JOHN (1894). Ants, bees and wasps. D. Appleton-Century Company, Inc., New York.

MICHENER, C. D. and M.H. (1951). American social insects. D. Van Nostrand Co., Inc., New York.

PECKHAM, G. W., and E. G. (1905). Wasps, social and solitary. Houghton Mifflin Company, Boston.

PLATH, O. E. (1934). Bumblebees and their ways. Macmillan Company, New York.

RAU, PHIL (1933). The jungle bees and wasps of Barro Colorado Island. Phil Rau, Kirkwood, Mo.

————, and RAU, N. (1918). Wasp studies afield. Princeton University Press, Princeton, N. J.

RICHARDS, O. W. (1953). The social insects. Philosophical Library Inc., New York.

SALT, G. T. (1929). A contribution to the ethology of the Meliponinae. *Trans. Ent. Soc. Lond.* **77** (2).

SMITH, M. R. (1928). Observations and remarks on the slave-making raids of three species of ants found at Urbana, Ill. *Journ. N. Y. Ent. Soc.* **36** (4). *Bibliography.*

SNYDER, T. E. (1924). Adaptations to social life; the termites. *Smiths. Misc. Coll.* **76** (12).

———— (1926). The biology of the termite castes. *Quart. Rev. Biol.* **1** (4). *Literature.*

———— (1948). Our enemy the termite. Comstock Publishing Company, Ithaca, New York.

THOMPSON, C. B. (1917). Origin of castes of the common termite, *Leucotermes flavipes. Journ. Morph.* **30** (1).

TURNER, C. H. (1910–1917). Literature on the behavior of spiders and insects other than ants. *Journ. Animal Behavior* **1–6.**

WHEELER, W. M. (1910). Ants, their structure, development and behavior. Columbia University Press, New York.

———— (1926). The natural history of ants from an unpublished manuscript in the archives of the Academy Sciences Paris by Réaumur. Alfred A. Knopf, New York.

———— (1923). Social life among the insects. Harcourt, Brace and Company, New York.

———— (1928). Foibles of insects and men; the termitodoxa, or biology of society. Alfred A. Knopf, New York.

### III. Commensalism and Symbiosis.

ABBOTTS, J. F. (1912). An unusual symbiotic relation between a water bug and crayfish. *Amer. Nat.* **46** (549).

BANKS, N. (1911). Cases of phoresie. *Ent. News* **22** (5).

BEQUAERT, J. W. (1922). Ants in their diverse relations to the plant world. *Amer. Mus. Nat. Hist. Bull.* **45.** *Bibliography.*

CLEVELAND, L. R. (1926). Symbiosis among animals with special reference to termites and their intestinal flagellates. *Quart. Rev. Biol.* **1** (1). *Literature.*

COUCH, J. N. (1938). The genus Septobasidium. University of North Carolina Press, Chapel Hill.

FROST, S. W. (1936). Ants attending various insects. *In,* Ancient artizans. Van Press, Boston, *Bibliography.*

HEINDEL, R. L. (1905). Ecology of the willow cone gall. *Amer. Nat.* **39** (468). *Bibliography.*

JONES, C. R. (1929). Ants and their relation to aphids. *Colo. Agric. Coll. Bull.* **341.**

LEACH, JULIAN GILBERT (1940). Symbiosis. *In,* Insect transmission of plant diseases. McGraw-Hill Book Company, Inc., New York.

MANN, W. M. (1910–1917). Literature on the behavior of ants and myrmecophiles. *Journ. Animal Behavior* **1–7.** *References.*

PEARSE, A. S., *et al.* (1936). Parasites and commensals. *In,* Ecology of *Passalus cornutus* Fab. a beetle which lives in rotten logs. *Ecol. Mongr.* **6** (4). *Bibliography.*

WALSH, B. D. (1864). On the insects, coleopterous, hymenopterous, dipterous, inhabiting the galls of certain species of willows. *Proc. Ent. Soc. Phil.* **3.**

WHEELER, WM. M. (1908). Studies on Myrmechophiles. *Journ. N. Y. Ent. Soc.* **16** (2, 3, 4). *Literature.*

——— (1913). Ants, their structure, development and behavior. Columbia University Press, New York.

# CHAPTER XIII

## SOLITARY INSECTS*

Solitary insects contrast with social insects such as honeybees, termites, ants, and paper wasps. All insects that are not social are, in a sense, solitary but entomologists designate certain characteristics for solitary insects. They are intermediate in habits between the majority of the free-feeding insects and the social insects. The females construct nests and provide food for the young without a cohort of workers. Social insects on the other hand are characterized by division of labor, parental care, and the caste system. Solitary and social insects provide food, in some special manner, for their young.

Incidentally, all insects provide food for their offspring for they generally lay their eggs within, upon, or near the food on which the larvae feed. Parasitoids usually lay their eggs upon their hosts; leaf-mining insects, with rare exceptions, lay their eggs in or upon the leaves in which the larvae mine; and gall insects lay their eggs in or upon the plants on which the larvae eventually form galls. In only a comparatively few cases are the eggs of insects laid far from the source of their food. Many of the Scarabaeidae lay their eggs in the ground but, when they hatch, the young grubs find food close at hand. Walkingsticks drop their eggs indiscriminately upon the ground. When they hatch,

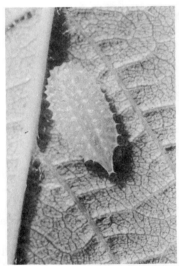

Fig. 201.—Larva of Limacodes (Eucleidae). This larva, like hundreds of others, leads an individual life; the adult provides food by laying its eggs upon the essential food plant.

the nymphs must seek their food plants. Blister beetles lay their eggs in the ground and the larvae are compelled to travel a considerable distance before they locate their hosts.

Although solitary insects are found chiefly among the bees and wasps, there are other solitary species. The tumble beetles of the genus *Copris*

* By some termed subsocial.

259

have been described by many entomologists as social insects because the adults guard the eggs until they hatch. Since these beetles practice mass provisioning and lack many of the characteristics of social insects, they approximate more closely the habits of the solitary insects. At least they represent intermediate forms. It is very evident that the three

CHARACTERISTICS OF THE MAJOR HABITS OF INSECTS

| Nonsocial and non-solitary insects | Solitary insects | Social insects |
|---|---|---|
| No definite provision for food for the young | Mass provisioning usually practiced† | Progressive provisioning usually practiced |
| No definite nest constructed by the adult* | A definite nest constructed | A definite nest constructed |
| No caste system or division of labor | No caste system or division of labor | A caste system and division of labor |
| No parental care | Little or no parental care | Definite parental care |
| Includes the majority of free-feeding insects | Carpenter bees, leaf-cutting bees, potter wasps, etc. | Honeybees, ants, termites, paper wasps, etc. |

* Larvae such as the ugly-nest caterpillar, *Cacoecia cerasivorana*, often construct nests.
† A few species such as *Bembex*, *Lyroda*, and *Ceramius* feed their young from time to time.

groups mentioned in the preceding table intergrade and that there are no distinct dividing lines.

The earth-boring beetles, *Geotrupes*, certainly belong to the solitary insects. The females burrow into the earth or beneath dung, which they gather in quantities. An egg is laid in each ball of dung which serves as food for the young. The female gives the young no further attention.

The curculionid beetles of the subfamily Attelabinae provide food for their young in a remarkable manner. They make compact thimbleshaped rolls from the leaves of the trees upon which they feed. A single egg is laid in each roll. When the egg hatches, the larva feeds on the inner layers of this roll; when mature, it enters the ground to transform.

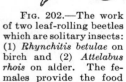

FIG. 202.—The work of two leaf-rolling beetles which are solitary insects: (1) *Rhynchitis betulae* on birch and (2) *Attelabus rhois* on alder. The females provide the food capsules for the larvae.

The Hymenoptera contain the majority of the truly solitary insects. There are in general two groups: the solitary wasps and the solitary bees. A summary of these groups is given in the table on page 261.

**Food of Solitary Hymenoptera.**—The solitary Hymenoptera have two sources of food: vegetable and animal. The solitary bees, Andrenidae and Megachilidae, provide their young with a paste of pollen and nectar. Some of the solitary Diploptera, namely the Masarinae and Zethinae,

SYSTEMATIC GROUPING OF THE SOLITARY HYMENOPTERA

| Family, subfamily, and tribe | Habits | Food of larvae |
|---|---|---|
| *Vespoidea:* | | |
| Pompilidae (spider wasps)............. | Burrowing,* occasionally making mud cells | Spiders, rarely orthoptera |
| Vespidae (solitary vespids): | | |
| Eumeninae........................ | Burrowing, constructing mud cells, mining pith, etc. | Caterpillars |
| Zethinae......................... | Cells of mud, gum, etc. | Pollen |
| Masarinae........................ | Burrowing or constructing cells | Pollen and honey |
| *Sphecoidea:* | | |
| Ampulicidae....................... | Unknown | Roaches |
| Sphecidae: | | |
| Larrinae: | | |
| Astatini........................ | Burrowing | Homoptera |
| Larrini......................... | Burrowing, or mining in brambles | Homoptera or Orthoptera |
| Dinetini........................ | Burrowing | Crickets, spiders |
| Trypoxyloninae.................... | Mining in brambles or mud cells | Spiders |
| Sphecinae (mud daubers, etc.)........ | Constructing mud cells | Spiders |
| Pseninae......................... | Burrowing or mining in pith | Homoptera |
| Bembicinae (cicada killer, etc.)........ | Burrowing | Ants, bees, beetles, Homoptera |
| Crabroninae...................... | Burrowing or mining pith, wood | Diptera |
| Andrenidae: | | |
| Halticus......................... | Burrowing | Pollen and nectar |
| Anthophora...................... | Burrowing | Pollen and nectar |
| Andrena......................... | Burrowing | Pollen and nectar |
| Ceratina (small carpenter bee)........ | Wood borer | Pollen and nectar |
| Xylocopa (large carpenter bee)........ | Wood borer | Pollen and nectar |
| Megachilidae: | | |
| Megachile (leaf-cutting bee).......... | Miner in pith or appropriate site | Pollen and nectar |
| Osmia........................... | Cells of many materials, mud cells | Pollen and nectar |
| Chalicodoma (wall bee).............. | Mud cells | Unknown |

* Burrowing used to indicate species that construct nests in the soil; mining used to indicate species that construct nests in wood, pith, etc.

also feed their young upon pollen.	A species of *Ceramius* is said to feed
its larva on a paste that appears to be dried honey.	The other solitary

FIG. 203.—Contents of a single cell of the mud dauber, *Sceliphron coementarium.*
Eighteen spiders, chiefly Misumena, can be seen as well as the wasp grub which has already
devoured some of the stored food.

FIG. 204.—A portion of the contents of a single cell of *Sceliphron coementarium,* showing
some of the spiders used as provision in the nest, to one of which the wasp egg is attached.

wasps provide the young with various species of insects.	The spider
hunters, Pompilidae, capture a single large spider which furnishes
sufficient food for one wasp larva.	The mud daubers of the genera

*Trypoxylon* and *Sceliphron* use small spiders and frequently select species of the genus *Misumena*. Twenty or more spiders are placed in a single cell.

The Hymenoptera, in providing food for their young, generally show preference for certain species. As stated before, the mud daubers generally select spiders of the genus *Misumena*. Other solitary insects select from groups of insects, such as grasshoppers, caterpillars, or spiders. Some are exceedingly particular and choose but a single species; *Monobia quadridens* takes only cutworms, the European *Larra anthema* selects only mole crickets, *Fertonia luteicollis* uses only certain species of ants, and the cicada killer, *Sphecius speciosus*, takes nothing but cicadas.

Fig. 205.—*Monobia quadridens*, one of the larger species of Eumeninae, which builds its nest in the burrows of other insects.

Solitary insects place food in cells or convenient places where the eggs are laid and the young obtain their nourishment. The female usually provides sufficient food for the entire development of the young, thus mass provisioning is the rule. This method contrasts with progressive provisioning, which generally takes place among the social insects, where the young are fed small portions of food from day to day. A few solitary insects, such as *Bembex*, *Lyroda*, and *Ceramius*, provide food for their young from day to day.

**Nests of Solitary Hymenoptera.—** The majority of the solitary Hymenoptera build nests in the ground. These are termed burrowing species. The terms mining and boring are reserved for species that excavate galleries in the pith of plants. Some dig individual cells in the ground in which a single host is placed. Others dig communal nests or burrows with

Fig. 206.—Mud cells of *Trypoxylon rugifrons* on the trunk of a tree, Panama.

lateral chambers. The stems of plants with pithy centers, such as rose, blackberry, sumac, and elder, are utilized by many solitary insects. The carpenter bee, *Ceratina calcarata*, is unable to make an entrance through the tough stems of sumac and similar plants but readily appropriates the stems of these plants when they are cut or broken. These small carpenter bees utilize any hole or burrow they find. They frequently inhabit the cavities made by wood-boring insects, build nests in cracks between shingles, or construct their cells in crevices in stone walls.

A third group of solitary Hymenoptera construct nests of mud or of vegetable material. These are fastened to the trunks of trees, to rocks, or to other surfaces. The mud daubers build nests of many cells. The potter wasps construct individual cells or small groups of cells (see Fig. 176).

FIG. 207.—The tarantula hawk, *Pepsis*, Panama.

**Spider Wasps, Pompilidae.**—The Pompilidae are comparatively large wasps, generally black or iridescent in color with spiny legs. Some of the largest Hymenoptera belong to this group, as for example the tarantula hawk, *Pepsis*, of the Tropics. The majority of the Pompilidae, however, are of moderate size. They are known as spider wasps because they provision their nests with spiders. Most of the species make their nests in the ground. The wasp first locates a spider and stings it until it is paralyzed. Then it digs a burrow which is enlarged at the lower end to form a cell for the reception of the spider. The spider is placed in the cell and an egg is attached to it. The passageway is closed with earth and the female leaves the egg, attached to the paralyzed spider, to hatch and the young to shift for itself.

It is interesting to note that the solitary habit approaches the parasitic habit and it is often difficult to differentiate between the two. For example, the Scoliidae are parasitic upon the white grubs. They sting their hosts in much the same manner as do the females of the Pompilidae and they attach an egg to the host. The habits of the scoliid wasp differ, however, since the female does not bury its host or construct a definite cell or burrow. The scoliid wasp merely locates its host and fastens an egg upon it.

Not all of the Pompilidae are digger wasps as species of the genus *Pseudagenia* make thimble-shaped cells of mud which are provisioned with spiders. These are located beneath stones, beneath bark, or in crevices in stone walls.

**Eumenidae.**—The wasps of the family Eumenidae have diverse habits. Some burrow in the ground, some appropriate the burrows of other insects, some mine in pith, and others build nests of mud. *Eumenes* *fraterna* is one of the common North American species. They construct earthen jugs upon the branches of trees. (See Fig. 176.) These structures are scarcely half an inch in diameter. They are neatly fashioned with a delicate liplike margin and a small opening which is eventually closed. The female provisions each jug with a number of caterpillars and frequently selects cankerworms. The caterpillars are paralyzed before they are placed in the earthen cells. The egg is suspended from the ceiling of the jug by means of a delicate silken thread so that the caterpillars will not injure the egg before it hatches. The caterpillars supply the young wasp with sufficient food to mature. Transformation takes

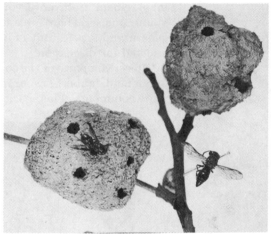

Fig. 208.—Nests of *Stenodynerus anormis* Say, and its builder. (*Champlain and Kirk.*)

place in this earthen cell and the adult breaks through the side of the little jug.

*Odynerus pedestris* builds its nest in the ground, usually selecting a bank so that a gallery can be sunk more or less perpendicular to the surface of the ground. Two or three side chambers are constructed from this gallery. Each chamber is provisioned with caterpillars and an egg is suspended from the ceiling of each cell.

*Odynerus geminus* also burrows in the ground but, instead of discarding the pellets of earth as it burrows, it uses them to build a turret or tower over the mouth of the burrow. These are cemented together in a very regular fashion. After the nest has been completed and provisioned, and the eggs have been laid, this tower is torn down stone by stone and the material is used to close the opening of the burrow. When the wasp has finished, she levels the ground so that it is difficult to tell where the nest is located.

Other species of *Eumenes* build nests in whatever convenient cavity or opening they can find. Comstock remarks, "One year these wasps plastered up many of the keyholes in our house, including those of the bureaus."

**Sphecidae.**—The typical solitary wasps belong to the family Sphecidae. The majority of the species provision their nests with sufficient food to enable the young to develop to maturity; however, *Bembex* and a few others feed their young from day to day. The sphecid wasps have the habit of stinging their prey to paralyze them so that they may remain fresh until consumed by the young wasp. There are a large number of genera and species but only a few will be selected for discussion.

In certain parts of the country the pipe-organ mud dauber, *Trypoxylon albitarsis*, is quite common. This species is shiny black

Fig. 209.—Nest of the pipe-organ mud dauber, *Trypoxylon albitarsis*. The cells are laid end to end in each tube.

Fig. 210.—Nests of the mud dauber, *Sceliphron coementarium*. Three cells lying flat are completed and covered with mud. Two cells standing on end are completed but unprovisioned and unsealed.

with white tarsi from which it derives its specific name. Tubes are constructed of mud which are divided by partitions into several cells. Three or four tubes may be placed side by side. The cells are provisioned with paralyzed spiders and an egg is attached to one of the spiders in each cell before it is sealed. Pupation takes place within the earthen cells. When the adult is ready to emerge, it makes a hole through the exposed side of the earthen tube.

The nests of another mud dauber, *Sceliphron coementarium,* are commonly seen upon the rafters of barns or attics. The species is brown in color with yellow spots and legs. It constructs a nest that

Fig. 211.—The mud dauber, *Sceliphron coementarium.*

looks like a blotch of mud thrown upon the side of the building. The nest consists of several cells, each of which is packed full of small paralyzed spiders. A single egg is attached to one of the spiders in each of the cells before it is sealed. (See Fig. 204.) At first the cells are delicately ridged on the outside, the ridges indicating where individual pellets of mud have been plastered. Later the cells are smeared with a mass of mud. The insects mature within these cells and break through the sides of the mud walls.

One of the largest and best known solitary wasps is the cicada killer, *Sphecius speciosus.* It makes burrows in the ground and provides each burrow with a cicada that has been paralyzed. An egg is laid on the side of the cicada. The egg hatches in two or three days and the grub feeds upon the internal organs of the cicada. It attains its growth in about a week and spins a white silken cocoon in the burrow. When

Fig. 212.—Larva of the mud dauber, *Sceliphron coementarium.*

the adult matures, it gnaws its way out of the cocoon and works its way through the burrow to the surface of the ground.

Species of *Bembex* usually burrow in sandy places. They are stout-bodied wasps, often black with greenish or greenish-yellow bands. They provision their nests with flies. At first, a single fly is placed in a burrow, to which the wasp attaches an egg. After the egg hatches and the young grub commences to feed, the female gathers flies from day to day and feeds them to the young wasp. She closes the nest each time she leaves and seals it permanently when the time arrives for pupation. The care of the young and the habit of feeding progressively are unusual among solitary insects.

**Megachilidae.**—The leaf-cutting bee, *Megachile latimana*, makes its nest in various situations. They have been found in burrows made by

FIG 213.—Leaf-cutting bee, *Megachile latimana*, its nest in wood, and rose leaves from which pieces have been cut to line the nest. (*After Comstock.*)

other insects, in tunnels in the ground, under stones, between the shingles of roofs, in lead pipes, and in the tubular leaves of the pitcher plants. The nests are generally found in the pith of various brambles. Bees of the genus *Megachile* are found in all parts of the world. They look like bumblebees but are much smaller. When a suitable burrow has been found or excavated, the bee cuts several circular pieces of leaves from a near-by plant, usually, rose, ash, or blackberry, which are forced down into the bottom of the burrow to form the base of a capsule. Then the bee cuts several oblong pieces of leaves which are used to line the excavation and make a thimblelike capsule. This is filled with a paste of nectar and pollen in the midst of which an egg is laid. Then several circular pieces of leaf are cut slightly larger than the diameter of the burrow and forced down into the open end of the cavity to make a tight-fitting plug. Several cells are made in a similar manner one above the other until the cavity is filled. The bee then deserts her nest and the eggs are left to hatch. The young grubs feed upon the paste of pollen and nectar, of which there is sufficient for their development.

A British leaf-cutting bee, *Megachile albocincta,* is said to take possession of earthworm burrows in the ground. If the burrow is too long, the bee cuts it off by means of a barricade of foliage. In other respects it is quite similar to our species.

Certain European species of *Osmia* are known to build their capsules of pieces of flower petals. Others build their cells in empty snail shells. One species covers the snail shell with a mound composed of fragments of grass or of pine needles.

The mason bees, *Chalicodoma,* make cells of mud which are plastered on stone walls. Comstock describes the nests of the Egyptian species, *C. muraria,* on the walls of the temple of Dendera. The temple, buried by shifting sands long ago, has been excavated by modern archaeologists but the inscriptions are being buried again beneath a layer of cementlike nests of this bee.

**Andrenidae.**—To this family belong a large proportion of the genera and species of our bees. A few are parasitic but the majority are solitary in habit.

The most obvious species of Andrenidae is the large carpenter bee, *Xylocopa virginica.* During the early spring, one's attention is often attracted by these large bees with shiny black abdomens. They hover about old unpainted buildings, or occasionally come drifting with a low musical hum in an open window. They are frequently mistaken for bumblebees but can be distinguished because the bumblebees are more hairy and often clothed with yellow and red hairs. During the middle of the summer, the carpenter bees frequently visit flowers in search of nectar and pollen. In the spring, however, they are busy excavating burrows for their nests. Soft, solid wood is selected. A hole about a quarter of an inch in diameter is made in a piece of wood which continues for scarcely an inch. At right angles to this gallery, a long burrow is excavated sometimes reaching a foot in length. Across the grain the boring is slow and it takes several days to progress an inch; once the bee turns and bores with the grain, progress is rapid and the bee is soon lost from view. The gallery is divided into several chambers, each marked by constrictions in the walls of the gallery. When the excavating has been completed, the female starts provisioning the cells with a paste of nectar and pollen. The lower cell is filled first, an egg is laid upon this food, and a wall is built across the gallery with small chips of wood which are securely cemented together. The remaining chambers are provisioned in succession until the gallery is completely filled. Sometimes nine or ten chambers are formed each with its supply of nectar and pollen and each containing an egg. The grubs that hatch from these eggs feed upon the food they find at hand. When they have consumed all the food, they change to pupae and remain until the following spring before coming forth as bees.

The small carpenter bee, *Ceratina dupla*, builds its nest in sumac, dead twigs, or the pithy stems of various plants. It is the most common bee encountered in this latitude. The bees are small, about half an inch long, and metallic in color. Excavations are always made in twigs with soft pith which can be readily hollowed out. At the bottom of the

Fig. 214.—The larger carpenter bee, *Xylocopa virginica*, and its burrow in wood. The opening to the burrow is slightly smaller than the burrow itself.

gallery, the female places a paste of pollen and nectar. She then lays an egg and builds a partition of chips of pith which serves as a roof to the cell below and a floor to the cell above. She continues to build cells and provision them until the gallery is nearly full. Then she takes her position in the space above the last cell and waits until her young hatch. The lowest egg hatches first, the bee attains its full growth, and, after tearing down the partition forming the roof of its cell, waits until the one above has matured. This continues until all the bees have matured. Then the mother bee leads the whole family from the nursery. The bees return again and clean out the old gallery and another brood is started by one of the young female bees. The parental care exhibited by this species places it close to the social insects.

Another little carpenter bee, *Ceratina calcarata*, is closely related to the preceding species but has somewhat different habits. It inhabits chiefly sumac and elder, but occasionally works in sassafras and rose. The bee bores into the soft pith of these plants but cannot gain entrance through the tough walls of the stem. It must have a cut or broken end in which to make a start. Phil Rau wanted to study a colony of these bees so he "provided building sites for the home seekers." This he did by cutting off a sumac patch. In about a month, 75 per cent of the cut twigs were occupied by new tenants. The cells are made and provisioned as in the preceding species, starting at the bottom and adding one cell above the other until the gallery is nearly full. The adult bee spends most of the time in the space just above the upper cell. Apparently she does not assume this position because she is brooding over her young, but is there because she is continuously adding new cells and the inhabitant of the bottom cell matures before the cells are all completed. Unlike the preceding species, *Ceratina calcarata*, when mature, does not wait for those above to emerge but tears down the ceiling over its head and pushes its way upward. It continues to cut away the partitions of the cells above. By this time the bees in these cells have matured and they begin to do the same. In turn, all begin to work upward, tearing down the partitions and depositing the bits at the bottom of the burrow. It takes the first bee about eight days to gain its freedom. The others follow in turn.

Fig. 215. —Nest of one of the smaller carpenter bees, *Ceratina dupla*.

The remaining species of Andrenidae are known as mining bees and belong to the genera *Halictus, Anthophora,* and *Andrena.* These genera include many species; all are comparatively small. They frequently build their nests close together forming large villages. *Halictus* constructs a burrow in the ground along the sides of which short galleries are made leading to cells. Each cell is provisioned with a paste of nectar and pollen upon which an egg is laid. When completed, the cells are closed. Several bees use the same corridor as a passageway to their cells. The corridor is constricted at the outer end and guarded by a sentinel whose head nearly fits the opening. The rightful owners are permitted to pass but enemies are driven away.

The nests of *Anthophora* are similar to those of *Halictus* but the exit to the nest is guarded by a cylindrical tube of clay extending outward and downward from the exit.

## BIBLIOGRAPHY

BEESON, C. F. C. (1938). Carpenter bees. *Indian Forester* **64** (12).

DONAHOE, H. C. (1938). Nests of leaf-cutting bees in dried figs. *Proc. Ent. Soc. Wash.* **40** (1).

FROST, S. W. (1944). Notes on the habits of *Monobia quadridens* (Linn.). *Ent. News* **55** (1).

HURD, P. D. (1955). The carpenter bees of California. Univ. Calif. Press.

KROMBEIN, K. V. (1936). Biological notes on some solitary wasps (Hymenoptera: Sphecidae). *Ent. News* **47** (4).

MALYSHEY, S. J. (1936). The nesting habits of solitary bees. *Eos Madrid* **11.**

MOTHERSOLE, H. (1924). Some observations of leaf-cutting bee (*Megachile*) and its parasite (*Collioxys*). *Essex Nat.* **21.**

MURRAY, W. D. (1940). *Podalonia* (Hymenoptera: Sphecidae) of North America and Central America. *Ent. Amer.* **20** (1 and 2). *Bibliography.*

NININGER, H. H. (1916). Studies of the life histories of two carpenter bees in California, with notes on certain parasites. *Journ. Ent. Zool. Claremont, Calif.* **8.**

PACKARD, A. S. (1897). *Megachile, Ceratina* and certain *Xylocopa*. In Notes on transformations of higher Hymenoptera. *Journ. N. Y. Ent. Soc.* **5** (3).

PARKER, J. B. (1915). Notes on the nesting habits of some solitary wasps. *Proc. Ent. Soc. Wash.* **17** (2).

PECKHAM, G. W., and E. G. (1898). On the instincts and habits of the solitary wasps. *Wisc. Geol. & Nat. Hist. Survey* **2** n. s. l.

——, and —— (1905). Wasps, social and solitary. Houghton Mifflin Company, Boston.

RAU, PHIL (1928). The nesting habits of the little carpenter bee, *Ceratina calcarata*. *Ann. Ent. Soc. Amer.* **21.**

—— (1933). The jungle bees and wasps of Barro Colorado Island. Phil Rau, Kirkwood, Mo.

—— (1940). Some mud-daubing wasps of Mexico and their parasites. *Ann. Ent. Soc. Amer.* **33** (3).

——, and RAU, N. (1916). Notes on the behavior of certain solitary bees. *Journ. Animal Behavior* **6** (5).

——, and —— (1916). The biology of the mud-daubing wasps as revealed by the contents of their nests. *Journ. Animal Behavior* **6** (1).

——, and —— (1918). Wasp studies afield. Princeton University Press, Princeton, N. J.

SANDHOUSE, G. A. (1940). A review of the Nearctic wasps of the genus *Trypoxylon* (Hymenoptera: Sphingidae). *Amer. Midl. Nat.* **24** (1).

SCHAFER, G. D. (1949). The ways of a mud dauber. Stanford Univ. Press.

SCHWARZ, H. F. (1934). The solitary bees of Barro Colorado Island, C. Z. *Amer. Mus. Novit.* **722.**

WHEELER, W. M. (1930). Demons of the dust. W. W. Norton & Company, Inc., New York.

# CHAPTER XIV

## SCAVENGERS, PREDATORS, AND PARASITES

The scavenger, predatory, and parasitic habits are somewhat closely related and may represent steps in specialization. As a matter of fact, the origin of the predator and parasite is often traced to the scavenger, for it is believed that insects first attacked dead plants and animals and eventually acquired a taste for living plants and animals. This theory seems to be substantiated by certain groups of insects such as the dipterous family Sarcophagidae,* which has a wide range of food varying from dead to living plants and animals.

**Saprophagous Insects (Scavengers).**—A scavenger is defined as a species that feeds upon waste material. Waste is usually considered as debris resulting from the decomposition of plant or animal matter or the excrement of animals. Saprophagous insects may be divided into *phytosaprophagous* species, those that feed upon dead plant material, *zoosaprophagous* (*sarcosaprophagous*) species, those that feed upon dead animal material, and *scatophagous* (*coprophagous*) species, those that feed upon excrement. There are other types of waste upon which insects feed. The ant-loving crickets, Myrmecophilinae, and

FIG. 216.—A pair of tumble beetles, *Copris*, rolling a ball of dung.

other myrmecophilous insects feed upon an oily substance on the bodies of ants and upon the walls of their burrows. Many ants feed upon honeydew, a sort of waste material produced by aphids. Other insects feed upon the waxy secretions of scale insects. The larvae of *Bradypus cuculliger* feed upon the backs of sloths, possibly eating the debris or algae that grow upon the fur of these animals.

There are no particular modifications for the scavenger habit such as one finds in the highly specialized boring, leaf-mining, and parasitic

* Now considered by some as a part of the family Metopidae.

insects. Some modifications of the alimentary canal are no doubt essential to utilize waste material which is low in nutriment. It is well known that the herbivorous insects have long alimentary canals, and no doubt the scavengers have also. Scavengers are found chiefly among the primitive insects, such as the Thysanura, Collembola, Blattidae, and the lower forms of Coleoptera and Diptera. The scavenger habit is not common in the Lepidoptera. *Aphomia sociella* is said to be a scavenger in the nests of social insects and *Aglossa suprealis* is apparently a scavenger. On the whole, the scavenger habit is uncommon in the Hymenoptera. Ants, of course, are well-known exceptions.

Scavengers are mandibulate and, as a rule, only the larvae or immature forms feed upon waste material. This is particularly true of the Diptera and Siphonaptera. The adults of a few Coleoptera and Mecoptera are saprophagous.

Scavengers can obtain a living in almost any environment. They exist even where terrestrial plants are absent. Darwin in his "Voyage of the Beagle" records a beetle belonging to the genus *Quedius* that feeds on the dung of birds on the barren St. Paul's Rocks, which are 540 miles from the coast of South America—an island upon which not a single plant, not even a lichen, grows. Parasites of animals and scavengers that feed upon the dung of higher animals are found closer to the poles than any other insects. Plant life is almost nonexistent as one approaches the South Pole; however, Byrd records Collembola and a wingless chironomid fly at Antarctica. The Coleoptera, Thysanura, and Collembola are the richest in number of world-wide species largely because they feed upon decayed organic material which is abundant nearly everywhere.

Some scavengers feed upon any kind of waste regardless of its origin. The ants, roaches, and fleas for example are general feeders. The ants gather sap exuding from trees, parts of dead insects and plants, and other plant and animal debris. Roaches are practically omnivorous. They feed upon dead animal matter, cereal products, and food material of all kinds. They eat woolens and leather. They often deface the covers of cloth-bound books, seeking the sizing used in their manufacture, and they eat gum or paste from the bindings of books. The larvae of fleas feed upon animal and vegetable matter found about the sleeping quarters of infested animals.

Certain groups of insects show mixed food habits. Although the Sarcophagidae are largely parasitic, some feed upon decayed animal and plant material; others feed upon excrement. The Scarabaeidae feed upon living plants, decayed plant material, or excrement. The Staphylinidae feed largely upon decayed animal or vegetable matter although a few species are predacious. The majority of the syrphid larvae are predacious; a few feed upon plants. Volucella larvae inhabit the nests of wasps and bumblebees and feed upon excrement and dead insects. The larvae of the Lepidoptera with rare exceptions are phytophagous. Although Silphidae are almost exclusively carrion feeders, a few species feed upon fungi or vegetables and an occasional species is predacious. The Cordyluridae, with the exception of a few plant feeding species and a few parasitic species, feed on excrement.

**Scatophagous Insects.**—The scatophagous habit is clear-cut. It is probable that these insects, as well as many of the sarcosaprophagous and the phytosaprophagous insects, feed upon microorganisms that are exceedingly abundant in the media in which they live.

Only a comparatively few species feed upon the fecula of insects. The young of the sticktight flea, *Echidnophaga gallinacea*, feed upon the fecula of the adults. Some of the minute brown scavenger beetles apparently feed upon the waste products of insects. Large numbers of *Melanophthalma americana* have been found in the abandoned nests of *Cacoecia fervidana* where the discarded fecular pellets of the caterpillars were abundant. The larvae of *Volucella* have been previously mentioned as scavengers in the nests of social insects.

Most of the scatophagous species feed upon the dung of higher animals. Some oviposit in fresh dung, others lay their eggs in material that is older, and still others utilize only well-rotted manure. The Coleoptera and the Diptera are the most frequent visitors to excrement. Among the beetles, certain of the Scarabaeidae, namely species of the genera *Geotrupes*, *Copris*, *Phanaeus*, *Canthon*, and *Aphodius*, are commonly found in or about dung. Some of the Histeridae also occur in dung. The dipterous larvae most commonly found in excrement are Scatopsidae, Cordyluridae, Helomyzidae, Anthomyiidae, Sarcophagidae, Syrphidae, Borboridae, Sapromyzidae, and Muscidae. A considerable proportion of the syrphid larvae are scavengers; some live in liquid media, others in solid material. Short-tailed varieties belong to the genera *Xylota*, *Syritta*, and *Tropidia*. Long-tailed varieties, or rat-tailed maggots, belong to the genera *Eristalis* and *Helophilus*. The Scatopsidae, Thysanura, and Collembola are found commonly in the compost of mushroom beds.

**Zoosaprophagous (Sarcosaprophagous) Insects.**—There are comparatively few insects that feed upon dead insects. The larvae of the water scavenger beetles, Hydrophilidae, feed to some extent upon insects that fall upon the water. Other insects, such as *Volucella*, feed upon dead insects and other debris in the nests of social insects. *Anthrenus* commonly destroys dried insects as well as other animal remains. The wax moth, *Galleria mellonella*, feeds upon nitrogenous material in the comb consisting of molted skins and other materials associated with the bees.

Fig. 217.—Scavenger beetles on dead mouse. *Silpha americana* on body of mouse, *Necrophorous americana* on tail, and *Silpha surinamensis* below. (*W. R. Walton.*)

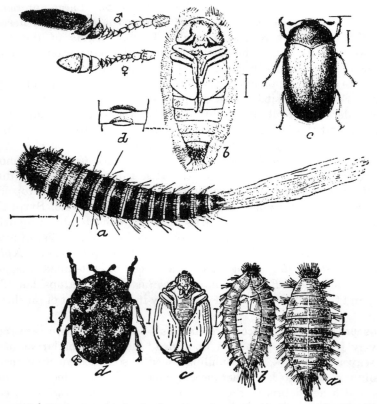

Fig. 218.—Feeders on dried animal matter: above, the black carpet beetle, *Attagenus piceus*, (*a*) larva; (*b*) pupa; (*c*) adult; (*d*) dorsal abdominal segments of pupa, antennae above; Below, the common carpet beetle, *Anthrenus scrophulariae;* (*a*) dorsal view of larva; (*b*) ventral view of larva; (*c*) pupa; (*d*) adult. (*After Riley.*)

The majority of the zoosaprophagous insects feed upon higher animals in various degrees of decay. The feeders of animal products can be divided into three classes: those that attack fresh or recently killed animals, those that attack decaying animals, and those that attack dried animal products including hair, feathers, and skins. Insects that attack living animals are known as parasites or predators and may cause the death of the animal. Then follow the species that oviposit on freshly killed meat, as, for example, many of the Sarcophagidae and the blow flies, Calliphoridae. When decomposition sets in, other insects are attracted. During the butyric fermentation a lepidopteron, *Aglossa suprealis*, may be attracted. During the ammoniacal fermentation, phorid and anthomyiid flies are attracted. As the stage of liquidation approaches, the carrion beetles, the Silphidae and the Histeridae (*Hister* and *Saprinus*) and certain Staphylinidae are attracted and lay their eggs. It is generally believed that these insects feed upon microorganisms that teem in these media. Finally the animal begins to dry and nothing but skin and bone remain. Animals in this condition attract another series of insects, *i.e.*, the dermestids (Dermestidae), the skin beetles (Nitidulidae, Trogiidae, and Corynetidae), and the cheese skipper, *Piophila casei*. The Trogiidae are frequently found about the refuse of tanneries and upon the hoofs and hair of decaying animals. Although carpets, woolens, and fur coats are not waste material, these articles are nevertheless attacked by the same insects that feed upon dried animal remains. The carpet beetle, *Anthrenus scrophulariae*, the black carpet beetle, *Attagenus piceus*, and the clothes moths are the species most commonly concerned.

**Phytosaprophagous Insects.**—Phytosaprophagous insects feed in much the same manner as the zoosaprophagous insects, *i.e.*, upon the rich supply of microorganisms that develop upon decomposed plant material. The Thysanura, Collembola, and Dicellura feed largely upon decomposing vegetable material. The naiads of the Ephemerida are said to be phytosaprophagous. The larvae of numerous Diptera, especially Tipulidae, Stratiomyiidae, Dolichopodidae, and Lonchopteridae, feed upon decayed plant material.

Decomposing wood shows a slightly different fauna. There are three steps in the decay of a tree: the dying tree, the dead tree, and the decayed tree. True scavengers do not enter until the third stage is reached and much depends upon the degree of the decomposition. Among the beetles, *Passalus*, *Cupes*, certain Scarabaeidae and Elateridae are common feeders. Among the Diptera, Tipulidae, Rhagionidae, Therevidae, and Asilidae are frequently found in decayed wood. The larvae of many other species of Diptera may be found in decayed wood but they are frequently predacious forms seeking other larvae. Some

tropical termites live in decayed wood.    Termites, ants, powder-post
beetles, and others commonly attack wood that is dried or cured.

A Comparison of Insects Feeding on Dried Animal and Vegetable Matter

| Species Feeding on Dried Animal Matter | Species Feeding on Dried Vegetable Matter |
|---|---|
| Dermestidae: | Dermestidae: |
| *Attagenus* | *Attagenus* |
| *Anthrenus* | *Anthrenus* |
| *Trogoderma* | *Trogoderma* |
| Blattidae (roaches) | Blattidae (roaches) |
| Siphonaptera (fleas) | Siphonaptera (fleas) |
| Orthoptera (house cricket) | |
| Coleoptera: | Coleoptera: |
| *Necrobia* (ham beetle) | Tenebrionidae (mealworms) |
| | Ptinidae (spider beetles) |
| | *Tribolium* (grain beetle) |
| | *Tenebroides* (grain beetle) |
| | *Silvanus* (grain beetle) |
| | *Sitophilus* (weevils) |
| | *Bruchus* (pea weevils) |
| | Bostrichidae (powder-post beetle) |
| Lepidoptera: | Lepidoptera: |
| Tineidae (clothes moth) | *Sitotroga* (flour moth) |
| | *Ephestria* (flour moth) |
| | *Plodia* (flour moth) |
| | *Pyralis* (flour moth) |
| | Termitidae (termites) |

**Insects Attacking Living Animals.**—There are three classes of insects
that attack living animals: predators, parasites, and parasitoids.   In

Fig. 219.—The rabbit botfly, *Cuterebra cuniculi.*   (*Kirk.*)

a broad sense, any organism, plant or animal that lives in, or on, or a
the expense of another is a parasite.   This definition of course include
predators, parasites, and parasitoids.

**Predators.**—A predator (*episite*) is an organism that lives at the expense of another, attacking and devouring the vital tissues. It generally lays its eggs near but not upon the host and usually attacks a succession of victims feeding from the outside. Predators are characteristically temporary for they accept a succession of hosts and sometimes consume hundreds of individuals.

Fig. 220.—Larvae of the rabbit botfly, *Cuterebra cuniculi*. (*Kirk.*)

The habits of the predator and the parasitoid intergrade to such an extent that the lines of demarcation are sometimes obscure. For example, the habits of the predator, *Promachus vertebratus*, resemble very closely the habits of the parasitoid, *Tiphia popillivora* (see Fig. 238). The larvae of both species feed externally upon white grubs. *Promachus*

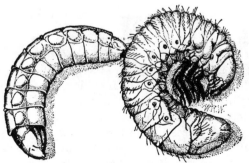

Fig. 221.—*Promachus vertebratus*, a predator attacking a white grub. The dipterous larva may shift its position on the host.

lays its eggs in crevices in the soil and the young larvae must find the host. *Tiphia* lays its eggs on separate hosts. When they hatch, the larvae are permanently attached. The larvae of *Promachus* often shift their position on the body of the hosts or may leave them and wander to others. Upon such slight differences our conception of predator and parasitoid often rests.

The outstanding examples of predatism are found among the Orthoptera, Neuroptera, Odonata, Plecoptera, Heteroptera, Coleoptera, and Diptera. The habit occurs to some extent in other orders.

Certain conspicuous modifications occur in the structure of predatory insects to fit them for the episitic habit. Both mandibulate and haustellate species are represented. The beaks of the predacious species are stronger and sharper than those of the phytophagous species. In the mandibulate species, the jaws are sharp and generally without molar surfaces. The sicklelike jaws of the Dytiscidae have openings at their tips through which the juices of their victims are sucked. In *Psychopsis elegans* (Neuroptera), the mandibles have on their ventral surfaces

Fig. 222.—The wheel bug, *Arilus cristatus*, a typical predacious heteropteran.

grooves that receive the corresponding maxillae. Each mandible is curved and untoothed but the edges of each are serrated at the tip. Juices pass between the mandibles and the maxillae in entering the mouth. In *Chrysopa*, the mandibles and the maxillae are also feebly grooved and fit together to form channels through which the liquid food is sucked. No matter whether the insect has mandibulate or haustellate mouth parts, it imbibes the juices of the host. The proboscides of the Heteroptera not only serve as feeding organs but are useful in holding the prey while the juices are being sucked out. Pentatomids and other Heteroptera frequently rest in an inverted position with their prey impaled on the beak, swung free from the plant, in which position the bugs can more conveniently suck the juices from their victims.

The naiads of the Odonata have a strong lower lip which is modified to seize their prey and hold it during the feeding process.

Raptorial legs reach their highest development in predacious insects. The front legs are generally modified for holding the prey. These are well illustrated in the Mantidae and the Mantispidae. The tarsi fold back against the tibia and both are armed with strong spines. In the ambush bugs, Phymatidae, the tibiae are conspicuously enlarged and the tarsi are very slender· both are armed with spines. In the scorpion

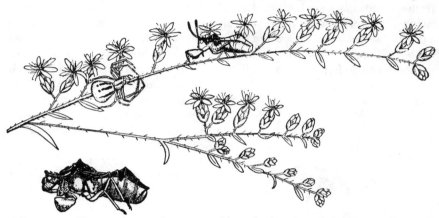

FIG. 223.—Two predacious forms on goldenrod; a crab spider, *Misumena*, and an ambush bug, *Phymata erosa*. The inset shows the enlarged front tibia of the ambush bug for grasping its prey.

FIG. 224.—Fore parts of the praying mantis, showing the powerful raptorial front legs.

flies, *Bittacus*, the front legs are generally used for support and the hind legs for seizing prey. A parasitic wasp, *Brachymeria fonscolombei*, which attacks numerous species of flies, also has the hind legs modified for grasping. The femora are expanded and armed along the inner margins with sharp teeth. The tibia folds against the femur making a remarkable raptorial organ which grasps the dipterous larva and holds it while the ovipositor is inserted into the host.

The Orthoptera, an order composed conspicuously of phytophagous species, contains one family, the Mantidae, which is exclusively predacious. A few grasshoppers and crickets are carnivorous. Among the mantids, *Stagmomantis carolina* is the only species native to the United States. It seldom occurs farther north than Maryland. The males and some of the females are grayish brown in color except the feet, which are green. Some of the females are wholly green in color. Two species have been introduced into the United States. The European, *Mantis religiosa*, is brown or green in color. The Asian, *Paratenodera sinensis*, can be distinguished because the front wings are brown with broad green margins. The females of these species consume a tremendous amount of food. One investigator states that a mantid consumed a few dozen flies, several grasshoppers, two young frogs, and stripped a lizard three times as large as itself.

The larvae of the Neuroptera are almost entirely predacious. The spongilla flies (Sisyridae) and the mantispids (Mantispidae) alone are parasitic. A few species of Neuroptera, for example, *Corydalis* and *Sialis*, are aquatic but the majority of the species are terrestrial. The aphis lions, Chrysopidae, are the most conspicuous

FIG. 225.—Larva of the ant lion, showing the strong sickle-shaped mandibles. (*Champlain and Kirk*.)

forms. There are more than 50 species easily distinguished from other insects because of their lacy green or yellowish wings. The larvae attack aphids and other small insects. The ant lions are another group of predacious Neuroptera. They are particularly interesting because of their habit of building pits or traps in sand. The larva lives at the bottom of the pit completely concealed by the sand except for its long mandibles which protrude at the lower end of the funnellike pit. When an ant or some unsuspecting insect comes too near the edge of the pit, the sand gives way and the insect rolls to the bottom and is quickly seized by the myrmeleonid and devoured.

The naiads of the Odonata and the Plecoptera are predacious. In the Odonata, the lower lip is tremendously enlarged and armed with sharp teeth. When at rest, the lip is folded back and conceals the mandibles. It is a powerful weapon of defense and an efficient organ for obtaining food. The naiads attack many aquatic insects and even prey upon small fish and frogs. The adults are likewise predacious but take their food upon the wing. They are the flycatchers of the insect world. They are extremely agile upon the wing, can poise in the air, and turn with extraordinary ease, which enables them to pursue and catch flying insects.

A few species of thrips (Thysanoptera) are predacious. They belong to the suborder Tubulifera which attack aphids and the eggs of mites and insects. The predacious species are, as a rule, larger

Fig. 226.—Two predacious pentatomids: left, *Brochymena arborea;* right, *Apateticus cynicus.*

and stronger than the phytophagous species. Some are giants compared with the other members of the order. The majority of the species of this order are phytophagous, however, and include many of our most serious plant pests, such as the pear thrips, onion thrips, and gladiolus thrips.

A large number of the Heteroptera are predacious. The Corixidae, Notonectidae, Nepidae, and Belostomatidae are semiaquatic in habit. They frequent aquatic locations, often dive below the surface of the water, and prey upon insects and small animals. Other Heteroptera, notably the Reduviidae, Cimicidae, Phymatidae, Pentatomidae, and Miridae, are terrestrial. The assassin bugs and the bedbugs are important enemies of man and frequently carry diseases. The Phymatidae, Miridae, and Pentatomidae are the most important species from the standpoint of biological control. The

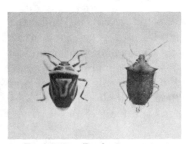

Fig. 227.—Predacious pentatomids: left, *Perillus bioculatus*; right *Podisus maculiventris.*

Pentatomidae have a wide range of hosts and seldom show specific preference. Species of the genera *Euschistus, Apateticus, Brochymena, Acrosternum, Podisus,* and *Perillus* commonly attack pests upon fruit and vegetable crops and thus become a valuable means of natural control. *Perillus bioculatus* is a striking species with several color forms.   It is generally yellow or red with a black Y-shaped mark on

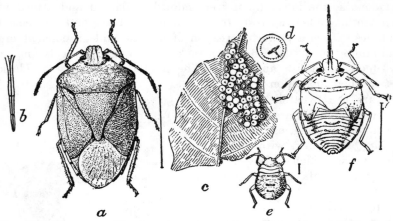

Fig. 228.—The green soldier bug, *Acrosternum* (*Nezara*) *hilare*: (*a*) adult; (*b*) proboscis or sucking mouth parts; (*c*) eggs; (*d*) cap of egg; (*e*) young nymph; (*f*) mature nymph. (*After Chittenden, U.S. Department of Agriculture.*)

the pronotum and two prominent black spots on the thorax.   It is predacious on cutworms and other caterpillars and is especially

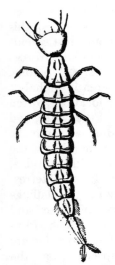

effective against the larvae of the Colorado potato beetle.   *Apateticus bracteatus* is predacious upon the larvae of the argus potato beetle.   *A. cynicus* often feeds upon the larvae of the oak tortricid, *Cacoecia fervidana*.   The adults of this species are brown in color and are inconspicuous when hidden among the dead, brown oak leaves.   *Crytorhinus mundulus* (Miridae) is a native of Australia, Java, the Philippine Islands, and the Fiji Islands, and is reported as an efficient agent in destroying the eggs of the sugarcane leafhopper.   Certain species of the genera *Acrosternum* and *Euchistus* have a liking for plant juices and frequently cause serious damage to plants.   *Acrosternum* (*Nezara*) *hilare* attacks the fruit of peach trees, causing considerable injury.

The Phymatidae are commonly seen in the fall resting on the flowers of goldenrod.   They do not hesitate to seize insects much larger than themselves.

Fig. 229.—Larva of a diving beetle, *Dytiscus*, showing the sickle-shaped mandibles adapted for a predacious life.

The Coleoptera have a wide range of food.   Many are predacious upon other insects.   Five families, Dytiscidae, Cicindelidae, Carabidae, Coccinellidae, and Meloidae, are worthy of mention.   The Dytiscidae and the Cicindelidae are of little importance as a means of biological

control.   The Dytiscidae are aquatic and both larvae and adults prey
upon other insects.   The larvae are known as water tigers because they are

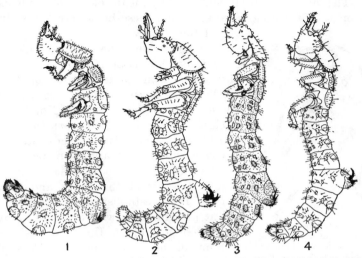

Fig. 230.—Larvae of tiger-beetles, Cicindelidae: (1) *Amblychila cylindriformis;* (2) *Omus
californicus;* (3) *Tetracha carolina;* (4) *Cicindela limbalis.* (*After Hamilton.*)

extremely voracious.   Their mandibles are long and slender, with a small
opening at the tip of each through which the juices of the host are sucked.

The Cicindelidae are predacious
in both larval and adult forms.   The
larvae construct vertical burrows in
the ground.   Some species live upon
sandy beaches and the adults are
almost white in color.   Other species
select roadsides or ploughed ground
that has become hard and dried.
Our common species, *Cicindela
sexguttata,* is iridescent blue or green
in color with six prominent white
or cream-colored spots on the elytra.
Other species are brown or metallic
in color.   All are active insects and
difficult to capture.   The larvae rest
in their burrows with their heads
flush with the ground.   They are
queer looking creatures and have
the appearance of being disjointed.   The head, extremely large in proportion to the rest of the body, is bent at right angles with the body

Fig. 231.—Cicindelid beetle attacking a hymenopterous insect. (*Champlain and Kirk.*)

and makes a neat plug which fits the opening of the burrow. Its powerful jaws extend upward ready to seize the first insect that walks over this trap. The larva is securely anchored by means of two sharp hooks on a hump on the fifth abdominal segment; these hooks are thrust into the wall of the burrow.

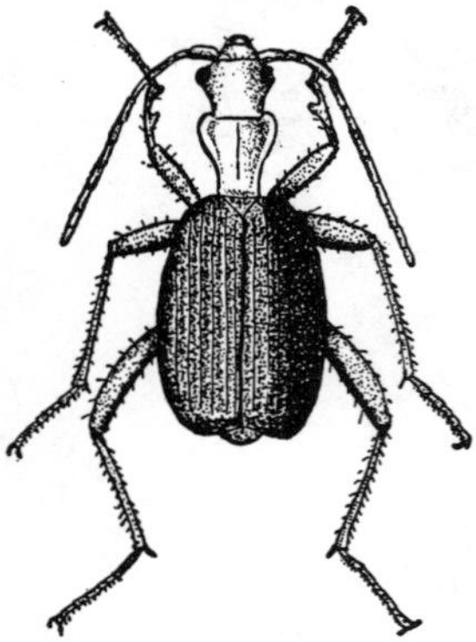

The larvae and the adults of most of the ground beetles (Carabidae) are predacious. They lurk about under stones and trash and usually come forth at night to seek their prey. There are approximately 1,200 North American species. The searcher, *Calosoma sycophanta*, is a striking example. It is a large iridescent green species with reddish margins on the outer edges of the elytra. This species was introduced from Europe to control the gypsy and brown-tail moths. It has established itself in this country and at present has a wide distribution over the eastern portions of the United States.

FIG. 232.—*Brachynus americanus* Lec., one of many predacious Carabidae. When alarmed this species shoots a puff of smoke from the posterior end of the body.

Some of the smaller species of Carabidae are commonly found beneath stones. Many are black or brown in color and have long legs. The bombardier beetles, *Brachinus*, are especially interesting. There are about 25 North American species. The head, thorax, and legs are reddish brown in color and the elytra are greenish or bluish. At the posterior end of the body are glands that secret a bad-smelling fluid which is used as a means of defense. These beetles shoot the fluid at their enemies in the form of little puffs of smoke.

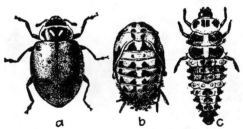

*a*     *b*     *c*

FIG. 233.—The convergent ladybird beetle, *Hippodamia convergens;* (*a*) adult; (*b*) pupa; (*c*) larva.

They are able to emit explosions five or six times in succession and elude their enemies by laying smoke screens.

The ladybird beetles, Coccinellidae, with few exceptions, are predacious in the larval and adult stages. They attack numerous scale insects, aphids, and the eggs of many insects, and are therefore considered as one of the most important groups of beneficial insects. They are oval or rounded beetles that vary considerably in color and markings. In general, they are red or yellow with black spots or black with yellow or red spots. The larvae are blackish grubs, often spotted or banded with red or yellow and armed with distinct spines. In many respects the larvae and the adults resemble superficially those of the leaf beetles, Chrysomelidae. When full grown, the larva suspends itself by the hinder end from some convenient support and either pushes the larval skin upward where it remains in a little wad about the posterior end, or it

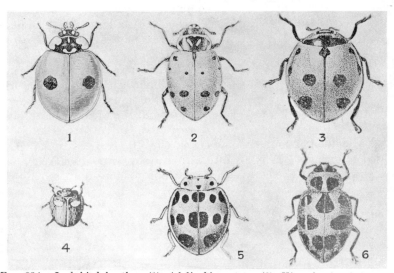

Fig. 234.—Ladybird beetles: (1) *Adalia bipunctata;* (2) *Hippodamia convergens;* (3) *Coccinella novemnotata* (4) *Chilocorus bivulnerus;* (5) *Epilachna borealis;* (6) *Ceratomegilla maculata.* Both larvae and adults are predacious.

remains within the skin until the beetle is ready to emerge. The larvae of *Hyperaspis* are woolly in appearance and look somewhat like mealy bugs.

*Adalia bipunctata* is one of our common species. It is generally reddish brown in color with one distinct black spot on each elytron.

*Hippodamia convergens* is another species common in the Eastern United States. The head and thorax are generally black, the latter bearing two conspicuous converging white bars from which the species receives its name. The elytra are reddish brown in color with five spots on each and a central spot adjacent to the thorax.

The fifteen-spotted ladybird beetle, *Anatis 15-punctata,* is somewhat larger than most species of Coccinellidae. The elytra are reddish brown with seven distinct spots on each elytron and one spot common to both elytra. The color often varies to gray or dirty brown.

Another striking and somewhat common species is the twice-stabbed ladybird beetle, *Chilocorus bivulnerus*. This species is almost entirely black, somewhat smaller than the other species, and has a reddish spot on each elytron.

The spotted ladybird beetle, *Ceratomegilla maculata*, is exceedingly common and often congregates in large numbers in the fall to hibernate. The adult is generally reddish brown in color and more elongate than other common species of Coccinellidae. There are two large black spots on the thorax and six spots on each of the elytra.

Fig. 235.—A dragonfly captured by an asilid fly, Panama.

Unfortunately, the family Coccinellidae includes two injurious species: the squash beetle, *Epilachna borealis*, and the Mexican bean beetle, *E. varivestis*.

The Meloidae, popularly known as the blister beetles, have been discussed elsewhere. It is an interesting group with about 200 North American species. The larvae are predacious but the adults are injurious to foliage.

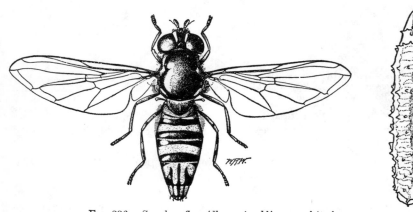

Fig. 236.—Syrphus fly, *Allograpta obliqua*, and its larva.

Many predacious forms occur in the Diptera. The black flies (Simuliidae), the horseflies (Tabanidae), the mosquitoes (Culicidae), and other similar species, attack man. Some species prey upon insects. The larvae and adults of the robber flies (Asilidae) are especially pugnacious and frequently attack insects much larger than themselves. The adults have the habit of resting in a sunny spot and waiting for other insects to enter the area, when they dart forth and capture them. Odonata, in spite of their size, are often attacked.

The Syrphidae are the most important group of predatory Diptera. The larvae of many species attack aphids. Their white eggs are laid in the midst of colonies of aphids. When they hatch, the maggots consume an enormous number of insects.

Some of the mecopterous larvae are predacious. The habits of the predacious adults are better known.

A comparatively small number of the Lepidoptera attack insects. Some are cannibalistic, others are predacious, and still others are parasitic. Balduf (1940) lists 20 families and 141 cannibalistic species of which 70 belong to the family Noctuidae. The predatory forms occur in the families Blastobasidae, Cosmopterygidae, Heliodinidae, Noctuidae, Pyralididae, Tineidae, and Tortricidae. Only five predacious species are known from North America; all attack Coccidae. Twenty-one parasitic species are known but only three occur in North America. These attack Fulgoridae and Coccidae.

The Hymenoptera include numerous predacious species. Many of the solitary and social wasps capture insects and spiders which they use to provision their nests. Common among these are the spider hunters, Pompilidae, and certain vespid and sphecid wasps. The cicada killer is the largest and most conspicuous species in this latitude.

**Parasites.**—A parasite is a species that feeds on the body of another without killing it and generally attacks only a single host. The eggs are usually laid upon the host. Among the insects, there are two general groups of parasites: those that attack vertebrates and those that attack insects and their near relatives. The former include the true lice (Anoplura), the bird lice (Mallophaga), some of the louse flies (Hippoboscidae), and the fleas (Siphonaptera). Parasites of insects are uncommon, the Strepsiptera and the bee louse, *Braula*, are about the only examples.

Parasites may be classified in various ways. They are divided, on the basis of host relations, as temporary or stationary. True lice, Anoplura, are more or less permanent parasites upon vertebrates; mosquitoes and horseflies feed upon their hosts for comparatively short periods. Parasites may be classified, on the position occupied on the host, as external (ecto) and internal (endo) parasites. Insects are generally ectoparasites. On the manner of attack, they may be facultative or obligatory. Most parasites are obligatory, that is, they are limited to a parasitic existence. A few are facultative or accidental parasites and normally live free, but can exist within a host. The larvae of certain Diptera are occasionally found in the human intestinal tract. Some species cause intestinal myasis. Two species of *Fannia* have been recorded in·this role. On the number of host species attacked, parasites may be monophagous, oligophagous, or polyphagous.   The species of Anoplura are generally limited to single hosts, for example, *Haematopinus asini* attacks the horse and *H. suis* attacks the hog.   On the other hand the Mallophaga show a tendency to select several hosts.   *Trichodectes bovis* attacks domestic cattle.

**Parasitoids.**—A parasitoid* is intermediate in habits between the parasite and the predator.   At first, it operates as a parasite avoiding the vital organs of the host; later, it operates as a predator and consumes the host.   It generally lays its eggs within or upon the host; the larvae usually attach themselves permanently and feed within or upon the host. For example, the scoliid wasp attaches its egg to the exterior of the host, which is a white grub or larva of the   Scarabaeidae.   When the eggs

Fig. 237.—*Pteromalus puparum*, a common chalcid parasitoid of the cabbage worm; left, female; right, male.

hatch, the larva feeds externally upon the host.   Most parasitoids, however, feed within the host.

Parasitoids occur only among insects with complete metamorphosis where the larva is adaptable to internal feeding and the active adult has well-developed senses to locate the host.   These species represent the highest development of insects that attack living animals.   They include most of the species that attack insects and are found abundantly in the Diptera and the Hymenoptera but rarely in the Coleoptera and the Lepidoptera.   They attack only in the larval stage.   Predators on the contrary attack in the larval, nymphal, and adult stages.   The Odonata, Heteroptera, Coleoptera, and some Diptera are predacious in the adult as well as the immature stages.   The Plecoptera attack only as naiads   the Hymenoptera often attack as adults.

Parasitoids are often *monophagous, i.e.,* limited to a single host. Under natural conditions, *Heterospilus cephi* attacks only the European wheat sawfly.   *Thersilochis conotracheli,* a native species, attacks only the plum curculio.   *Ascogaster carpocapsae* attacks only the codling moth.   Predators are seldom monophagous although the pentatomid, *Perillus bioculatus,* feeds almost exclusively on the Colorado potato beetle.

* Parasitoids are considered as parasites by earlier workers.

Some parasitoids are *oligophagous, i.e.,* have a restricted range of food. The Scoliidae and Pyrgotidae are restricted to white grubs. *Cryptochaetum* (Agromyzidae) is restricted to scale insects. The genus *Opius* is restricted to the Diptera.

On the other hand, many parasitoids are *polyphagous. Trichogramma evanescens* attacks the eggs of more than 100 host species. *Compsilura concinnata,* an introduced species, has more than 125 hosts, chiefly the larvae of Lepidoptera.

Parasitoids attack all stages of insect life—eggs, larvae, nymphs, pupae, and sometimes adults. The forms attacking eggs are numerous. The Trichogrammidae, Proctotrupidae, Eulophidae, and Chalcididae are best known. *Tetrastichus asparagi* attacks the eggs of the asparagus beetle. The larvae mature entirely within the eggs of the host. This species apparently has no male. The females have the peculiar habit of puncturing the host eggs with their ovipositors to cause the juices upon which they feed to flow. Other species lay their eggs in those of the host but they do not hatch until the host has reached the larval stage.

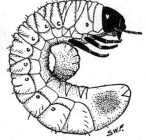

The larvae and pupae of insects are attacked by a large number of parasitic species but the adults are infrequently attacked. The best known adult parasitoid is *Centeter cinerea,* a tachinid that attacks the Japanese beetle. The white eggs of this species are deposited on the

Fig. 238.—*Tiphia popilliavora* Rohwer, an ecto-parasitoid on the grub of the Japanese beetle, *Popillia japonica* Newm.

upper side of the thorax of the beetle; the young larvae have sharp jaws to make entrance holes through the body wall of the host where the larvae mature and pupate.

The parasitoid and the solitary habits sometimes approach each other. For example, the solitary spider wasps, Pompilidae, and the parasitic white grub wasps, Tiphidae, are quite similar in habits. Both sting their prey to paralyze it and lay an egg on the side of the host. They differ in the following respects. The tiphia adult does not construct a cell or burrow for the reception of the host; the pompilid wasp prepares a cell in the ground in which she places her host.

The Hymenoptera take the lead as enemies of other insects. More than 50 families are represented. The more important families, from the standpoint of biological control are the Ichneumonidae, Braconidae, Trichogrammidae, Pteromalidae, Chalcididae, Scoliidae, and Tiphiidae. All but the Chalcididae are strictly parasitic.

*Megarhyssa lunator* is one of the largest and most conspicuous of the Ichneumonidae. It attacks the wood-boring larva of the pigeon tremex,

*Tremex columba.* To reach the larva of this species, the female must drill a hole to the *Tremex* burrow, which is no small task. Sometimes the adult *Megarhyssa* gets her ovipositor wedged in the wood so tightly that she is unable to withdraw it and she is held prisoner.

The oriental fruit moth parasite, *Glypta rufiscutellaris*, also belongs to the Ichneumonidae. This species has been extensively reared and distributed as a means of biological control.

The Braconidae is a large family containing several hundred species. The genus *Apanteles* contains about 50 species; all attack larvae and

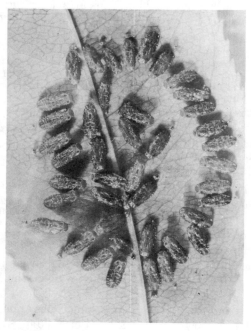

Fig. 239.—Pupae of *Comedo*, the larvae emerged from their host, a caterpillar, and arranged in a regular manner with a small pile of discarded fecular pellets at the posterior end of each.

are somewhat similar in habits. The small white cocoons, frequently seen upon the backs of the tomato hornworm, belong to this group. Some of the species, such as *Apanteles congregatus*, which attacks the sphingid caterpillar, and *A. militaris*, which attacks the armyworm, emerge from their hosts and spin their cocoons on the backs of the caterpillars. Other species, such as *A. glomeratus*, which attacks the cabbage worm, and *A. lacteicolor*, which attacks the gypsy moth, emerge from their hosts and spin their cocoons in a mass on some object adjacent to the abandoned caterpillar.

To the Braconidae belongs one of the commonest enemies of the oriental fruit moth and strawberry leaf roller, namely *Macrocentrus ancylivorus*. Many of the Braconidae attack lepidopterous larvae. The genus *Meteorus* contains about 10 species; one, *M. laphygme*, is a common enemy of the fall armyworm.

The Trichogrammidae is a small family with but two North American species. *Trichogramma minutum* is an important species and has a large number of hosts. It has been reared from the eggs of more than 125 species of insects and is decidedly polyphagous. This species has

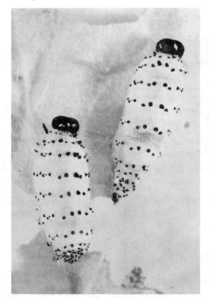

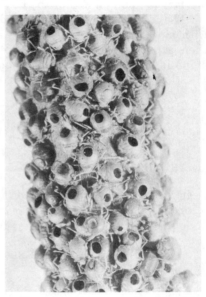

Fig. 240.—Larvae of the red-humped apple worm, *Schizura concinna*, attacked by *Hyposoter fugitivus*. The pupae are formed within the shriveled skins of the hosts.

Fig. 241.—A group of aphids attacked by *Lysiphlebus testaceipes*. Some show the emergence holes of the parasites; others show lids cut in preparation for emergence.

been widely used as a means of control of various injurious insects and has been propagated and liberated more than any other insect.

The Chalcididae are especially interesting because they have such a wide range of habits. The majority of the species are parasitic upon insects. They attack the eggs, larvae, pupae, and in a few cases the adults. A few species form galls and some feed within the seeds of various plants. *Lysiphlebus* attacks numerous species of aphids. The female deposits her eggs in the bodies of the aphids. After the aphid is killed, the body turns brown and becomes distended, and the parasite pupates within this parchmentlike skin. The adult emerges by cutting a hole through the body wall of the aphid.

The Scoliidae and Tiphiidae attack white grubs, the larvae of May beetles.

The Diptera stand next in importance to the Hymenoptera as enemies of insects. The Sarcophagidae and the Tachinidae are the most important groups although there are about 20 other families that attack insects. The habits of the Sarcophagidae, as previously noted, vary greatly but the majority are parasitoids. They attack Orthoptera, Neuroptera, Homoptera, Coleoptera, Lepidoptera, and Hymenoptera. As far as is known, all the species are larviparous and the female deposits her larvae upon or within the host.

The Tachinidae are a large and important group of beneficial insects with more than 1,400 species. All are parasitoids and the majority of them attack the larvae of Lepidoptera. They have diverse habits: some oviposit in the ground, some deposit their eggs upon foliage, some lay

Fig. 242.—Larvae of *Eulophus viridulus* attacking the larva of the European corn borer.

their eggs upon the host, and not a few are larviparous. *Centeter cinerea* is one of the comparatively few Diptera that oviposits upon adult insects. It attacks the Japanese beetle.

## BIBLIOGRAPHY

I. **Scavengers.**—See also references to insects related to plants.

ABBOTT, C. E. (1937).   The necrophilous habit in Coleoptera.   *Bull. Brooklyn Ent. Soc.* **32** (5).

BAUMBERGER, J. P. (1919).   A nutritional study of insects with special reference to microorganisms and their substrata.   *Journ. Exp. Zool.* **28** (1).

CLARK, C. U. (1895).   On the food habits of certain dung and carrion beetles.   *Journ. N. Y. Ent. Soc.* **3** (2).

HOWARD, L. O. (1900).   A contribution to the study of the insect fauna of human excrement.   *Proc. Wash. Acad. Sci.* **2**.

LENG, C. W. (1913).   Beetles of carrion and excrement.   *Journ. N. Y. Ent. Soc.* **21** (2): 168.

MOTTER, M. G. (1898).   A contribution to the study of the fauna of the grave. *Journ. N. Y. Ent. Soc.* **6** (4).

STEELE, B. F. (1927).   Notes on the feeding habits of carrion beetles.   *Journ. N. Y. Ent. Soc.* **35** (1).

WEISS, H. B. (1922).   A summary of the food habits of North American Coleoptera. *Amer. Nat.* **56** (643).

——— (1924).   Ratios between the food habits of insects.   *Ent. News* **35** (10).

——— (1925).   Notes on the ratios of insect food habits.   *Proc. Biol. Soc. Wash.* **38**.

**II. Predators.**

BALDUF, W. V. (1935). The bionomics of entomophagous Coleoptera. John S. Swift & Co., Inc., St. Louis.

—— (1939). The bionomics of entomophagous insects. John S. Swift & Co., Inc., St. Louis. *References.*

BEQUAERT, J. (1922). Predaceous enemies of ants. Part III, Ants of the Belgian Congo. *Bull. Amer. Mus. Nat. Hist.* **45.**

BHATIA, M. L. (1939). Biology, morphology and anatomy of aphidophagous syrphid larvae. *Parasitology* **31** (1). *References.*

BROMLEY, S. W. (1914). Asilids and their prey. *Psyche* **21** (6).

CLAUSEN, CURTIS P. (1940). Entomophagous insects. McGraw-Hill Book Company, Inc., New York.

DETHIER, V. G. (1939). Further notes on cannibalism among larvae. *Psyche* **46** (1).

FROST, S. W. (1936). Assassins, with bibliography and list of hosts of predacious insects. *In,* Ancient artizans. Van Press, Boston.

HUNGERFORD, H. B. (1917). Food habits of corixids. *Journ. N. Y. Ent. Soc.* **25** (1). *Bibliography.*

SLOUGH, W. W. (1940). The feeding of ground beetles (Carabidae). *Amer. Midl. Nat.* **24** (2).

STAGE, H. H., and YATES, W. W. (1939). Ground beetles predatory on the eggs of Aedes mosquitoes. *Proc. Ent. Soc. Wash.* **41** (6).

SWEETMAN, H. L. (1936). The biological control of insects. Comstock Publishing Company, Inc., Ithaca, N. Y. *Bibliography.*

TURNER, C. H. (1911). A note on the hunting habits of an American Ammophila. *Psyche* **18** (1).

WEBSTER, F. M. (1903). Notes on the food of predacious beetles. *Bull. Ill. State Lab. Nat. Hist.* **1.**

**III. Parasites and Parasitoids.**

BALDUF, W. V. (1939). The bionomics of entomophagous insects. John S. Swift & Co., Inc., St. Louis.

BROOKE, WORTH C. (1939). Observations on parasitism and superparasitism (Lepid.: Sphingidae: Braconidae, Chalcididae). *Ent. News* **50** (5).

BRUES, C. T. (1921). Correlation of taxonomic affinities with food habits in Hymenoptera, with special reference to parasitism. *Amer. Nat.* **55** (637). *Bibliography.*

CLAUSEN, CURTIS P. (1936). Insect parasitism and biological control. *Ann. Ent. Soc. Amer.* **29** (2).

—— (1940). Entomophagous insects. McGraw-Hill Book Company, Inc., New York.

DAVIS, J. J. (1919). Contributions to a knowledge of the natural enemies of *Phyllophaga. Bull. Ill. State Lab. Nat. Hist.* **13** (5). *Bibliography.*

EWING, H. E. (1929). A manual of external parasites. Charles C. Thomas, Publisher, Springfield, Ill.

FENTON, F. A. (1918). The parasites of leafhoppers with special reference to the biology of the Anteoninae. *Ohio Journ. Sci.* **18** (6, 7, 8).

HOWARD, L. O. (1889). A commencement of a study of the parasites of cosmopolitan insects. *Proc. Ent. Soc. Wash.* **1** (3). *Literature.*

—— (1910). On the habit with certain Chalcidoidea of feeding at puncture holes made by the ovipositor. *Journ. Econ. Ent.* **3.**

LAING, J. (1937). Host-finding by insect parasites. *Journ. Animal Ecol.* **6** (2). *References.*

RILEY, C. V. (1891–1893). Parasitism in insects. *Proc. Ent. Soc. Wash.* **2** (2 and 4).

SALT, GEORGE (1927). The effects of stylopization on aculeate Hymenoptera. *Journ. Exp. Zool.* **48** (1). *Bibliography.*

———— (1938). Experimental studies in insect parasitism. VI—Host suitability. *Bull. Ent. Res. Lond.* **29** (3). *Bibliography.*

SMITH, H. S. (1916). An attempt to redefine the host relationships exhibited by entomophagous insects. *Journ. Econ. Ent.* **9** (5).

SWEETMAN, H. L. (1936). The biological control of insects. Comstock Publishing Company, Inc., Ithaca, N. Y. *Bibliography.*

THOMPSON, W. R. (1930). The principles of biological control. *Ann. Appl. Biol.* **17** (2).

———— (1939). Biological control and the theories of interaction of populations. *Parasitology* **31** (3).

# CHAPTER XV

## ASSOCIATIONS OF PLANTS AND INSECTS

The associations between plants and insects may be mutual; the plants may be benefited or the insects may be benefited. In most cases, the insects obtain nourishment and the plants are injured or defoliated. A large part of economic entomology deals with the control of insects that injure plants.

Insects depend directly or indirectly upon plants for their existence. They are frequently attracted by the abundance of available food and by the shelter that leaves, stems, or other portions of the plant offer. Insects utilize many parts of the plant. They seek nectar, fruit juices, and sap. They feed upon green plants, dried plants, and decayed plants. They eat buds, leaves, shoots, branches, bark, wood, roots, flowers, pollen, seeds, and fruit. They often carry away portions of the plant such as leaves, bark, and plant fuzz which are used to construct cases, cocoons, or resting places.

Some insects have jaws to consume portions of plants. Others have haustellate mouth parts to suck the plant juices or to sip nectar. Bees have devices to gather and carry pollen. Caterpillars have legs

FIG. 243.—Oak leaf skeletonized by sawfly larvae.

or prolegs to cling to plants. Sphinx moths can hover while they drink nectar from flowers. Insects often have well-developed senses to locate flowers or to find the hosts upon which they lay their eggs. Thus they may be strongly attracted to plants.

On the other hand, insects are often repelled by certain physical characteristics of plants such as hairiness or toughness of leaves, toxic qualities, resinous or other undesirable properties of the plant. Economic entomologists are constantly seeking plants with these qualities, for they resist the attacks of injurious insects.

Fig. 244.—White stippling on the leaves of black cherry produced by the nymphs and adults of the lace bug, Corythuca.

About 50 per cent of the insect species are plant feeders. Weiss (1924–1939) gives the following figures:

### FOOD HABITS OF SOME INSECTS

| Locality and number of species considered | Per cent phytoph-agous | Per cent saproph-agous | Per cent harpact-ophagous | Per cent para-sitic | Per cent polleno-phagous | Per cent miscel-laneous habits |
|---|---|---|---|---|---|---|
| West Arctic coast of North America, 400 species.... | 47 | 27 | 14 | 10 | 2 | 0 |
| New Jersey, 10,500 species. | 49 | 19 | 16 | 12 | 2 | 2 |
| Death Valley, 557 species. | 37 | 20 | 25 | 14 | 4 | 0 |
| Connecticut, 6,781 species. | 52 | 19 | 16 | 10 | 3 | 0 |
| North Carolina, 9,249 species.................. | 46 | 17 | 22 | 11 | 4 | 0 |
| Mount Desert Is., Me., 5,177 species.......... | 52 | 17 | 14 | 15 | 2 | 0 |

Species that are not phytophagous feed upon other insects or animals which in turn feed upon plants, or they feed upon the remains of plants or insects. The Homoptera are strictly phytophagous. The larvae of the Lepidoptera, with rare exceptions, are plant feeders. The Orthoptera, with the exception of the Mantidae, and a few other species are herbivorous. The naiads of Ephemerida are almost entirely herbivorous. Certain groups of Coleoptera such as the Chrysomelidae, Cerambycidae, Elateridae, Buprestidae, and Curculionidae are wholly phytophagous. The larvae of the sawflies are plant feeders.

The Homoptera, Coreidae, and certain Pentatomidae are phytophagous in both nymphal and adult stages. Most lepidoptera feed upon plants as larvae; the adults sip nectar or have vestigal mouth parts and take no food at all. The blister beetles, Meloidae, have a strange combination of food habits. The larvae are beneficial, preying upon the eggs of certain grasshoppers or brood of bees, but the adults are injurious to flowers and foliage.

Predacious species seldom attack plants. There are, however, some exceptions. *Acanthocephala femorata* is predacious upon the cotton worm but frequently causes damage to the fruits of cherries. Several northern species of green stinkbugs of the genus *Acrosternum* are predacious but they occasionally attack peaches and cause a type of injury known as "cat facing."

FIG. 245.—The four-lined plant bug and its characteristic manner of feeding upon chrysanthemum.

Different types of plant feeding have led to groups such as leaf miners, borers, gall makers, seed feeders, and free feeders. In most cases, the feeding is detrimental to the plant. It is interesting to note that some insects take food without killing the plant or at least without causing pronounced ill effects. The severity of the attack determines the resulting damage. The columbine leaf miner is frequently abundant upon the leaves of Aquilegia but the leaves do not wilt and the flowers apparently show no serious effect of the attack. Many gall insects cause deformations of the plant but do not check growth or seed formation. It has been said that many of these galls increase the amount of available food for both insect and plant. Bees, of course, take honey and nectar without damage to the plant and, as we well know, pollinate the flowers. The fig *Blastophaga* is also distinctly beneficial to the plant as fruit production depends upon this tiny insect.

FIG. 246.—*Phytomyza miniscula* frequently disfigures the leaves of columbine by its mining operations. This is the common miner on Aquilegia in North America.

**Free Feeders.**—A large proportion of the insects are free feeders, that is, they do not mine, bore, or form galls. Some are subterranean, others are aquatic, but all feed upon the exterior of the host and are free

FIG. 247.—A typical forager or external foliage feeder, *Sphecodina abbotti*, a sphingid caterpillar on grape.

to wander about and accept food from a more or less wide area. They are foragers or hunters. All orders contribute to the free-feeding habit. Some are sucking insects, others are chewing insects. They attack plants and plant products, and, of course, other groups attack animals and animal products.

Grasshoppers, June beetles, cutworms, armyworms, apple tent tree caterpillars, webworms, and leaf beetles are typical foragers. They often operate in companies and cause conspicuous injury. Other species, such as the polyphemus larvae, feed more or less alone.

The foragers include the beneficial as well as the injurious. Many examples will be cited in discussing the relation of plants to insects. The ants, roaches, and other insect foragers are discussed under *saprophagous* insects (see page 273).

Undoubtedly the free or external feeders are the direct ancestors of other insects with more specialized habits. The free feeders are primarily mandibulate insects, the more primitive type of feeding. Most of the haustellate insects are really internal feeders. Although they rest upon the surface of the host, they have specialized mouth parts to reach into the plant to obtain their nourishment.

**Insect and Host Abundance.**—There is scarcely a plant that does not harbor some insect pest. Ginko is said to have few or no insect enemies. Ailanthus is almost without insect pests. Strangely, poison ivy is host to a considerable number of species of insects.

Fig. 248.—A community tree for insects, illustrating how numerous species feed upon white pine: (1) five species feed on foliage; (2) three species feed on buds; (3) three species feed on twigs; (4) two species bore in wood; (5) four species are cambium miners; (6) one species feeds on bark; (7) two species feed on roots.

A single plant usually supports more than one species of insect. Two hundred species have been reported from corn, 400 from apple, and more than 150 from pine. Twenty species, known to occur on white pine in North America, are adjusted to different parts of the tree so that they avoid, to some extent, competition with one another. Five species feed on foliage, three species feed on the buds, three species are twig borers, two species are wood borers, two species are root borers, four species are cambium miners, and one species is a bark borer. Oaks are attacked by at least 1,000 species of insects, 50 of which are leaf miners.

Insects may be restricted to a single plant (*monophagous*). This is especially true of leaf miners, borers, and gall makers that are particularly dependent upon certain characteristics of plants. The leaf-mining

Lepidoptera are particular in the selection of food plants. Nearly every species of *Nepticula, Phyllocnistis,* and *Lithocolletis* has its specific host. The boll weevil is restricted to cotton. The larva of the common copper butterfly, *Heodes hypophlaeas,* is restricted to *Rumex acetosella.*

On the other hand, insects often attack many plants or groups of related plants. The Colorado potato beetle is restricted to species of the family Solanaceae. The larvae of the milkweed butterfly feed only on species of *Asclepias.* The imported cabbage worm feeds chiefly on species of Cruciferae. The hessian fly attacks only species of the Gramineae. When a few species are attacked, they are said to be *oligophagous;* if many species are attacked, they are said to be *polyphagous.*

Some insects have a large number of hosts. The gypsy moth has 78 native hosts and is known to accept 458 species of plants in captivity. The Japanese beetle feeds upon more than 250 plants including grasses, weeds, field crops, flowering garden plants, truck and garden plants, shrubs, fruit and shade trees. Although the Japanese beetle has a wide range of food plants, there is a considerable list of plants they rarely or never attack: oaks (except pin and chestnut oak), evergreens (except cypress), spirea, lilac, chrysanthemum, petunia, and many others. The gypsy moth likewise attacks many evergreens but will not mature on cypress, arborvitae, or cedar.

It is not difficult to understand the food preferences of insects that feed upon related groups of plants, but it is hard to explain how some species feed upon plants belonging to unrelated systematic groups. Many apple insects feed upon oak, and vice versa; the oblique-banded leaf roller, *Cacoecia rosaceana,* the red-banded leaf roller, *Argyrotaenia velutinana,* and the tent feeder, *Cryptolechia tentoriferella,* are a few examples. In the case of the oblique-banded leaf roller and the red-banded leaf roller, this might be expected, because they are general feeders but the tent feeder, as far as known, attacks only apple and oak. Some strange food combinations might be cited. *Agrotoxa semipurpurana* feeds upon oak, cherry, and maple, three widely separated groups of plants. The twig pruner, *Hypermallus villosus,* bores in the wood of apple, grape, and maple. The codling moth is a pest primarily of apple but also attacks the husks of walnuts. The apple maggot, a pest principally of apple, also feeds on blueberries. The nun moth, *Liparis monacha,* which normally feeds on oak in this country, attacks pine in Germany.

As a rule, the food plant or plants of an insect are definitely fixed and most insects will die rather than accept another host. In recent experiments, however, it has been found that a species can be bred to accept a hitherto undesirable or unused host. Such breeding requires a large number of individuals and numerous generations. Perhaps only one out of a thousand larvae can be forced to accept a new food plant. In nature, a species may be compelled to accept an undesirable host before the plant upon which it normally oviposits is ready. For example, leaf beetles, Chrysomelidae, often emerge from their hiding places early in the spring before the foliage of the preferred host has expanded. They therefore select other plants as temporary hosts. This may explain, in a manner, how certain insects have developed strange diets.

**Feeding Activities.**—Insects assume numerous characteristic feeding attitudes for convenience or for protection. The feeding of borers, leaf miners, gall makers, and leaf rollers has been described elsewhere. This discussion is applicable chiefly to free feeders such

FIG. 249.—A typical haustellate insect, an aphid with mouth parts inserted into the tissues of the plant.

as grasshoppers, caterpillars, the larvae of sawflies, and certain nymphs.

In general, haustellate insects feed much alike. The stylets of the mouth parts are forced into the tissues of the plant and the juices are sucked from the interior. These insects perform a neat trick; they remain on the surface of the plant but feed upon the clean juices within.

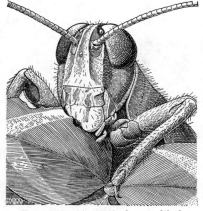

FIG. 250.—A typical mandibulate insect, a grasshopper feeding on clover (*Illinois Natural History Survey.*)

Thus, as we know, sucking insects cannot be killed by stomach poisons. They usually remain motionless while feeding. As a matter of fact, their mouth parts are frequently embedded so securely in the plant tissue that it is often difficult for the insect to remove them quickly when disturbed. Similarly, butterflies drinking in moist sand frequently work their maxillae so deep that they can be approached quite closely because they are unable to free themselves except with difficulty. Predacious pentatomids often hold their prey at the point of their proboscides, swinging them free from the foliage while sucking the juices from their victims.

Chewing insects employ several methods of feeding. Some dig their mandibles downward, removing patches of the surface of the leaf. Flea beetles usually eat small holes. Cabbage worms eat large holes or remove entire portions of the leaf. The pear slug, the Japanese beetle, and the elm leaf beetle skeletonize the leaf. On the other hand many caterpillars, sawflies, and other insects feed at the edge of the leaf.

Certain of the notodontid larvae have perfected the habit of feeding on the edge of the leaf. They combine this method of feeding with protective coloration. One of the best examples is *Nerice bidentata*, which feeds on elm. The larva is wholly green in color. Each segment of the body is produced dorsally into two points which resemble the teeth on the margin of the leaf. This is one of the most difficult caterpillars to locate because it is so well protected by its resemblance to the edge of the leaf. Other notodontids feed on the edges of the leaves of hickory, oak, and black walnut.

FIG. 251.—*Nerice bidentata* feeding on the edge of an elm leaf.

These larvae are green with brown markings which resemble injured portions of the leaves.

A number of sawfly larvae feed upon the edges of leaves or eat holes in the leaves and feed on the edges of the holes thus formed. The larvae often assume S positions which conform somewhat with the curved outlines of the holes in which they feed. When feeding in this manner, the larvae are difficult to recognize and are easily overlooked by the beginner.

Maggots have mouth hooks which can be turned so that the larvae can feed in a horizontal or a vertical position. They generally feed upon soft or semisoft media.

Special methods of feeding occur in some larvae. Sick or dying larvae of the clover leaf weevil, *Hypera punctata*, encircle the stem of the plant on which they feed. Abbott's sawflies, *Neodiprion pinetum*, use the posterior ends of their bodies to grasp the slender needles of the pine and thus maintain their position. Adult curculios have snouts of various lengths to make incisions in the fruit or other portions of the plant in which they lay their eggs. Bill bugs feed upon corn leaves before they have expanded,

FIG. 252.—Sawfly larvae feeding in characteristic manner on the edge of a leaf.

resulting in regular series of holes when the leaves finally open.

Some larvae excavate deep cavities in the fruit, stems, or roots of plants and almost completely bury themselves. The corn ear worm, *Heliothis obsoleta*, is an excellent example. The rose leaf beetle, *Nodonota puncticollis*, has a similar habit but many of the beetles work together in a single cavity.

The type of feeding is often so characteristic that the order, family, genus, or even species producing the injury can be recognized. For example, bill bug injury or feeding by the four-lined leaf bug is distinctive.

**Mutual Associations.**—In many cases the relationship between insects and plants is mutual. Insects visit flowers to obtain pollen and nectar and in so doing unconsciously pollinate the flowers. Although

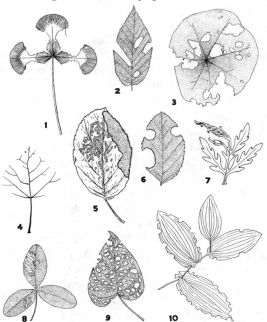

Fig. 253.—Typical leaf injuries by haustellate insects: (1) clover weevil, *Hypera punctata* feeding before leaflets opened; (2) feeding by sawfly larvae on birch; (3) injury to nasturtium by the imported cabbage worm; (4) advanced stage of skeletonizing by chrysomelid larvae; (5) feeding areas by the adults of *Baliosus ruber* and mine produced by the larvae of the same; (6) work of a megachilid bee; (7) ragweed leaf collapsed by a leaf-mining larva; (8) feeding by the green clover worm, *Plathypena scabra;* (9) Tortoise beetle injury on morning glory; (10) *Polygonum* leaflets injured by immature sawfly larvae.

bees and wasps are outstanding pollinators, the beetles, flies, and other insects play a part. Very often the pollination of flowers is a simple process; the insect brushes against the anthers of the flowers and becomes dusted with pollen and then carries it to the next flower. In other cases, the insect or the flower is especially adapted for these visitations. Although numerous insects visit flowers, only a few are of vital importance. Many secure nectar illegitimately. Bumblebees puncture the nectaries of snapdragon, columbine, and trumpet creeper and sip the nectar. Hymenoptera of the genera *Odynerus* and *Xylocopa*, and perhaps others, cut through the corolla, making a hole opposite each nectary and take the sweet floral juices without fertilizing the flower.   Certain moths force their proboscides between the petals of flowers and sip nectar without entering the flowers.

**Adaptations of Flowers for Insect Visitation.**—The simplest devices that attract insects to flowers are color, odor, and form.   Color and odor attract many insects but the form of the flower may be such that only certain species may enter.   Nectar lying at the base of deep corollas is available, normally, only to long-tongued insects.

Some flowers are so constructed that insects cannot obtain the nectar without carrying pollen away and they cannot enter the next flower of the same kind without leaving some pollen on the stigma of this flower.   The blue flag is one of the most remarkable flowers in this respect.   Three sepals form the floor of a passageway leading to the nectar, which is located in deep pockets in the flower.   The liplike stigma is attached to the corolla of the flower and is arched above the sepals forming a narrow passageway leading to the nectaries.   When a bee, covered with pollen, enters a flower, it brushes against the outer sticky surface of the stigma, to which a considerable amount of the pollen, obtained from the flower visited previously, adheres.   As the bee pushes its head into the flower to obtain nectar, its body rubs against the anther and is dusted with pollen.   When the bee leaves the flower, it may encounter the stigma again but this time it brushes against the inner surface which is not sticky and no pollen is deposited.   Cross pollination is thus ensured.

Fig. 254.—B l u e flag showing adaptation for cross pollination. The stigma is located on the upper petal and a narrow passage above the anther leads to the nectar well. (*After Folsom.*)

The milkweeds have special pollen masses (*pollinia*) which are so arranged that, when certain insects step upon the edge of the flower to sip the nectar, their legs slip into a fissure in front of the anther.   As the insect draws its legs upward, the claws, hairs, or spines frequently catch in the V-shaped slit in the pollinia.   The pollinia become fastened securely to the insect's legs and are carried to the next flower. In struggling, the insect generally frees itself from these pollinia which are left in the stigmatic chamber of the newly visited flower.   Sometimes the insect loses a leg or is permanently trapped by the flower.   Although numerous bees, wasps, flies, and certain butterflies carry pollinia, the red milkweed beetle, *Tetraopes tetrophthalmus*, is apparently so well adapted to living on milkweed that it is able to avoid the fissures that lead to the pollinia.   Charles Macnamara states, in writing, that he has never seen the milkweed beetle carrying pollinia.

**Adaptations of the Insect for Flower Visitation.**—Often the insect is especially modified for flower visitation. Common adaptations are found in pollen baskets (*corbiculae*) of pollen-carrying species. The hairs of bees also gather much pollen. These hairs are dense, often twisted, branched, or barbed. Caudell states that, of the 200 species of bees, 23 have branched hairs and these are pollen gatherers. The pollen is combed from the hairs of the body by means of pollen combs on the inner surface of the hind tarsi and is transferred to the pollen baskets on the hind tibiae. In the honeybees, only the workers possess these structures. In the bumblebees, the females start the new colonies and they possess pollen baskets. The antennal combs on the front legs of bees are used to clean pollen and other dirt ' from the antennae. The pollen-feeding beetles of the genus *Euphoria* are very hairy and the mouth parts are especially well clothed with dense hairs. Pollen adheres to these hairs when the beetles visit flowers.

Many insects have long tongues or suitable mouth parts to reach nectar located in deep flowers. The bees are thus divided into long- and short-tongued species. *Euglossa cordata*, peculiar to tropical America, has a tongue that exceeds the length of the

FIG. 255.—A fly caught in the pollinia traps of the common milkweed, *Asclepias syriaca*. Only a single flower of the umbel is shown. Normally the insect carries the pollinia away on its legs. (*Charles Macnamara.*)

body and undoubtedly feeds in flowers with long nectar tubes. *Macrosila cruentis*, a sphingid from Madagascar, has a proboscis which is 9½ inches long.

*Euphoria fulgida*, like many other insects, visits flowers to obtain nectar but also drinks the honeylike secretion that fills the glands at the base of the peach leaves. When the secretions from these glands run low, the beetle uses one of the sharp spines on the outer edge of the front tibia to lacerate the glands and cause the juice to flow.

The yucca moths, *Tegeticula* (*Pronuba*), exhibit an extraordinary adaptation for flower visitation. The yuccas are dependent upon these insects for pollination. The female is uniquely adapted for this purpose. The proboscis is not long and curled, as is usually the case in the Lepid-

Fig. 256.—A tabanid with the pollinia of the common milkweed, *Asclepias syriaca*, attached to the feet and mouth parts. (*Charles Macnamara.*)

optera, but is short, spiny, and modified to collect pollen. After gathering pollen from several flowers, the moth thrusts her long lancelike ovipositor into the ovary and lays several eggs. Then she ascends the pistil and thrusts the pollen mass into the stigma. The ovules develop into seeds; some are consumed by the larvae of the yucca moth, others are left and perpetuate the plant. Riley states that the yucca never produces seeds unless the *Tegeticula* are present.

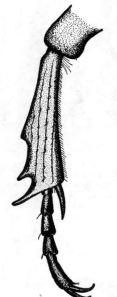

Another remarkable association exists between the fig and a chalcid. *Blastophaga psenes* is essential for the propagation of the Smyrna fig, the flowers of which are exclusively female and produce no pollen. Furthermore, the Smyrna fig owes its flavor and development to the number of ripe seeds which it contains and these develop only when the flowers are pollinated. The wild or caprifig produces pollen and for centuries has been used to fertilize the Smyrna fig. The *Blastophaga* develops in small galls in modified, infertile female flowers at the base of the wild or caprifig. The male of the *Blastophaga*, unique among insects, is wingless. It lives in and upon the fig from which it hatches and never wanders far away. When a male issues from a seedlike gall, it seeks a female gall in the interior of the same fig and gnaws a hole through the exterior and inserts its abdomen into the hole to fertilize the female. In emerging from the

Fig. 257.—Spine on the tibia of *Euphoria fulgida* often used to cut the glands on the petiole of peach to cause the juices to flow.

caprifig, the female becomes covered with pollen. She flies to another fig to oviposit. If by chance she enters a Smyrna fig, she walks among the flowers of the interior of the undeveloped fig seeking a proper place to lay her eggs. She does not lay her eggs in the Smyrna fig but in this futile search to find a place to lay her eggs, she dislodges sufficient pollen, which has been brought from the caprifig, to fertilize the female flowers of the Smyrna fig.

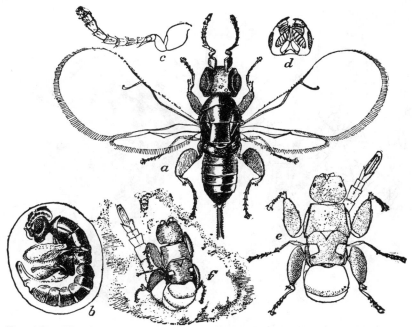

Fig. 258.—The fig wasp, *Blastophaga psenes;* (*a*) adult winged female; (*b*) female enclosed in pupal case and within gall of fig flower; (*c*) antenna of female; (*d*) lower surface of head of female; (*e*) and (*f*) adult males. (*After U. S. Department of Agriculture.*)

The most interesting cases of mutual relationship between insects and plants are found among tropical ants. Wheeler in his "Ants of the Belgian Congo" cites many examples of such associations. As a matter of fact, the hollow stems of numerous plants serve as homes for ants and have been termed ant plants. A species of ant, *Pseudomyrma*, from Nicaragua and the Amazon Valley inhabits the hollow thorns of *Acacia*. The ants gain entrance to the thorn by boring a hole near the apex. They obtain shelter within the thorn and feed upon a sugary secretion which exudes at the base of the petioles of the leaves. The ants raise their brood in these hollow cavities. They drive away leaf-cutting or leaf-eating species and thus protect the tree from injurious insects.

The cecropia tree of Central America provides an excellent dwelling place for ants and scale insects. The trunk of this tree is hollow, much

like bamboo, with sections divided by nodes.    The ants gain entrance to
one of these sections by making a small hole through the exterior of the
plant.    Then they cut holes through the nodes or partitions so that they
have the freedom of several sections.    One of these sections is used by
the ants as a nursery.    Here the eggs are laid and cared for until they

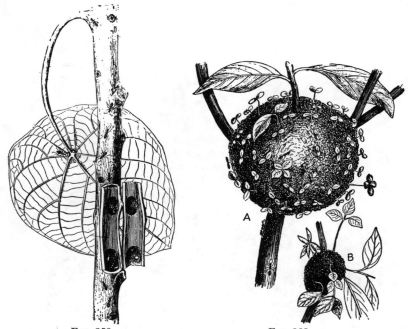

Fig. 259.                          Fig. 260.

FIG. 259.—A case of mutual relation between plant and insect.    The stem of *Endospermum formicarium* inhabited by ants, *Camponotus quadriceps*.    Openings to the nest appear on the upper part of the stem; an ant is feeding at one of the nectaries at the junction of the petiole and the leaf.    (*After Wheeler.*)

FIG. 260.—Ant gardens of the Amazons: (*A*) a large spherical ant garden covered with seedling plants; (*B*) smaller garden on Cordia.    (*After Ule.*)

hatch.    When the eggs have hatched, they are carried to the section
below which is reserved for the developing larvae.    When the larvae are
ready to pupate, they are transferred to another section.    In the mean-
time, certain scale insects are introduced and propagated in a fourth
section of the stem.    The scale insects feed upon the inside of the plant
and secrete a sweet substance which is appropriated by the ants.    The
ants also feed upon tender rounded bodies which grow at the base of each
leaf petiole.    The ants drive the invaders away and thus protect the
cecropia tree.

An interesting type of symbiosis exists between certain scale insects
and fungi.    Fungi of the genus *Septobasidium* are entirely dependent

upon scale insects for their development.    Twenty-four species of that genus are known to have a definite association with scale insects.    This habit has been observed in connection with 19 species of scale insects, including three common species, *Chionaspis corni*, *Aspidiotus perniciosus*, and *Lepidosaphes beckii*.    The young scale insects become infected with the fungus by crawling over the bud cells when emerging from the parent scale, previously infected.    These young scales settle upon the host and insert their beaks into the substratum of the host plant to feed. Shortly afterward, these scales become surrounded with a dense growth cf stroma (*hyphae*).    The stroma are attached through the setal pores to fungous coils (*haustoria*) which are inward growths of the fungous organism within the body of the scale.    Insects thus attacked live through the dormant season.    In the spring, the young scales are produced within the mass of stroma covering the parent scale.    Many of the scales clear a space beneath the stroma and settle down to feed; others make an opening to the outside, wander off, and start new infestations.    The fungus forms homes for the scale insects and protects them, to some extent, from parasitic and predacious enemies.    In return, the scale insects provide food for the fungus and serve as a means of distribution.    As a rule, the scales live over winter and succeed in laying young before they are killed by the fungus.    There is a delicate balance between these two forms that seems to perpetuate both insect and fungus.

**Insectivorous Plants.**—In some cases, plants receive direct benefit from insects but the insects suffer. Conspicuous are the numerous plants that trap insects.    There are a

Fig. 261.—Sundew, *Drosera*, showing the numerous sticky hairs that entrap small insects.

number of types of insectivorous plants.    The pitcher plant, a native of North America, is best known.    The leaves and petioles are modified to form hollow receptacles or "pitchers."    Glands on the inside of the leaf secrete a substance that attracts insects to the rim of the pitcher.    As the insects attempt to reach this material, they lose their balance and fall within the pitcher and are unable to escape because they are trapped by numerous downward pointing sticky hairs. The insects fall into a liquid at the bottom of the pitcher.    They are digested by this liquid and the food material is absorbed through the walls of the pitcher.

The sundew, *Drosera*, has specialized leaves designed to catch small insects. These leaves have numerous sticky hairs that trap insects. A fluid is secreted by the leaves which digests the insects.

The Venus's-flytrap, *Dionaea muscipula*, occurs along the coast of North Carolina. The leaves have flattened petioles, terminated by two semicircular lobes, the edges of which are armed with numerous spines. When an insect alights upon a leaf of this plant, it closes like a trap imprisoning the insect. The insect is then digested by a secretion poured from numerous glands opening upon the surface of the leaf.

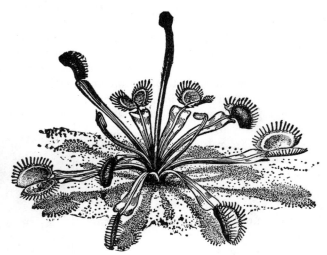

FIG. 262.—Venus's-flytrap, *Dionaea muscipula*, showing some leaves expanded and others closed. (*After Brown, courtesy of Ginn and Company.*)

Certain aquatic plants also trap insects. *Utricularia vulgaris* captures many mosquito larvae and other minute aquatic insects and animals. This plant bears small bladders which act as traps. When set, the walls of these traps are rigidly compressed. Larvae enter the outer vestibule and by their movements stimulate the valve which suddenly opens; the side walls expand and the inrushing water carries the insect within the trap. The valve then closes and the insect is held fast and is digested within these bladderlike chambers. A plant seven inches long may capture 150,000 crustaceans and other small animals.

A strange plant from Ceylon, *Aristolochia*, traps flies and holds them until the pollen covering their bodies has fertilized the stigmas of the flowers. The peculiar odor and color of the flowers attract the flies. Once they have entered the narrow opening of the flower, the recurved hairs at the entrance prevent them from emerging until the plant, so to speak, is ready to release them. The flies can be distinctly heard buzzing

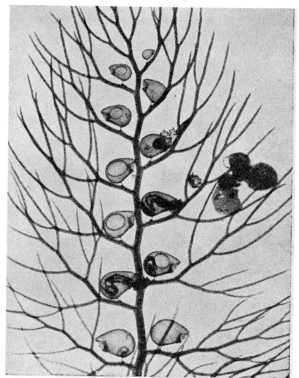

Fig. 263.—*Utricularia vulgaris*, an aquatic plant that traps small animals.   Several of the vesicles contain dipterous larvae.   (*After Matheson.*)

Fig. 264.—Salvinia and Riccia crowding the surface of a pool so completely that i⁻ find it difficult to reach the surface to obtain air.

in the bulbous portion of the flower and can be released artificially by slitting the base of the flower.

Insects form food for plants in other ways. Numerous fungi grow upon and obtain their nourishment from insects. *Cordyceps* are especially abundant upon wireworms and white grubs. Species of *Empusa* attack grasshoppers and flies. More than 40 genera of fungi have been reported to attack insects. The Lepidoptera, Coleoptera, and Homoptera seem to be the most outstanding hosts.

Insects may also transmit plant diseases, sucking insects, such as leafhoppers and aphids, being the common transmitters. Many viruses and bacterial wilts are transmitted by these insects. A few mandibulate insects, notably the Coleoptera, are also responsible for the transmission of plant diseases. The white-pine beetle is associated with the white-pine blister rust, and certain bark beetles are closely associated with the Dutch elm disease.

## BIBLIOGRAPHY

See also references to gall insects, leaf-mining insects, and boring insects.

### I. Food Relations between Insects and Plants.

BRUES, C. T. (1920). The selection of food-plants by insects, with special reference to lepidopterous larvae. *Amer. Nat.* **54** (633).

———— (1923). The choice of food and numerical abundance among insects. *Journ. Econ. Ent.* **16** (1).

———— (1924). The specificity of food plants in the evolution of phytophagous insects. *Amer. Nat.* **58** (655). *Literature cited.*

———— (1930). The food of insects viewed from the biological and human standpoint. *Psyche* **37** (1).

———— (1936). Aberrant feeding behavior among insects and its bearing on the development of specialized food habits. *Quart. Rev. Biol.* **11** (3). *Literature.*

CHAPMAN, ROYAL N. (1931). Nutrition. *In*, Animal ecology. McGraw-Hill Book Company, Inc., New York. *References.*

DIKMAN, A. (1931). Studies on the intestinal fauna of termites with reference to the ability to digest cellulose. *Biol. Bull.* **61** (1).

EVANS, A. C. (1938). Physiological relationships between insects and their host plants. *Ann. Appl. Biol.* **25** (3).

ᵀOWARD, N. F. (1941). Feeding of the Mexican bean beetle larva. *Ann. Ent. Soc. Amer.* **34** (4).

ᶜH, E. M. (1938). Food-plant catalogue of the aphids of the world. *Me. Agric. ᵀxp. Sta. Bull.* **393.**

ᵀ, A. S. (1939). Nutrition. *In*, Animal ecology. McGraw-Hill Book Com-ᵥ, Inc., New York.

PAUL (1913). Handbuch der Pflanzenkrankheiten. Die tierischenfeinde ᵣ Classe Insekten Band III Bearbeit von L. Reh.

P. (1928). Insects and metabolism. *Trans. Ent. Soc. Lond.* **31.** *Liter-*

ᵗ921). A summary of the food habits of North American Hemiptera. ᵧₙ *Ent. Soc.* **16** (5).

———— (1922). A summary of the food habits of North American Coleoptera. *Amer. Nat.* **56** (643).

———— (1924). Insect food habits and vegetation. *Ohio Journ. Sci. Columbus* **24** (2).

———— (1924). Ratios between the food habits of insects. *Ent. News* **35** (10).

———— (1925). Insect food habit ratios in Death Valley and vicinity. *Ohio Journ. Sci.* **25** (5).

———— (1925). Notes on the ratios of insect food habits. *Proc. Biol. Soc. Wash.* **38**.

———— (1926). The similarity of insect food habit types on the Atlantic and Western Arctic Coasts of America. *Amer. Nat.* **60** (666).

———— (1939). Insect food habit ratios of North Carolina and Mount Desert Island, Maine. *Journ. N. Y. Ent. Soc.* **47** (2). *References.*

**II. Resistance of Plants to Insect Attacks.**

CHESTER, K. S. (1933). The problem of acquired physiological immunity in plants. *Quart. Rev. Biol.* **8**. (2–3). *Literature.*

DAHMS, R. G., and FENTON, F. A. (1939). Plant breeding and selecting for insect resistance. *Journ. Econ. Ent.* **32** (1).

DeLONG, D. M. (1934). Insecticidally induced immunity in plants to sucking insects. *Science* **80** (2075). *Literature.*

FELT, E. P. (1931). Developing resistance or tolerance to insect attack. *Journ. Econ. Ent.* **24** (2).

FLINT, W. P., and BIGGER, J. H. (1938). Biological control of insects through plant resistance. *Can. Ent.* **70** (12).

HOWITT, J. E. (1924). A review of our knowledge concerning immunity and resistance in plants. *Quebec Soc. Prot. Plants 16th Ann. Rept.* 1923–1924. *Literature cited.*

JEWETT, H. H. (1933). The resistance of leaves of red clover to puncturing. *Journ. Econ. Ent.* **26** (6).

KOFOID, C. A., et al. (1934). Termite resistant materials. *In,* Termites and termite control. University of California Press, Berkeley.

MARTIN, HUBERT (1936). Host resistance. *In,* The scientific principles of plant protection. Edward Arnold & Co., London.

MUMFORD, E. P. (1931). Studies in certain factors affecting the resistance of plants to insect pests. *Science n. sr.* **73** (1880). *Literature.*

PAINTER, R. H. (1936). The food of insects and its relation to resistance of plants to insect attack. *Amer. Nat.* **70** (731). *Literature cited.*

TREHERNE, R. C. (1917). The natural immunity or resistance of plants to insect attack. *Agric. Gaz. Canada* **4** (10).

WARDLE, R. A. (1929). Host resistance. *In,* The problems of applied entomology. McGraw-Hill Book Company, Inc., New York.

**III. Feeding Punctures and Scars Caused by Insects.**

CARTER, W. (1939). Injuries to plants caused by insect toxins. *Bot. Rev.* **5** (5).

DAVIDSON, J., and HENSON, H. (1929). The internal condition of the host plant in relation to insect attack, with special reference to the influence of pyridine. *Ann. Appl. Biol.* **16** (3).

FULTON, B. B. (1920). Insect injuries in relation to apple grading. *Geneva Agric. Exp. Sta. Bull.* **475**.

HORSFALL, J. L. (1923). The effects of feeding punctures of aphids on certain plant tissues. *Pa. State College Bull.* **182**. *Bibliography.*

JOHNSON, C. G. (1937). The biology of *Leptobyrsa rhododendri* Horvath (Hemiptera, Tingitidae). The rhododendron lacebug. II. Feeding habits and the histology of the feeding lesions produced in rhododendron leaves. *Ann. Appl. Biol.* **24** (2). *References.*

KING, W. V. (1932). Feeding punctures of mirids and other plant-sucking insects and their effect on cotton. *U. S. Dept. Agric. Tech. Bull.* **296.**

KNIGHT, H. H. (1918). An investigation of the scarring of fruit caused by apple red-bugs. *Cornell Univ. Agric. Exp. Sta. Bull.* **396.** *Bibliography.*

—— (1922). Studies on insects affecting the fruit of the apple with particular reference to the characteristics of the resulting scars. *Cornell Univ. Agric. Exp. Sta. Bull.* **410.** *Bibliography.*

LEES, A. H. (1926). Insect attack and internal condition of the plant. *Ann. Appl. Biol.* **13** (4). *Bibliography.*

PORTER, B. A., et al. (1928). Some causes of cat facing in peaches. *Ill. Nat. Hist. Survey Bull.* **17** (6).

PUTNAM, WM. C. (1941). The feeding habits of certain leafhoppers. *Canad. Ent.* **78.** *References.*

SMITH, F. F. (1933). The nature of the sheath material in the feeding punctures produced by the potato leaf hopper and the three-cornered alfalfa hopper. *Journ. Agric. Res.* **47** (7). *Literature cited.*

——, and POOS, F. W. (1931). The feeding habits of some leaf hoppers of the genus *Empoasca*. *Journ. Agric. Res.* **43** (3). *Literature.*

SMITH, K. M. (1920). Investigations of the nature and cause of the damage to plant tissue resulting from the feeding of capsid bugs. *Ann. Appl. Biol.* **7** (1). *Bibliography.*

—— (1926). A comparative study of the feeding methods of certain Hemiptera and of the resulting effects upon the plant tissue, with special reference to the potato plant. *Ann. Appl. Biol.* **13** (1). *Literature cited.*

TATE, H. D. (1937). Method of penetration, formation of stylet sheaths and source of food supply of aphids. *Iowa State Coll. Journ. Sci.* **11** (2).

**IV. Relation between Insects and Fungi.**
See also food relations of wood borers in Chap. XIX.

BAILEY, I. W. (1920). Some relations between ants and fungi. *Ecology* **1** (3). *Literature.*

BEQUAERT, J. C. (1922). Ants in their diverse relations to the plant world. *Amer. Mus. Nat. Hist. Bull.* **45.** *Bibliography.*

FROST, S. W. (1936). Leaf cutting and fungus growing ants, ambrosia or fungus growing beetles, fungus growing termites. *In*, Ancient artizans. Van Press, Boston. *Bibliography.*

HENDEE, E. C. (1935). The role of fungi in the diet of the common damp-wood termite. *Hilgardia* **9** (10). *Literature.*

LEACH, J. G., et al. (1940). Observations on two ambrosia beetles and their associated fungi. *Phytopathology* **30.**

WEBER, N. A. (1940). The biology of fungus-growing ants. Part IV. *Revista de entomologia, Rio de Janeiro, Brazil* **11** (1/2).

—— (1941). The biology of the fungus-growing ants. Part VII. The Barro Colorado Island, Canal Zone species. *Revista de entomologia, Rio de Janeiro, Brazil.* **12** (1/2). *Literature cited.*

WEISS, H. B. (1921). A bibliography on fungus insects and their hosts. *Ent. News* **32** (2).

—— (1922–1924). Notes on fungous insects. *Can. Ent.* **54** (9), **55** (9); also *Psyche* **31** (5).

——, and WEST, E. (1920–1921). Fungus beetles and their hosts. *Proc. Biol. Soc. Wash.* **33–34.**

**V. Insects and Pollination.**

CAMMERLOHER, H. (1931). The mutual relations between flowers and insects. Blutenbiologie I Wechselbezieh ungen zruischen Blumen u. Insekten. Berlin Gebruder Bornstaeger.

DARWIN, CHARLES (1885). The habits of insects in relation to the fertilisation of flowers. *In*, The effects of cross and self fertilization in the vegetable kingdom. D. Appleton-Century Company, Inc., New York.

—— (1903). On the fertilization of orchids by insects. 2d ed. D. Appleton-Century Company, Inc., New York.

HENSLOW, G. (1907). On some remarkable adaptations of plants to insects. *Journ. Roy. Hort. Soc.* **32.**

KIRCHNER, O. VON (1911). Blumen und Insekten. Leipzig und Berlin.

KNOLL, F. (1921–1926). Insekten und Blumen. *Abh. zool.-bot. Ges. Wien.* **12.**

KNUTH, PAUL (1909). Handbook of flower pollination. Trans. by J. R. A. Davis. Oxford University Press, New York.

LEACH, JULIAN GILBERT (1940). The interrelationships between plants and insects. *In*, Insect transmission of plant diseases. McGraw-Hill Book Company, Inc., New York. *Bibliography.*

MÜLLER, H. (1883). The insects which visit flowers. *In*, The fertilization of flowers. Macmillan & Company, Ltd., London. *Bibliography.*

PARKER, R. L. (1926). The collection and utilization of pollen by the honeybee. *Cornell Univ. Agric. Exp. Sta. Mem.* **98.**

PEARSON, J. F. W. (1933). Concerning the apparent adaptations of certain bee species to plant communities, Part IV. Studies on the ecological relations of bees in the Chicago region. *Ecol. Monogr.* **3** (3). *Bibliography.*

PHILLIPS, E. F. (1933). Insects collected on apple blossoms in Western New York. *Journ. Agric. Res.* **46** (9).

PLATH, O. E. (1925). The role of bumblebees in the pollination of certain cultivated plants. *Amer. Nat.* **59** (664). *Literature cited.*

RILEY, C. V. (1873). The pronuba moth. *Trans. Acad. Sci. St. Louis* **3.**

—— (1880). Further notes on the pollination of yucca and on Pronuba and Prodoxus. *Proc. Amer. Assoc. Adv. Sci.* **29.**

—— (1892). The yucca moth and yucca pollination. *3rd Ann. Rept. Mo. Bot. Gardens.*

ROBERTSON, CHARLES (1924). Phenology of entomophilous flowers. *Ecology* **5** (4). *Literature cited.*

—— (1929). Flowers and insects. Science Press Printing Co., Lancaster; also *Ecology* **9** (4). *Literature.*

VANSELL, G. H. (1931). Nectar and pollen plants of California. *Calif. Agric. Exp. Sta. Bull.* **517.**

WILSON, G. F. (1929). Pollination of hardy fruits. Insect visitors to fruit blossoms. *Ann. Appl. Biol.* **16** (4).

VI. Symbiotic Relations of Insects and Plants.

BEQUAERT, J. C. (1922). Ants in their diverse relations to the plant world. *Amer. Mus. Nat. Hist. Bull.* **45.** *Bibilography.*

COUCH, J. N. (1938). The genus Septobasidium. Univ. North Carolina Press. *Bibliography.*

DYAR, H. G. (1928). Water bearing plants of Panama which harbor mosquitoes, with a synopsis of *Wyeomyia* (Diptera: Culicidae). *Proc. Ent. Soc. Wash.* **30** (6).

HOWARD, L. O. (1900). Smyrna fig culture in the United States. *Yearbook U. S. Dept. Agric.*

HUBBARD, H. G. (1896).   Some insects which brave the dangers of the pitcher plant. *Proc. Ent. Soc. Wash.* **3** (5).

KNAB, FREDERICK (1913).   Larvae of Cyphonidae in bromeliads.   *Ent. Mo. Mag.* **49** 2d ser. 24.

KUMM, H. W. (1933).   Mosquitoes breeding in bromeliads at Bahia, Brazil.   *Bull. Ent. Res.* **24** (4).   *References.*

LEACH, JULIAN GILBERT (1940).   The interrelationships of plants and insects.   *In,* Insect transmission of plant diseases.   McGraw-Hill Book Company, Inc., New York.   *Bibliography.*

NUTTALL, G. H. F. (1923).   Symbiosis in animals and plants.   *Amer. Nat.* **57** (652).

WHEELER, W. M. (1913).   Ants, their structure, development and behavior.   Columbia University Press, New York.

**VII. Insectivorous Plants.**

DARWIN, CHARLES (1883).   Insectivorous plants.   D. Appleton-Century Company, Inc., New York.

FOLSOM, J. W. (1934).   Entomology, with special reference to its ecological aspects. 4th ed.   Revised by R. A. Wardle.   The Blakiston Company, Philadelphia. *Bibliography.*

HEIM DE BALSAC, F., and HEIM DE BALSAC, H. (1932).   Captures d'insectes par certaines plantes.   *5th Congress Inter. Ent. Paris.*

JONES, F. M. (1921).   Pitcher plants and their moths.   *Nat. Hist.* **21.**

LEACH, JULIAN GILBERT (1940).   Entomophagous plants.   *In,* Insect transmission of plant diseases.   McGraw-Hill Book Company, Inc., New York.   *Bibliography.*

LLOYD, F. E. (1935).   *Utricularia. Biol. Rev.* **10** (1).

—— (1942).   The carnivorous plants.   *Chronica Botan.,* Waltham, Mass.

MATHESON, R., and HEINMAN, E. H. (1928).   *Chara fragilis* and mosquito development.   *Amer. Journ. Hygiene* **8** (2).   *Bibliography.*

——, and —— (1931).   Further work on *Chara* spp. and other biological notes on Culicidae (mosquitoes).   *Amer. Journ. Hygiene* **14** (4).   *Bibliography.*

RABAUD, E. (1921).   La capture des insectes par les plantes.   *Bull. Ent. Soc. France.*

SMITH, J. G. (1893).   Recent studies of carnivorous plants.   *Amer. Nat.* **27** (317).

WILLIAMS, AMY (1913).   Carnivorous plants of Ohio.   *Ohio Nat.* **13** (5).

**VIII. Insect Vitamins.**

BACOT, A. W., and HARDEN, A. (1922).   Vitamin requirements of *Drosophila* I. *Biochem. Journ.* **16.**

CHAPMAN, ROYAL N. (1931).   Animal ecology.   McGraw-Hill Book Company, Inc., New York.

CRAIG, R., and HOSKIN, W. M. (1940).   Vitamins.   *In,* Insect biochemistry.   *Ann. Rev. Bio. Chemistry* **9.**   *References.*

IMMS, A. D. (1937).   Vitamins.   *In,* Recent advances in entomology.   The Blakiston Company, Philadelphia.

McCAY, C. M. (1940).   Insect life without vitamin A.   *Science* **92** (2387).

RICHARDSON, C. H. (1926).   A physiological study of the growth of the Mediterranean flour moth, *Ephestia kuehniella* Zeller.   *Journ. Agric. Res.* **32** (10).

SWEETMAN, M. D., and PELMER, L. S. (1928).   Insects as test animals in vitamin research.   *Journ. Biol. Chem.* **77** (1).

UVAROV, B. P. (1928).   Vitamins.   *In,* Insects and metabolism.   *Trans. Ent. Soc. Lond.* **31.**

# CHAPTER XVI

## LEAF-MINING INSECTS

Leaf miners, like aquatic insects, form a natural rather than a taxonomic group of species which are modified for an existence in a special type of environment. As a matter of fact, the two environments are somewhat similar because leaf-mining larvae live in a semiaquatic situation. The cell sap of the leaves has about the same consistency as the body fluids of the insect, and the mines frequently contain much moisture that is derived from ruptured cells.

**Leaf Miner Defined.**—A leaf miner is a species, the larva of which lives and feeds, for a part or all of its existence, between the two epidermal layers of a leaf. The mining habit converges on the one hand with the borer and the gall maker and on the other hand with the external feeder and the scavenger. There is little difference between a borer and a miner except that a boring insect feeds deep in the tissues and a miner feeds just below the surface of the portion of the plant attacked. The codling moth, a typical borer in fruit, occasionally comes to the surface and mines for a short distance just beneath the skin of the fruit.

Fig. 265.—Plum leaf bearing a number of mines of *Nepticula slingerlandella.* (*After Crosby.*)

*Marmara arbutiella*, of British Columbia, normally mines in the cambium of twigs but, when crowded, extends its operations into the leaves. As a matter of fact, the species of *Marmara* show a diversity of habits. Many mine in cambium; some mine in leaves or the joints of cacti and a few mine beneath the skin of fruit. The Agromyzidae also show habits intergrading between leaf mining and boring. *Agromyza youngi* mines in the petioles and flower stalks of dandelion as readily as it does in the leaves.

319

Although the leaf-mining and gall-making habits are usually distinct, the same species may do both. *Agromyza laterella* produces a gall on

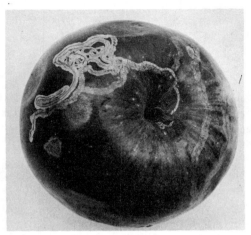

Fig. 266.—*Marmara pomonella* mining beneath the skin of apple.

the young unfolding leaf but a mine on the mature leaf of iris. The following table shows the interrelation of habits in the genus *Agromyza*.

RELATION OF LEAF MINING TO OTHER HABITS IN THE GENUS *Agromyza*

| Species | Habits | Food plants |
|---|---|---|
| *Agromyza tiliae*............. | Gall maker | Citrus |
| *Agromyza schineri*.......... | Gall maker | Poplar |
| *Agromyza websteri*.......... | Gall maker | Wisteria |
| *Agromyza simplex*.......... | Stem borer | Asparagus |
| *Agromyza aeniventris*........ | Pith borer | *Trifolium, Ambrosia, Helianthus* |
| *Agromyza virens*............. | Pith borer | *Ambrosia, Helianthus, Nabalus* |
| *Agromyza amelanchieris*...... | Cambium miner | *Amelanchier canadensis* |
| *Agromyza aceris*............. | Cambium miner | *Acer rubrum* |
| *Agromyza pruinosa*.......... | Cambium miner | *Beta nigra* |
| *Agromyza* sp............... | Cambium miner | *Baptisia tinctoria* |
| *Agromyza laterella*.......... | Leaf miner and gall maker | Iris, mine on mature leaves, gall on young leaves |
| Remaining species.......... | Largely leaf miners | Herbaceous and woody plants |

The Cecidomyiidae also have many forms intergrading between leaf mining and gall making. The touch-me-not gall, produced by *Cecidomyia impatientis*, shows unmistakable proliferation of cells and modification of the plant tissues. On the other hand, the tulip spot gall, *Thecodipolosis liriodendri*, is a mere blister on the leaf which is only slightly thickened. The tar-spot gall, produced by species of *Asteriomyia*, shows even less thickening of the leaf and is scarcely more than a leaf mine. The boxwood leaf miner, *Monarthropalpus buxi*, makes a definite mine but the tissues proliferate within the mine and the leaf is distinctly thickened. This species presents a combination of the leaf-mining and gall-making habits. None of the Cecidomyiidae produce typical leaf mines.

The habits of surface feeders may approximate those of the leaf miners. Certain chironomid larvae dig channels in leaves but remove the upper epidermis and close the opening with a layer of frass pellets. A southern caterpillar, *Homaledra heptathalama*, feeds upon the palmetto in a similar manner. It excavates a burrow in the surface of the leaf and covers the channel with frass pellets webbed together with silk. Neither of these species is a true leaf miner because it does not feed between the two epidermal layers.

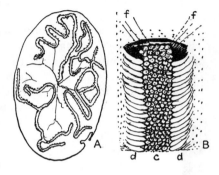

**Economic Importance of Leaf Miners.**—Succulent leaves such as ragweed, parsley, and delphinium often wilt as the result of the attack by leaf miners. The leaves of woody plants usually remain flat and turgid in spite of their operations. Even the leaves of columbine do not wilt, although the extensive, scriptlike, tortuous mines may cover nearly the entire surface of the leaf. The greatest distortion results from the

Fig. 267.—Channels resembling mines in the upper surface of leaves: above, (*A*) galleries produced by the larvae of *Chironomus braseniae* in the leaf of water shield, *Brasenia*; (*B*) details of the gallery showing pellets of frass forming a roof over same (*after Leathers*); below, a gallery of *Homaledra heptathalama* with its frass covered roof on palmetto. (*After Busck.*)

operations of the tentiform miners, which spin silk within their mines.

Miners frequently attack the leaves of vegetables, fruit, or ornamental plants, causing serious injury. The spinach and beet are mined by *Pegomyia hyoscyami*. Sprays are usually ineffective against these miners. Growers have learned to recognize the chalky white eggs of this species, which are laid on the undersurfaces of the leaves. When they determine that the flies are beginning to oviposit, they cut the spinach prematurely to avoid injury. Several generations occur during the summer, and one must be alert to avoid injury by the later generations.

Nasturtium, delphinium, and columbine are often disfigured by the mines of the Agromyzidae. The oaks are attacked by more than 50 species of leaf miners; none except *Brachys ovata* has been found abundant enough to prove serious.

The hispid, *Chalepus dorsalis*, causes serious damage to locusts by its feeding and by its mining operations. The adults at first do considerable damage feeding upon the foliage; later, the larvae mine in the leaves. This species often causes the trees to appear by fall as though fire had swept through them. Fortunately, the feeding habits of the adults allow ample time to kill them with arsenate of lead before they lay their eggs and the larvae start mining the leaves.

The birch leaf miner, *Fenusa pumila*, is an introduced species which has caused serious damage in the New England States. The larvae make blotch mines on the tender, terminal leaves of the gray birch, the paper birch, and the European white birch. Several larvae may unite to mine or remove the entire contents of a leaf. Nicotine sulphate and summer oil emulsion apparently penetrate the thin, parchmentlike surfaces of the mines and kill the larvae.

The boxwood leaf miner, *Monarthropalpus buxi*, is known to attack 11 varieties of *Buxus sempervirens* and to inflict serious damage. The dwarf English boxwood is seldom infested and the varieties, *pendula* and *argentovariegata*, are apparently immune. The loss is especially serious when boxwoods fifty or seventy-five years old are attacked. Cyanide fumigation and hot-water treatments have been used with more or less success in reducing this pest. Plants dug for transportation are usually treated with water at a temperature of 120°F. for ten minutes in July and August and for five minutes in October or March to May.

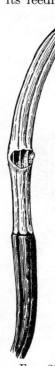

Fig. 268. Leaf of tamarack mined by the larch casebearer, *Coleophora laricella*. Only the head and anterior portion of the larva enter the leaf. (*After Herrick.*)

The larch casebearer, *Coleophora laricella*, is also a leaf miner. It frequently damages the young tender leaves of tamaracks. In the fall, the larvae move their minute cases to the branches, where they remain during the winter. As soon as the buds begin to push in the spring, these larvae migrate to the new unfolding leaves. The larva attaches its case to a leaf, cuts a hole through the epidermis, and mines out the tissues as far on each side of the entrance hole as it can reach. During this time it does not let go its hold on the case but grasps it securely by its posterior pair of prolegs. When the larva has eaten as much of one leaf as it can reach, it moves to another. It has been estimated that a single larva may destroy 129 leaves. During the feeding of the larva, the case is enlarged by adding small pieces of leaves to the sides. The larva pupates late in May and the adult emerges shortly afterwards. The female lays her small, sculptured, brown eggs singly upon the leaves. There is only one brood a year.

The European elm sawfly, *Kaliofenusa ulmi*, often causes conspicuous injury to American elms. Twenty or more larvae may mine in a single leaf and the mines frequently coalesce to form a large blotch, sometimes occupying the interior of the whole leaf. The pest can be effectively controlled by spraying the foliage, as soon as the mines begin to appear, with nicotine sulphate and soap or summer oil emulsion.

**Extent of the Leaf-mining Habit.**—Leaf miners attack nearly all families and groups of plants. They mine in plants with milky juices, in plants poisonous to higher animals, and occasionally in aquatic plants.

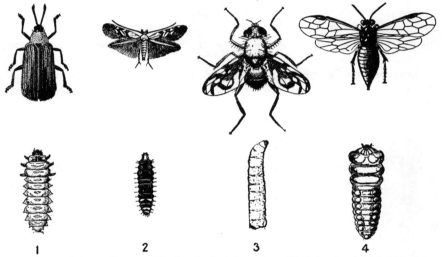

1      2      3      4

FIG. 269.—Adults and larvae of the four leaf mining orders: (1) Coleoptera; (2) Lepidoptera; (3) Diptera; (4) Hymenoptera.

Leaf miners are most numerous in the Tropics but have a fair distribution in the temperate climate. They are apparently rare or nonexistent on high mountain tops and in the Arctic zone even where plants are abundant. They appear to be absent in the Katmai district of Alaska but occur in Labrador and are numerous in southern Finland.

Insects of four orders have developed the leaf-mining habit: Coleoptera, Lepidoptera, Diptera, and Hymenoptera. The habit occurs only in orders with complete metamorphosis, *i.e.*, in orders that generally show the highest degree of specialization or development. The adults are winged, active, and able to select the proper food plants; the larvae are adaptable to the unusual manner of feeding. The larvae and pupae may occupy the interior of leaves but the adults are not fitted for this type of living. The adults are comparatively small but are often gorgeous creatures clad in silver, gold, ermine, and jet.

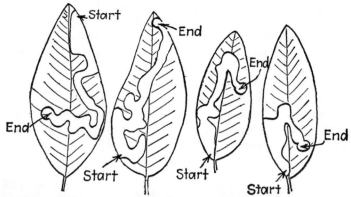

FIG. 270.—Mines of *Pachyschelus laevigatus* on the leaves of Desmodium. (*After Weiss and West.*)

**Leaf-mining Coleoptera.**—There are about 50 North American species of leaf-mining beetles, widely scattered in three families: Buprestidae, Chrysomelidae, and Curculionidae. The wood-boring habit is predominant in the Buprestidae and the mining habit is more or less limited to the genera *Trachys*, *Brachys*, and *Pachyschelus*. The mines of the Buprestidae are usually of the blotch type although some species make very broad linear mines. The eggs are, as a rule, deposited on the lower surface of the leaf and are covered with a transparent secretion which glistens conspicuously long after the larvae have matured and left the leaf. They generally mine the leaves of trees and shrubs although *Taphrocerus* mines the leaves of sedges and a few species mine the leaves of Leguminosae. The larvae usually spin cocoons in the mine or cut circular pieces from the upper and lower epidermis of the leaf, fastening these together by means of silk. The species are minute, often less than a millimeter in length. They contrast conspicuously with the tropical buprestid, *Euchroma goliath*, which is more than two and a half inches long. This species, is, however, a borer.

FIG. 271.—Feeding scars of adult and mine of *Baliosus ruber* on an apple leaf.

The Chrysomelidae are primarily leaf feeders and the leaf-mining habit fits here naturally. The majority of the leaf-mining species occur in the Hispinae and in the genera *Chalepus* and *Microrhopala*. A few species occur in the genera *Stenispa, Anoplitis, Baliosis,* and *Uroplata.* The adults are somewhat similar in shape and have been called wedge-shaped beetles. The typical species are often conspicuously spiny but the North American species are never prominently spined although they are delicately ridged and sculptured. The larvae likewise are quite similar in form. They are flattened, have well-developed short legs, and closely resemble the Chrysomelidae that feed externally on the foliage. They mine the leaves of weeds, grasses, or occasionally trees, and make large, irregular, blotch mines, where pupation generally occurs.

Other species of leaf-mining Chrysomelidae belong to the sub-family Halticinae. One of the outstanding species is *Dibolia borealis* which makes a long tortuous linear mine on the leaf of the broad-leaved plantain. When these insects are abundant, the leaves of plantain are completely mined and the plant withers as a result of the attack of this species.

Another somewhat common chrysomelid is *Hippuriphila modeeri*, which makes a small blotch mine on the leaves of *Rumex.* This species is minute and brown in color.

The leaf-mining Curculionidae belong primarily to the genus *Orchestes.* The species is more numerous in Europe than in America. They mine the leaves of oaks, alders, and willows. The apple flea weevil, *Orchestes pallicornis,* is somewhat injurious to fruit in the eastern part of the United States. It has been taken on the leaves of elm, willow, and shadbush and has been found mining the leaves of alder, cherry, and apple.

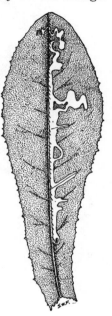

Fig. 272.—Linear mine of *Phytomyza lactuca* on the underside of the leaf of wild lettuce, *Lactuca scariola integrata.*

A single species of the genus *Prionomerus* occurs in Northeastern North America. *Prionomerus calceatus* mines the leaves of sassafras and tulip. The larvae make conspicuous blotch mines which cover nearly one-half the area of the leaf. The adults feed considerably on the leaves before they lay their eggs, and make round holes in the leaves somewhat like the feeding of flea beetles. The cocoon is formed in the mine and surrounded by a mass of fecula which causes the mine to bulge conspicuously.

**Leaf-mining Diptera.**—There are about 200 North American species of leaf-mining Diptera. The habit is limited chiefly to the Agromyzidae and Anthomyiidae. A single species of Tipulidae, a few species of Trupaneidae, several Cecidomyiidae, and a few other Diptera also mine leaves. Various types of mines are produced by these species. Some of the Agromyzidae make linear mines. The mines of *Phytomyza obscurella nigritella* on peach and the mines of a species of *Agromyza* on nasturtium are excellent examples. Blotch mines are not so abundant in the Diptera

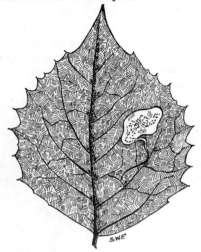

as generally believed. *Agromyza coronata* produces a small blotch on goldenrod and *A. posticata* produces a large blotch on the same plant. Linear blotch mines are produced by species such as *A. aristata* on elm or *A. borealis* on jewel weed.

**Leaf-mining Hymenoptera.**—Leaf-mining Hymenoptera, with few exceptions, occur in a compact group known as the sawflies, Tenthredinidae. They are predominantly free feeders; about 15 North American species mine leaves; a few bore in the stems of plants. The leaf-mining species fall in five subfamilies: Phyllotominae, Holocampinae, Dineurinae, Scolioneurinae, and Schizocerinae. They are conspicuously blotch miners.

FIG. 273.—Linear-blotch mine produced by *Agromyza melampyga* on the upperside of the leaf of Syringa.

There is an interesting similarity of form between the sawfly larva, the hispid larva, and the primitive lepidopterous larva. All are conspicuously

SUMMARY OF THE HABITS OF THE SUPERFAMILY TENTHREDINOIDEA

| Family | Habits | Approximate number of genera and species | Distribution |
|---|---|---|---|
| Blasticotomidae...... | Stem miners | 1 genus, 1 species | European |
| Xyelidae............. | Free feeders | 7 genera, 25 N. Am. species | Chiefly N. Am. |
| Tenthredinidae....... | Free feeders, gall makers, leaf miners | 58 genera, many species | Europe and N. Am. |
| Pamphiliidae......... | Leaf rollers, casebearers | 9 genera, 55 N. Am. species | Peculiar to Northern Hemisphere |
| Megalodontidae...... | Leaf rollers, casebearers | 4 genera, 35 species | Europe, Asia, N. Africa |
| Xiphydriidae......... | Borers | 4 genera, 8 N. Am. species | Europe, N. Am. |
| Siricidae............. | Borers | 5 genera, 60 species | Largely Northern Hemisphere |
| Cephidae............ | Borers | 14 genera, 16 N. Am. species | Intercontinental |

flattened and have greatly reduced legs.    Even the mines are strikingly alike, for they are of the blotch type.

**Leaf-mining Lepidoptera.**—The leaf-mining habit reaches its highest development in the Lepidoptera.    There are about 400 North American species scattered in about 20 families.    The majority of the species occur in the families Nepticulidae, Tischeriidae, Gracilariidae, and Gelichiidae.

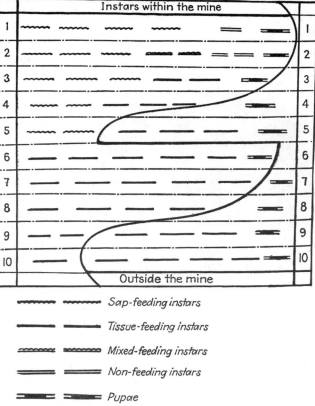

Fig. 274.—A summary of the methods of mining in the Lepidoptera.

They are so numerous that the early workers, such as Swammerdam, DeGeer, and Réaumur, discussed these species almost to the exclusion of other orders.    Within the Lepidoptera one finds all the habits and variations found in other leaf-mining orders, as, for example, types of mines, methods of feeding, frass disposal, and larval forms.    The life cycles are often complicated and the species show extreme specialization. A glance at Fig. 274 shows something of the feeding habits of the larvae. The habits of many of the species intergrade with gall making and leaf rolling.

The larvae of the leaf-mining Lepidoptera are more specialized than those of other orders and depart widely from related free-feeding forms. The legs, antennae, and eyes are often absent or considerably reduced. The head is often held in a horizontal position and is frequently tele-scoped into the thorax. The mandibles of the sap feeders have numerous teeth for cutting the cells and causing the sap to flow. These are highly developed in the genus *Phyllocnistis* and in certain *Lithocolletis*.

The Lepidoptera present numerous types of mines but the linear mine predominates. Shallow linear mines, with little or no frass, are characteristic of the sap-feeding *Phyllocnistis*. The trumpet mine is a distinct type but not common. The best example is the apple leaf trumpet miner, *Tischeria mali-foliella*. Short broad linear mines with prominent central frass lines are common among the Nepticulidae. Small blotch mines are procured by the larvae of *Cosmopteryx*, and certain of the Nepticulidae, Lyo-netiidae and Gracilariidae. These mines are generally preceded by short scriptlike linear mines; the linear portion is usually obliterated in the completed mine. Large blotch mines are somewhat rare in the Lepidoptera. The larvae of *Lithocolletis aceriella* often cooperate to form a large blotch mine on the maple leaf. Many unique mines are produced by the Lepidoptera, as for example, the intricate mine produced by the larva of *Chrysopora* on *Chenopodium* which turns back and forth to form a peculiar looped mine.

Fig. 275.—Mine of *Phyllocnistis popul-iella* on poplar, showing a prominent central frass line.

*Paraclemensia acerifoliella* produces characteristic elliptical cases on the leaves of maple. This species mines only in the early stages. The young larva cuts an oval piece of leaf, places it over its back, and fastens it down with silk around the edges. This forms a shelter beneath which it lives. As the larva grows, it reaches beyond the edge of the original case and cuts a new oval piece slightly larger than the original piece and fastens it to the outer edge of the smaller piece, the larva taking its position between the two pieces. It then turns the case over so that the smaller piece is upon the leaf and the larger piece is upon its back. The larger piece is then fastened at intervals to the leaf by means of silk. The larva rests between the two pieces and feeds by protruding its head and taking leaf tissue that it can reach without leaving the case. When it wishes to obtain fresh food, it cuts the case loose and moves to a new location on the leaf. When walking with the oval case, it looks like a tiny turtle.

One of the most remarkable species is the basswood leaf miner, *Lithocolletis lucetiella*. This species has received the common name, polygon leaf miner, because the mines are small and often rectangular in form. The larva feeds from the outside toward the middle of the mine. Thus, the larva must mark the final area to be mined before it begins to feed; otherwise, it would not have sufficient food to mature. During the first two instars, it is a sap feeder working on the lower surface of the mine, feeding upon the liquid contents of the cells, and voiding little or no fecula. After the third moult, the larva becomes a tissue feeder but does not extend the range of the mine. It goes back over the rectangular area previously indicated and strips all the cells between the upper and the lower epidermal layers. The fecula is deposited in minute pellets at the outer edge of the mine and the larva, always feeding toward the center of the mine, has fresh food uncontaminated by fecal discharges. This species contrasts with another, *Lithocolletis lucidicostella* (*aceriferella*), that makes polygonal mines on the leaves of maple but deposits the fecula in the center of the mine. Later the larva constructs a tentiform mine by the customary method of spinning silk within the mine.

Two comprehensive works cover the field of leaf-mining insects. Dr. Martin Hering treats the European species in "Die Ökologie der blatt-minierenden Insektenlarven," published in 1926. Needham, Frost, and Tothill discuss the North American species in "Leaf-mining Insects," published in 1928.

**Eggs of Leaf Miners.**—The eggs of leaf miners differ little from those of other species and are generally laid upon or within the leaf. However, the ermine moth, *Yponomeuta malinellus*, lays its eggs upon a twig, and the young larvae migrate to the leaves in spring. Probably all the

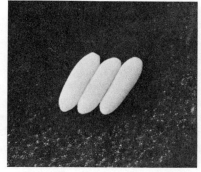

Fig. 276.—Eggs of *Pegomyia vanduzei*, a common miner on the leaves of Rumex.

Buprestidae lay their eggs upon the surfaces of leaves and they are frequently covered with fecula or regurgitated food. The leaf-mining

Chrysomelidae also lay their eggs upon the surfaces of the leaves. As far as known, the Curculionidae lay their eggs singly within leaves. The Hymenoptera, with few exceptions, insert their eggs within leaves. They are oval, flattened, and usually transparent or pale in color. They swell considerably in size after they are laid and often discolor the leaf, rendering it more conspicuous than when the eggs were first laid. The eggs of the Diptera are laid within or upon the surfaces of the leaves.

**Oviposition and Entrance of Leaf Miners into Leaves.**—It makes little difference whether the egg of the miner is laid upon or within the leaf. The Agromyzidae, Trupaneidae, Tenthredinidae, and the specialized Lepidoptera have highly developed ovipositors and insert their eggs within the leaves of their hosts. The European *Phytomyza varipes* has an ovipositor which is as long as the remainder of the abdomen. The

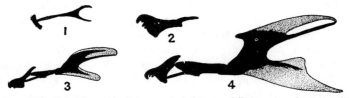

Fig. 277.—Mouth hooks and pharyngeal skeletons of *Pegomyia calyptrata*: (1) of first instar; (2) and (3) of second instar; (4) of third instar. The first instar has sawlike mandibles to cut its way through the egg and into the leaf.

Curculionidae generally chew holes into which they lay their eggs. The Anthomyiidae, Chrysomelidae, and the primitive Lepidoptera have generalized membranous ovipositors and, not being able to insert their eggs within the tissues, lay them on the surface of the leaves. The larvae of these species have well-developed mandibles by which they cut their way through the egg shells and gain entrance into the leaves. The Anthomyiidae show the most remarkable development. The mouth hooks of the first instar larvae are sawlike and adapted for cutting their way into leaves. After the first moult, these hooks are replaced by normal, triangular mouth hooks.

**Larvae of Leaf Miners.**—The larvae of leaf miners may spend their entire existence in the leaf or feed there for only a few instars. Some mine for one or more instars, then emerge and feed externally. Leaf rolling or gall making may follow leaf mining. Occasionally, stem borers migrate to leaves, where they mine.

Leaf-mining larvae are usually flat in form; the legs and setae are usually absent or reduced. The head is often rotated to a horizontal position and the posterior margin is frequently obsolete and telescoped into the thorax. There is a tendency for the formation of a wedge-shaped head, which is an efficient device in separating the two epidermal layers of the leaf as the larva moves forward. The antennae and eyes are often reduced. In the highly specialized forms, the eyes may be arranged along the sides of the head. In *Laverna* they are placed almost in a straight line; in *Tischeria eikbladhella* they form a straight line with the first three and the last three grouped. The jaws are usually sharp and provided with powerful muscles. In the sap-feeding larvae, the mandibles are platelike with many sharp teeth which are well adapted for cutting through plant cells and causing the sap to flow. Sclerotized

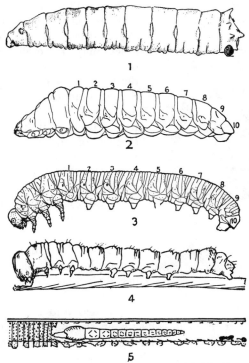

Fig. 278.—Leaf mining larvae: (1) *Hylemyia* (Diptera); (2) *Metallus rubi* (Hymenoptera); (3) external feeding sawfly; (4) external feeding caterpillar; (5) leaf-mining caterpillar.

plates or tubercles are often developed to aid the larva in holding itself within the mine as it feeds.

**Pupae.**—When the time for pupation arrives, leaf miners have two choices: to transform within the mine or to make their exit and transform elsewhere. In a few cases, the naked pupa rattles like a seed within the dried walls of the mine. In other cases, the pupa is attached to the interior surface of the mine by means of frass, silk, or sharp spiracles which pierce the leaf surface. Silken cocoons are often spun in the mine. This is especially true of Lepidoptera, Hymenoptera of the genus *Phyllotoma*, and Coleoptera of the genera *Pachyschelus* and *Orchestes*.

*Prionomerus* incorporates considerable frass in its cocoon and makes a large ball that causes the leaf to bulge. *Coptodisca* and certain of the Buprestidae cut neat, circular pieces from the epidermal layers of the leaf and fasten these together with silk to form pupal chambers. Other species make their exit through holes in the leaf and pupate elsewhere.

Larvae may escape from the leaves in different ways. Sometimes the parchment like surface of the mines crack and permit the larvae to

FIG. 279.—Pupal skin of a species of *Gracilaria* protruding from an empty tentiform mine on dogwood.

escape. The larvae often make their exit by means of holes cut through the surfaces of the leaf. The exit holes may be circular, semicircular, or straight slits. Each species has a definite manner of making its escape. Some emerge through the upper epidermis but the greater number make their exit through the lower surface of the leaf. The pupae are then formed on the surface of the leaf, beneath rubbish on the ground, or in the soil.

**Food of Leaf Miners.**—The leaf-mining species have adjusted themselves to a specialized method of obtaining their share of leaves, which form the food of the majority of insects. They secure sustenance and shelter within the confines of a single leaf and sometimes in a small portion of that. The food consists of the palisade cells, the paren-chyma, or sap rasped from these cells. Veins sometimes impede the progress of miners or confine their operations to small areas of the leaf and determine, to some extent, the type of mine produced On the other hand, many of the species cut through the veins and reach out into the whole area of the leaf. If latex cells are cut, the secretions that pour forth may drown the larvae. *Agromyza pusilla*, mining the leaves of milkweed, avoid these cells. Resins make mining difficult for some species. Alkaloids and other toxic principles apparently do not deter the activities of most leaf miners, for larvae are found in the leaves of poison ivy and other plants poisonous to higher animals.

Some species have developed the habit of entering new leaves when their food supply is exhausted or when the leaves in which they are mining wilt or become undesirable for occupation. This habit is highly developed in certain Diptera and occasionally occurs in the Lepidoptera and Hymenoptera. Species of the genus *Pegomyia*, especially those that mine the leaves of beet, spinach, *Chenopodium* and *Rumex*, frequently enter fresh leaves. Their mines are large; the leaves are succulent and often wilt before the larvae mature, forcing the larvae into new leaves. A few Lepidoptera of the genera *Acrocercops*, *Argyresthia*, *Scythris*, and *Aphelosetia* also have this habit. *Aphelosetia orestella*, for example, mines the basal overwintering leaves of grass, *Hystrix hystrix*, and transfers in the spring to new leaves. The *Coleophora* move freely from one

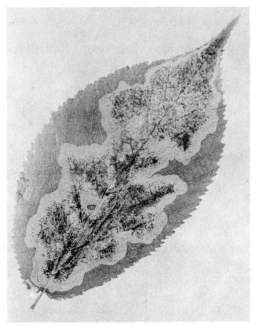

Fig. 280.—Mine of *Lithocolletis ostryaefoliella* on hop hornbeam, showing frass smeared on the lower wall of the mine. The periphery is unsoiled and reserved for feeding.

leaf to another because they have the habit of feeding from cases and can reach only a small area of the leaf.

Blotch miners tear down and consume quantities of cells. Some feed on the palisade cells, others feed on the parenchyma. *Lithocolletis ostryaefoliella* feeds only on the palisade cells. The whole area of the leaf may be excavated but the mine is exceedingly shallow. When the mine is completed, the parchmentlike upper epidermis can be removed like a sheet of paper.

Linear miners generally cut the cells and feed upon cell sap. They often produce mines that are exceedingly shallow. *Phyllocnistis populiella*, for example, makes a beautiful, glistening, white mine on poplar, with a contrasting black central frass line. Another species of *Phyllocnistis*, which mines the leaves of tulip, makes a much shallower mine which is silvery white because of the film of air separating the upper epidermis from the bottom of the mine and because the frass is sparse. Some Diptera also feed in this manner. *Phytomyza lactuca* produces a shallow mine on wild lettuce with little or no evident frass.

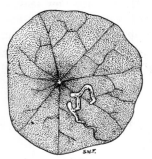

FIG. 281.—A linear mine on nasturtium produced by *Agromyza propepusilla*.

Within the genus *Lithocolletis*, there are two well-marked groups of species: the *Lithocolletis* group and the *Cameraria* group. Each is characterized by a distinct type of feeding. In the *Lithocolletis* group, the larvae are flat in form until the fourth larval instar after which they become cylindrical in form with a head like an ordinary free-living caterpillar. With this change in form, there is a change in feeding habits from a sap feeder to a cell feeder. In the *Cameraria* group, the larvae retain their flat form and continue as sap feeders through the later instars. In the last two instars, the larvae cease feeding and prepare for pupation. The *Lithocolletis* group represents a distinct type of hypermetamorphosis for there is more than one form of larva.

The larvae of the Coleophoridae never actually enter the leaf. Many live in small cases and push their heads into the leaves as far as their bodies can reach without letting go of the cases. Naturally their feeding range is limited and, to obtain new plant tissue, they move their cases to another portion of the leaf or to other leaves. A leaf thus mined shows several small mines, each with a hole through which the larva has fed.

**Forms of Mines.**—Each species makes a characteristic type of mine. "They write their signatures in the leaves," a habit which greatly assists in determining the species. There are two general types of mines: linear and blotch. Modifications of these are the linear-blotch, the trumpet, the digitate, and the tentiform mines. In the linear-blotch, the larva at first produces a linear

FIG. 282.—The linear-blotch mine of *Agromyza aristata* on elm, showing double, central frass lines.

mine, then suddenly expands it into a blotch. This type of mine occurs commonly among the Diptera as, for example, the mine of *Agromyza melampyga* on syringa and *A. borealis* on *Impatiens*. If the mine is broadened gradually, a trumpet mine is formed. An excellent example is the trumpet leaf miner of apple, *Tischeria malifoliella.* The tentiform mines are produced by larvae which spin silk within the mine. When this silk shrinks, the mine is thrown into a fold. This habit is particularly striking in the genus *Parornix.*

**Position of the Mines.**—Mines may occur on the upper or lower surface of the leaf or they may alternate from one surface to the other. Some mines are visible from both surfaces. As previously mentioned, the nature of the feeding determines to some extent the position of the mine. Species that feed upon the palisade cells are visible only from the upper side of the leaf; those that feed on the parenchyma are visible

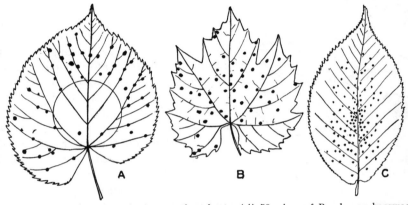

FIG. 283.—Distribution of mines on three hosts: (*A*) 50 mines of *Brachys* on basswood; (*B*) 50 mines of *Antispila viticordifoliella* on grape; (*C*) 100 mines of *Lithocolletis ostryarella* on hophornbeam.

only from the lower side. Species that eat both the parenchyma and the palisade cells are visible from both sides of the leaf. Some mines occur near the center of the leaf, others only along the edge of the leaf. Figure 283 shows the distribution of *Brachys, Antispila,* and *Lithocolletis* on three different food plants.

**Frass Disposal.**—The matter of frass disposal is one of the most serious problems confronted by leaf miners, confined as they are between the two epidermal layers of the leaf. External-feeding larvae let their fecula drop but leaf miners must devise some means of getting rid of it. Although the majority of the leaf-mining larvae discard their fecula as worthless, a few species use it in the construction of pupal cases. The pupa of *Prionomerus* is enclosed in a large ball of black fecula. Some Lepidoptera mine and then construct pupal cases of frass.

There are three general methods of disposing of the waste material: it may be laid down in the mine, it may be pushed from the mine, or the larva may abandon the mine and start a fresh one in another leaf. The primitive miners make blotch or blotchlike mines and the frass is often spread in an even layer on the lower surface of the mine. In such cases the larva feeds at the outer edge of the mine, where the food is uncontaminated. This is well illustrated in *Lithocolletis ostryaefoliella*, which makes a large blotch mine on hop hornbeam. The completed mine may cover the entire area of the leaf. The larvae feed only on the palisade cells, the upper epidermis becomes parchmentlike, and the black frass is smeared over the entire floor of the mine.

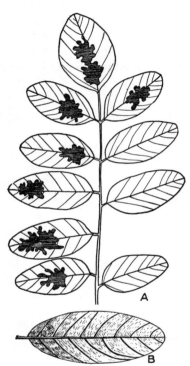

The highly specialized species dispose of the frass in a more elaborate manner. The larvae of *Scaptomyza adusta* make digitate mines on the leaves of Cruciferae and deposit the frass in the fingerlike prolongations of the mines. The digitate miner on locust, *Parectopa robiniella*, disposes of its frass in another manner. In addition to the characteristic mine on the upper surface of the leaf, it constructs small subcuticular mine on the lower surface of the leaf, into which the cast skins and fecula are deposited.

*Lithocolletis lucetiella* makes rectangular-shaped mines on the leaves of basswood. The larva feeds toward the center of the mine and deposits its sparse fecula at the outer edges of the mine. On the other hand, *L. lucidocostella* feeds from the center of the mine to the outside and deposits its frass in the center.

The linear and serpentine miners have other means of disposing of their fecula.

Fig. 284.—Mines of *Parectopa robiniella* on locust: (*A*) digitate mines on the uppersides of the leaflets; (*B*) small mine on underside of leaflet into which frass and cast skins are deposited.

As the larvae feed, they progress forward leaving the frass behind and find a clean supply of food in front of them. In some species, the frass is laid down in a single continuous central line. This is prominent in *Phyllocnistis*, in Nepticulidae, and in certain of the Agromyzidae. Other larvae form interrupted central frass lines. This is well illustrated by the mine produced on peach by *Phytomyza nigritella*. Still other larvae deposit the frass in two parallel lines, one on each side of the mine. In these cases, the larva swings the posterior end of the body from one side of the mine to the other as it moves along. The larva of the elm leaf miner, *Agromyza aristata*, feeds in this manner.

Many species push their fecula out of holes made through the surface of the mine. *Parectopa robiniella* has already been mentioned, but this species pushes its frass into another portion of the leaf. *Lyonetia speculella*, a species that commonly mines the leaves of apple, cuts several holes to the exterior and pushes the frass out of the mine. The frass emerges in small pellets which adhere to one another like minute links of sausage. Some linear mines follow the edge of the leaf and the

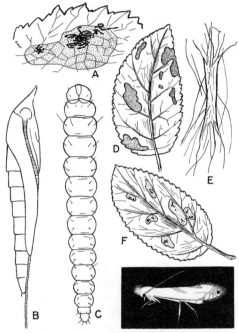

Fig. 285.—The apple leaf miner, *Lyonetia speculella*: (*A*) a bit of leaf showing frass extruded from holes through lower side of mine; (*B*) pupa; (*C*) larva; (*D*) upper side of leaf showing mines; (*E*) white silken cocoon; (*F*) lower side of leaf showing appearance of mines.

larva cuts holes at regular intervals and pushes the frass from the edge of the leaf.

**Sequence of Leaf Miners.**—The majority of leaf mines are produced during the middle or toward the end of the summer. As a rule, early attacks on leaves lead to gall formation. When meristematic tissue is attacked, the cells tend to proliferate and form galls rather than leaf mines. Leaf mines on the contrary occur on mature leaves although a few leaf miners appear early in the spring. The mines of *Agromyza aristata* develop upon elm very shortly after the leaves unfold. There is only one brood. A few weeks after the leaves develop, larvae can no longer be found in the mines. The dock leaf miner, *Pegomyia calyptrata*, appears in the early spring, and several generations are produced during the summer. Beetles do considerable feeding on the foliage before they oviposit. This delays the appearance of the mines. The locust leaf miner, *Chalepus dorsalis*, does not produce its mines until late in the summer. Other species persist as long as green leaves remain in the fall. Thus, some of the grass-mining Agromyzidae may continue their activity in sheltered places as late as November or December. *Aphelosetia orestella* mines in the leaves of *Hystrix hystrix* until late fall, then hibernates as larvae in the old leaves. It migrates to new leaves in the spring and continues its mining operations. The most remarkable habit is that of an undescribed species of *Coleophora* that feeds on laurel, *Kalmia*. Green leaves are available throughout the entire year. The larvae feed until cold weather causes them to cease. Then they withdraw into their cases and wait for more favorable weather. Sometimes they migrate to the twigs but more often they remain on the leaves all winter. They feed to some extent during warm periods of winter and finish their growth early in the spring. There is only one brood a year. Pupation takes place early in May and the adults emerge in June.

**Ecological Aspect of Leaf Miners.**—Leaf miners, like aquatic or subterranean insects, fill an ecological niche unoccupied by other species. They compete with other leaf-feeding insects but take their food in a different manner. A surface feeder and a leaf miner may exist upon the same leaf without interfering with each other, although the free-feeding form, being the larger, usually survives at the expense of the miner. Species like *Nepticula* require but a small portion of the leaf in which to mature; such areas seem to be avoided by the free-feeding larvae.

Leaf-mining insects serve as excellent material for ecological studies. They are abundant and easily collected, and the mines can be conveniently preserved. A mine often reveals the complete life history cf the insect. It is often possible to determine from it the position of the egg, the type of feeding, and the number of moults. The cast larval and pupal skins often remain to tell the complete story. These insects not only write their signatures in the leaves but leave their fingerprints.

Leaf-mining insects present some of the most interesting and the most complicated of insect life histories.

### BIBLIOGRAPHY

BRISCHKE, C. G. A. (1882). Der Blattminirer in Danzig's Umgebung. *Deut Naturf. u. Arzte ent. u. bot. Vers.* **53.**

FROST, S. W. (1924). A study of the leaf-mining Diptera of North America. *Cornell Univ. Mem.* **78.** *Bibliography.*

—— (1924). The leaf-mining habit in the Coleoptera. *Ann. Ent. Soc. Amer.* **17.** *Bibliography.*

———— (1925). The leaf-mining habit in the Hymenoptera. *Ann. Ent. Soc. Amer.*

**18.** *Bibliography.*

———— (1925). Convergent development in leaf-mining insects. *Ent. News* **36.**

———— (1930). Collecting leaf miners on Barro Colorado Island, Panama. *Scientific Mo.* **30.**

———— (1931). Habits of leaf-mining Coleoptera of Barro Colorado Island, Panama. *Ann. Ent. Soc. Amer.* **24.**

HENDEL, FRIEDRICH (1918). Die Palaartischen Agromyziden (Dipt.) *Archiv. fur Naturgeschichte* **84** (7).

HERING, MARTIN (1920–1934). Minenstudien. Parts I to XIV. Parts I to III *Deut. ent. Zeit.*, Parts IV, VI, VII, and VIII *Zeit. f. Morph. u. Oekol.*, Parts V and XI *Zeit. f. Wiss. Insektenbiol*, Part IX *Zool. Jahrb.*, Part X *Zeit. f. Angew. ent*, Parts XII, XIII, and XIV, *Zeit. f. Pflanzenkrankheiten.*

———— (1926). Die Ökologie der blattminierenden Insektenlarven. Verlag v. Gebruder Bornstaeger, Berlin.

HERING, M. (1951). Bestimmungstabellen der Blattminen von Europa. Parts I, II and III. Dr. W. Junk, s'Gravenhage.

———— (1957). Biology of leaf miners. Dr. W. Junk, s'Gravenhage.

LINNANIEMI, R. A. F. (1913). Zur Kenntnis der Blattminirer. *Soc. Fauna et Flora Fennica Acta* **37.**

NEEDHAM, J. G., et al. (1928). Leaf-mining insects. The Williams & Wilkins Company, Baltimore. *Bibliographies.*

TRÄGÅRDH, IVAR (1915). On what depends the ability of leaf miners to preserve the green color in Autumn mines? Swedish. *Skogvardsforeningens Tidskrift* **3.**

WATT, M. N. (1921–1924). Leaf-mining insects of New Zealand. Parts I to IV. *Trans. N. Z. Inst.* **53–55.**

# CHAPTER XVII

## LEAF-ROLLING INSECTS

René A. F. de Réaumur described leaf-rolling caterpillars in "Mémoires pour servir à l'histoire des insectes," published in 1737. Since that time the habits of this interesting group have scarcely been discussed except as individual species. The term leaf roller is applied to numerous economic forms, for example, the fruit tree leaf roller, the basswood leaf roller, and the strawberry leaf roller. Unfortunately, the expression has a vague meaning and is used to cover many different types of foliage distortion.

**Leaf Roller Defined.**—Insects distort leaves in various ways. Leafhoppers and aphids often cause them to pucker or curl by interfering

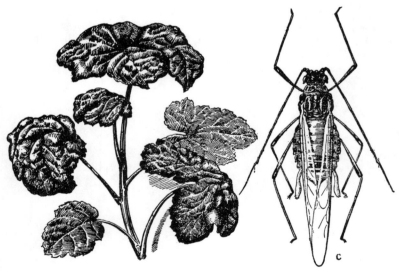

Fig. 286.—The currant aphid and the typical distortion, "leaf rolling," produced by this species. (*After U. S. Department of Agriculture.*)

with the cellular structure of the leaf. Although Rennie (1845) speaks of "leaf-rolling aphids," deformations caused by these insects are not generally considered in the category of leaf-rolling insects. A leaf roll, in the true sense, is produced by a larva that spins silk which is used to twist or distort the leaf.

**Evolution of the Leaf-rolling Habit.**—There are several evolutionary steps in the development of the leaf-rolling habit. The larva of Harris's sphinx, *Lapara bombycoides*, rests, head downward, in a group of pine needles. The leaves are not curled or drawn together but are naturally bunched and the larva is protected by its striped color pattern, which resembles the pine needles. Other species live in leaves tied together with silk forming protective shelters. For example, the larvae of the pine tube moth, *Argyrotaenia pinatubana*, loosely fastens a number of pine needles together, in which it rests and from which it protrudes its head to feed upon the tips of the needles composing the tube. However, this species is not a true leaf roller.

The larvae of many of the skippers, Hesperiidae, rest in folded leaves and emerge to feed. They spin no silk but the larvae take advantage of curled leaves. The larvae of *Cryptolechia tentoriferella* spin dense mats or sheets of silk on the undersurfaces of leaves beneath which the larvae rest and from which they protrude their heads to feed. These sheets of silk usually cause slight bulges upon the upper sides of the leaves and thus they approximate the leaf rollers. The larvae of *Papilio troilus* go a step further and spin dense mats of silk in folded leaves upon which they

FIG. 287.—Larva of Harris's sphinx, *Lapara bombycoides*, resting in a group of pine needles. (*Redrawn after Comstock.*)

rest and from which they emerge to feed. The next step is attained by the true leaf rollers that bend the edges of the leaves into definite rolls by means of silk.

Many insects utilize rolled or curled leaves but few "roll their own." A typical leaf roller is an insect whose larva rolls a single leaf as a shelter from which it feeds and in which it rests, transforms, and may pupate. An excellent example is the basswood leaf roller, *Pantographa limata*.

FIG. 288.—Larva of a typical skipper, *Epargyreus tityrus*.

**Extent of the Leaf-rolling Habit.**—The leaf-rolling habit is most conspicuous in the Lepidoptera, where silk spinning is paramount and useful for the leaf-rolling, leaf-folding, and leaf-tying habits. At least 17 of the families of the Lepidoptera show the habit in one form or another. The Tortricidae, Pyralididae, Gelechiidae, and the Gracilariidae are especially conspicuous. The habit also occurs in a few Coleoptera of the subfamily Attelabinae, in the web-spinning sawflies, *Pamphilius,* and is represented in the Orthoptera by the unique leaf-rolling grasshopper, *Camptonotus carolinensis.*

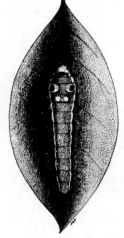

FIG. 289.—*Papilio troilus* Linn., at rest upon a mat of silk in a folded leaf.

**Method of Rolling Leaves.**—The methods of rolling and folding leaves are essentially the same. The chief qualification of a leaf roller is the ability to spin an abundance of silk, for this is the mechanism by which the leaf is manipulated. Threads of silk are spun across the portion of the leaf to be folded or rolled. If the roll is to be lengthwise, the strands of silk are spun perpendicular to the midrib of the leaf; if the roll is to be cross-wise, the strands of silk are spun parallel to it. As the strands of silk dry, they shrink and pull the edges of the leaf inward. New and shorter strands are then spun which in turn shrink and pull the edges of the leaf closer together. This is continued until the edge of the leaf is drawn completely over and is fastened with other strands of silk. Apparently the larvae do not use pressure to cause the leaf to curl but at times the growth or expansion of the leaf may assist the larvae in their operations. The leaf is sometimes cut and a small flap rolled into a cone. This is the common type of leaf-rolling found in the Gracilariidae. Leaf folders bend the leaf at the

FIG. 290.—Silken sheet spun by the larva of *Cryptolechia tentoriferella* on the underside of an apple leaf.

FIG. 291.—Leaf rolled by a Central American species, showing conspicuous bands of silk.

midrib or along one of the principal lateral veins. The silk is always spun on the upper side of the leaf, and the leaf naturally bends more easily in this direction.

**Purpose of Leaf Rolling.**—Leaf rolling is often associated with special methods of feeding. The basswood leaf roller, *Pantographa limata*, feeds upon the inner portions of the roll. The buckthorn tortricid, *Anchylopera braunae*, feeds upon the ends of the roll. Many species protrude their heads and feed upon adjacent foliage or fruit. The dusky leaf roller, *Amorbia humerosana*, a pest upon apple, feeds in this manner.

The leaf-rolling Coleoptera are really solitary insects, for they provide food for their young in a peculiar manner. The adult of the Attelabinae rolls a portion of the leaf into a compact thimble-shaped capsule in which it lays an egg. The larva, upon hatching, feeds upon the inner part of the roll and, when mature, enters the ground to pupate. *Attelabus rhois* rolls the leaf of alder; *Rhynchites betulae*, a British species, rolls the leaf of birch (see Fig. 202).

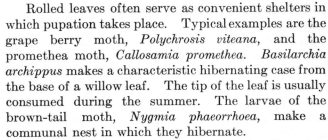

Fig. 292.—Leaf roll of *Anchylopera braunae* on the buckthorn. (*After Busck.*)

Rolled leaves protect the insects during the feeding period and provide some relief from parasites and predators. The leaf-rolling grasshopper uses its case solely for protection and abandons it during the daytime to feed upon aphids. The larvae of certain hesperids and papilios also use folded leaves as retreats.

Rolled leaves often serve as convenient shelters in which pupation takes place. Typical examples are the grape berry moth, *Polychrosis viteana*, and the promethea moth, *Callosamia promethea*. *Basilarchia archippus* makes a characteristic hibernating case from the base of a willow leaf. The tip of the leaf is usually consumed during the summer. The larvae of the brown-tail moth, *Nygmia phaeorrhoea*, make a communal nest in which they hibernate.

Fig. 293.—Cocoon of *Callosamia promethea* in a folded leaf securely attached to its support.

After leaf rollers abandon their leaf rolls, other insects and small animals often take possession. Certain scavengers, particularly mites and small beetles, feed upon the fecula left by the leaf rollers. On one occasion the writer found a large number of minute brown scavenger beetles, *Melanophthalma americana*, feeding in the ugly nests of *Cacoecia fervidana*, which were abundant on oaks. Many insects use the discarded leaf rolls as hiding places. During cool damp weather, leaf rolls are sure to be occupied by many different species. They make excellent places to collect beetles and spiders. For example, one cool afternoon, at least a dozen chestnut weevils were taken from an equal number of leaf rolls on witch hazel. The presence of spiders can usually be detected by the abundance of silk spun in these rolls.

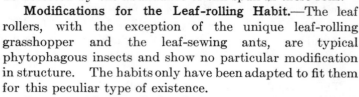

**Modifications for the Leaf-rolling Habit.**—The leaf rollers, with the exception of the unique leaf-rolling grasshopper and the leaf-sewing ants, are typical phytophagous insects and show no particular modification in structure. The habits only have been adapted to fit them for this peculiar type of existence.

**Guests and Parasites of Leaf Rollers.**—Numerous insects and their near relatives enter leaves rolled by various insects. Some live as guests much as the inquilines of gall insects. The ugly-nest tortricids are especially attractive to parasites and predators. Many Braconidae, Chalcididae, and Tachinidae have been reared from leaf rollers. Numerous predacious forms such as syrphid larvae, carabid larvae, capsid and pentatomid nymphs attack leaf rollers. The nymphs and adults of *Apateticus cynicus* are particularly fond of the larvae of the ugly-nest oak tortricid. This pentatomid is brown in color and is well protected when resting among the dead leaves of the nest. Ants have also been reported to attack leaf rollers. Mites and predacious thrips are often found in leaf rolls.

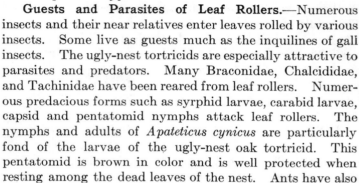

Fig. 294. Rolled leaf in which the viceroy, *Basilarchia archippus*, feeds and later hibernates.

**Typical Leaf Rollers.**—Examples of insect larvae that roll single leaves are not numerous. Many of the tortricids have this habit when the larvae are young; the older larvae tie the leaves together to make ugly nests. This is particularly true of the species of *Cacoecia* and *Argyrotaenia*. A few species of *Exartema* make typical leaf rolls; others sew leaves together. *Exartema inornatanum*, a common feeder upon dog-wood and wild cherry, rolls the leaf crosswise. *E. connectum*, another feeder on dogwood, rolls the leaf lengthwise. Several of the gelechiids make typical leaf rolls. *Telphusa latifasciella* and *T. querciella* roll the leaves of oak; *T. betulella* rolls the leaves of birch; *T. belangerella* rolls the leaves of alder.

Fig. 295.—Basswood leaf roller, *Pantographa limata*.

The basswood leaf roller, *Pantographa limata*, is one of the best examples of a typical leaf roller. Basswood trees often show the operations of these insects late in the summer. Single leaves seldom bear more than one or two rolls. These are large and conspicuous.

The larva makes a cut about halfway across the leaf and then rolls the piece towards the midrib. A tube or roll is formed of several thicknesses of leaf. A bright green larva, with shiny black head and thoracic shield lives within this tube. The larva feeds upon the inner rolls of the tube and produces copious fecular pellets, which gather at the lower end of the roll. When full grown, the larva leaves this roll and makes a smaller one which is a mere fold of one edge of the leaf. This fold is lined with silk and here the larva passes the winter in the fallen leaf. Pupation does not take place until the middle of the following year and the adults emerge about the first of August to start a new generation.

Several typical leaf rollers feed upon the ends of their cases. Heinrich describes the habits of *Anchylopera braunae*, which attacks the buckthorn (see Fig. 292). The young larva makes a characteristic fold near the tip of the leaf. Feeding takes place beyond the fold toward the tip. As it continues feeding, the larva extends the fold backward toward the base of the leaf.

The gelechiid, *Anacampsis innocuella*, exhibits a variation of the leaf-rolling habit. It makes a cylindrical roll on the leaf of poplar and usually cuts the petiole early in the last larval stage and finishes its feeding in the decaying leaf upon the ground.

Another gelechiid, *Brachmia hystricella*, rolls the blade of grass, *Hystrix hystrix*, upon which it feeds. The last generation passes the winter in one of these rolls.

The young larvae of the Gracilariidae make blotch mines in the leaves. Many of the species emerge from their mines and fold over

FIG. 296.—*Gracilaria negundella* on box elder, is first a leaf miner, then a leaf roller.

the edge of the leaf or make conical rolls from which they continue their feeding. *Parornix anglicella*, for example, makes a slender conical roll on the edge of the leaf of thorn or strawberry. A tortricid, *Episimus argutanus*, has a similar habit. It rolls the edge of the leaf of *Rhus* or of witch-hazel into a small cone.

The bean leaf roller, *Eudamus proteus*, like many of the hesperiids, makes a retreat in a folded leaf. The larva constructs this retreat by folding over a flap of the leaf made by cutting along two lines converging from the margin. The larva emerges only to feed and seals the edge of the flap when the time for moulting approaches.

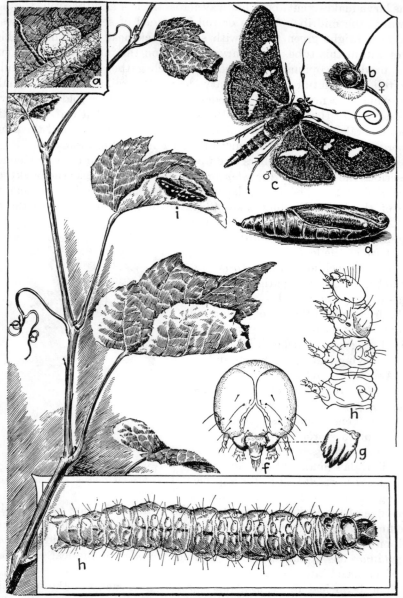

FIG. 297.—The grape leaf folder (*Desmia funeralis*): (*a*) egg; (*b*) head of adult female; (*c*) adult male; (*d*) pupa; (*e*) folded leaf serving as a retreat for the larva; (*f*) head of larva; (*g*) mandible of same; (*h*) larva; (*i*) adult on leaf. (*U. S. Department of Agriculture.*)

The leaf-rolling grasshopper also makes a typical leaf roll although it is not a plant feeder.

The cecidomyiid larva, *Camptoneuromyia rubifolia*, produces a definite marginal roll on the leaf of *Rubus*. The injury, however, approaches a gall formation. The same species is said to make a spot gall on the leaf of smilax.

**Leaf Folders.**—Leaf folders, in contrast to leaf rollers, fold the edges of the leaves instead of twisting them into rolls. They are sometimes called leaf sewers. The apple leaf sewer, *Anchylopera nubeculana*, is a typical example and is quite common in the apple orchards of the Eastern United States. It lives in a single folded leaf and spins an abundance of silk within the leaf. Numerous species of the Tortricidae, Gelechiidae, Pyralididae, and Oecophoridae also have this habit.

The grape leaf folder, *Desmia funeralis*, is another typical leaf folder. The species has long been known as a pest of grape vines and is widely distributed in the United States. The adult is a handsome moth measuring nearly an inch in width when the wings are expanded. The wings are dark brown in color narrowly bordered with white. Each of the fore wings has two conspicuous white spots; each of the hind wings of the male has a single spot, that of the female, two spots. The body is black with two narrow white crossbands in the female and one in the male. The moths emerge from their overwintering pupae during the early part of May. The eggs are deposited singly on the under surfaces of the leaves. The newly hatched larvae are unable to fold the leaves, so they seek some protected place and start feeding upon the upper epidermis of the leaf. The larvae are at first almost colorless; later the head and thoracic shield become light brown and the first two thoracic segments develop light brown lateral spots. When about two weeks old, the larva starts the construction of a fold in the

Fig. 298.—Six leaf rolls on a small branch of witch hazel, produced by the larvae of *Cacoecia rosaceana*.

leaf. This is done by spinning strands of silk from side to side across a portion of the leaf. When the silk dries, it contracts and pulls the edge of the leaf over. Each successive strand of silk is made shorter until the edge of the leaf is gradually drawn over and fastened with shorter bands of silk. The larva skeletonizes the leaf within this fold. When full grown, it generally drops to the ground where it transforms among leaves and trash. A second brood occurs in some parts of the country.

The witch-hazel leaf folder, *Episimus argutanus*, is very abundant in Pennsylvania where nearly every leaf upon a bush may show the work of this species. A single leaf may bear several folds. The fold is usually made along one of the principal lateral veins and the feeding is done from the inside.

Many species of *Anchylopera*, *Ancylis*, and *Olethreutes* fold individual leaves. The apple leaf sewer, *Anchylopera nubeculana*, and the strawberry leaf roller, *Ancylis comptana*, are common examples. The rose budworms, *Olethreutes nimbatana* and *O. cyanana*, also fold leaves and the larvae bore into the flower heads as well.

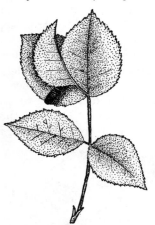

Another rose leaf folder, *Argyrotoxa bergmanniana*, has been found abundant at Ithaca, N. Y. The larva matures in a single folded leaf where it also pupates. The pupa forces its way partly from the folded leaf before the adult emerges.

The sweetpotato leaf roller, *Pilocrocis tripunctata*, is actually a leaf folder. It is a native of the West Indies and occurs to some extent in the Southern United States. The larva lives in a folded leaf where the pupa is also formed.

Fig. 299.—The rose leaf folder *Argyrotoxa bergmanniana*, pupa, protruding from folded leaf.

Several species of the Gelechiidae fold leaves. *Dichomeris eupatoriella* folds the leaves of *Eupatorium*, and *Aristotelia roseosuffusella* folds the leaves of clover. Both are common species.

A few of the cecidomyiid larvae fold leaves. The clover leaf midge, *Dasyneura trifolii*, makes a characteristic fold along the midrib of the leaf. The larvae feed gregariously and from one to twenty larvae may be found in a single folded leaflet.

**Leaf Tyers.**—Leaf tyers differ from leaf rollers and leaf folders in that more than one leaf is involved. The habit varies from species that sew two leaves together to those that draw in large numbers of leaves and even incorporate flowers, fruits, and other parts of the plant. They include the webbing insects and those that make ugly nests.

The hydrangea leaf tyer, *Exartema ferriferanum*, sews two terminal leaves together enclosing the flower bud. A single larva occupies such a nest and feeds upon the developing flower as well as the inner surfaces of the two leaves, which continue to grow and eventually form a bladderlike pouch on the terminal. These dwarfed terminals cause conspicuous injury to the plant.

A species of *Agonopteryx*, still undescribed, feeds upon *Hypericum* in a similar manner sewing together the two terminal leaves and enclosing the bloom upon which the larva feeds.

The poplar leaf tyer, *Melalopha inclusa*, lays a batch of small, spherical, red eggs on the poplar leaf. The young larvae are gregarious and at first tie two leaves together between which they feed. This forms a bladderlike nest to which they retreat when not feeding. Later they

Fig. 300.—The poplar leaf tyer, *Melalopha inclusa*. The larvae congregated before a moult.

draw in other leaves and construct a larger nest. When mature, the larvae drop to the ground and spin irregular cocoons under leaves and other trash.

The larva of the pine tube moth, *Argyrotaenia pinatubana*, draws a group of needles together and ties them with silk to form a tube. Here it lives and feeds on the tips of the needles forming the tube. It may make two or three such tubes during its development but eventually transforms in the last one. *Epinotia nanana* feeds in a similar manner on the needles of spruce.

Another excellent example of leaf tying is presented by the silver spotted skipper, *Eparavreus tityrus*. The larva feeds upon various leguminous plants and is common on the locust. It makes a nest by fastening together the leaflets of the compound locust leaf. Within the nest it remains concealed and emerges only to feed.

**Leaf Webbers and Ugly-nest Builders.**—The habit of webbing and tying leaves together is common in the Lepidoptera. It occurs conspicuously in the Tortricidae, Geometridae, Notodontidae, Arctiidae, Oecophoridae, Gelechiidae, Pyralididae, and Yponomeutidae. The family Tortricidae contains many economic forms such as the cherry tree ugly-nest tortricid, *Cacoecia cerasivorana*, the oak ugly-nest tortricid, *C. fervidana*, and the fruit tree ugly-nest tortricid, *C. argyrospila*. All have somewhat similar habits. The eggs are laid in compact masses on the bark of the tree. The larvae are gregarious. They attack the opening

FIG. 301.—The pine tube builder, *Argyrotaenia pinatubana.*

buds and roll and fasten leaves together until a large, ugly nest is formed.

FIG. 302.—The red-banded leaf roller, *Argyrotaenia velutinana,* showing larva feeding between fruit and leaf.

They may roll individual leaves at first but, as the larvae develop, this habit changes to the leaf-tying habit. Pupation takes place in these nests and the pupae work themselves part way from the nests before the adults emerge. The genus *Argyrotaenia* (Eulia) contains numerous species that tie leaves together in somewhat the same manner, although they frequently operate independently.

The larva of a geometrid, known as the scallop shell moth, *Calocalpe undulata*, webs the leaves of wild cherry together in much the same

FIG. 303.—*Calocalpe undulata*, a common leaf tyer on black cherry.

manner as the cherry tree ugly-nest tortricid. The eggs are laid in a cluster on a leaf near the tip of the twig. The larvae tie the leaves together in a tight roll and, as they grow in size, incorporate new leaves toward the base of the twig. Eventually the leaves die and turn brown, making a very unsightly appearance. There are two generations during the summer. The first generation pupates in the nest. When the second generation matures, the larvae drop to the ground, where they remain as pupae until the following spring.

There is a tendency for the gelechiid larvae to feed upon the terminal growth and tie the leaves together. For example, *Gelechia nundinella* feeds upon the terminal growth of *Solanum carolinense, Gnorimoschema axenopis* feeds upon the terminal growth of *Artemesia canadensis,* and *Trichotaphe nonstrigella* feeds upon the terminal growth of asters.

The apple and cherry ermine moths, Yponomeutidae, often cause serious damage to fruit trees. The habits of these insects are especially interesting. Oviposition occurs during the middle of the summer. The eggs hatch in autumn and the young larvae remain sheltered during the winter beneath the old egg shells. The following spring, the larvae bore into the leaves and mine for a short period, then emerge and feed externally, tying the leaves together in conspicuous tents. The cocoons are formed in masses in these nests.

The Pyralididae often web the leaves together in a characteristic manner. *Pyrausta futilalis* is one of the common species. Sometimes, like its relative *P. nubilalis,* it bores in the stems of milkweed but more often it feeds upon dogbane or milkweed, tying the leaves together to form a tent. The greenhouse or celery leaf tyer, *Phlyctaenia rubigalis,* is another common leaf tyer which draws the leaves and flower buds together and ties them with much silk. *P. theseusalis* rolls the tips of fern leaves.

Various webworms belonging to the genera *Loxostege, Pachyzancla,* and *Crambus* feed on grasses at or below the surface of the ground. The larvae live in tubular nests constructed of silk, bits of earth, and vegetable matter. Their habits differ greatly from the leaf-rolling and leaf-webbing species.

The Oecophoridae have habits similar to the Gelechiidae. The parsnip webworm, *Depressaria heracliana,* feeds upon the flower heads and seeds more often that it does upon the leaves.

The leaf-rolling habit is rare and unusual among the Noctuidae. *Brephos notha,* a European species, feeds upon aspen, drawing two or three leaves together. It pupates in a hole bored in soft wood or bark, with a double silken covering over the exit.

The fall webworm, *Hyphantria cunea,* stands out as one of the most conspicuous of the webbing species. It is the only arctiid that has this habit. There is little or no rolling of leaves but much spinning and tying. It represents the extreme development of the leaf-tying habit. The tent tree caterpillars build large nests in trees but use their tents only as retreats and emerge from them to feed.

**Leaf Crumplers.**—Leaf crumplers differ little from the leaf tyers and, as a matter of fact, represent a special type of them. The name has been given to the species of the genus *Acrobasis.* The larvae live in silken tubes which are conical in shape, much curved and twisted, and densely covered with fecula. These tubes are often large and conspicuous, small at one end and considerably larger at the open end. From them the larvae feed, drawing the leaves in and tying them together in more or less compact masses. They hibernate in these cases and resume feeding in the spring. Several species are commonly seen. *Acrobasis comptoniella* feeds on sweet fern, *A. rubrifasciella* on alder, and *A. caryae* is a serious pest on pecan.

The apple leaf crumpler, *Mineola indigenella*, is often abundant and causes serious injury upon young trees, especially those that are unsprayed. The adults come forth early in June and lay their eggs upon the leaves. The eggs hatch in a few days and larvae make trumpet-

Fig. 304.—Case of the apple leaf crumpler, *Mineola indigenella*.

shaped feeding tubes which are covered with an abundance of black fecula. The larvae remain in these tubes and push their heads out to feed upon adjacent fruit and foliage. The leaves are drawn in by means of strands of silk. Before the end of the summer, the leaves are gathered together in conspicuous masses and the silken tubes containing the larvae are obscured from view. By fall, the larvae are about one-half grown. They fasten their cornucopia-shaped cases securely to the twigs and hibernate in this manner. In the spring, they cut their cases from the winter resting places and wander to the opening leaves. The mature larvae are brown or green in color with black roughened heads. They pupate within their silken cases.

Fig. 305.—The false pine web-worm, *Itycorsia zappei*. Silk, spun by the larvae, binds the needles together and incorporates a great abundance of frass.

The web-spinning or leaf-tying habit is somewhat unusual in the Hymenoptera. The web-spinning sawfly, *Pamphilius inconspicuus*, attacks plums and cherries. The eggs are laid in a compact chainlike mass along the under side of the midrib of a leaf. When the larvae hatch, they begin to draw the leaves together. A fully formed nest may be 10 or 12 inches long. The full-grown larvae drop to the ground and make cells in which flimsy cocoons are spun. Here they spend the winter. The greater part of their life is spent in the ground.

The larvae remain in the ground from July of one year until about the end of May of the second year.    The pupae are formed toward the end of May.

The false pine webworm, *Itycorsia zappei*, is another rather common web-spinning sawfly.    It is reported from Connecticut on Austrian pine and has been found somewhat common in Pennsylvania on red pine. The crescent-shaped eggs are laid on the needles late in June or early in July.    The larvae are olive green in color with shiny black heads and short but distinct pairs of setalike appendages on the posterior segment. As the larvae mature, they take on a reddish color; when dropped into hot water, they turn brilliant red in color.    The larvae tie together bunches of needles and incorporate great masses of brown fecular pellets.

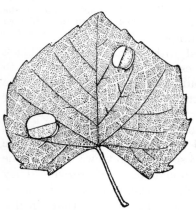

FIG. 306.—Cocoons of the grape berry moth, *Polychrosis viteana* Clem.

Several larvae live together in a single nest and feed upon the needles.    They feed until cold weather approaches, when they go to the ground and form earthen cells.    Here they remain until spring, when they transform to pupae. The adults emerge in the latter part of June.

Leaf-tying ants have been described by Wheeler and others. Several species of *Camponotus, Oecophylla*, and *Polyrhachis* of Africa, Singapore, and Australia spin silken nests. This habit is extraordinary, because no adult insects are known to spin silk.    The ants, however, seize larvae in their jaws and work them back and forth over the edges of the leaves like little shuttles until the leaves have been properly joined (see Fig. 194).

Web-spinning psocids have been described but these insects use silk only to cover the eggs.    Similar habits occur in other species.

**Manipulation of Leaves in Other Manners.**—Leaves are often cut and shaped to make cases or for other purposes.    The leaf-cutting ants and termites take pieces of leaves to their nests and make mushroom beds.    The leaf-cutting bees construct thimbles or capsules of pieces of leaves which are placed in hollow stems and form cells in which their young are raised.    Many larvae cut pieces of leaves and form cases to rest during the moulting process, during transformation, or during hibernation.    Conspicuous examples are the midges (Chironomidae), the caddis worms (Trichoptera), the basket worms (Psychidae), and the casebearers (Coleophora).    See also the discussion of casemaking insects, Chap. XXII.

## BIBLIOGRAPHY

No separate texts have been published on this subject but there are innumerable papers dealing with individual species, of which a few are selected.

BALDUF, W. V. (1938). Bionomic notes on *Exartema ferriferanum* (Lepid., Olethreutidae). *Journ. N. Y. Ent. Soc.* **46.**

BUGNION, E. (1923). La fourmi fileuse de Ceylon (*Oecophylla smaragdina* F., sous fam. Camponotinae For.). *Riviera Scient.* **10.**

CAUDELL, A. N. (1904). An orthopterous leaf roller. *Proc. Ent. Soc. Wash.* **6** (1).

CHITTENDEN, F. H. (1912). The larger canna leaf roller. *U. S. Dept. Agric. Bur. Ent. Circ.* **145.**

DONISTHORP, H. J. H. (1914). Ant larvae at sewing. *Trans. Ent. Soc. Lond.* **2.**

FROST, S. W. (1925). The red-banded leaf roller. *Pa. State College Bull.* **197.** *Bibliography.*

———— (1926). The dusky leaf roller. *Pa. State College Bull.* **205.**

HERRICK, G. W. (1915). The fruit-tree leaf roller. *Cornell Bull.* **367.**

JONES, T. H. (1917). The sweet-potato leaf-folder. *U. S. Dept. Agric. Bull.* **609.**

MCATEE, W. L. (1908). Notes on an orthopterous leaf roller. *Ent. News* **19** (10).

SEVERIN, H. C. (1920). The webspinning sawfly of plums and sandcherries. *S. D. Agric. Exp. Sta. Bull.* **190.**

STRAUSS, J. F. (1916). The grape leaf-folder. *U. S. Dept. Agric. Bull.* **419.**

WALDEN, B. H. (1912). A new sawfly pest of blackberry, *Pamphilius dentatus* (a leaf roller). *36th Rept. Conn. Agric. Exp. Sta.*

WEBSTER, R. L. (1909). The lesser apple leaf folder. *Iowa Exp. Sta. Bull.* **102**

———— (1918). The strawberry leaf roller. *Iowa Exp. Sta. Bull.* **179.**

WEIGLE, C. A., *et al.* (1924). The greenhouse leaf tyer, *Phlyctaenia rubigalis* (Guenee) *Journ. Agric. Res.* **29** (3).

WHEELER, W. M. (1906). The habits of the tent-building ant, *Cremastogaster*. *Bull. Amer. Mus. Nat. Hist.* **22.**

# CHAPTER XVIII

## GALL INSECTS

Galls are better known than the organisms that produce them. Nearly everyone is familiar with plant deformities produced by gall insects. They are abundant, colorful, and often grotesque. The causative organisms are, on the other hand, frequently small and difficult to identify.

The formation of galls was probably the first insect habit to be observed by man, although for a long time he was unaware that most of them were produced by living organisms. The nature of gall formation was not discovered until comparatively recent times. Pliny, about A.D. 60, refers to their habits and speaks of their rapid growth. Malpighi, 1686, a physician and a botanist, first explained their formation in a treatise, "De Gallis." As an article of trade, their value has been known for centuries. Theophrastus, 372–286 B.C., refers to their medical and curative properties.

**The Gall Defined.**—The term gall is applied, in a broad sense, to any abnormal growth of plant or animal caused by another organism. They are, however, particularly characteristic of plants; thus, a gall is defined as an abnormal growth of plant tissue produced by a stimulus external to the plant itself. The stimulus may be a mechanical irritation such as the rubbing of two limbs, a fungus growth such as the organisms that produce witches'-broom, or an animal such as a nematode, mite, or insect.

Insects and mites are the chief causative agents of plant galls. They are horticulturists, for they develop new plant forms. Normally the blackberry stem is slender and the oak leaf is flat. The gall maker changes the stem of the blackberry into a woody knot and the oak leaf into an "apple."

It is often difficult to determine where boring ends and gall making begins. *Cryptorhynchus lapathi* is a typical borer of poplar; nevertheless, the bark often thickens at the point of injury to create a gall-like formation. Likewise it is difficult, at times, to differentiate between leaf mining and gall making. Both habits may be exhibited by the same species. *Agromyza laterella* produces a gall on the young developing leaves of iris but a mine on the mature leaves. The tendency of plants to overcome injury often results in callousing which, in a sense, is a gall formation. Thus the crescent egg scars of the curculio on plum, cherry, and apple, or the calloused areas produced by treehoppers on apple twigs, are in reality galls. Many leaf mines show traces of similar development.

**Economic Importance of Galls.**—Tannic acid is one of the chief products obtained from galls. This is particularly interesting to the economic entomologist because tannic acid is used in the production of nicotine tannate, an insecticide applied against the codling moth and other insects. Tannins are obtained from various sources but the greatest supply comes from galls. A gall produced by an Eurasian cynipid yields 65 per cent of tannic acid. The North American species yield less. Our native sumac gall yields about 50 per cent of tannic acid.

Dyes are obtained from numerous galls. The "mad apple" of Asia Minor yields Turkey red. The natives of East Africa use galls as a source of a dye for tattooing, the only record of savage people making use of galls. No galls, so far as known, were ever used by the American Indians for any practical purpose. Pliny mentions the use of the Aleppo gall to dye the hair black.

The best permanent inks have for years been made from galls. The Aleppo gall, produced by *Cynips gallae-tinctoriae* on several species of oaks in Eastern Europe and Western Asia, is one of the chief sources of the ingredients for inks. In some places, the law requires that permanent records be made with ink derived from gallnuts. Such inks were known and used by the monks of the ninth and tenth centuries. The Aleppo gall has been specified in formulas for inks used by the United States Treasury, the Bank of England, the German Chancellery, and the Danish Government.

Galls have frequently been used in medicine. The ancients had many superstitions concerning them. Galls were supposed to contain a maggot, a fly, or a spider. If they contained a maggot, famines were sure to come; if they contained a fly, war would result; if they contained a spider, pestilence resulted. One of the most powerful vegetable astringents is obtained from the Aleppo gall. The use of galls in medicine is mentioned in various pharmacopoeias. During the eighteenth century, they were used in France to control fevers. Today they have little use in medicine although they are still used in the preparation of *Unguentum galae*.

Galls have occasionally been used as food. In the Near East, a gall on *Salvia pomifera*, produced by a species of *Aulax*, has been an article of commerce. It is known as the gall of sage or "pomme de sauge." It is esteemed for its aromatic and acid flavor and is said to be especially delicious when prepared with honey and sugar. The galls of another cynipid, *Aulax glechomae*, which attacks *Glechoma hed-*

*eracea*, were used as food in France. It is said that they have an agreeable taste and the sweet odor of the host plant.

In Missouri and Arkansas, a black cynipid gall has been used as food for domestic animals. This minute gall, which resembles a kernel of wheat and is produced by a species of *Callirhytis*, has unquestionably high food value. An analysis shows that it contains 63.6 per cent carbohydrate and 9.34 per cent protein.

On the other hand, gall insects may be exceedingly injurious to crops. The hessian fly, clover leaf midge, chrysanthemum midge, and pear leaf blister mite are a few of our common gall pests.

**Extent of the Gall-making Habit.**—The great majority of the plant galls are produced by nematodes, mites, and insects. We are concerned primarily with the insect galls. The gall-making habit is scattered throughout six orders of insects: Coleoptera, Lepidoptera, Homoptera, Thysanoptera, Diptera, and Hymenoptera. Felt (1917) lists 1,440 North American species, including the mites, Eriophyidae. The distribution is as follows: 162 Eriophyidae, 12 Coleoptera, 17 Lepidoptera, 60 Homoptera, 701 Diptera, and 488 Hymenoptera. His figures for 1940 are slightly different but total approximately the same. Four hundred and forty-four of the Hymenoptera are cynipid wasps and 682 of the Diptera are midges, Cecidomyiidae. Certain Thysanoptera make galls on *Callestemon*, *Bursaria*, and *Acacia*. One hundred and seventy-seven gall-making thrips are known from Asia and the Dutch East Indies.

Gall insects attack nearly all parts of the plant: buds, leaves, petioles, flower heads, stems, bark, and roots. The host and the portion attacked are often characteristic and are a great aid in identifying the species. Gall makers usually produce the same type of gall on different hosts. There are exceptions; for example, *Camptoneuromyia rubifolia* produces a spot gall on smilax but a leaf roll on *Rubus*. Different species working on the same plant usually produce entirely dissimilar types of galls.

More than one-half of the families of plants are attacked by gall insects. The Cecidomyiidae alone attack 69 families and 202 genera. They are especially numerous on the willows, oaks, roses, legumes, and composites. One hundred and fifty-one, or about one-sixth of the species, attack the Compositae. The cynipid wasps attack chiefly the oaks; they work to some extent on the rose and rarely make galls on other plants. The gall-making plant lice live on a variety of hosts. Although the Phylloxera are conspicuous on the hickories, the grape Phylloxera is perhaps the best known species. The mites form their galls largely on woody plants, especially the willows, maples, birches, beeches, and roses.

Galls are often exceedingly numerous upon individual trees or plants. Felt estimates that more than 500,000 cynipid wasps were living at the expense of one oak tree. Another cynipid was so abundant on a large pin oak that the sweet exudation from the galls attracted hosts of bees and flies. The twigs of these oaks were literally covered in part with minute, tubular galls.

Psyllid, coccid, and aphid galls are often minute and tend to be exceedingly numerous on their hosts. This is especially true of *Phyllocoptes quadripes* on maple, Phylloxera on hickory or grape, and similar species. The spangle and blister galls produced by the midges may be equally as abundant.

**The Gall as a Dwelling Place.**—The larval and often the pupal period is spent in the gall, which serves as a shelter and as a source of food. The insect obtains some relief from parasites and predators, and is protected from desiccation. The inner walls of the gall are often richer in protein than the portion of the plant upon which it grows and thus the larva or nymph finds an abundance of concentrated food there.

Many inquilines utilize insect galls. Some feed upon the extra food available, others are transient visitors. The pine-cone willow gall has been studied in considerable detail. Hendel lists 31 dwellers in addition to the gall maker: 10 inquilines, 16 parasites, and 5 transients. In North America, the larvae of Lepidoptera, sawflies, cynipid wasps, and midges are common guests, and the eggs of the meadow grasshopper are frequently found between the scales of the pine-cone willow gall. Many Coleoptera inhabit galls. Leng lists at least a dozen species of Curculionidae found in the galls of cynipid wasps.

**Types of Galls.**—In general there are two types of galls: open and closed. The open galls are produced by haustellate forms: aphids, psyllids, coccids, and mites. Typical examples are the pear leaf blister mite, *Eriophyes pyri*, the witch-hazel cone gall maker, *Hormaphis hamamelidis*, and the elm cockscomb gall maker, *Colopha ulmicola*. These forms at first feed from the outside but later cause the leaf to fold or grow inward to produce a pocket in which the insects live and feed. Reproduction occurs within the galls and the young escape through a small opening on the lower surface of the leaf. The tubular galls made by the midges resemble cone galls but are of the closed type.

The witch-hazel cone gall maker, *Hormaphis hamamelidis*, is but one of the many species of Homoptera that make open galls. The life history of this species is somewhat complicated and, like that of most aphids, reveals alternation of generations, parthenogenesis, and the production of wingless generations. These insects are so modified in form that they hardly resemble the free-feeding aphids one is accustomed to see. The eggs of the witch-hazel cone gall maker are laid upon the branches and twigs of the witch hazel late in the fall.   They hatch early in the spring and the stem mothers migrate to the lower surface of the leaves.

FIG. 307.—Elm cockscomb gall, *Colopha ulmicola.*

In some way, the feeding nymphs cause the upper surface of the leaf to thicken and to develop into conical growths.   The interior of the cones are hollow and openings remain to the exterior on the lower surface of the leaf. The females produce their young within these galls.   When the galls become crowded, agamic winged forms are produced which leave the galls and fly to birch.   The second and subsequent generations on birch show a remarkable reduction of legs, antennae, and beaks.   The body becomes broadly oval and aleyrodiform in shape.   The edges of the body are decorated with rods of glistening white wax which are forced through numerous wax pores arranged about the entire

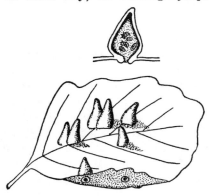

FIG. 308.—The witch-hazel cone gall, *Hormaphis hamamelidis,* an open type of gall.   Above, a cross section of gall showing young within.   (*After Pergande.*)

margin of the insect.   The sixth generation is winged and these individuals return to the witch hazel.   They give birth to a seventh generation, which consists of wingless males and females. Each female lays about 10 winter eggs on the witch hazel.

The spiny witch-hazel gall aphid, *Hamamelistes spinosus*, produces another typical open gall. The life history of this species differs remarkably from that of the witch-hazel cone gall maker. The second generation migrates to the birch. These aphids produce young which feed for a short time and then settle down close to the leaf buds, where they hibernate. The fourth generation is produced on birch early the next spring. The nymphs cause the upper surface of the leaf to fold and produce pockets which are open below. The fifth generation is produced in these pseudo galls. The sixth generation is winged and migrates

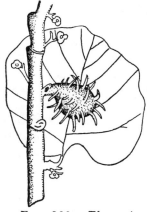

FIG. 309.—The spiny witch-hazel gall, *Hamamelistes spinosus*. An open type of gall. (*After Pergande*.)

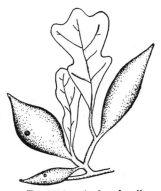

FIG. 310.—A closed gall, *Amphibolips ilicifoliae*, on oak.

back to the witch hazel early in the summer. The seventh generation, which is produced later in the summer, consists of males and females which mate and lay their winter eggs on the branches of the witch hazel.

Closed galls are made by the larvae of mandibulate insects: Coleoptera, Lepidoptera, Diptera, and Hymenoptera. None produces young in the galls. The majority of the galls are simple (*monothalamous*), that is, they contain but a single larva. A few are compound (*polythalamous*), that is, each gall contains several larvae in separate cells or chambers. The oak apples, produced by species of *Amphibolips*, are examples of the former; the oak hedgehog gall is an example of the latter.

The spherical goldenrod gall, produced by *Eurosta solidaginis*, is a typical and common example of a closed gall. In some localities nearly every goldenrod stem bears one or two galls. They are found only on the stems of *Solidago canadensis*. The fly, a member of the family Trupaneidae, is a pretty insect with pictured wings. The female lays her eggs upon the surface of the goldenrod stems. The larvae wander somewhat before they enter the stems but eventually they bore into the stems and cause the formation of galls. New galls become apparent about the first of June. By fall, the galls are about the size of hickory nuts, round, and of much the same texture as the stems themselves.

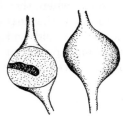

The insects remain in the galls during the winter. Some transform in the fall but the majority of them pass the winter as larvae and transform in the spring. Numerous guests may inhabit the thickened walls of these galls. Most common of these is the mordellid beetle, *Mordellistena unicolor*. At least 12 other insects have been found in these galls; some are parasites, others are true inquilines.

FIG. 311.—The spherical goldenrod gall produced by *Eurosta solidaginis* (Trupaneidae).

Galls can be further classified according to their form and texture. The blister or spot galls are very shallow and are usually produced by the Cecidomyiidae. The spangle galls have greater thickness and look like shallow saucers on the lower surfaces of the leaves. They occur commonly on oaks and hickories and are produced by the Cecidomyiidae and rarely by the Cynipidae. The button galls are the work of the Cynipidae. The tube galls, frequently seen on grape and hickory, are produced by the Cecidomyiidae. The oak bullet galls and the oak apples are the results of the attacks of the Cynipidae. Irregular woody galls on rose, oak, and blackberry are formed by the larvae of the cynipid wasps. Galls found on the flower heads are usually produced by the Cecidomyiidae but occasionally by the Eriophyidae, Cynipidae, or Psyllidae.

FIG. 312.—Oak button galls, *Neuroterus umbilicatus*. (*After Felt.*)

A number of galls are produced by the egg-laying habits of certain insects, as for example the snowy tree crickets, cicadas, tree hoppers, or leafhoppers. These insects frequently insert their eggs in the twigs of trees, making more or less conspicuous scars. During the second year, the egg scars often become calloused and assume the characteristics of galls.

**Adaptations for the Gall-making Habit.**—The most conspicuous adaptations are found in the plant rather than the insect. Certain remarkable changes occur in the plant, following the attack by the gall maker. Tannin and other unpalatable substances are formed which antagonize feeding by other species. Hairs and spines are often developed on the surface of the gall. The walls of the gall are sometimes much thicker and harder than the plant itself. The portion of the plant attacked is frequently modified in color and shape to resemble flowers, fruits, seeds, or fungi.

The gall-making habit occurs in groups of insects that are typically phytophagous and no particular modification of the larva is necessary. Some mechanism of the egg, larva, or adult, still little known, is essential to start the formation of the gall. Some of the groups, as for example, the Cecidomyiidae and the Agromyzidae, show intermediate forms between the gall-making and the leaf-mining habits. The Cecidomyiidae are of mixed habits: scavengers, fungus feeders, predators, and leaf miners. On the other hand, the Cynipidae are almost entirely gall makers or inquilines.

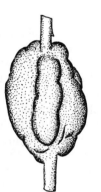 

The larvae and nymphs of gall-making species usually show a reduction of antennae and legs. In the mites, the antennae are minute and the legs are reduced to two, instead of the customary four pair. The bodies of the gall-inhabiting species are also elongate. The Hymenoptera are typically legless. The Homoptera show considerable reduction in antennae and legs. The gall-inhabiting *Phylloxera* have minute legs and antennae, and the root-inhabiting forms have well-developed legs and antennae. The group of Coleoptera containing the gall-making species is legless by nature.

Fig. 313.—An irregular woody gall on blackberry, produced by *Diastrophus nebulosus.*

Fig. 314.— The cypress flower gall, produced by a Cecidomyid, *Itonida anthici.* (*After Felt.*)

The alternation of generations in the Cynipidae and the Aphididae is not peculiar to the gall-making habit but leads to a special method of hibernation that occurs in no other gall-making species.

**Origin of Galls.**—The gall-making habit has undoubtedly developed independently in widely separated groups. There are only three outstanding groups of gall makers: the eriophyids, the cynipids, and the midges. The habit may have developed in different ways in these groups. Felt believes that it had its origin through the fungus-feeding or predacious habit. This is probably true in the Cecidomyiidae, where both types of feeding are represented. There are also indications that the habit may have arisen through the leaf-rolling habit. This may have taken place in the Eriophyidae, where many species roll leaves and others form blisterlike galls. Hering believes that the leaf-mining and the gall-making habits are closely associated. In many of the leaf mines there is some proliferation of cells, not sufficient to be conspicuous but enough to indicate a tendency to gall formation. The spot galls produced by the Cecidomyiidae are borderline forms. *Monarthropalpus buxi* is usually considered as a leaf miner; nevertheless, the leaf shows considerable development of cells and a thickening about the injury.

**Formation of Galls.**—The gall insect lays its egg upon the host or inserts it within the tissues of the plant. Upon hatching, the larva soon finds its way to the cambium or to some portion of the plant that is capable of growth. There is usually no distortion of the plant until the egg hatches; then the gall grows with the insect. The insect feeds upon the abundant food produced by the gall. Mandibulate species tear down the cells; haustellate species suck out their contents, feeding on the inside of the gall. An enzyme, produced by the insect, changes the starch of the plant cells into sugar thus rendering the same function as the enzyme of the plant that normally changes starch to sugar. The production of excess food material stimulates the activity of the plant protoplasm and causes the cells to multiply and produce what is known as a gall. Both the plant and the insect derive food by this association.

The physiology of gall formation is still obscure. It is generally believed that the stimulus for gall formation is a secretion from the larva and that it is not due to a mechanical or a chemical irritation produced by the female at the time the egg is laid. Ping has shown that, in the case of the goldenrod gall and other similar species, the gall does not start to develop until the larva has done considerable wandering and feeding. It is well known that galls are formed only on meristematic tissue. Galls are the result of cell multiplication, which cannot take place after a leaf or stem has become mature or fixed. A large proportion of the galls begin their development in the bud and many are inflorescent. Galls are more abundant during spring and early summer although they may persist until fall or even remain on the trees or plants during the winter. Leaf miners, in contrast, are more abundant during the middle or the latter part of the summer. Active tissues may be found in certain parts of the plant during the entire season. The cambium is always present in living trees. Leaves often retain small areas of meristematic tissue after they have expanded and hardened. These areas are capable of gall formation. It will always be a question how such remarkable formations are produced, as the woolly oak galls with embedded seedlike grains.

**Alternation of Generations.**—The alternation of generations is common among the Cynipidae and the Aphididae. This habit occurs in insects other than the gall makers but is well adapted to the gall-making habit. Thus two types of reproduction are represented in the gall insects: parthenogenetic, in which only agamic females are produced; and sexual, in which males and females are produced. The agamic generation is the overwintering form; the sexual generation is the summer form.

This differs somewhat from the normal aphid development in which the species reproduces agamically all summer long and sexes are produced only at the approach of fall to mate and lay winter eggs.

The insects as well as the galls of these two generations are so unlike in form and appearance that they are often mistaken for different species. As a matter of fact, several gall insects have been described as separate genera and species which were later found to be different generations of the same species.

Two classical examples of alternation of generations are the oak hedgehog gall, *Acraspis (Andricus) erinacei,* and the spruce cone gall, *Adelges abietis.* The latter is an exceedingly injurious species and has a very complicated life history with generations on the spruce and on several secondary hosts and with one- and two-year cycles. The summer generation of *Acraspis erinacei* produces the familiar hedgehog gall on the midrib of the leaf of white oak. These galls are hard, thickly walled, and contain many larvae. The sexual generation is produced in spring. These individuals make galls on the tip of the buds and on the bud scales. They are small, inconspicuous, reddish in color, thin walled, and egg shaped, and each gall contains a single larva.

**Classification of Galls by Orders.**—The species of insects producing galls are too numerous to discuss in detail. There are about 1,440 North American species.

*Diptera.*—This order includes the largest number of the gall-making insects. Felt (1917) lists 701 species of which 682 belong to the family

Fig. 315.—The hedgehog gall of oak, produced by *Acraspis erinacei.*

Cecidomyiidae. The hessian fly, chrysanthemum midge, and clover leaf midge are perhaps best known. Next in importance are the Trupaneidae. The majority of these species make solid galls on the buds, stems, and roots of goldenrod. The Oscinidae produce inconspicuous galls on various grasses. The Agromyzidae are predominantly phytophagous and are composed largely of leaf-mining species, with some cambium miners, pith and stem miners, and a few gall makers. Four species of Agromyzidae produce definite galls: *Agromyza tiliae* on linden and lime, *A. schineri* on populus, *A. websteri* on wisteria (Japan), and *A. laterella* on iris.

*Hymenoptera.*—The Hymenoptera constitute one of the most important groups of gall-making insects. It is exceeded in number of species only by the Diptera. The majority of the species belong to the family Cynipidae. About 86 per cent of these occur on oaks. **The** cynipid galls vary considerably in form: some are bulletlike, some fruit-like, some woody, some mossy, and some woolly. Both simple and compound galls are produced by the cynipid wasps and a few species produce both types. Aside from the Cynipidae and *Euura* (Tenthredinidae), there are no Hymenoptera that make distinct galls. The Tenthredinidae show a tendency towards leaf mining and boring. The Eurytomidae, Torymidae, and Chalcididae are seed feeders. The Cephidae bore in the stems of plants, dwarf and stunt their growth, and sometimes cause gall-like enlargements.

*Coleoptera.*—Gall-making Coleoptera tend to intergrade strongly with the boring species and cause much confusion to the observer who

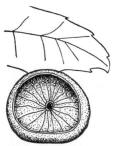

Fig. 316.—The wool sower, *Callirhytis seminator*, showing remarkable development of the stem of the oak.

Fig. 317.—Oak apple produced by *Amphibolips confluens*.

attempts to describe the type of injury produced by them. Scarcely more than 12 North American species of beetles make definite galls. These are scattered among the families Curculionidae, Buprestidae, and Cerambycidae. The Curculionidae include several economic species: the pine gall weevil, *Podapion gallicola*, the grape cane gall maker, *Ampeloglypter sesostris*, and the virginia creeper stem gall, *A. ater* are well-known examples. The Buprestidae likewise contain many injurious species: the red-necked cane borer, *Agrilus ruficollis*, the raspberry cane borer, *Oberea bimaculata*, and the bronzed birch borer, *Agrilus anxius*, are common forms. The latter does much boring as well as gall making. The Cerambycidae contain a few gall-making species, although wood boring is the predominant habit in this group. The gall-making maple borer, *Xylothrechus aceris*, often becomes a pest. *Saperda concolor* and *S. populnea* produce galls on the branches of poplar and willow.

*Lepidoptera.*—The largest number of gall-making Lepidoptera belong to the genus *Gnorimoschema*. Most of these species make galls on the stems of goldenrod. One species produces galls on the roots of asters.

Next in importance are the Tortricidae and Elachistidae. None appears to be of economic importance. Occasional gall-making species are found in the families Aegeriidae, Tineidae, Olethreutidae, Lavernidae, and Pyralididae. The lima bean vine borer, *Monoptilota pergratialis* (*nubilella*), is perhaps the best known. It is said that the injury inflicted

Fig. 318.—A section of blackberry cane showing a gall caused by the larva of the red-necked cane borer, *Agrilus ruficollis* (*After U. S. Department of Agriculture.*)

Fig. 319.—Gall of the solidago gall-moth, *Gnorimoschema gallaesolidaginis.*

varies with the position of the gall and the thriftiness of the vine. When the larva forms its gall in a sturdy stalk, no injury results; when it works in a small stalk near the tip, the terminal portion often dries and wilts.

*Homoptera.*—Forty-seven Aphidae, eleven Psyllidae, and two Coccidae are recorded as gall makers. All produce open galls, usually of the pocket or cone type. The galls are often small and thickly placed on the leaves, as for example the grape phylloxera. This interesting species produces one form of gall on the leaves and another on the roots. The elm cockscomb gall, *Colopha ulmicola*, produces a large pocket gall on the upper side of the leaf. The woolly apple aphid, *Eriosoma lanigerum*, produces nodules on the roots which are frequently termed galls. Several species of Psyllidae produce galls on the leaves and stems of hackberry.

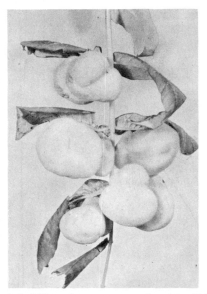

Fig. 320.—A fruitlike gall on sumac, produced by *Melaphis rhois*.

## BIBLIOGRAPHY

BEQUAERT, JOSEPH (1922). Gall insects, in ants of the Belgian Congo. *Amer. Mus. Nat. Hist. Bull.* **45.**

BEUTENMULLER, WM. (1904). The insect galls of the vicinity of New York City. *Journ. Amer. Mus. Nat. Hist.* **4.**

CHADWICK, G. H. (1908). A catalogue of the "Phytoptid" galls of North America. *N. Y. State Mus. Bull.* **124.**

COCKERELL, T. D. A. (1890). The evolution of insect galls. *Entomologist* **23.**

CONNOLD, E. T. (1902). British vegetable galls. E. P. Dutton & Company, Inc., New York.

COOK, M. T. (1902–1904). Galls and insects producing them. Parts I to IX. *Ohio Naturalist*, Vols. 1–4. *Bibliography.*

—————— (1910). The insect galls of Michigan. *Mich. Biol. Survey* **1.**

COSENS, A. (1912). A contribution to the morphology and biology of insect galls. *Toronto Stud. Insect Biol. Ser.* **13.**

—————— (1913). Insect galls. *Can. Ent.* **45** (11).

FAGAN, MARGARET (1918). The uses of insect galls. *Amer. Nat.* **52** (614). *Bibliography.*

FELT, E. P. (1916). American insect galls. *Ottawa Natural* **30.**

—————— (1917). Key to American insect galls. *N. Y. State Mus. Bull.* **200.**

—————— (1925). Key to gall midges. *N. Y. State Mus. Bull.* **257.**

——— (1940). Plant galls and gall makers. Comstock Publishing Company, Inc., Ithaca, N. Y.

GARMAN, H. (1883). The Phytopti and other injurious plant mites. *Rept. Ill. State Ent.* 12.

HASSAN, A. S. (1928). Biology of Eriophyidae. *Univ. Cal. Pub. in Ent.* 4 (11). *Literature cited.*

HENIDEL, R. L. (1905). Ecology of the willow cone gall. *Amer. Nat.* 39 (468).

HERING, MARTIN (1927). Gall und mine. *Ent. Jahrb. v. Kranches.*

HODGKISS, H. E. (1930). Eriophyidae of New York II, the maple mites. *Geneva Agric. Exp. Sta. Tech. Bull.* 163.

HOUARD, C. (1908). Les Zoocecides des plantes d'Europé et du Basin Méditerranée. Vols. I and II.

——— (1933). Les Zoocecides de l'Amerique du Sud et de l'Amerique Central.

JARVIS, T. D. (1936–1937). Insect galls of Ontario. *Ent. Soc. Ont. Repts.* 37 and 38.

KINSEY, A. C. (1929). Gall-wasp genus Cynips, a study of the origin of species. *Ind. Univ. Studies* 16.

NALEPA, A. (1910). Zur Öekologie der Gallmilben; *In, Eriophyiden Gallenmilben, Zoologica, Heft* 61.

——— (1927). Zur Phanology und Entwicklungsgeschichte der Gallenmilben. (Phenology and life history of mite galls.) *Marcellia* 24 (87–98).

——— (1928–1929). Neuer Katalog der bisher beschriebenen Gallenmilben, ihrer Gallen und Wertzpflanzen. (New catalogue of the gall mites described to the present date, their galls and host plants.) *Marcellia* 25.

NIERENSTEIN, M. (1930). Galls. *Nat. Lond.* 125 (3149).

PARROTT, P. J., *et al.* (1906). Eriophyidae I, the apple and pear mites. *Geneva Agric. Exp. Sta. Bull.* 283.

SEARS, P. B. (1914). The insect galls of Cedar Point and vicinity. *Ohio Naturalist* 15 (2).

STEBBINS, F. A. (1910). Insect galls of Springfield, Massachusetts and vicinity. *Springfield Mus. Nat. Hist. Bull.* 2.

THOMPSON, M. T. (1916). An illustrated catalogue of American insect galls. Supplemental list of American gall insects, Nassau, N. Y.

TOWNSEND, H. T. (1894). Notes on the Tenthredinid gall of *Euura orbitalis* on *Salix* and its occupants. *Journ. N. Y. Ent. Soc.* 2 (3).

WALSH, D. B. (1864). On the insects, coleopterous, hymenopterous and dipterous inhabiting the galls of certain species of willows. *Proc. Ent. Soc. Phil.* 3.

WELD, L. H. (1928). Cynipid galls of the Chicago area. *Trans. Ill. Acad. Sci.* 20.

WELLS, B. W. (1916). The comparative morphology of the Zoocecidia of *Celtis occidentalis*. *Ohio Journ. Science* 16 (7).

# CHAPTER XIX

## BORING INSECTS

Insects bore in many substances such as plants, animals, and soil. The borers in animals (parasitic insects) and the borers in soil (subterranean insects) are discussed elsewhere. The present chapter deals with the plant borers (subcutaneous feeders) exclusive of the leaf miners and gall makers.

**Economic Importance of Borers.**—There is perhaps no group of insects more injurious than borers. The destruction caused by many species occurs so suddenly that control methods are difficult to administer. The injury is generally evident long before the presence of the insect is noted. The squash and the cabbage, for example, wilt rapidly when

attacked by borers. The sun helps to aggravate the injury and hasten the wilting. The conspicuous hole in the seed, stem, fruit, or wood, or the abundant borings protruding therefrom, indicate that the damage is done and that the larva or the adult has escaped. Poplars and willows suffer from the attack of the poplar borer long before the average person is aware that the tree is infested by grubs. Half the terminals of a peach tree may wilt in a few days' time following an attack by the larvae of the oriental fruit moth. Many

FIG. 321.—Injury to telephone pole caused by cerambycid larvae; Riobomba, Ecuador. Eucalyptus trees in background.

of the boring insects attack and kill trees after several years of growth, and the loss is increased by the necessity of replacing the trees. The table on page 371 will give some idea of the losses caused by boring insects. It is difficult to summarize such losses because figures are not available for consecutive years.

Borers are a difficult group of insects to control because they feed beneath the surface of the plant where it is often impossible to kill them with the ordinary stomach poisons. The larva of the oriental fruit moth bores in the twigs and fruit of peach and allied plants. It even casts aside the first borings as it cuts its way through the epidermis.

370

LOSSES IN DOLLARS BY SOME COMMON BORING INSECTS*

| Insect | Loss | Cost of control | Total loss | Value of crop |
|---|---|---|---|---|
| Cotton boll weevil.......... | $118,083,000 | $ 3,200,000 | $121,283,000 | $500,000,000 |
| Codling moth.............. | 13,500,000 | 17,500,000 | 31,000,000 | 110,000,000 |
| Seed-corn maggot.......... | 2,600,000 | .......... | 2,600,000 | |
| Peach borers............... | .......... | 1,000,000 | 1,000,000 | 46,550,500 |
| European corn borer........ | 865,000 | .......... | 865,000 | 102,571,072 |
| Wireworms................ | 40,000,000 | 108,000 | 40,108,000 | |

* Figures from *U. S. Dept. Agric. Bur. Ent.* E-444, 1938.

Once beneath the surface, it cannot be poisoned. Nicotine will kill the young larva before it has a chance to dig into the tissues. Five or six applications are necessary to control the pest but this is generally expensive and impracticable. Bait traps have been tried, with some

FIG. 322.—Termite colony showing winged adults, workers and injury.

success, to dispose of the adults before they lay their eggs. Predacious and parasitic insects have not reduced the pest appreciably. Thus the control of this injurious insect is still much of a problem.

On the other hand, the habits of the peach tree borer lend themselves well to control methods. The application of a "gas," popularly known as P.D.B., about the base of the infested trees gives almost complete control. As a matter of fact, it excels the old method of hand grubbing for it kills the young larvae before they have a chance to do much damage.

**Extent of the Boring Habit.**—Borers attack many parts of the plant: buds, leaves, stems, cambium, pith, bark, wood, roots, fruit, and seeds.

Fig. 323.—Method of application of P.D.B. (Paradichlorobenzene) for the control of the peach tree borer.

Some bore in algae and fungi, others in manufactured products such as furniture and cigars. This heterogenous group will be discussed under such more or less popular headings as bud borers, fruit borers, nut borers, seed borers, stalk borers, cambium borers, and wood borers.

The boring habit occurs chiefly in five orders of insects: Isoptera, Coleoptera, Lepidoptera, Diptera, and Hymenoptera. As a rule, only the larvae of boring insects are modified for this particular habit. The adults usually leave the burrows and live free lives. A few species, however, spend a large part of the time in the burrows and these species show remarkable development. This is true of certain Coleoptera, especially species of the families Bostrichidae, Ptinidae, and Ipidae. The body form of these insects is cylindrical; the antennae are short and can be withdrawn into grooves in the head; and the legs are short and fit closely to the body. Some of the species rotate while boring through wood and make perfect tubular galleries. They send forth a stream of borings like a drill or auger. These beetles excavate burrows in which the eggs are laid and this unusual boring habit is essential for the development of the species.

Fig. 324.—A plant-boring cricket, *Cylindrodes kochi* from Australia. (*Redrawn after Tillyard.*)

Several species of ants inhabit twigs but none is so highly developed as the neotropical species *Pseudomyrmix filiformis*. It is adapted for life in hollow twigs of small diameters. The body is cylindrical in form, the head is elongate, the antennae are comparatively short, the peduncle and first gaster segment are elongate and without a distinct scale, and the abdomen is unusually long and slender.

To these may be added a few Orthoptera such as the unique Australian wood-boring cricket, *Cylindrodes kochi*, and the North American wood-boring roach, *Cryptocerus punctulatus*.

**Significance of the Boring Habit.**—The boring habit has three functions: a source of food, a means of protection during the larval or pupal period, and a home or place for rearing the broods of certain social and solitary species.

**Eggs of Borers.**—The egg-laying habits of boring insects are particularly fitted for this peculiar type of living. The curculionid uses its proboscis to excavate a hole in fruit, in which the egg is laid. The snout is often elongate and can reach deep into the tissues. In the chestnut weevil, the proboscis is twice as long as the rest of the body. This is used by the female to drill a hole into the spiny bur of the chestnut. The Scolytidae engrave galleries in the cambium with niches for the reception of the eggs. Many of the Lepidoptera and Coleoptera lay their eggs on the surface of the plant and the larvae find their way to the interior of the host. The majority of the beetles deposit their eggs on the surface of the plant or in crevices in the host. Ants and termites make tunnels through the wood, some of which serve as egg chambers. Solitary wasps dig their own burrows or utilize those made by other insects. The eggs are laid in cells together with a provision of food. The sawflies have sharp lances to cut incisions in the host for the reception of the eggs. The ovipositor of the pigeon tremex, *Tremex columba*, is extremely sharp and stiff and is used to drill holes in solid wood. This is a strenuous task. The ovipositor frequently becomes wedged in the wood, and the female is trapped and dies. The seed-infesting wasps, Chalcididae, have long sharp ovipositors which penetrate the flesh of various fruits and reach the seeds in which they lay their eggs. Diptera that bore in fruits, succulent stems, or roots lay their eggs in, on, or near the host. The Anthomyiidae usually lay their eggs in the ground near the host plant, and the Trupaneidae generally insert their eggs in the fruit or plant. The Mediterranean fruitfly lays a group of eggs in a small pocket just beneath the skin of citrus fruit. The apple maggot lays its eggs singly just beneath the skin of the fruit. The Lepidoptera usually lay their eggs on the exterior of the host; however, the yucca moth, *Tegeticula (Pronuba) yuccasella*, has a long extensile ovipositor that is used to insert the eggs into the ovaries of the yucca. Other specialized Lepidoptera have ovipositors adapted to insert eggs within the tissues of the plant.

**Modifications of the Larvae for the Boring Habit.**—The larva of the boring insect is typically cylindrical in form, legless, with the head telescoped within the thorax, and with the antennae reduced.  Boring insects reach their maximum development in the Buprestidae, Cerambycidae, and Siricidae.  The species of these families bore in solid wood, and extreme specialization is evident.  In other groups there is often considerable variation from the typical boring form.  Legs are absent in the larvae of most of the boring Coleoptera; however, they are present but minute in the Bruchidae, Bostrichidae, Anobiidae, Mordellidae, and Brentidae.  The larvae of *Phyllotreta*, which bore in the roots of various plants, are elongate forms with much reduced legs.  They are quite different in appearance from the typical external-feeding chrysomelid larvae.  Thoracic legs are well developed in the Scarabaeidae, Lucanidae, and Elateridae, forms that bore chiefly in the soil or in decayed wood.

Fig. 325.—Apple showing the mining of a codling moth larva beneath the skin and a plug of frass at the exit hole of the boring.

Legs are present or absent in the Lepidoptera.  In *Cossus*, they are well developed and in *Tegeticula* (*Pronuba*) they are absent.  In the Hymenoptera, legs are absent or minute.  Legless larvae usually have roughened areas or scansorial warts on various body segments which aid in holding the larva in position while it is feeding.  These are highly developed in the Cerambycidae and Buprestidae.  They take the form of transverse rows of papillae, ambulatory setae, in the Diptera.  The larvae of the boring insects are usually white or cream in color; however, the larvae of a few species such as the hop borer, *Gortyna immanis*, and the leopard moth *Zeuzera pyrina*, are distinctly spotted.

**Elimination.**—The larvae of all boring insects are confronted with the problem of the disposal of the waste material.  This is of more serious concern than in the case of leaf-mining insects, where the portion of the plant confining the larva is thin and can readily be cut or punctured to eject the frass. The squash vine borer packs its burrow with much of the soft green fecular pellets and feeds in the opposite direction so that clean food is always available. The larva of the plum curculio packs its fecula in a mass in a cavity within the fruit. The codling moth and the oriental fruit moth larvae push their borings and waste materials from their burrows by means of their heads. Ants, termites, and bostrichids remove the excavated material from their burrows by means of their jaws. Many of the wood-boring larvae, at first, push their borings to the exterior but later pack them in the burrow to serve as material for pupal chambers. The anal fork, used so frequently by free-feeding larvae to dispose of their fecula, is usually absent in the boring forms. However, the larvae of a few, such as the oriental fruit moth, have well-developed anal combs.

**Life Cycle of Boring Insects.**—The species that bore in fruits or succulent plants have a short life cycle, at least one generation a year. The oriental fruit moth has four broods in Connecticut, five broods in Pennsylvania, and seven broods in Georgia. Borers in solid wood develop more slowly and require at least two or three years to mature. The Cossidae require three years and the Cerambycidae from one to three years. Wood borers, that ordinarily mature in one or two years, under normal conditions, may require from 15 to 20 years if the timber in which the larva is feeding is cut and converted into furniture or other manufactured products. This is due to the lack of more suitable nutriment and humidity.

**Emergence.**—Species that bore in fruits or succulent plants seldom have any difficulty in emerging; those that bore in solid wood are confronted with a more serious problem. Many larvae abandon their boring quarters before they pupate; some go to the ground and others seek rotten wood. The Coleoptera and the Hymenoptera have strong jaws with which they can gnaw their way to freedom but the Lepidoptera must follow other means. The pupae of the carpenter moth and the leopard moth are formed just beneath the bark; when ready to emerge, they work their way partly out of the burrows before the adults emerge. The pupa frees itself by means of well-developed, backward-projecting spines on the abdominal segments which permit it to move only forward. The mature currant stem borer, *Synathedon tipuliformis*, has a similar habit but the larva eats a hole to the exterior and spins a silken web across the hole before it pupates. Normally, seed feeders wait until the fruit has decayed before the adults emerge from holes previously cut by the larva almost through the seed coat.

**Food of Boring Insects.**—Plant borers can be divided, on the basis of nutrition, into two general groups: feeders on living tissue and feeders on dead or decaying tissue. The former includes buds, terminal growth, cambium, stalks, roots, fruits, seeds, and fungi. The food of the borers of living tissue is similar to that of other plant feeders. It consists of proteins and soluble carbohydrates. The feeders upon dead plant tissue include the borers of pith, dry wood, and decayed wood. Our knowledge of the food of wood-boring species is still meager. It is well known that termites depend upon protozoa, that occur in the alimentary tract, to convert cellulose into available food. The larvae of certain Tipulidae, Cerambycidae, and Buprestidae also depend upon symbiotic organisms for preparing their food. Insects that feed upon decayed wood utilize the juices resulting from decay or consume fungi and other organisms prevalent. The ambrosia beetles and leaf-cutting ants cultivate fungi in their burrows upon which they feed. How some insects derive nourishment from cork and dried wood is still a problem. It has been previously pointed out that such insects often require many years to develop.

**Bud Borers.**—Bud boring is often a temporary habit. Numerous larvae bore into buds to obtain nourishment before the leaves open. These species may feed upon leaves during the remainder of the season. The habit is especially well developed in the Lepidoptera; the eye-spotted budmoth, *Spilonota ocellana*, is an excellent example. Other insects confine their feeding to bud and terminal growth. The pine tip moth, *Rhyacionia* (*Evetria*) *frustrana*, the European pine shoot moth, *R.* (*Evetria*) *buoliana*, and the white-pine weevil, *Pissodes strobi*, are common examples. All cause a similar type of injury, check the growth of the terminals, and misshape the tree. The peach twig borer, *Anarsia lineatella*, and the oriental fruit moth, *Grapholitha molesta*, work in somewhat the same manner although the latter bores in the fruit as frequently as in the twigs. The eggs of the oriental fruit moth are laid on the newly formed tip leaves and the larvae bore into the terminals.

Fig. 326.—Hole in apple bud and frass resulting from the boring of the larva of the eye-spotted bud moth, *Spilonota ocellana* Schiff.

**Cambium Borers.**—Cambium borers or miners reach the peak of their development in the Lepidoptera. The habit occurs frequently in the Opostegidae and in the genera *Marmara* and *Euzophera*. In the Aegeriidae and in certain of the Coleoptera, the larvae often start feeding in the cambium but later migrate to the sapwood and finally finish in the heartwood. The pitch pine moth, *Parharmonia pini*, works almost entirely in the cambium. The Scolytidae also feed characteristically in the cambium. Among the Diptera, there are a few species of the genus *Agromyza* that mine strictly in the cambium.

**Stalk Borers.**—Stalk borers are found in the Lepidoptera, Diptera, Coleoptera, and Hymenoptera. They intergrade to some extent with the root borers. The Pyralididae (Lepidoptera), the Oscinidae and Ephydridae (Diptera), and the Cephidae (Hymenoptera) contain the largest number of stalk borers. They attack the grasses and more succulent plants. Many extremely injurious species are included, as

Fig. 327.—Masses of gum on pine produced by the larvae of the pitch pine borer, *Parharmonia pini.* The larvae live and eventually pupate in these masses.

for example, the European corn borer, the squash vine borer, and the wheat stem sawfly.

**Root Borers.**—Root boring reaches its highest development in the Lepidoptera. Vegetables and ornamentals frequently suffer from their attack. The iris borer, *Macronoctua onusta,* and the columbine borer, *Papaipema purpurifascia,* are noteworthy. In the Diptera and the Coleoptera, the larvae frequently feed from the exterior and tunnel into the roots but are not strictly borers. The cabbage maggot, the corn root borer, and the tobacco and potato flea beetle larvae are examples.

**Fruit Borers.**—Fruit borers are conspicuous among the Diptera, Coleoptera, and Lepidoptera. The fruitflies (Trupaneidae), the curculios (Curculionidae), and various fruitworms (Olethreutidae) include the most important species. This group embraces some of the most serious pests of fruit crops, such as the apple and cherry maggots, the oriental fruit moth, the plum curculio, and the codling moth. The codling moth has been a problem for more than two centuries and will probably continue to annoy fruit growers for many years to come.

**Nut Borers.**—The larvae of the Curculionidae are the outstanding borers in nuts. They lay their eggs in the young nuts, using their long proboscides to excavate tunnels. The larvae generally abandon the nuts before they pupate. One of the most interesting species is the borer in vegetable ivory, tagua. Fifty larvae of this tropical species, *Dryocoetes dactyloperda*, were taken from a single nut. The nut when mature is as hard as animal ivory. The insect apparently secretes a substance that softens the nut so that it can feed. A few Lepidoptera, chiefly the larvae of the *Laspeyresia*, also bore in nuts.

**Seed Borers.**—There are two general groups of seed borers: species that feed in green or living seeds and those that attack dry seeds. The Chalcididae, most important in the first group, attack the seeds of a large number of plants. The eggs are deposited in the seeds by means of a long ovipositor that penetrates the flesh of the fruit. The adults usually emerge after the fruits have decayed. A few Carpocapsidae are also seed feeders. The common feeders upon dried seeds are known as weevils. *Bruchus brachialis* feeds upon the seeds of vetch. The bean weevils, *Acanthoscelides obtectus* and *Bruchus rufimanus*, feed upon green beans as well as upon dried beans. The movements of the so-called Mexican jumping beans (*Sabestiania*) are caused by the larvae of *Grapholitha*

FIG. 328.—Excavations of the clover root borer, *Hylastinus obscurus* (Marsham). (*After Webster, U. S. Department of Agriculture.*)

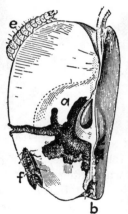

FIG. 329.—The codling moth, *Carpocapsa pomonella:* (*a*) borings of larva; (*b*) place where egg was laid; (*e*) larva; (*f*) adult. (*U. S. Department of Agriculture.*)

*saltitans*, a species somewhat closely related to the codling moth. The larvae throw themselves from one wall of the seed to the other and so cause the seeds to jump.

**Fungus Borers.**—The larvae of many Diptera and Coleoptera bore in fungi. The larvae of Lepidoptera and Hymenoptera are rarely found in fungi. Among the Diptera, the Mycetophilidae are the most outstanding group and are therefore called the fungus flies. A few

Tipulidae and Cecidomyiidae are also fungus feeders. The Coleoptera include the largest number of fungus feeders. The more important groups are the minute tree fungus beetles (*Cioidae*), the pleasing fungus beetles (Erotylidae), the handsome fungus beetles (Endomychidae), the hairy fungus beetles (Mycetophagidae), and the silken fungus beetles (Cryptophagidae).

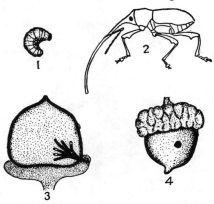

**Wood Borers.**—Wood boring is conspicuously the habit of certain Coleoptera. Some Lepidoptera and Hymenoptera bore in solid wood; few or no Diptera have this habit, although the larvae of many predacious Diptera are found in decayed wood seeking other insects.

FIG. 330.—Acorn borer, *Balaninus rectus*: (1) larva; (2) adult; (3) egg in gallery in acorn; (4) exit hole of larva.

Twig boring is a special type of wood boring. These forms have

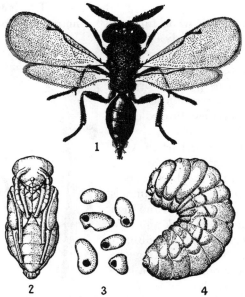

FIG. 331.—Clover seed chalcid, *Brucophagus gibbus:* (1) adult; (2) pupa; (3) clover seeds showing emergence holes of adults; (4) larva. (*After Urbahns, U. S. Department of Agriculture.*)

been discussed to some extent under the bud borers. Some, such as the oriental fruit moth, cause the terminals to wilt. Others cause a proliferation of cells and the formation of gall-like swellings. The red-necked cane borer, *Agrilus ruficollis,* and the bronzed birch borer, *A. anxius,* bore in spiral manner beneath the bark and produce conspicuous ridges on the surface of the plant.

Another group of wood borers is known as girdlers or pruners, because the larvae or adults girdle the twig and check the growth of the plant. The female currant stem girdler, *Janus integer,* girdles the stem with her knifelike ovipositor just above the position of the egg. The larva of *Oncideres cingulatus* feeds on woody plants particularly oak, pecan, pear, and peach. The female girdles the stem before she lays her egg beneath the bark. The oak pruner, *Hypermallus villosus,* lays her egg beneath the bark of the host. The larva eats out the inside of the twig and, when full grown, enlarges the burrow suddenly so that the branch is nearly severed.

FIG. 332.—*Eburia quadrigeminata.* The larva bores in the wood of hickory, ash, and locust.

The branch, thus weakened, readily breaks and falls to the ground. The larva remains in this piece of branch until spring when it transforms to an adult. The raspberry cane is girdled by the larva of *Hylemyia rubivora.* The flies appear in the latter part of April and the female deposits a large white egg near the tip of the raspberry shoot. When the egg hatches, the larva tunnels within the interior of the shoot and then works its way out to the bark and tunnels around the shoot, thus girdling it from the inside. After checking the growth in this manner, the larva burrows downward to the base of the plant and finally forms its puparium within the stalk. Here it remains until the following spring when the fly emerges.

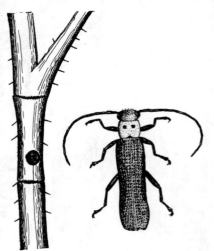

FIG. 333.—The raspberry cane girdler, *Oberea bimaculata,* and section of cane showing egg scar and girdling by adult.

**Decayed Wood Borers.**—Decayed wood is frequented by the larvae of numerous species of Coleoptera and Diptera. The larvae of the

Lepidoptera and Hymenoptera are seldom found in decayed wood except when pupating. The Scarabaeidae, Lucanidae, and Elateridae are the common feeders in decayed wood. The food is to some extent the same as that found in fungi. As a matter of fact, many of the species feed upon the fungi that grow on decayed wood.

**Dry Wood Borers.**—Dry wood attracts comparatively few species of insects. The powder-post beetles, the deathwatch beetles, the drug store beetles, and the timber beetles are practically the only species that attack dry wood. Ants and termites often make their nests in dry wood. Furniture and other products manufactured from wood may contain other coleopterous larvae which normally feed in green wood.

**Frass of Wood-boring Larvae.**— The *frass,** *i.e.*, the refuse left by wood-boring larvae, or the manner in which it is disposed of, often serves as a means of recognizing the insect concerned in producing certain types of injury. The excess chips or shreds of wood cut by the wood-boring larvae are frequently mixed with fecula, which is generally composed of cellulose or undigested lignum that passes through the alimentary canal. It is usually pushed from the burrow. The white-spotted sawyer, *Monochamus scutellatus*, the poplar borer, *Saperda calcarata*, and the carpenter worm, *Prionoxystus robiniae*, keep at least a portion of their burrows free from frass and other waste material.

Fig. 334.—The parandra borer, *Parandra brunnea* which attacks various shade and fruit trees. Bark removed to show larvae and abundant frass. (*Pennsylvania Department of Agriculture.*)

The buprestid, *Dicerca prolongata*, keeps its burrow comparatively free of frass by pushing it out of the openings through the bark. On the other hand, the pigeon tremex, *Tremex columba*, maintains no opening to the exterior but packs its frass in the rear of the burrow as it advances.

Frass varies considerably in form, texture, and color, depending upon the nature of the wood utilized by the insect. Soft woods such as hemlock and locust cut more readily into long shreds. Hard woods such as

* Consult p. 144, the fecula of insects.

oak and hickory cut more readily into chips. Gougelike mandibles cut broad chips, and acute mandibles produce narrow shredded fibers. When the locust borer works in the bark or cambium, the borings are brown in color; when it reaches the heartwood, the borings are yellow in color.

The roundheaded apple tree borer, *Saperda candida*, the locust borer, *Cyllene robiniae*, and other cerambycid larvae cut excelsiorlike borings which are used to plug the outer end of the pupal cells. The ribbed pine borer, *Rhagium lineatum*, forms a rounded pupal cell of long woody fibers just beneath the bark.

The buprestids usually cut chips which are eventually used to form pupal chambers. The flatheaded apple tree borer, *Chrysobothris femorata*, is a common species which attacks numerous forest trees and also a variety of orchard trees.

The powder-post beetles, Anobiidae and Ptinidae, burrow in seasoned wood and produce meallike borings. Ants make borings which are intermediate in size and resemble sawdust.

Many borers leave their signatures in the wood or stems in which they bore. The most evident telltale mark is the hole left by the larva, pupa, or adult that has emerged, or the abundant borings that are often pushed from the burrows. The shot-hole borers leave numerous small shotlike holes in the bark when they emerge. If the bark is stripped off, shallow engravings are evident that continue into the wood. Larger, more irregular cavities in the bark, extending into the wood, may be the work of the Buprestidae. Deep cavities in the wood are frequently produced by the larvae of the Cerambycidae, the larvae of the carpenter and leopard moths, and occasional hymenopterous larvae.

Injuries by insects often remain in lumber and later appear in manufactured objects. The smaller beetles, Anobiidae and Ptinidae, produce small holes or narrow burrows in the wood. The Cerambycidae and other larvae produce deep cavities in the wood. Pitch or gum pockets result from previous injuries by various insects. Blue stain follows the attack of *Dendroctonus* beetles. Black check is a defect resulting from the staining of the sapwood or heartwood of hemlock, fir, and spruce by the larvae of certain flies and moths. Pith flecks are caused by the larvae of certain Diptera.

**Systematic Classification of Borers.**—The boring habit occurs chiefly in insects with complete metamorphosis and insects with mandibulate mouth parts. The Coleoptera, Lepidoptera, Diptera, and Hymenoptera include the outstanding boring insects.

*Coleoptera.*—On the whole, the larvae of Coleoptera are the most conspicuous forms found in wood or the products of trees. This is a comparatively large group and includes many injurious species. The Buprestidae and the Cerambycidae are the typical wood borers although the Brentidae and a few Curculionidae also attack the wood of living trees.

The Cerambycidae or long-horned beetles show great diversity of feeding habits. Some larvae bore exclusively in decayed wood, others in dead dry wood, and many work under the bark in dead or dying trees, either completing their growth between the bark and the wood or going deep into the sapwood or heartwood.

The Buprestidae, because of their preference for woody plants and the tendency to metallic coloration, have been called the metallic wood

FIG. 335.—The roundheaded apple tree borer, *Saperda candida*, at rest upon the bark of a tree. (*Champlain and Kirk.*)

borers. The larvae of most of the species feed beneath the bark, in pith, or in solid wood. A few are leaf miners. In the wood-boring forms, the larvae are legless, the prothorax is exceedingly broad and flat, and the succeeding segments are slightly flattened. Because of the flattened nature of the thorax, they have been called flatheaded borers. The red-necked cane borer, *Agrilus ruficollis*, and the flatheaded apple tree borer, *Chrysobothris femorata*, are common examples; both are exceedingly injurious.

The Curculionidae stand somewhat alone as borers chiefly of fruit and nuts. The apple, plum, and quince curculios and the various nut weevils are common examples. A few species have other habits: the white-pine weevil, *Pissodes strobi*, attacks the growing shoots of pine; the rhubarb curculio, *Lixus concavus*, bores in the stalks of *Rumex;* the boll weevil, *Anthonomous grandis*, infests the cotton bolls.

The bark beetles, Scolytidae, make galleries, for the most part, in the cambium of dead or dying trees but their injuries are not confined to trees. The clover root borer, *Hylastinus obscurus*, bores in plants of the second year's growth and destroys them.

The Lyctidae, Anobiidae, Bostrichidae, and Lymexylidae bore in dry wood and often attack articles manufactured from lumber.

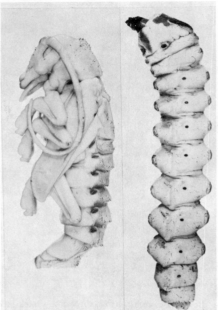 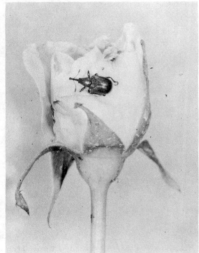

Fig. 336.—Pupa and larva of the roundheaded apple tree borer, *Saperda candida.* (*Craighead.*)

Fig. 337.—The rose hip beetle, *Rhynchites bicolor.* The adults feed upon the rose petals, the larvae feed in the seeds.

The larvae of many of the Scarabaeidae, Passalidae, Lucanidae, and Elateridae feed in decayed wood. They show little resemblance in form and habit to the true boring species and resemble more closely the subterranean species.

The Mycetophagidae, Cryptophagidae, Cioidae, Platystomidae, Endomychidae, and Erotylidae are borers in fungi. However, one of the erotylids, *Languria mozardi*, bores in the roots of clover.

The Chrysomelidae are principally free feeders or leaf miners, but a few species bore in the roots of plants. The corn rootworms, *Diabrotica longicornis* and *D. duodecimpunctata*, numerous flea beetles of the genus *Phyllotreta*, and the grape rootworm, *Fidia viticida*, are common examples.

*Lepidoptera.*—The Lepidoptera constitute one of the most extensive groups of boring insects. Eighteen families are represented, of which only a few are outstanding. Larvae of the primitive Hepialidae feed upon wood or bark and are found chiefly at the roots or within the stems of plants; *Sthenopis* bores in the roots of submerged plants and *Hepialus* bores in the stalks of ferns.

Fig. 338.—Galleries made by the Monterey pine engraver, *Ips radiatae* Hopk., It attacks living, dying, and recently killed Monterey and other pines of California. (*Pennsylvania Department of Agriculture.*)

The Incurvariidae have various habits, the most important of which is leaf mining; a few species exhibit the boring habit. The larvae of the yucca moths of the genus *Tegeticula* burrow in the seeds of the yucca. The larvae of the bogus yucca moth, *Prodoxus*, feed upon the yucca, burrowing in the flower stems and in the fruit.

The Cossidae are borers in solid wood; the leopard moth, *Zeuzera pyrina*, and the carpenter worm, *Prionoxystus robiniae*, are both general feeders. The leopard moth attacks more than 125 species of woody trees. It lays its salmon-colored eggs in groups of four or more in crevices in the rough bark. The larvae burrow into the heart or pith of the twigs and into the heartwood of the larger branches. Six or eight borers commonly work in a single tree but occasionally larger numbers are found. The larvae transform in their burrows two years after the eggs hatch. No cocoons are spun and no silk or wooden chips are employed to make pupal cases. On the head of the pupa, there is a sharp spine that aids the pupa in pushing its way out of the burrow preparatory to the emergence of the moth.

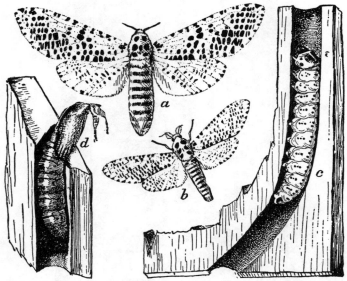

Fig. 339.—The leopard moth, *Zeuzera pyrina*: (*a*) female; (*b*) male; (*c*) larva; (*d*) empty pupal case. (*After U. S. Department of Agriculture.*)

The Tineidae contains a single fungus-boring genus, *Scardia*.

The majority of the larvae of the Gracilariidae are leaf miners but many of the species of the genus *Marmara* alternate between leaf mining and cambium mining. A few species excavate shallow channels beneath the skin of fruit. Apples and citrus fruit often show the operations of these larvae.

The Opostegidae are strictly cambium or bast miners.

The larvae of the Yponomeutidae display an unusual variety of habits. *Acrolepia incertella* ties and skeletonizes the leaves of smilax and bores in the bulbs of lilium.

The Aegeriidae constitute one of the most outstanding groups of borers; some excavate twigs, others wood. This family includes some of the most serious pests, as for example, the peach tree borer, *Conopia exitiosa*, the raspberry root borer, *Bembecia marginata*, the imported currant borer, *Synanthedon tipuliformis,* the lilac borer, *Podosesia syringae,* the squash borer, *Melittia satyriniformis,* and others. All deposit their

Fig. 340.—Peach tree borer, adult male.

eggs on the surface of the host. The larvae work their way into the host and generally pupate in their burrows near the exit holes. The squash

borer works in the stems of squash, pumpkin, and other cucurbits.    The
larvae bore through the stems, cause them to rot at the affected points,
and eventually kill the plants.    Unfortunately the presence of the pest
is not evident until the larva is nearly mature and the damage is completed.    The borer then manifests itself by coarse yellowish-green fecula
which is forced from the burrows and by the wilting of the leaves.    One
to six larvae may inhabit a single stem; in some instances 50 or more
individuals have been taken from a single plant.    The eggs are oval,
dull red in color, and minutely sculptured.    They are laid on all parts of
the vine but more particularly on the stems.    The larvae are at first
green; later they become white or cream colored.    The heads are shiny
black and the thoracic legs are well developed.    When mature, the larvae

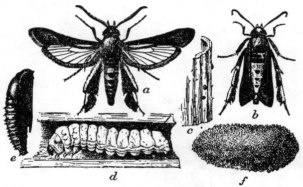

Fig. 341.—The squash vine borer, *Melittia satyriniformis*: (a) male; (b) female; (c) eggs on a
bit of squash stem; (d) larva; (e) pupa; (f) pupal cell.    (*After Howard.*)

leave the stems and enter the ground burying themselves to a depth of
one to two inches.    They spin silken cocoons which are covered with
small particles of earth.    There is apparently but one generation in the
northern part of the United States.    The adults are rather striking,
clear-winged moths.    The front wings are opaque and olive brown in color
with a metallic sheen.    The hind wings are transparent.    The abdomen
is conspicuously marked with red, black, and bronze.    The hind legs are
fringed with long hairs which are red on the outer surface and black on
the inner surface.    This species is closely related to the peach tree borer
and the pine pitch moth.

The pine pitch moth, *Parharmonia pini*, lays its eggs on the bark of
the tree.    The larvae mine in the cambium causing an abundant flow of
gum which hardens in masses on the exterior of the tree trunk.    At
least two years are required for the larvae to mature.    They pupate in
the mass of gum, and the pupae work themselves partly out of it before
the adults emerge (see Fig. 327).

The Olethreutidae contains numerous boring species, some of which are serious pests. All the species of the genus *Bactra* bore in rushes. The *Proteoteras* are borers exclusively in the Sapindaceae. The larvae of *Carpocapsa* bore in fruit and rarely in nuts. *Carpocapsa pomonella* bores

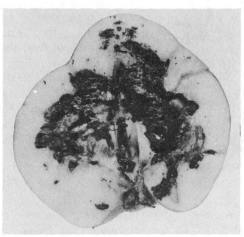

FIG. 342.—Quince showing excavations made by the larva of the oriental fruit moth, *Grapholitha molesta* Busck.

in the fruit of apple or in the husks of walnuts. The *Grapholitha* are borers in fruit, seeds, and stems. The oriental fruit moth, *Grapholitha molesta*, is a serious pest on peach, quince, cherry, and other fruits. The larvae of *Grapholitha saltitans* bore in the seeds of *Sabestiania*, commonly known as Mexican jumping beans. They hibernate in the seeds, where pupation occurs. The adults emerge in the spring through small circular holes. The larvae of the genus *Epiblema* bore in the stalks of *Solidago*, *Ambrosia*, and *Bidens*. The larvae of *Rhyacionia* (*Evetria*) feed on the young shoots of pine or in pitch nodules, or bore in the cones of spruce. Occasional borers are found in the genera *Eucosma*, *spilonota Epinotia* and *Thiodia*.

FIG. 343.—The adult cranberry borer, *Mineola vaccinii*, and injury to fruit.

The Carposinidae bore exclusively in fruit. The currant fruit worm, *Carposina fernaldana*, feeds within the fruit, attacking both seeds and pulp.

The larvae of the Pyralididae exhibit a wide range of food habits: leaf rolling, leaf crumpling, leaf tying, external feeding, stored-food feeding, plant boring, and preying upon other insects. The majority of the borers are found in the stems of succulent plants. Common examples are the European corn borer, *Pyrausta nubilalis*, the cornstalk borer, *Diatraea zeacollis*, the sugarcane borer, *Diatraea saccharalis*, the cranberry girdler, *Crambus hortuellus*, the corn root webworm, *Crambus caliginosellus*, the melon stem borer, *Eudioptis nitidalis*, and many others. All lay their eggs upon the stalks of the plant; the larvae usually feed at the base of the plant or below the ground. The larvae of *Crambus* frequently feed from the outside and girdle the plant.

FIG. 344.—Larvae of the European corn borer feeding in corn cobs. (*After U. S. Department of Agriculture.*)

Two other pyralids have very unusual habits. The larvae of the wax moth, *Galleria mellonella*, tunnel through the wax of beehives. A species of *Euzophera* is a bast borer.

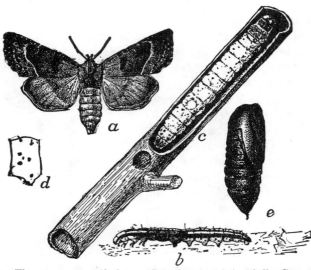

FIG. 345.—The common stalk borer, *Papaipema nebris nitella* Guenee: (*a*) adult; (*b*) half-grown larva; (*c*) full-grown larva; (*d*) side view of first segment of larva; (*e*) pupa. (*After Chittenden, U. S. Department of Agriculture.*)

The larvae of a few Noctuidae bore in plants. To the genus *Papaipema* belong several species especially injurious to the roots and stalks of flowering plants; *Papaipema purpurifascia* is known as the columbine root borer; *P. nebris nitela* is a general feeder. The larvae of this species are found in the stalks of numerous weeds and cultivated plants, especially corn, tomato, and eggplant. Young corn is most susceptible to injury. The infested plants wilt and break down and the larvae attack new plants thus increasing their destructiveness. The small, light-brown eggs are deposited on the stalks of the various hosts. They hatch in early spring into brownish larvae with two conspicuous darker brown lateral stripes, which are broken in the region of the first, second, and third abdominal segments. As the larvae mature, the color fades and varies from light to dark brown. The larvae pupate in the stalks and the adults emerge late in the fall. They resemble most noctuids in appearance; the fore wings are a grayish brown marked with white lines and the hind wings are paler in color.

The larvae of *Bellura* bore in the stems of pond lilies. Those of *Arzama* and *Archanara* bore in the stalks of cattails. The corn ear worm, *Heliothis obsoleta*, and the tobacco budworm, *H. virescens*, are well-known species. *Gortyna immanis* often cause serious damage to hops by boring into the stalks. The iris borer, *Macronoctua onusta*, is another serious pest.

Fig. 346.—One of the horn tails, the pigeon tremex, *Tremex columba* (Linn). (*After Jordan and Kellogg.*)

*Hymenoptera.*—The boring habit is somewhat restricted in the Hymenoptera. The Siricidae bore in solid wood. One of the best known species is the pigeon tremex, *Tremex columba*. The larvae of this species infest elm, maple, beech, oak, sycamore, apple, and pear. The eggs are inserted about half an inch into the solid wood. The larvae bore in the wood and transform within their burrows. They construct cocoons composed of silk and small chips of wood.

The Cephidae are borers in the stems of grasses and include many injurious species, as for example, the wheat stem sawflies, *Cephus pygmaeus*, *C. cinctus*, and *C. tabidus*. The females have sharp ovipositors and insert their eggs within the hosts. The currant stem girdler, *Janus integer*, bores in woody or pithy stems. After laying her egg in the cane, the female moves up the stem a short distance and girdles the cane with her ovipositor. This causes the tip of the cane to wilt and checks the growth of the plant so that it does not crush the egg.

The family Xiphydriidae is composed of a small number of species which bore in dead or decaying trees.

The Chalcididae include a large number of insects with widely different habits; many are parasitic or hyperparasitic, others are phytophagous. Some bore in the stems of grasses. Important among these are the wheat jointworm, *Harmolita tritici*, and the wheat straw worm, *H. grandis*. Not a small number of the Chalcididae bore in seeds, as for example the apple seed chalcid, *Callimome druparum*, and the clover seed chalcid, *Bruchophagus gibbus*. Crosby (1909) describes the habits of several species of chalcid flies that feed upon the seeds of sorbus, rose, grape, Virginia creeper, and sumac.

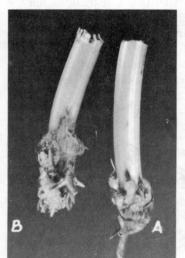

Fig. 347.—Wheat stems severed, after boring by (*A*) *Cephus tabidus* Fab.; (*B*) *Cephus pygmaeus* (Linn). Note smooth cut in one case and rough cut in the other. (*Hill and Udine.*)

Many of the free-feeding sawflies retreat to the ground to pupate and some utilize rotten wood in which to transform. The dogwood sawfly refuses to pupate unless it can find dead or decayed wood in which to transform.

Among the solitary bees and wasps, the adults do the excavating. Conspicuous examples are the carpenter bees, *Xylocopa* and *Ceratina*, and the leaf-cutting bees, *Megachile*. The carpenter bees bore in solid wood; the other species bore in pith or appropriate the burrows of other insects.

A few social Hymenoptera build nests in wood. The bees and wasps frequently use cavities in trees. The ants and termites often excavate their burrows in wood.

*Diptera.*—The boring habit is not well developed in the Diptera. As a matter of fact, few or none of them bore in solid wood. The larva of *Tanyptera frontalis* bores in maple wood that is almost solid. Of the many dipterous larvae that are found in decayed wood, some feed on the juices resulting from the processes of decay, others feed upon microorganisms in this medium, and still others prey upon insects that inhabit decayed wood. The larvae of the Therevidae, Asilidae, Stratiomyiidae, and Dolichopodidae prey upon other insects that live in these situations. The larvae of certain Tipulidae, Rhagionidae, and Anisopidae undoubtedly feed upon the products of decay.

The fungus gnats, Mycetophilidae, and certain Helomyzidae, Platypezidae, and Borboridae live and feed in fungi.

A large number of the boring Diptera attack the roots of plants. They often feed from the exterior, girdle the roots, and eat burrows in

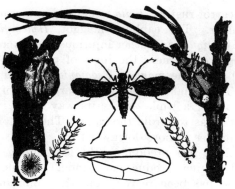

Fig. 348.—The resin gnat, *Retinodiplosis resinicola*, showing lumps of resin on pine in which the larvae live. Adult, and empty pupal cases protruding from the resin lumps. (*After Comstock.*)

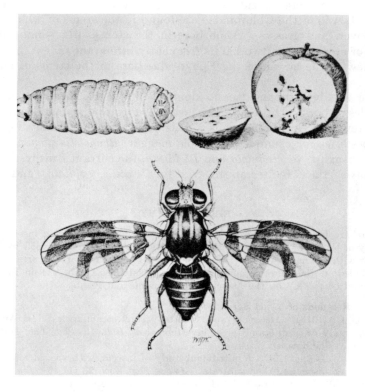

Fig. 349.—The apple maggot, *Rhagoletis pomonella* illustrating larva, injury to apple by larva, and adult.

the stalks.    The carrot rust fly, *Psila rosae,* and numerous Anthomyiidae feed in this manner.    The cabbage maggot, *Hylemyia brassicae,* the onion maggot, *H. antiqua,* and other injurious species are included here. They lay their eggs on or near the roots of plants.    A closely related species, the raspberry cane maggot, *Pegomyia rubivora,* feeds in woody or pithy stems.    The female lays her eggs near the tip of the shoot.    The larva, after feeding for a while, works its way nearly to the bark and tunnels the shoot so as to girdle the cane.    The top of the plant wilts and the larva continues to bore downward.

The Agromyzidae are predominantly leaf-mining insects.    A few species are parasitic, others produce galls, and a few mine in the cambium; *Agromyza pruinosa* has been reared from the cambium of river birch, *A. aceris* from the cambium of maple, and *A. amelanchieris* from the cambium of serviceberry.    The larvae of the cambium miners depart from the usual agromyzid type and are unusually long and slender.

The larvae of the Chloropidae are found in the stems of wheat, oats, rye, clover, and grasses.    Some bore in the stems, others mine in the leaves or stems.    The two habits frequently intergrade.

A few of the larvae of the Ephydridae bore in the stems of aquatic plants.

The family Trupaneidae no doubt contain the most outstanding examples of boring Diptera.    The larvae are fruit borers and leaf miners, and a few species produce galls.    This family includes many injurious fruit pests, as for example, the apple maggot, *Rhagoletis pomonella,* the cherry maggots, *R. cingulata* and *R. fausta,* the currant fruitfly, *Epochra canadensis,* the Mediterranean fruitfly *Ceratitis capitata,* and many others.

## BIBLIOGRAPHY

There are innumerable papers dealing with the habits of boring insects.    The majority of these are concerned with the digestion or utilization of wood by boring forms or with the habits of individual species.    A few of the more important papers are included in the following bibliography.    Consult also Chap. VII on Immature Insects and Chap. XV on the Associations of Insects and Plants.

I. Food Relations of Wood Borers.
ADAMS, C. C. (1915).    Decaying wood and tree trunk communities.    *In,* An ecological study of prairie and forest invertebrates.    *Bull. Ill. State Lab. Nat. Hist.* **11** (2).    *Bibliography.*
BAUMBERGER, J. P. (1919).    A nutritional study of insects, with special reference to microorganisms and their subtrata.    *Journ. Exp. Zool.* **28** (1).
BLACKMAN, M. W., and STAGE, H. H. (1924).    On the succession of insects living in the bark and wood of dying, dead and decaying hickory.    *N. Y. State Forestry College Tech. Pub.* **17.**    *Bibliography.*
CARTRIGHT, K. ST. G. (1938).    A further note on fungus association in the Siricidae. *Ann. Appl. Biol.* **25** (2).    *References.*

CHAPMAN, ROYAL N. (1931). Nutrition. *In*, Animal ecology. McGraw-Hill Book Company, Inc., New York.

CLEVELAND, L. R. (1924). The physiological and symbiotic relationships between the intestinal protozoa of termites and their host, with special reference to *Reticulitermes flavipes* Kollar. *Biol. Bull.* **46** (5).

—— (1925). The ability of termites to live perhaps indefinitely on a diet of pure cellulose. *Biol. Bull.* **58.**

—— (1934). The wood-feeding roach *Cryptocerus*, its protozoa and the symbiosis between protozoa and roach. *Mem. Amer. Acad. Arts and Sci.* **17.**

DAVIS, W. T., and LENG, C. W. (1912). Insects on a recently felled tree. *Journ. N. Y. Ent. Soc.* **20** (2).

GRAHAM, S. A. (1925). The felled tree trunk as an ecological unit. *Ecology* **6** (4).

HENDEE, E. C. (1935). The role of fungi in the diet of the common damp-wood termite. *Hilgardia* **9** (10).

INGLES, L. G. (1933). The succession of insects in tree trunks as shown by collections from various stages of decay. *Journ. Ent. Zool. Claremont* **25.**

LEACH, JULIAN GILBERT (1940). Symbiosis between insects and microorganisms. *In*, Insect transmission of plant diseases. McGraw-Hill Book Company, Inc., New York.

MANSOUR, K., and MANSOUR-BEK, J. J. (1934). On the digestion of wood by insects. *Journ. Exp. Biol.* **11** (3).

PARKIN, E. A. (1936). A study of the food relations of the *Lyctus* powder post beetles. *Ann. Appl. Biol.* **23** (2). *References.*

SAVLEY, H. E. (1939). Ecological relations of certain animals in dead pine and oak trees. *Ecol. Monogr.* **9** (3).

SCHWARTZ, E. A. (1901). "Sawdust" of wood borers. *Proc. Ent. Soc. Wash.* **4** (4).

TRÄGARDH, IVAR (1929). Investigations of the fauna of a decaying tree. *Trans. 4th Intern. Congress Ent.*

—— (1938). Survey of wood-destroying insects in public buildings in Sweden. *Bull. Ent. Res. Lond.* **29** (1).

UVAROV, B. P. (1928). Insect nutrition and metabolism. *Trans. Ent. Soc. Lond.* **31.** *References.*

YOUNG, C. M. (1938). Recent work on the digestion of cellulose and chitin by invertebrates. *Sci. Progr. Lond.* **32.**

**II. Coleoptera That Bore in Wood or Other Substrata.**—See also Chap. IV, references to forest and shade tree insects.

BEUTENMÜLLER, WM. (1896). Food-habits of North American Cerambycidae. *Journ. N. Y. Ent. Soc.* **4** (2).

CHAMBERLIN, W. J. (1939). The bark and timber beetles of North America north of Mexico. O. S. C. Cooperative Assoc. Oregon.

HILL, C. L. (1927). Marine borers and their relation to marine construction on the Pacific coast. University of California Press, Berkeley.

HOPKINS, A. D. (1904). Insect injuries to hardwood forest trees. *Yearbook U. S. Dept. Agric.* 1903.

—— (1905). Insect injuries to forest products. *Yearbook U. S. Dept. Agric.* 1904.

KAMPE, O., and ISLEY, D. (1936). Notes on the biology of nut infesting weevils. *Journ. Kan. Ent. Soc.* **9** (1). *Literature.*

KNOWLTON, G. F., and THATCHER, T. O. (1936). Notes on wood-boring insects. *Proc. Utah Acad. Sci. A. L.* **13.**

KRAUS, E. J., and HOPKINS, A. D. (1911). A revision of the powder-post beetles Lyctidae of the United States and Europe. Appendix, notes on habits and distribution with lists of described species. *U. S. Dept. Agric. Bur. Ent. Tech. ser.* **20** Part III.

PARKIN, E. A. (1933). Larvae of some wood-boring Anobidae (Coleoptera). *Bull. Ent. Res. Lond.* **24** (1). *References.*

PECHUMAN, L. L. (1937). An annotated list of insects found in the bark and wood of *Ulmus americana* in New York State. *Bull. Brooklyn Ent. Soc.* **32** (1).

PIERCE, W. D. (1907). On the biologies of the Rhynchophora of North America. *Nebr. State Bd. Agric. Bibliography.*

SCHWARTZ, E. A. (1882). Wood-boring Coleoptera. *Amer. Nat.* **16** (9).

TOWNSEND, C. H. T. (1886). Coleoptera found in dead trunks of *Tilia americana* L. in October. *Can. Ent.* **18** (4).

TRÄGÅRDH, IVAR (1938). Survey of wood-destroying insects in public buildings in Sweden. *Bull Ent. Res.* **29** (1).

WADE, J. S. (1935). A contribution to a bibliography of the described immature stages of North American Coleoptera. *Bur. Ent. and Quarantine U. S. Dept. Agric. E-358*, mimeographed.

WALSH, B. D. (1867). Bark-borers. *Practical Entom.* **2** (6).

**III. Borers in Metal.**

BURKE, H. C., *et al.* (1922). Lead cable borer or "short-circuit beetle" in California. *U. S. Dept. Agric. Ent. Bull.* **1107.**

FROGGATT, W. W. (1917). A lead boring beetle *Xylothrips gibbicollis. Agric. Gaz. N. S. W. Sydney* **28** (11).

LAING, F. (1919). Insects damaging lead. *Ent. Mo. Mag.* **55.**

LITTLER, F. M. (1909). Notes on *Lyctus causalisulatus* F. *Entom.* **42.**

RENDELL, E. J. P. (1930). Depredations to lead-covered aerial cables by beetles in Brazil. *Proc. Ent. Soc. Wash.* **32** (6).

SNYDER, T. E. (1927). Insect metal workers. *Nat. Mag.* **8** (5).

**IV. Wood-boring Bees and Wasps.**

ACKERMAN, A. J. (1916). The carpenter bees of the United States of the genus *Xylocopa. Journ. N. Y. Ent. Soc.* **24.**

ASHMEAD, W. H. (1894). The habits of aculeate Hymenoptera. *Psyche* **7.**

BEESON, C. F. C. (1938). Carpenter bees. *Indian Forester* **64** (12).

HANSON, H. S. (1939). Ecological notes on *Sirex* wood wasps and their parasites. *Bull. Ent. Res.* **30** (1). *References.*

HICKS, C. H. (1929). Notes on the habits of *Anthidium collectum* Howard. *Can. Ent.* **61** (4).

―――― (1929). The nesting habits of *Anthidium mormonum fragariellum* Ckll. *Ent. News* **40** (4).

MOTHERSOLE, H. (1924). Some observations on leaf-cutting (*Megachile*) and its parasite (*Coelioxys*). *Essex Nat.* **21.**

NININGER, H. H. (1916). Studies in the life histories of two carpenter bees of California, with notes on certain parasites. *Journ. Ent. Zool. Claremont, Calif.* **8** (4).

PACKARD, A. S. (1897). "*Megachile, Ceratina* and *Xylocopa.*" *In,* Notes on transformations of higher Hymenoptera III. *Journ. N. Y. Ent. Soc.* **5** (3).

RAU, PHIL (1928). The nesting habits of the little carpenter bee, *Ceratina calcarata. Ann. Ent. Soc. Amer.* **21** (3).

YUASA, HACHIRO (1922). A classification of the larvae of the Tenthredinoidea. *Ill. Biol. Mon.* **7**, No. 4.

**V. Miscellaneous Boring Insects.**

BARBER, H. S. (1913). Notes on a wood-boring syrphid. *Proc. Ent. Soc. Wash.* **15** (2).

BIRD, HENRY (1902). Boring noctuid larvae. *Journ. N. Y. Ent. Soc.* **10** (4).

CROSBY, C. R. (1909). On certain seed-infesting chalcis-flies. *Cornell Univ. Agric. Exp. Sta. Bull.* **265.**

FLINT, W. P., and MALLOCH, J. R. (1920). The European corn borer and some similar native insects. *Bull. Ill. State Lab. Nat. Hist.* **13** (10).

FROST, S. W. (1936). Australian cricket, *Cylindrodes campbelli* that bores in wood. *In,* Ancient artizens. Van Press, Boston.

GRAHAM, S. A. (1918). The carpenter ant as a destroyer of solid wood. *Minn. State Ent. Rept.* **17.**

HARVEY, F. L. (1895). Mexican jumping beans. *Amer. Nat.* **29** (344).

HUTCHINGS, C. B. (1924). The life history, habits and control of the lesser oak carpenter worm. 16*th Ann. Rept. Quebec Soc. Prot. Plants. Bibliography.*

PICER, J. L. (1908). Life history of the carpenter ant. *Marine Biol. Lab. Woods Hole Bull.* **14** (3). *Bibliography.*

RILEY, C. V. (1891). Mexican jumping bean, the determination of the plant. *Proc. Ent. Soc. Wash.* **2** (2).

———— (1891). Further note on *Carpocapsa salitans* and on a new *Grapholitha* producing jumping beans. *Proc. Ent. Soc. Wash.* **2** (2).

Consult also references to boring insects found in *U. S. Dept. Agric. Div. Ent.* as follows: The locust borer, *Bull.* **787**; The Parandra borer, *Bull.* **262**; The round-headed apple borer, *Bull.* **847**; The spotted apple borer, *Bull.* **886**; The clover stem-borer, *Bull.* **889**; The timothy stem borer, *Bull.* **95.** In the *Journ. Agric. Research*, the following references will be found valuable: The oak sapling borer, Vol. XXVI, No. 7, 1924; The cambium curculio, Vol. XXVIII, No. 4, 1924; The three-lined fig tree borer, Vol. XI, No. 9, 1917.

# CHAPTER XX

## SUBTERRANEAN INSECTS

A subterranean insect is one that spends a part or all of its existence beneath the surface of the soil. There are two points of view: surface insects seek the soil for various reasons, and subterranean insects seek the surface for various reasons. The Japanese beetle spends almost 11 months in the soil as egg, larva, and pupa, then emerges to feed for a short time and to mate and returns to the soil to lay its eggs. Lepidoptera, on the other hand, frequently enter the soil to spend but a few days during pupation.

The subterranean habit approaches the *cavernicolous* (the cave inhabiting habit), the *subcutaneous* (leaf-mining or boring habit), and the *subaqueous* (the aquatic habit). All these species live in environments where the light is reduced or absent, the moisture is generally high, the temperature is somewhat constant, and the parasitic and predacious hazard is somewhat reduced.

**Attending Soil Factors.**—Soil is a mixture of air, water, and earth and is therefore a composite medium intergrading with the cover. The line of demarcation between the air and the soil is not so well defined as that between air and water. The study of soil insects should take into consideration the stratification of the soil, especially the cover of leaves and debris, the loam, the subsoil, clay, and hardpan. The texture of the soil and the type of the cover often determine what species may be present. Shelford has shown that different species of tiger beetles, Cicindelidae, choose different types of soil in which to oviposit. Some species oviposit in moist sandy places, others in soils composed of sand and

Fig. 350.—Interior of a formicary, showing the classification of the eggs, larvae, and pupae according to stages. A subterranean colony. (*After Andre.*)

loam, and still others along roadsides where there may be considerable vegetation.

SOIL PREFERENCE OF COMMON SPECIES OF TIGER BEETLE LARVAE

| Type of Soil Inhabited | Typical Species of *Cicindela* Occupying Such Soil |
|---|---|
| Dry sand containing some humus | *C. scutellaris* |
| | *C. tranquebarica* |
| Sand or clay with some humus | *C. sexguttata* |
| Shifting sands | *C. lepida* |
| Moist clean sand | *C. hirticollis* |
| Sandy margins of pools and streams | *C. repanda* |
| Bare rocky soil, steep wooded hillsides | *C. unipunctata* |
| Hard dry soil, usually with humus, often in clumps of grass | *C. punctulata* |
| Clay or humus | *C. duodecimguttata* |
| Clay on steep banks | *C. limbalis* |
| Moist black soil especially in grassy ravines | *C. purpurea graminea* |

Soils rich in decaying organic material attract saprophagous and scatophagous insects, especially certain dipterous larvae, white grubs,

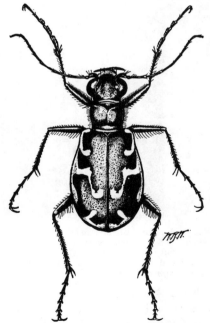

FIG. 351.—A tiger beetle larva, Cicindelidae. (*Kirk.*)

FIG. 352.—A tiger beetle, *Cicindela repanda* Dej. (*Walton.*)

Collembola, and Thysanura. Loose soil is desirable for pupating insects, for emergence, and for permitting insects such as wireworms and white grubs to move about readily. Sandy soil is desirable for species that make traps or pitfalls, as for example, the ant lions. Loose porous soil is suitable for burrowing larvae. Clay is preferable for ants, termites, and solitary insects because it does not cave in readily. Many of the

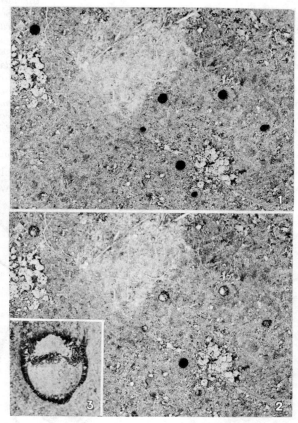

Fig. 353.—A colony of tiger beetle larvae: (1) Seven openings to the burrows of tiger beetle larvae. Two of them show piles of recently excavated earth. (2) Same group as above with all but one of the openings to the burrows plugged by the heads of tiger beetle larvae. (3) Head of a larva closing the opening of a burrow, much enlarged. (*After Macnamara, Arnprior, Canada.*)

mining bees, Andrenidae, and solitary wasps, Eumenidae, select such locations. Carabidae and similar insects lurk beneath stones or beneath trash upon the surface of the ground. An Australian termite is said to burrow in a stone. Custer describes a bee which digs a hole in sandstone and collects its food from the cactus. Banks records a rock-boring mite. The soil is thus an ecological niche where many species escape the over-crowded conditions existing above the surface.

Temperature and moisture determine to a large extent the abundance of insects in the soil. The temperature varies most at the surface, the variation decreasing with the depth until a constant zone is reached. Many insects seek the soil as fall approaches and avoid the severe winter conditions. Decaying humus may supply heat for some species.

Excess moisture often drives insects from the soil. Ants and termites, seeking dry places for their brood, frequently build their nests above the ground and avoid unfavorable conditions. The average insect avoids moisture at the time of pupation, and insects such as ant lions require dry soil or sand in which to construct their pitfalls. It is said that the pupae of the Sphingidae are not affected by extreme dryness in the soil. The grape phylloxera prefers dry soil and is extremely sensitive to moisture. In France, submersion has been used as a means of control. On the other hand, many insects seek moisture. If the soil becomes too dry, some insects dig deeper. Corn borer larvae can resist submersion for 48 days. The cranberry worm, *Sparganothis sulfureana*, can withstand submersion, under normal winter conditions, for five and a half months.

Fig. 354.—A group of ant lion pits in sandy soil, Panama.

Insects are affected indirectly by the acidity and alkalinity of the soil. These factors act upon decomposing material and organisms in the soil which in turn react upon the insects. The acid water of bogs is generally unfavorable for insect development. Tannins and other toxins, such as those derived from black walnut trees, may render the soil unfavorable for certain insects as it does for certain plants. Heavy alkaline soils also tend to paucity of insect life.

The coverage of grass, leaves, vegetation, and snow not only regulates the physical nature of the soil but determines to a large extent what species of insects may be present. More insects will be found beneath a forest floor than in a desert soil. The soil cover also determines the character of the flora and hence the accompanying fauna.

**Food of Subterranean Insects.**—The food of subterranean insects consists of living plants and animals, dead plants and animals, excrement, bacteria, and fungi. In short, it is little different from that of the terrestrial species. Some insects, such as the ant lions and tiger beetles, live near the surface and take their food from the surface. Others feed in true subterranean fashion.

**Extent of Soil Occupation.**—The majority of the subterranean insects occupy the soil for only a part of their life history, as eggs, larvae, nymphs, pupae, or adults.   One or several stages may be passed in the ground. The subterranean habit occurs in most of the orders.   In some, the habit is almost completely absent as, for example, the Anoplura, Mallophaga, Siphonaptera, Strepsiptera, Plecoptera, and Trichoptera.   It never occurs as a dominant habit in any order but exists in individuals, small groups, or families.

Only occasionally do adults occupy the soil.   The Thysanura, Dicellura, and Collembola live commonly in the debris and trash on the surface of the soil.   The mole cricket and the sand cricket spend most of the time in their burrows.   The adults of many of the small shore bugs (Saldidae), the toad bugs (Gelastocoridae), the burrowing bugs (Cydnidae), and the minute, little known Schizopteridae, burrow in the soil.   Subterranean species of aphids and scale insects may also live almost continuously in the soil.   Ants, termites, and social wasps and bees, as well as their guests, spend much time in the soil.   The most remarkable insects, in this respect, are certain Carabidae that live continuously in small pockets in the ground and never come to the surface.   They are so evasive that special means must be employed to collect them.

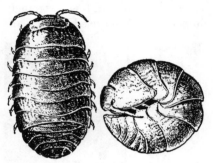

Fig. 355.—Sow bugs, crustaceans commonly found in the soil.

Numerous insects lay their eggs in the soil.   This is especially true of the grasshoppers, earwigs, beetles, flies, and certain species of social and solitary insects.   The eggs occasionally remain in the ground over winter; more often they are in the ground only for short periods during the summer.   In either case, they are somewhat protected from parasites, predators, and desiccation.   However, they have their enemies. The eggs of grasshoppers often fall prey to the wandering meloid larvae. Carabid larvae also prey upon the eggs of many insects laid in the ground.

The larvae of Coleoptera, Diptera, and Hymenoptera are commonly found in the ground.   Among the Coleoptera, the Cicindelidae, Carabidae, Scarabaeidae, Meloidae, Elateridae, and Curculionidae are most outstanding.   The tiger beetle larvae, Cicindelidae, live in vertical burrows in the ground and prey upon passing insects.   The Carabidae generally live beneath stones or rubbish; they are primarily insects of the soil cover.   The larvae of the blister beetles, Meloidae, search in the soil for grasshopper eggs or the larvae of certain solitary bees.

The Scarabaeidae, with the exception of a few species that live in decayed wood, are dwellers in the soil and feed upon living or decayed plant material. These and the Elateridae are the most conspicuous larvae found in the soil.

Among the Diptera, the craneflies (Tipulidae), March flies (Bibionidae), long-legged flies (Dolichopodidae), snipe flies (Rhagionidae), dance flies (Empididae), robber flies (Asilidae), bee flies (Bombyliidae), and the Anthomyiidae are the most common soil dwellers. Some feed upon plant material, some upon decomposed material, and others prey upon insects in the soil. The Tipulidae live largely in the soil or in decayed wood; the majority of the species are scavengers. The Bibionidae feed upon decayed material or upon the roots of plants. *Bibio albipennis* is frequently encountered; the adults emerge in large numbers during March, hence the common name. The Dolichopodidae occupy various situations; some live upon decomposing material, others are found in the burrows of wood-boring insects. The Rhagionidae thrive in the soil or in decaying wood; some are scavengers, others are predacious. The Asilidae live in the soil or in decaying wood and the majority of the species prey upon the larvae of beetles. *Promachus vertebratus* is especially abundant and attacks the larvae of Scarabaeidae. A few of the species apparently feed upon plants. The Bombyliidae are parasitic upon numerous species of subterranean insects. About 500 species have been described. The adults feed upon nectar and pollen; the larvae are predacious or parasitic. The species of *Anthrax* are predacious upon the larvae of Lepidoptera. Other species are parasitic upon the eggs of grasshoppers, the larvae of beetles, or the larvae of Lepidoptera. The Anthomyiidae have various habits but a large proportion of the larvae live in the ground and feed upon the roots of plants.

The larvae of the Lepidoptera are seldom found in the soil except immediately preceding pupation. However, certain species of cutworms feed upon roots and tender shoots of herbaceous plants and often hibernate as larvae in the soil.

Certain of the Mecoptera also live in the ground. *Panorpa klugi*, *P. communis*, and *P. rufescens* spend most of their larval life in burrows in the ground. They are somewhat similar in habits to the Cicindelidae, for their heads fill the openings to their nests when they are at rest. If disturbed, they retreat to the depths of their tubes. Other species of Mecoptera larvae live largely beneath stones.

The larvae of Hymenoptera are seldom found in the soil except as parasites or predators, or during pupation. Broods of solitary and social insects commonly occur in the soil.

Comparatively few nymphs are found in the soil. Certain root-feeding aphids and Coccidae are outstanding exceptions. The cicada is a classic example. The naiads of Odonata and Ephemerida often burrow in the mud at the bottom of ponds.

Pupae are numerous in the soil and nearly every order except the parasitic groups, Anoplura and Mallophaga, contribute many species.

**Numerical Relations.**—In desert areas, subterranean insects are scarce, in open country they are few, but in forest areas they are abundant. In an Illinois forest, the subterranean insect population was reported from 1,000,000 to 65,000,000 per acre (Windsor). Other figures are more conservative. Thompson reports 4,500,000 insects per acre, and Morris 673,000 insects per acre in unmanured soil. Of these 379,000 were Collembola and 163,000 were Coleoptera. Morris also records from 164,983 to 198,653 elaterid larvae per acre. In infested areas, Japanese beetle larvae average 175 per square yard of sod. As many as 1,531 larvae have been taken per square yard in a heavily infested golf course. One thousand Asiatic beetle larvae have been taken per square yard in lawns and 4,500 per square yard in sod. Two hundred white grubs varying from one-half to one inch in length have been taken from the base of a single corn plant. Trägardh gives the following figures for insects and related forms taken from the soil cover.

INSECTS AND MITES FROM DRY LEAVES, SWEDEN

| Depth, centimeters | Number per kilogram dry substance | |
| --- | --- | --- |
| | Mites and Collembola | Insects |
| 0–2 | 15,500 | 950 |
| 3–4 | 11,500 | 810 |
| 5–6 | 1,700 | 310 |

INSECTS AND MITES FROM SOIL COVER, SHORTLEAF PINE, NORTH CAROLINA

| Depth, centimeters | Number per kilogram dry substance | |
| --- | --- | --- |
| | Mites and Collembola | Insects |
| 0–2 | 14,500 | 400 |
| 3–4 | 4,600 | 265 |

The effect of different types of soil upon the insect population is plainly evident from figures on white grubs by Fluke, Graber, and Koch.

Number of White Grubs per Acre, Southern Wisconsin
(Fluke, Graber, and Koch)

| Type of Soil | Number of Insects |
|---|---|
| Nonvirgin sod previously heavily cropped | 218,000 |
| Thin turf | 205,800 |
| Hillside, thin dry soil | 194,000 |
| Blue grass adjacent to sweet clover | 148,000 |
| Virgin sod | 102,000 |
| Thick, heavy turf | 70,600 |
| Hillside, narrow valleys and shielded ravines | 52,000 |
| Sweet clover | 28,000 |

**Degrees of Subterranean Existence.**—This subject will be discussed as to duration and as to penetration. Soil insects may be occasional, transient, intermittent, or constant dwellers. Some species visit the soil only occasionally. Many of the staphylinid beetles have the habit of dropping to the ground when disturbed and quickly burying them-

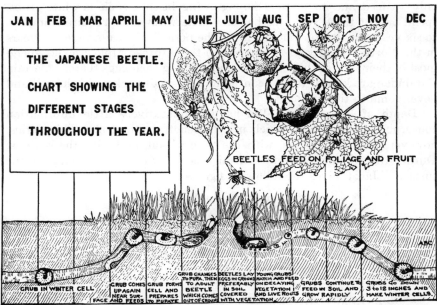

Fig. 356.—Diagrammatic illustration of the life history of the Japanese beetle, indicating the excessive amount of time spent in the soil. (*After Pennsylvania Department of Agriculture.*)

selves beneath leaves or rubbish. The burying beetles, *Necrophorus*, may bury a dead bird or mouse before they oviposit, or they may lay their eggs on the underside of the carrion without burying it.

The adult June beetles are intermittent visitors to the soil or to the dense covering upon it. At night, they feed upon the foliage of various

trees such as apple and nut.   At the approach of daylight, they migrate
to the bases of the trees, burrow in the ground, or hide in trash upon the
ground.   Many cutworms have similar habits.   During the night they
feed upon foliage; some species even climb trees to obtain their food.
During the daytime, many of them hide beneath brush or burrow beneath
the soil.   Certain sand-inhabiting beetles that normally live in the soil
emerge during the cool evening to feed but, with the approach of daylight, they return to the soil where they avoid the high temperatures
on the surface.   Solitary and social insects that nest in the ground make
intermittent trips to obtain provision for their young.

Most insects are transients in the soil for they spend only a part
of their life history there; some spend a few days as eggs or as
pupae; others spend nearly their whole life in the ground.   The
oriental beetle, *Anomala orientalis*, and many of its near relatives spend
almost 11 months in the soil as eggs, larvae, and pupae, and emerge
only to feed and to mate.   Slow-growing insects, such as white grubs,
spend two or three years in the soil.   The wireworms require from two
to six years for their development.   The periodical cicada takes 17 years
in the North and 13 years in the South to mature.   Mole crickets spend
most of their time in the soil, mating and laying their eggs there.   Certain
Carabidae are known to live continually in small pockets in the soil and
never come to the surface.

**Depth of Penetration.**—We have comparatively little information
concerning the depth to which insects penetrate the soil.   This varies,
however, with the species, with the type of soil, and with the seasons.
During the summer, insects generally occur within four inches of the
surface.   Jacot, gives the following figures.

SOIL POPULATION PER SQUARE FOOT

| Depth, Inches | Number of Insects |
|:---:|:---:|
| 1–3 | 500 |
| 2–4 | 148 |
| 5–7 | 32 |
| 10–13 | 22 |

During the winter, insects are found at greater depths.   The larvae
of the Japanese beetle feed during the summer just beneath the surface;
as fall approaches, they descend to a depth of 6 to 12 inches and remain
there during the winter.   About the first of April, they migrate toward
the surface again and remain until they pupate about the end of May
or first of June.

Some of the ground beetles penetrate the soil for only a fraction of an
inch.   On the other hand, insects often go to great depths.   The burrow
of the cicada killer may extend 13 inches into the ground.   Most of

the tiger beetle larvae have extensive burrows; those of *Cicindela gratiosa* penetrate the soil for about 44 inches; the burrows of *C. lepida* penetrate the soil for 58 to 72 inches. Certain white grubs penetrate to a depth of 5 feet during the winter; the cicada is said to descend to a depth of 18 feet.

**Purpose of the Subterranean Habit.**—Some insects obtain food from the soil, others store food below the ground, and many species obtain shelter or protection.

Many insects find food in the ground. This ranges from the roots or shoots of living plants to decomposed vegetable matter and from living insects and animals to decomposed animal matter and excrement. The Scarabaeidae are typical soil dwellers whose food habits show a

Fig. 357.—An ant lion trap for catching small insects that wander too close to the edge (*After Lutz*.)

wide range of adaptability. The larvae of all the species live in the ground or in rotten wood. The Japanese beetle and other closely related species feed upon the roots of living plants. The oriental beetle, *Anomala orientalis*, may live and grow in soil containing only decayed sod. *Cotinis nitida*, the green June beetle, feeds upon vegetable mold in the soil. The tumble beetles, *Canthon*, *Copris*, and *Phanaeus*, the dung beetles, *Aphodius*, and the earth-boring beetles, *Geotrupes*, feed upon excrement.

Many insects store food in the ground. Certain species of solitary and social wasps place food in cells in the ground for their young. The ants store nectar, pollen, portions of plants, seeds, and other material. The burrowing Eumenidae store caterpillars. The Pompilidae store spiders. *Sphecius speciosus* digs a hole and places a cicada therein.

A few species construct traps in the soil by which they obtain their food. Tiger beetle larvae, Cicindelidae, live in vertical burrows in the soil. The larva takes a position at the opening of the burrow with the head bent at right angles to the body to form a plug that closes the entrance. In this attitude, the larva waits until some insect passes, then seizes it and drags it into the burrow to consume it.

The larvae of several species of ant lions, Myrmeleonidae, make pitfalls in sandy soil and feed upon small insects that fall to the bottom of the pits. These traps have been described elsewhere. The larvae of the snipe flies of the genus *Vermileo* have similar habits.

Fig. 358.—Ant lions, Myrmeleonidae: above, adults; below, larva and cocoons. (*Champlain and Kirk.*)

The soil affords an excellent place to obtain protection from certain parasitic and predacious insects or from unfavorable weather conditions. The largest number of insects pupate in the soil. Often no silken cocoon is necessary as the larva forms a cell of dirt or mud. The Lepidoptera seldom construct silken cocoons in the soil. The regal moth, *Citheronia regalis*, the imperial moth, *Basilona imperialis*, and probably all members of this family have naked pupae. The same is true of the Sphingidae.

Eggs laid in the soil are protected from drying and to some extent from enemies. All of the species of the Scarabaeidae and Elateridae and most of the Carabidae lay their eggs in the soil or in rotten wood. The same is true of many of the Diptera. The onion maggot, the cabbage maggot, and the carrot rust flies are common examples.

Although some insects retreat to the soil and avoid sunshine and high temperatures, others leave the ground and find dry places. Termites often enter houses and burrow in timber where moisture is low. Ants seek dry places where they rear their brood and store their food. Mound-building ants construct their nests above the ground and avoid moisture.

The soil is an ideal place for hibernation since insects can descend below the frost line or at least below the zone of critical temperature. Many mechanical devices aid the insect in withstanding winter conditions. Even the presence of moisture raises the temperature of the surrounding soil when freezing occurs. Under such circumstances, insects may hibernate in any stage of development: eggs, larvae, pupae, or adults.

**Modifications for the Subterranean Habit.—** Certain modifications have taken place in the structure and habits of insects that fit them for this type of existence. Some adults have developed fossorial legs for digging. One of the best examples is the mole cricket. The tibiae of the fore legs are broadened to form shovellike organs. The femora and the coxae are also expanded and the tarsi are reduced. The Scarabaeidae have legs that are flattened and

Fig. 359.—Longitudinal section of a cicada turret.

spiny. These are most highly developed in *Copris* and *Phanaeus*. The middle and posterior tibiae are dilated. In *Phanaeus*, the fore tarsi are wanting. The spider wasps, Pompilidae, the cicada killer, *Sphecius speciosus*, and similar wasps have long spiny legs which are used in digging. The fore and middle legs are used to loosen the soil and the hind legs are used to throw the soil from the burrow. Wasps also use their jaws to manipulate stones or obstructions. In the burrowing bugs, Cydnidae, all the tibiae are expanded and fitted for digging. They are thickly set with strong setae and long hairs except the anterior pair where the spines and hairs are confined to the cephalic edges of the broad apexes. In *Glyptocombus saltator*, the anterior pair of legs are conspicuously broadened. The front legs of the cicada nymphs are tremendously modified for digging.

Although many grasshoppers oviposit on plants, the short-horned grasshoppers, Acridiidae, have an ovipositor adapted for laying their eggs in the soil. It is short and quite different from the saberlike ovipositor of other Orthoptera. The female makes a hole in the ground which is sometimes deep enough to accommodate the entire abdomen. The eggs are deposited in the bottom of the hole and covered with a varnishlike material.

May beetles and, in fact, most of the Scarabaeidae enter the ground to oviposit. The Japanese beetle scoops out a small pocket in the soil and deposits a few eggs in this cavity.

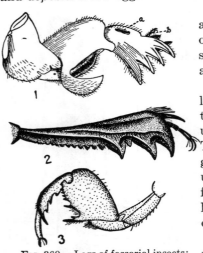

Fig. 360.—Legs of fossorial insects: (1) mole cricket, *Gryllotalpa*; under surface showing (*a*) ear and (*b*) tarsus; (2) Canthon; (3) "harvest fly," *Tibicen linnei.*

Certain damselflies use their long abdomens to insert eggs in moist earth or in mud. The majority of the species, however, lay their eggs upon aquatic foliage.

The bee flies of the family Bombyliidae are parasitic upon hymenopterous and lepidopterous larvae and upon the pupae and eggs of Orthoptera. They naturally deposit their eggs in the ground. *Anthrax anale* is parasitic upon the larvae of the tiger beetles. The female sights the hole of the cicindelid larva and drops her eggs at the opening of the hole.

Subterranean larvae usually have well-developed legs, especially when they must travel through the soil. The Carabidae, Staphylinidae, and Meloidae seek their prey. They are exceedingly active and have long legs. Although the Elateridae have extremely short legs, they are cylindrical in form and slip through the soil readily. They are usually found close to their food and it is not necessary for them to travel far in the soil. The dipterous larvae are legless; however, the eggs are usually laid close to the food plant. The larvae of the cabbage maggot and of the radish maggot will perish if the eggs are not laid close to the plant. This habit has been utilized in the control of these maggots. Disks of paper are placed about the base of the plant to prevent the flies from laying their eggs adjacent to it.

Subterranean pupae are usually provided with numerous rows of short spines, especially upon the dorsum of the abdomen. These assist the pupae in working their way to the surface of the ground. In the puparia of Diptera, minute transverse ridges usually take the place of spines. Similar structures are found in other insects.

Another subterranean habit has been developed in the mole cricket. Baumgartner states that both the male and the female of this species stridulate. This is unique for, as a rule, only the male produces sounds. It is supposed that this habit developed so that the sexes could locate each other when hidden from view in their earthen cells.

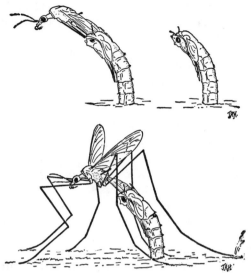

FIG. 361.—Various stages in the emergence of the smoky cranefly, *Tipula infuscula*, from the ground. (*After Hyslop.*)

**Problems Associated with the Subterranean Habit.**—Under subterranean conditions, moisture often becomes so excessive that drowning may occur. Cicadas when emerging during wet seasons are known to build turrets over their exit holes, apparently an attempt on the part of the insect to overcome excessive moisture. Wireworms and white grubs are commonly attacked by fungi and bacteria, and soil conditions are especially favorable for the development of these diseases. Although the soil eliminates certain parasitic and predacious enemies, it introduces new enemies such as moles and mice.

Other insects suffer from the lack of moisture. Certain of the tiger beetles plug the openings to their

FIG. 362.—Entrance to an ant's nest showing a pile of particles of uniform size brought from the burrows. Panama.

burrows when the soil becomes too dry and thus prevent evaporation. Various steps are taken to ward off enemies. Stingless bees build a funnellike opening of wax to their nest. This is coated within with a sticky material and closed at night to prevent intruders from entering. Some of the Eumenidae build turrets over the entrances to their nests as a protection against intruders while they are at work. Later, these are torn down and the material is used to close the nests. Some of the burrowing bees stand guard at the entrances of the nests, plugging the openings with their heads.

The disposal of the earth excavated from the burrows of insects often becomes a serious problem. It is generally discarded or disposed of in various ways. Ants, characteristically, and certain wasps, such as the cicada killer, form mounds about the entrance of their burrows. Many wasps scatter the excavated pellets about the entrance of the burrows. *Odynerus dorsalis,* for example, strews its pellets in this manner. The black hunting wasp, *Priononyx atratum,* makes a pile of pellets near the entrance of the burrow and later uses them to fill the burrow. Some of the eumenids construct turrets above the opening of the burrow which are later torn down and the material used to close the openings of the nests. The large sand wasp, *Chlorion (Proterosphex) ichneumoneum,* makes a pile of pellets about four inches from the opening of its nest. Tiger beetle larvae throw the excavated pellets a considerable distance from the openings of their burrows, otherwise these pellets would interfere with the proper functioning of the traps. The insects upon which they prey would probably be diverted by these pellets and would not approach the openings of the burrows where the tiger beetle larvae lie in wait to seize them.

The earth is sometimes compressed against the walls of the burrow. This is the method employed by the larvae of the wireworms and white grubs and certain other larvae.

**Burrows as a Means of Identifying Insects.**—Insect burrows in the soil, also those in wood, may be used as a means of identifying the insects that produce them. Bryson summarized the important characteristics of burrows as follows: size of burrow, which should include diameter and length; shape and characteristic turns in the burrow; location or placement of the burrow, especially in relation to the nearness of the food supply; size, weight, and measurement of the soil particles at the entrance; amount, structure, and type of the soil excavated; character of the interior of the burrow; the quantity and nature of the materials stored in the burrow.

Open burrows are somewhat uncommon among insects although spiders generally have them. Traps made by ant lions, which are not true burrows, are open. The larvae conceal themselves at the bottom of funnellike pits. The burrows of *Bolbocerosoma* (Scarabaeidae) are entirely open. Many insects use their heads to close the openings of their burrows. This is characteristically the habit of the tiger beetle larvae. Bees of the genus *Halictus* build communal nests; each bee has its own series of cells along the main corridor, the outer end of which is guarded by a sentinel whose head nearly fits the opening.

Other insects take great pains to close the openings of their burrows. The cicada killer, *Sphecius speciosus*, plugs its nest only on completion.

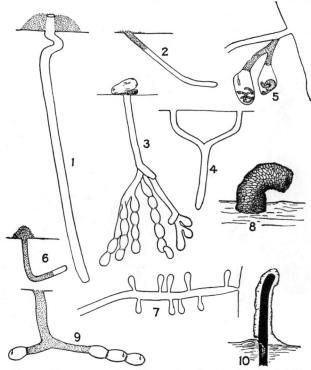

FIG. 363.—Types of burrows produced by various insects: (1) cicada killer; (2) *Bembex;* (3) *Odynerus annulatus;* (4) *Bolbocerosoma;* (5) *Odynerus pedestris;* (6) *Harpalus caliginosus;* (7) *Halictus;* (8) *Odynerus geminus;* (9) *Pterochilus 5-fasciatus;* (10), cicada turret. (1, 2, 4, and 6 *after Bryson;* 3 and 9 *after Isley;* 7 *after Comstock;* 5 and 8 *after Rau;* 10 *after Marlatt.*)

Some wasps close their nests every time they go in search of material or provision for the nest. The ground beetle, *Harpalus caliginosus*, the pompilid wasps, and the burrowing eumenid wasps generally fill the unoccupied portions of their burrows completely with soil or mud. The bembicine wasps usually plug the upper portion of their burrows.

Mounds are constructed above the openings of the nests of numerous ants and the type of the mound often reveals the species of the maker. The beetles, *Harpalus caliginosus, Geopinus fluviaticus,* and *Canthon lecontei,* and the cicada killer, *Sphecius speciosus,* also produce distinct mounds over their burrows.

Turrets are constructed by various insects above the openings of their burrows.  The seventeen-year cicadas often make turrets if the season is wet at the time they emerge.  These earthen chimneys unfortunately are seen only at long intervals.  Some of the eumenid wasps construct turrets which are removed when the nests are completed and the materials are used to close the openings of the burrows.

The nature of the vegetation about the openings of insect burrows varies considerably.  Some species select open places void of.grass and other foliage.  The burrows of *Odynerus dorsalis,* for example. are

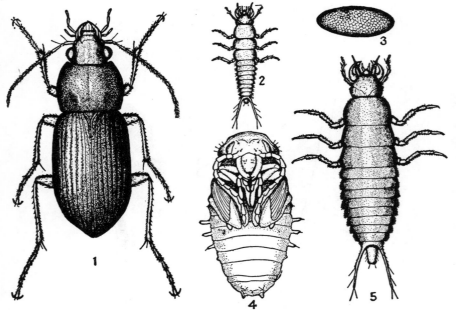

Fig. 364.—A typical ground beetle, *Chlaenius impunctifrons*: (1) adult; (2) first instar larva; (3) egg; (4) pupa; (5) mature larva.  (*After Claussen.*)

generally found on barren spots.  The bembicine wasps also select open places.  The Raus state that *Bembex nubilipennis* nested four consecutive summers on the same barren spot in a vacant lot near St. Louis.

Certain ants and possibly certain wasps deliberately remove the vegetation before they construct their nests.

Bumblebees and yellow jackets may conceal the opening of their nests in the ground.  The eumenid wasp, *Odynerus annulatus,* conceals the opening of its nest in a clump of grass or beneath a small stone.  Many wasps protect their nests during construction by building turrets over them or by closing the nest before they leave to hunt for food.

Solitary bees and wasps frequently construct their nests in communities. This may be due in part to the type of soil suitable for their operations. Bees of the genera *Halictus, Anthophora,* and *Andrena* are

CHARACTERISTICS OF SOME REPRESENTATIVE INSECT BURROWS*

| Insect | Burrow constructed by | Type of burrow | Habitat | Direction of burrow | Diameter of burrow, inches | Length of burrow, inches | Weight of excavated soil, grams |
|---|---|---|---|---|---|---|---|
| Hymenoptera: | | | | | | | |
| Bembicine wasps | Adult | Closed, simple | Clay and loam | 45° angle | 0.25 | 10.0 | 25.2 |
| *Bembex* spp | | | | | | | |
| Cicada killer | Adult | Closed, simple, or branched | Usually in packed soil | Chiefly vertical | 0.83 | 13.0 | 60.0 |
| *Sphecius speciosus* | | | | | | | |
| Eumenid wasps | Adult | Closed, simple, or branched | Compact soil | Vertical | | | |
| *Odynerus* spp. | | | | | | | |
| Mining bees | Adult | Open,† branched | Clay | Horizontal | | | |
| *Halictus* spp | | | | | | | |
| Mining bees | Adult | Open, branched | Compact soil | Horizontal | | | |
| *Anthophora* spp | | | | | | | |
| Mining bees | Adult | Open, branched | Compact soil | Vertical | | | |
| *Andrena* spp | | | | | | | |
| Ants | Adult | Open, branched | Various | Vertical and horizontal | | | |
| Formicidae | | | | | | | |
| Coleoptera: | | | | | | | |
| Ground beetles | Larva | Open, simple | Soils rich in humus | Horizontal on surface | 0.125 | 3–10 | 3.5 |
| Carabid spp. | | | | | | | |
| Ground beetle | Larva | Closed, simple | Loam | Curved | 0.25 | 6–7 | 14.25 |
| *Harpalus caliginosus* | | | | | | | |
| Ground beetle | Larva | Closed, simple | Sand dune | 45° angle | 0.31 | 3–4 | 6.07 |
| *Geopinus fluviaticus* | | | | | | | |
| Scarabid beetles | Larva | Open, Y shaped | Usually in packed clay | Chiefly vertical | 0.25 | 10 | 20.0 |
| *Bolbocerasoma* spp. | | | | | | | |
| Tumble beetle | Larva | Closed, simple | Sand | Vertical | 0.20 | 1.5 | 1.3 |
| *Canthon lecontei* | | | | | | | |
| Tiger beetles | Larva | Open,† simple | Clay or sand | Vertical | 0.20 | 12–15 | 3.7 |
| Cicindelidae | | | | | | | |
| June beetles | Chiefly larva | Open, simple | Dry loam | Chiefly horizontal | | | |
| *Phyllophaga* spp. | | | | | | | |
| Elater beetles | Larva | Open, simple | Dry or moist loam | Tortuous | | | |
| Elateridae | | | | | | | |
| Homoptera: | | | | | | | |
| Cicada | Nymph | Open or closed, simple | Dry or moist loam | Tortuous | 0.75 | | |
| Diptera: | | | | | | | |
| Crane flies | Larva | Open, simple | Dry or moist loam | Tortuous | 0.20 | | |
| *Tipula* spp. | | | | | | | |
| Spiders | Adult | Usually open | Loam or sand | Vertical | 0.50 | 10 | 15.0 |

* Taken largely from Bryson, 1939.
† Opening usually closed by head of insect.

gregarious. One hundred colonies of *Halictus* have been found in a small area on a clay bank. The bembicine wasps, including the well-known cicada killer, are also gregarious. Tiger beetle larvae are frequently gregarious. A Kansas species, *Amblychila cylindriformis,* occurs in colonies of two to eleven. A colony is usually enclosed within a radius of 10 inches. Seventy to a hundred of the small shallow burrows of ground beetles have been found in one square yard of ground surface. The slant of the burrow is often characteristic. The tiger beetle larvae usually construct their burrows perpendicular to the surface of the ground although a few species build in the sides of banks. (*Ctenostoma* lives in dead wood among colonies of ants.) Many of the bembicine wasps and the pompilid wasps as well as beetles of the genus *Geopinus* construct burrows at an angle of 45 degrees. Some of the carabid larvae construct horizontal burrows on the surface of the ground.

**Insects Dwelling in Animal Burrows.**—Numerous insects have been taken from the burrows of gophers, squirrels, rabbits, and other animals. The Diptera are most commonly found under these conditions; the majority of them feed upon excrement or other waste material in the animal burrows. Many Staphylinidae, Pselaphidae, Diptera, and Hymenoptera are found in the nests of termites and ants. They live as guests or are parasitic upon social insects.

**Cave Insects.**—Many sections of North America have caves capable of supporting an endemic fauna of cavernicolous species or troglodytes.* Caves are generally found in limestone sections and are especially common in Eastern and Southern United States. Limestone forms the bulk of the Mississippian system and the Arizona Plateau, underlies nearly all of the Great Plains, and outcrops frequently in the Appalachian region. Within these strata, caves are often formed. The larger caves, such as Carlsbad Cavern, Mammoth Cave, and the numerous caves of Yucatan, yield the most interesting examples of cave insects although the smaller caves have their peculiar fauna.

**Cave Environment.**—Caves are portions of the subterranean environment but they present conditions quite different from the soil habitat. The uniform temperature, the absence of light, and the scarcity of food are three distinctive characteristics of the cave environment. The temperature of the larger caves is more or less uniform and approximates the annual mean temperature for the locality. In the deeper caves, the temperature varies less than 2°C. throughout the year. The moisture in caves is generally high and constant. Dry caves are seldom inhabited by insects. Universal darkness prevails except at the entrances, where the fauna is always different. Darkness is one of the most important factors in attracting and holding cave insects. The food is often scanty and its supply is frequently irregular. Ultimately all the food comes

---

* Correctly speaking, Troglodytes is the genus of wrens that nest in houses or holes. The term is derived from the *trogle*, hole, and *dytus* to creep into or to dwell. Recently the term has been used for cave-dwelling animals. The adjective cavernicolous is preferable although troglobic has its place.

from the outside. Green plants, of course, are absent in caves although roots of trees may extend through the roofs of smaller caves. The dung of larger animals, especially the guano of bats, is one of the chief sources of food. Bats are conspicuous inhabitants of most caves and the guano often accumulates in great heaps beneath their roosts. They leave the cave at night and return with their prey, discarding portions which also serve as food for insects. Streams may bring plant debris and micro-organisms. Fungi and bacteria grow on the plant debris which in turn form food for certain insects, especially the Diptera and the Silphidae. The Collembola and Thysanura feed upon decomposing vegetable matter. Some Collembola feed upon colloidal substances carried by the water that seeps into the cave. These insects often collect in large numbers on the stalactites and stalagmites to obtain food from their moist surfaces. The cave blattids are predacious upon Collembola, arachnids, and other small animals, which are frequently abundant. Numerous parasites have been taken from cave bats and other animals but in the strict sense these are not cavernicolous for they are brought in from the outside. An occasional dead animal serves as food for some cave species.

**Characteristics and Modifications of Cave Insects.**—Cavernicolous species usually occur among insects that normally live in the soil or are subterranean in habit. They are by nature small and photonegative, and often show a tendency to a wingless and eyeless condition. The Blattidae, for example, are exceedingly photonegative. Their near relatives the mole crickets (Gryllotalpinae), the pigmy crickets (Tridactylinae), the sand crickets (Stenopelmatinae), the Grylloblattidae, and the Hemimeridae are subterranean species. The members of the last three groups are even blind. In the Carabidae we find a similar condition. The members of this family, which form one of the largest groups of cave insects, are predominantly soil insects and are known by the common name ground beetles. The cavernicolous species have not essentially changed their habitat. In like manner, the Collembola and Thysanura are found characteristically in the soil and are everywhere photonegative. They are often blind even when not troglobic.

On the other hand, certain modifications have taken place in cave insects which generally distinguish them from the external forms. There is a tendency for cave insects to be lighter in color than terrestrial forms. Most cave beetles are melanistic but they are usually paler in color than their relatives living above the ground. Some of the European species, especially *Aphaenops*, have exceedingly thin integuments. The North American *Pseudonophthalmus* is straw color or reddish yellow. The eyes of cavernicolous species are often absent or reduced and the associated structures are degenerated. *Machaerites mariae* is with or without eyes depending upon the distance it lives from the entrance of the cave.    The appendages of cave insects are often long and attenuate.    The Carabidae usually have long legs and antennae, six long "flying" hairs on the elytra, and six long discal bristles.    The caverniculous Orthoptera have conspicuously long antennae.    There is a

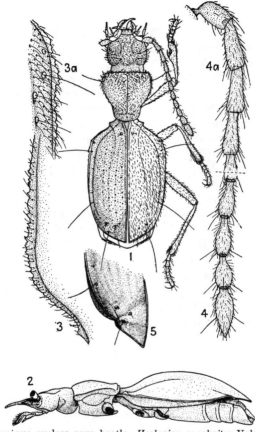

Fig. 365.—A unique eyeless cave beetle, *Horlogion speokoites* Valentine, (Carabidae) taken from a small cave in West Virginia.    (1) Dorsal view, right elytron showing pubescence, left, showing sculpture; (2) left side, mandible extended and antenna removed; (3) humeral margin of left elytron, ventral view; (3*a*) humeral margin of right elytron, dorsal view; (4) and (4*a*), apical and basal segments of antenna, respectively; (5) apex of right elytron, dorsolateral view.    (*After J. M. Valentine.*)

tendency for cave species to be more hairy and have better developed tactile organs.    The loss of sight has, in some cases, been replaced by a more acute sense of touch.    The legs and antennae of the beetle *Anthroerpon apfelbecki* are unusually hairy.    Cavernicolous insects are usually wingless.    This is especially noticeable in the Orthoptera and Mecoptera, where the majority of the terrestrial forms have well-developed wings.

In the cavernicolous Carabidae, the second pair of wings is absent and the elytra are fused down the back. It is interesting to note that cave species retain their thigmotropic response. It is well known that the Collembola, Thysanura, and Carabidae hide beneath fragments of wood or other trash even though they are in eternal darkness. J. M. Valentine reports that living cave beetles kept in artificial caves come out more frequently at night although they are eyeless.

**Insect Fauna of Caves.**—In general there are two groups of insects that inhabit caves: (1) species that are exclusively cavernicolous, *i.e.*, live permanently in caves, and (2) species that normally live outside but frequently invade caves. The latter usually inhabit small caves or live near the entrances of larger ones. Individuals are often more numerous than species although several hundred species have been described from the world. The literature dealing with the subject is tremendous. Packard's "Cave Fauna of North America" (1886) is a monumental piece of work and is one of the best general papers in English dealing with the subject.

According to the general rule in the animal world, the arthropods constitute the greater portion of the cave fauna. Call (1897) lists 38 species of animals from Mammoth Cave, of which 14 are insects. The infusoria and arachnids form the greater part of the balance. Giovannoli (1933) lists 35 species of insects from the same cave. Dearolf (1941) lists 79 species from 37 Pennsylvania caves. Since the Siphonaptera and many of the Diptera mentioned are not strictly cavernicolous, this leaves a population of about 20 species of true cave insects.

The springtails, Collembola, are the most frequently seen of cave insects. Four-fifths of a Moravian cave population was estimated to be Collembola. These insects probably form the chief food for the predacious species. Five species have been described from Mammoth Cave.

The bristletails, Thysanura, like the springtails are naturally subterranean in habit and readily adapt themselves to the cave habitat. Only two species are known from Mammoth Cave.

The largest number of cave species belong to the Coleoptera, chiefly the Carabidae. According to Valentine, at least eight tribes are represented by truly cavernicolous species. The tribe Trechini includes the great majority, at least 20, of cave-inhabiting genera, of which but two occur in Eastern North America, the others being mainly European. Cavernicolous Platynini are found on both sides of the Atlantic, but cave species of the Pterostichini, Scaritini, Nebriini, and Brachinini seem to be restricted to Europe and Africa. The Bembidiini, typically an external group, contains at least three North American genera which habitually live in caves, and the Pogonini is so far represented by but one genus on this continent.

Several genera of Silphidae are known to live in caves. Only one cave genus, *Adelops*, occurs in North America. These are the fungus beetles of caves. Numerous species of pselaphids feed on carrion and several staphylinids are known to feed on decayed vegetable and animal matter in caves.

The Orthoptera are somewhat numerous in caves. The species occur in the families Blattidae and Tettigoniidae. Cave blattids have apparently not been recorded from North America. Several species of wingless long-horned grasshoppers, known as cave crickets, occur in Mammoth Cave, Carlsbad Cavern, and the various caves of Indiana, Tennessee, Virginia, West Virginia, and other states.

Corrodentia and Diptera have been reported frequently from caves. Although about 12 species have been described, only 3 species are definitely cavernicolous.

Mecoptera, Trichoptera, Hemiptera, and Lepidoptera are apparently somewhat rare in caves and probably dwell near the entrances. Only the Lepidoptera have been found in North American caves. A tineid, closely related to the clothes moth, has been reported from several North American caves. The larvae feed upon the dung of larger animals.

**Isolation Problem of Caves.**—Although cavernicolous insects have been described from many parts of the world, caves are generally widely separated and the conditions for the existence of cave insects are somewhat limited. The northern limit of troglobic forms in Europe coincides with the southern limits of glaciation. The distance between caves is an important and interesting factor in accounting for the species found in them. Caves are generally more completely separated than islands. Troglobic insects are strongly photonegative and avoid dry atmosphere. The chances of a species wandering from one cave to another is slight unless the caves are very closely associated. Underground streams and crevices following rock strata sometimes connect caves and make this possible. It is quite noticeable that species of cave insects are usually confined to a single cave or to a few neighboring ones. In the case of the European Leptoderinae, nearly every cave has its own species. The Orthoptera show somewhat the same tendency. *Hadenoecus subterraneous* occurs only in Mammoth Cave and adjoining caves. *Ceuthophilus stygius* is rare in Mammoth Cave but is common in the caves of Indiana; *C. carlsbadensis* and *C. longipes* occur only in Carlsbad Cavern.

## BIBLIOGRAPHY

I. **General References to Soil Insects.**

ADAMS, C. C. (1915). Forest soil and undergrowth communities. *In,* An ecological study of prairie and forest invertebrates. *Bull. Ill. State Lab. Nat. Hist.* **11** (2). *Bibliography.*

BANKS, NATHAN (1907). A census of four square feet. *Science n. s.* **26.**

BIRD, HENRY (1921).    Soil acidity in relation to insects and plants.    *Ecology* **2** (3).
BRYSON, H. R. (1924).    Interrelation of insects and soil.    Thesis.    Kansas State
    College.
———— (1931).    The interchange of soil and subsoil by burrowing insects.    *Journ.
    Kans. Ent. Soc.* **4** (1).
———— (1933).    The amount of soil brought by insects to the surface of a watered and
    an unwatered plot.    *Journ. Kans. Ent. Soc.* **6** (3).
———— (1939).    The identification of soil insects by their burrow characteristics.
    *Trans. Kans. Acad. Sci.* **42.**
BUCKLE, P. A. (1921).    A preliminary survey of the soil fauna of agricultural land.
    *Ann. Appl. Biol.* **8** (3 and 4).    *Bibliography.*
CAMERON, A. E. (1917).    The relation of soil insects to climatic conditions.    *Agric.
    Gaz. Can.* **4** (8).
———— (1925).    A general survey of the insect fauna of the soil within a limited area
    near Manchester.    *Journ. Econ. Biol.* **8.**
CHAPMAN, R. N., *et al.* (1926).    Studies in the ecology of sand dune insects.    *Ecology*
    **7** (4).    *Literature cited.*
DAVIS, A. C. (1934).    Insects in the burrows of the California ground squirrel in
    Orange County, California.    *Bull. Brooklyn Ent. Soc.* **29** (2).
DOW, R. P. (1916).    Plaster-casting insect burrows.    *Psyche* **23** (3).
EDWARDS, E. E. (1929).    A survey of the insects and other invertebrate fauna of
    permanent pasture and arable land of certain soil types.    *Ann. Appl. Biol.* **16** (2).
    *References.*
GHILAROV, M. S. (1937).    The fauna of injurious soil insects of arable land.    *Bull.
    Ent. Res. Lond.* **28** (4).    *Bibliography.*
GLASGOW, J. P. (1939).    A population study of subterranean soil Collembola.    *Journ.
    Animal Ecol.* **8** (2).
HAMILTON, C. C. (1917).    The behavior of some soil insects in gradients of evaporat-
    ing power of air, carbon dioxide and ammonia.    *Biol. Bull. Woods Hole* **32.**
HESSE, R., and ALEE, W. C. (1937).    Subterranean animal life.    *In,* Ecological
    animal geography.    John Wiley & Sons, Inc., New York.
HUBBARD, H. G. (1894).    The insect guests of the Florida land tortoise.    *Insect
    Life* **6** (4).
———— (1895).    Additional notes on the insect guests of the Florida land tortoise.
    *Proc. Ent. Soc. Wash.* **3** (5).
———— (1899).    Insect fauna in the burrows of desert rodents.    *Proc. Ent. Soc. Wash.*
    **4** (3).
JACOT, A. P. (1936).    Soil structure and soil biology.    *Ecology* **17** (3).    *Literature.*
———— (1940).    Fauna of the soil.    *Quart. Rev. Biol.* **15** (1).    *Bibliography.*
KING, K. M. (1939).    Population studies of soil insects.    *Ecol. Monogr.* **9** (3).
KULASH, W. M. (1940).    Insects of the forest floor available as food for game animals.
    *Journ. Forestry* **38** (7).
LADELL, W. R. S. (1936).    A new apparatus for separating insects and other arthro-
    pods from the soil.    *Ann. Appl. Biol.* **23** (4).    *References.*
LINDQUIST, A. W. (1933).    Amounts of dung buried and soil excavated by certain
‾   Coprini (Scarabaeidae).    *Journ. Kans. Ent. Soc.* **6** (4).    *Literature cited.*
McCLURE, H. E. (1937).    Insect remains.    *Ent. News* **48** (1).
McCOLLOCH, J. W. (1926).    The rôle of insects in soil deterioration.    *Journ. Amer.
    Soc. Agron.* **18** (2).
————, and HAYES, W. P. (1922).    The reciprocal relation of soil and insects.    *Ecol-
    ogy* **3** (4).    *Literature cited.*

MAIL, G. A. (1930).   Winter soil temperatures and their relation to subterranean insect survival.   *Journ. Agric. Res.* **41** (8).

MORRIS, H. M. (1927).   Insect and other invertebrate fauna of arable land at Rothamsted Part II.   *Ann. Appl. Biol.* **14** (4).   *Literature.*

MOTTER, M. G. (1898).   A contribution to the study of the fauna of the grave.   *Journ. N. Y. Ent. Soc.* **6** (4).

PACKARD, A. S. (1894).   The origin of the subterranean fauna of North America.   *Amer. Nat.* **28** (333).   *Bibliography.*

SILVEY, J. K. G. (1936).   An investigation of the burrowing inner-beach insects of some fresh water lakes.   *Mich. Acad. Sci. Ann Arbor* **21.**

SNYDER, T. E., and SHANNON, R. C. (1919).   Notes on the insect fauna of bank swallows' nests in Virginia.   *Proc. Ent. Soc. Wash.* **21** (5).

TAYLOR, W. P. (1935).   Some animal relations to soils.   *Ecology* **16** (2).   *Literature cited.*

THOMAS, C. A. (1931).   Mushroom insects, their biology and control..   *Pa. State Coll. Agric. Exp. Sta. Bull.* **270.**   *Bibliography.*

———— (1939).   The animals associated with edible fungi.   *Journ. N. Y. Ent. Soc.* **47.**   *References.*

TRÄGÅRDH, IVAR (1929).   Studies in the fauna of the soil in Swedish forests.   *Trans. 4th Intern. Congress Ent.*

———— (1933).   Methods of automatic collecting for studying the fauna of the soil.   *Bull. Ent. Res.* **24** (2).   *References.*

UVAROV, B. P. (1931).   Insects and climate.   *Trans. Ent. Soc. Lond.* **97.**

VAN DERSAL, WM. R. (1937).   The dependence of soils on animal life.   *Trans. Second N. Amer. Wild Life Conf.*   *Bibliography.*

WENE, G. (1940).   The soil as an ecological factor in the abundance of aquatic chironomid larvae.   *Ohio Journ. Sci.* **40** (4).   *Bibliography.*

YOUNG, F. N., and GOFF, C. C. (1939).   An annotated list of the arthropods found in the burrows of the Florida gopher tortoise.   *Fla. Exp. Sta. Bull.* **22** (4).

## II. Some Important References to Burrowing Bees and Wasps.

CUSTER, C. P., and HICKS, C. H. (1927).   Nesting habits of some Anthidiine bees.   *Biol. Bull. Woods Hole* **52** (4).

DAVIS, W. T. (1920).   Mating habits of *Sphecius speciosus*, the cicada killing wasp.   *Bull. Brooklyn Ent. Soc.* **15** (5).

FABRE, J. H. (1915).   The hunting wasps.   Dodd, Mead & Company, Inc., New York.

HAMM, A. H., and RICHARDS, O. W. (1930).   The biology of the British fossorial wasps of the families Mellinidae, Goryidae, Philanthidae, Oxybelidae and Trypoxylidae.   *Trans. Ent. Soc. Lond.* **78.**   *Bibliography.*

HARTMAN, F. (1905).   Observations on the habits of some solitary wasps of Texas.   *Bull. Univ. Texas* **65.**

HUNGERFORD, H. B., and WILLIAMS, F. X. (1912).   Biological notes on some Kansas Hymenoptera.   *Ent. News* **23** (6).

ISELY, DWIGHT (1913).   The biology of some Kansas Eumenidae.   *Kans. Univ. Sci. Bull.* **8.**

MELANDER, A. L., and BRUES, C. T. (1903).   Guest and parasites of burrowing bees Halticus.   *Biol. Bull.* **5** (1).

NEWCOMER, E. J. (1930).   Notes on the habits of a digger wasp and its inquiline flies.   *Ann. Ent. Soc. Amer.* **23** (3).

PECKHAM, GEO. W., and E. G. (1905).   Wasps, social and solitary.   Houghton Mifflin Company, Boston.

RAU, PHIL (1926). The ecology of a sheltered clay bank. *Trans. Acad. Sci. St. Louis* **25.**

——— and RAU, NELLIE (1918). Wasp studies afield. Princeton University Press, Princeton, N. J.

SMITH, J. B. (1901). Notes on some digger bees. *Journ. N. Y. Ent. Soc.* **9,** (1 and 2).

TURNER, C. H. (1908). The homing of the burrowing bees (Anthophoridae). *Biol. Bull. Woods Hole* **15** (6).

——— (1923). The homing of the Hymenoptera. *Trans. Acad. Sci., St. Louis* **24,** No. 9.

TURNER, R. E. (1912). Studies in the fossorial wasps of the family Scoliidae. *Proc. Zool. Soc. Lond.*

WILLIAMS, F. X. (1914). Notes on the habits of some wasps that occur in Kansas with descriptions of a new species. *Kans. Sci. Bull.* **8** (6).

**III. References to Subterranean Coleoptera Exclusive of Cave Species.**

DAVIS, W. T. (1906). The burrows of *Cicindela rugifrons* and *Cicindela modesta*. *Can. Ent.* **38** (4).

FABRE, J. H. (1918). The sacred beetle and others. Trans. by A. T. Mattos. Dodd, Mead & Company, Inc., New York.

FLUKE, C. L., *et al.* (1932). Populations of white grubs in pastures with relation to the environoment. *Ecology* **13** (1). *Literature.*

FORBES, S. A. (1907): On the life history, habits and economic relations of white-grubs and may-beetles (Lachnosterna). *Bull. Ill. Agric. Exp. Sta.* **116.**

HAMILTON, C. C. (1925). Studies on the morphology, taxonomy and ecology of the larvae of Holarctic tiger beetles. *Proc. U. S. Nat. Mus.* **65** (17).

LINDQUIST, A. W. (1933). Amounts of dung buried and soil excavated by certain Coprini (Scarabaeidae). *Journ. Kans. Ent. Soc.* **6** (4).

SHELFORD, V. E. (1908). The life histories and larval habits of tiger beetles. *Journ. Linn. Soc. Zool.* **30.**

TRAVIS, B. V. (1939). Migration and bionomics of white grubs in Iowa. *Journ. Econ. Ent.* **32** (5).

WICKHAM, H. F. (1899). Habits of American Cicindelidae. *Proc. Davenport Acad. Nat. Sci.* **7.**

**IV. References to Miscellaneous Burrowing Insects.**

ALLARD, H. A. (1937). Some observations on the behavior of the periodical cicada, *Magicicada septendecim* L. *Amer. Nat.* **71** (737).

BANKS, NATHAN (1906). A rock-boring mite. *Ent. News* **17** (6).

BEYER, GUSTAV (1904). Insects breeding in adobe walls. *Journ. N. Y. Ent. Soc.* **12** (1).

CUSTER, C. P. (1928). The bee that works in stone, *Perdita opuntiae* Cockerell. *Psyche* **35** (2).

CUTRIGHT, C. R. (1925). Subterranean aphids of Ohio. *Ohio Bull.* **387.**

DOWDY, W. W. (1937). The hibernation of certain Arthropod fauna of the soil. *Proc. Mo. Acad. Sci.* **3** (4).

GLASGOW, J. P. (1939). A population study of subterranean soil Collembola. *Journ. Animal Ecol.* **8** (2).

ISELY, F. B. (1937). Seasonal succession, soil relations, numbers and regional distribution of northeastern Texas Acridians. *Ecol. Monogr.* **7** (3). *Literature cited.*

——— (1937). The relation of Texas Acrididae to plants and soils. *Ecol. Monogr.* **8** (4). *Literature cited.*

KOFOID, C. A., *et al.* (1934). Subterranean termites. *In,* Termites and termite control. University of California Press, Berkeley.

LANDER, B. (1895). Domed burrows of *Cicada septendecim.* *Journ. N. Y. Ent. Soc.* **3** (1).

LUTZ, F. E. (1928). Little "beasts of prey" of the insect world. *Nat. Hist.* **28** (2).

McCOOK, H. C. (1877). Mound-making ants of the Alleghanies, their structure and habits. *Trans. Amer. Ent. Soc.* **6.**

MARLATT, C. L. (1907). A successful seventeen-year breeding record for the periodical cicada. *Proc. Ent. Soc. Wash.* **9** (1).

NEEDHAM, J. G. (1920). The burrowing mayflies of our larger lakes and streams. *U. S. Bur. Fisheries* **36.** 1917–1918, *Document* **883.**

SNYDER, T. E. (1935). Our enemy the termite. Comstock Publishing Company, Inc., Ithaca, New York.

WHEELER, W. M. (1910). Ants, their structure, development and behavior. Columbia University Press, New York.

———— (1930). Demons of the dust. W. W. Norton & Company, Inc., New York.

**V. References to Cave Insects.**

BAILEY, VERNON (1928). Animal life of Carlsbad Cavern. The Williams & Wilkins Company, Baltimore.

———— (1933). Cave life of Kentucky mainly in the Mammoth Cave region. *Amer. Midl. Nat.* **14** (5).

BANTA, A. M. (1907). The fauna of Mayfield's cave. *Carnegie Inst. Wash. Pub.* **67.**

BARBER, H. S. (1929). Cave and other subterranean beetles. *Journ. Wash. Acad. Sci.* **19** (2).

BLATCHLEY, W. S. (1897). Indiana caves and their fauna. *21st. Ann. Rept. Ind. Dept. Geol. and Nat. Resources.* 1896.

———— (1899). Ten Indiana caves and the animals that inhabit them. *In,* Gleanings from nature. Nature Publishing Co., Indianapolis.

CALL, R. E. (1897). Some notes on the flora and fauna of Mammoth Cave, Kentucky. *Amer. Nat.* **31** (365).

CHOPARD, L. (1931). Biospéologica LVI Campagne spéologique de C. Bolivar et R. Jeannel dans l'Amérique du Nord, Orthoptères. *Arch. de Zool. Esp. et Gen.* **71** (3).

DEAROLF, K. (1937). Notes on cave invertebrates. *Proc. Pa. Acad. Sci.* **11.**

———— (1941). The invertebrates of 37 Pennsylvania caves. *Proc. Pa. Acad. Sci.* **15.** *References cited.*

FROST, S. W. (1953). Cave insects. *The American Caver Bull.* **15.**

GARMAN, H. (1892). The origin of cave fauna of Kentucky. *Science.*

GIOVANNOLI, L. (1933). Invertebrate life of Mammoth and other neighboring caves, Part VII. *In,* Cave life of Kentucky by Vernon Bailey. *Amer. Midl. Nat.* **14.** *Bibliography.*

HUBBARD, H. G. (1900). Insect life in Florida Caves. *Proc. Ent. Soc. Wash.* **4** (4).

IVES, J. D. (1927). Cave fauna with special reference to ecological factors. *Journ. Elisha Mitchell Soc. Sci.* **43.**

———— (1934). Notes on the fauna and ecology of Tennessee caves. *Journ. Tenn. Acad. Sci.* **9** (2).

JEANNEL, R. (1931). Biospéologica Campagne spéologique de C. Bolivar et R. Jeannel dans l'Amérique du Nord. Insectes Coléoptères et Révision des Trechinae de l'Amérique du Nord. *Arch. de Zool. Exp. et Gen.* **71.**

PACKARD, A. S. (1889). The cave fauna of North America, with remarks on the anatomy of the brain and origin of the blind species. *Memoir Nat. Acad. Sci. Wash.* **4** (1). *References.*

PEARSE, A. S., *et al.* (1938). Fauna of the caves of Yucatan. *Carnegie Inst. Wash. Pub.* **491.**

REESE, A. M. (1934). Fauna of West Virginia caves. *Bull. Soc. Sci. Cluj* **7.**

SCHWARTZ, E. A. (1890). A list of the blind or nearly eyeless Coleoptera hitherto found in North America. *Proc. Ent. Soc. Wash.* **2** (1).

――― (1898). A new cave-inhabiting silphid. *Proc. Ent. Soc. Wash.* **4** (2).

VALENTINE, J. M. (1932). A classification of the genus *Pseudonaphthalmus* Jeannel (Carabidae) with descriptions of new species and notes on distribution. *Journ. Elisha Mitchell Soc. Sci.* **47** (2).

――― (1932). *Horologion*, a new genus of cave beetles (Fam. Carabidae). *Ann. Amer. Ent. Soc.* **25** (1).

――― (1937). Anophthalmid beetles (Fam. Carabidae) from Tennessee caves. *Journ. Elisha Mitchell Soc. Sci.* **53** (1).

WOLF, B. (1937–1938). Animalium Cavernarum Catalogus. 's. *The Hague u. Gravenhage W. Junk.*

――― (1938). Fauna Fossils Cavernarum. 1. Fossilium Catalogus.

WICKHAM, H. F. (1897). A collection of Indiana cave beetles. *Ann. Rept. Geol. Ind.* (1896).

# CHAPTER XXI

## AQUATIC INSECTS

Aquatic insects are species that are more or less closely associated with water. They do not comprise a systematic group but are scattered through a large number of insect orders. We have reason to believe that all life was originally aquatic and that animals and plants emerged from the water long before insects were created. Insects have, therefore, readapted themselves to this ecological niche unoccupied by other insects. Since the line of demarcation between air and water is sharp, the strictly aquatic forms are unable to live in the air.

Fig. 366.—Pockets in the roots of trees, especially in Central America, often form breeding places for mosquitoes and other insects.

**Water as a Sphere of Life.**— Seventy-two per cent of the earth's surface is water and the mean depth of the sea is 3,795 meters, a most extensive realm for life. Nevertheless, only a small portion of the aquatic environment is occupied. Less than 5 per cent of the insects are aquatic and only a very few invade the oceans. Lakes, streams, and ponds sustain the greatest population of insect life.

The stability of water makes it an ideal medium for life. Temperature and light, at given depths, are more or less constant and the humidity is of course 100 per cent. Water may freeze or evaporate but eventually it returns to its original form. These and other factors determine the extent to which water has been invaded by insects.

The depth affects insects directly and indirectly. More silt and more gases are held at greater depths. The pressure increases about one atmosphere for each 33 feet. Insects are seldom found at great depths, although chironomid larvae have been taken 1,000 feet below the surface. The eggs of *Chironomus bathophilius* begin to sink as soon as they are laid and die if they do not hatch before they reach a depth of 16 to 36 meters. Most insects live in comparatively shallow water. Light determines, to a large extent, the depth to which insects descend.

Light penetrates to various depths depending upon the amount of silt suspended. In the Pacific it penetrates to 57 meters, in the Mediterranean to 42 meters, in Lake Cayuga, N. Y., to 5 meters, and in the Spoon River, Ill., at flood season, to a depth of only 0.013 meter. The amount of silt cuts the light penetration and consequently the ability of plants to grow. As plants decrease, the food and the oxygen supply decrease and insect life is less abundant. The bloodworms, Chironomidae, live in stagnant water where dissolved oxygen is deficient. They rely upon hemoglobin in their blood to extract oxygen from the water. On the other hand, silt may raise the temperature of the upper regions of water and increase the amount of available food.

Water currents affect insect life. Wave action often makes life untenable for insects and plants but there is a considerable population of insects between the low and high tide regions, especially where grass, rocks, or other protection is present. There is also a definite adjustment of species to the various currents of streams. The black flies and the net-winged midges select the swiftest part of the stream, where food and oxygen are abundant. Certain May fly and dragonfly naiads and other insects burrow into the mud at the bottom of the stream or inhabit sluggish waters. The adjustment of insects to water currents has been discussed under Rheotropism (page 215) and Thigmotropism (page 213).

Surface tension plays an important part in aquatic life. The forces involved are greater than those between air and any other liquid except mercury. These strong forces serve as a support for many insects that are heavier than water. The water striders and the whirligig beetles skip over its surface; the larvae of mosquitos, the nymphs of *Nepa* and *Ranatra* and other species suspend themselves from the surface film. The eggs of some insects are borne by or suspended from the surface film.

The nature of the bottom of the pond or stream determines to a considerable extent the species of insects that may be present. Some caddis flies prefer clear streams with sandy bottoms; others frequent ponds where sticks and vegetable material are abundant. Certain dragonfly and May fly naiads burrow in the mud. Stones and plants are often essential for the attachment of eggs or larvae. Accumulations of rich vegetable debris is necessary for the saprophagous species, such as the naiads of certain May flies.

The size and shape of bodies of water also determine the abundance of insect life. It is a well-recognized fact that the abundance of aquatic life is proportional to the length of the shore line. This depends upon the fact that plants grow more abundantly in shallow water along the edges of the pond. The plants provide the insects with hiding places, materials for cases and cocoons, places to lay eggs, anchorage, food, and oxygen.

**Extent of the Occupation of Water by Insects.**—About 5 per cent of the animals are aquatic; only 3 per cent of the insects are closely associated with water. Some live on or near the surface; others penetrate more deeply. Many obtain their oxygen from the air above (the semi-aquatic species);* others obtain their oxygen from the water (the truly aquatic species). These two groups are discussed under Respiration.

There are various degrees of adaptation to the aquatic habit. Few or no insects pass their whole life history in the water. China and Easki discuss species of the family Helotrephidae which occur in Lake Tanganyika in the Rift Valley of East Africa and which apparently live permanently submerged. A species of corixid lives 40 feet below the surface and several miles from the shores of the Great Lakes. It apparently comes to the surface only at long intervals. With these exceptions, insects have at most acquired gills or other means of obtaining oxygen only in the immature stages. In some cases, the aquatic life starts with the egg. The eggs of many aquatic insects are laid outside but close to the water. Other species insert their eggs in plants or drop them into the water where they may fall to the bottom or float freely. Many insects live in the water only as larvae or as naiads. Insects rarely pupate in the water: the Trichoptera and certain Diptera are notable exceptions. Few adults are strictly aquatic although the Gyrinidae, Dytiscidae, Hydrophilidae, certain Heteroptera, and other insects live on the water or dive below its surface.

Aquatic insects are occasionally found living under extreme conditions. They may exist in very small bodies of water. Certain species of mosquitoes breed freely in the water contained in a discarded tin can *Wyeomyia smithii* breeds in the liquid in the pitcher plant and often passes the winter frozen in ice cores in the leaves of this plant. Other species breed in the moisture held between the leaves of bromeliads Insects are found in cold glacial streams at considerable altitudes, in hot springs with temperatures as high as 150°F., in stagnant water, in swift water, in well aerated water, in fresh water, in salt water, and even in petroleum. Insects rarely live in extensive bodies of water such as the oceans. This is not due to salinity, for the larvae of certain Ephydridae live in Great Salt Lake which has a greater concentration of salt than most oceans. It is probably due to depth and lack of vegetation.

**Oceanic Insects.**—The water striders (Gerridae) of the genus *Halo bates* are truly pelagic, living at the surface of the ocean several hundred miles from the mainland. They are semiaquatic species for they obtain oxygen from the air above the water. They feed upon dead animal floating upon the surface of the ocean. Like other species of gerrids and

---

* Species, such as Collembola, which live in damp situations are often terme semiaquatic.

like their relatives the hydrometrids, they skip over the surface of the water and are frequently found among floating plants. They attach their eggs to debris found on the surface. These species are more numerous in tropical and subtropical regions. *Halobates micans* occurs off the coast of Florida and *H. sericeus* has been taken off the coast of California.

On the other hand, species of the genus *Pontomyia* (Chironomidae) are truly aquatic for they live completely submerged. An increasing number of species are being described from Samoa and Japan. *Pontomyia natans* occurs in the lagoons of Samoa. The males are active swimmers; the adult females are wingless and legless with closed spiracles. Both larvae and females live in tubes constructed between the leaves of submerged plants. These species are found far from land and are not to be confused with the benthic species such as those of *Thalassomyia*, *Halirytus*, and other genera that are common along the coasts.

**Insects of Hot Springs.**—Hot springs occur in many parts of the world, especially in areas where there is, or has been, volcanic activity. Famous thermal springs occur in Iceland, New Zealand, Algeria, and the Americas. Their waters differ noticeably from those of the usual stream or pond. The temperature varies from the boiling point to that which one finds in any small pond during the summertime. The important factor is the constancy of the temperature of a given spring or area. The oxygen content is low, for gases are driven off when water is heated. The concentration of salts is high. Sodium sulphate and calcium carbonate are usually abundant, in addition to sodium chloride which is found in many waters. Appreciable amounts of arsenic may be present. Owing to the presence of large quantities of salts, the specific gravity and the osmotic pressure are high. The hydrogen-ion concentration varies remarkably. Brues (1929) gives the range generally between 6.0 and 9.5. Plants are usually abundant as food for insects. Chlorophyll-bearing plants occur regularly at temperatures from 140° to 146°F. Nonchlorophyll-bearing plants may exist in water with temperatures as high as 155° to 160°F.

On first thought, it seems strange that insects should live in hot springs but some of them do not withstand temperatures greatly different from species that dwell in terrestrial situations close by. The temperature at the surface of the sands of the desert may be approximately the same as that of certain hot springs. Nymphs of mantids and grasshoppers have been found on the ground in Palestine where the temperature was 121°F. Graham found certain buprestid larvae living under bark at a temperature of 120°F.

With the preponderance of insect life in the world, we might expect insects to represent a fairly large proportion of the animal life of thermal springs. However, the insect fauna of hot springs is not abundant. Insects, no doubt, have frequently had a struggle to adapt themselves to the unusual conditions existing there. The inhabitants of these springs are usually found among those insects that are able to adjust themselves to brackish water. Hence there is a similarity between the insect fauna of hot springs and that living in the coastal areas and the sea itself. Practically all groups of aquatic insects contribute a few species known to inhabit hot springs. Among the truly aquatic species, the Diptera are the most outstanding. The larvae of Stratiomyiidae, Chironomidae, Culicidae, Ephydridae, and Tabanidae are most numerous. May fly and dragonfly naiads have also been found in thermal springs. The semiaquatic species such as the Coleoptera and the Heteroptera are able to withstand higher temperatures but the immature forms are never found living in hot springs.

TEMPERATURES AT WHICH TRULY AQUATIC INSECTS HAVE BEEN FOUND LIVING IN HOT SPRINGS

| Species or group | Temperature, degrees Fahrenheit | Locality |
|---|---|---|
| Mosquito *Culex pipiens*............... | 105 | Europe |
| *Chironomus*........................... | 124 | N. America |
| Stratiomyiidae....................... | 122* | N. America |
| Ephydridae........................... | 109 | N. America |
| Syrphidae............................ | 92 | N. America |
| Tabanidae............................ | 109 | N. America |
| Odonata.............................. | 104 | N. America |
| *Cloe diptera*........................ | 113 | Europe |

* Johnson (1895) cites a case of a stratiomyid larva at a temperature of 156°F., but this record has been questioned.

**Problems of the Aquatic Situation.**—Aquatic insects live in an environment quite similar to the densities of their own bodies and avoid the serious problem of desiccation so common to terrestrial insects. The few species that dwell in salt water live in a medium almost identical to the density of their own bodies. Such insects may be confronted with the problem of getting rid of excess salt, although most insects have the problem of acquiring it. Salt is apparently essential for the development of some insects. Haber (1926), Griffiths (1891), and others have shown that the insect's blood is usually alkaline and that salts of calcium and magnesium often occur in appreciable amounts in the blood.

Aquatic insects avoid competition with their overcrowded terrestrial neighbors. They obtain certain protection from sudden changes in temperature and avoid numerous parasitic and predacious enemies but acquire new enemies such as frogs, fish, and other predacious species.

The elimination of waste material is not a problem and requires no special adaptation as is the case with leaf-mining and boring insects. On the other hand, aquatic insects are beset with numerous problems in connection with food selection, respiration, oviposition, locomotion, anchorage, emergence, and the dangers of drying ponds.

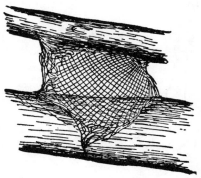

**Obtaining Food.**—As a rule, aquatic situations offer an abundance of food, chiefly in the form of plants. Many of the Chrysomelidae and Lepidoptera feed upon the portions of the plants that protrude above the water. Leaf miners occasionally attack the leaves of aquatic plants.

FIG. 367.—The nets spun by the larvae of many of the species of Trichoptera collect microorganisms which are gathered at irregular intervals by the larvae and eaten.

Hydrophilid beetles and many other species devour the submerged portions of plants. *Bellura* bores into the stems of lilies. *Donacia* larvae feed upon the roots of aquatic plants. Certain chironomid larvae have been found in kelp at sea, apparently obtaining their nourishment from the kelp or organisms therein. The drift, composed of debris that falls upon the surface of the water, serves as food for many insects as well as fish. The water striders feed largely upon insects that fall upon the surface of the water. The predacious Heteroptera—*Notonecta, Nepa,* and *Ranatra*—attack living insects and even turn upon fish and other small animals. Microscopic organisms, known as plankton, serve as food for numerous species of insects. Organic material is continually washing down from the uplands and is being deposited upon the bottoms of ponds and streams where many species obtain nourishment from this rich accumulation. Mosquitoes feed upon the solutes and organic waste present in the breeding pools. Sewage serves as food for a special group of insect larvae, the Spongilla flies, certain chironomids, and psychodids.

FIG. 368. —Head of naiad of damsel fly showing the elongate labium unfolded (*After Sharp*).

Aquatic insects are modified in various ways to obtain food. Often the developments are similar to those of terrestrial insects, as for example, prehensile front legs. These are well developed in *Nepa* and *Ranatra*. The naiads of the Odonata have a strong lower lip which is an efficient organ for seizing and holding its prey. The net-spinning Trichoptera construct minute silken nets which are placed in the streams to collect microorganisms. Other Trichoptera strain these organisms from the water as it passes through their cases. The larvae of the black flies have rakelike appendages which sweep in quantities of diatoms and other microscopic animals.

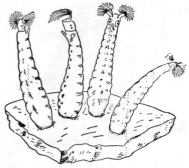

Fig. 369.—Larvae of black flies, *Simulium.* The comblike rakes situated near the mouth gather food. (*After Miall.*)

**Respiration in Semiaquatic Insects.**—Respiration forms one of the most interesting and the most complex problems in connection with the aquatic habit. The rich supply of oxygen usually present in streams, ponds, and other bodies of water supplies free air for all that have learned the art of obtaining it. Insects can be divided, on the basis of respiration, into two groups: semiaquatic and truly aquatic species.

The semiaquatic insects dwell for the most part on or near the surface of the water. They are air breathers and submerge for only comparatively short periods. All have tracheal systems with open spiracles. The spiracles of *Dytiscus* are dorsally located and are exposed when the beetle breaks through the surface film and protrudes the tip of its abdomen. To this semiaquatic group belong the adults of numerous beetles, the nymphs and adults of certain Heteroptera, the larvae of a few Diptera, and, rarely, the adults of Hymenoptera.

The semiaquatic species have various ways of obtaining oxygen. The water striders skip on the surface of the water, rarely descend below its surface, and seldom wander far from its vicinity except to locate new ponds during breeding seasons. They breathe like terrestrial insects. *Hebrus* is found in moist earth at the margins of ponds and runs out upon the water when disturbed. The Hydrometridae creep slowly upon the surface of the water or live on floating plants. The Hydrophilidae and Dytiscidae dive below the surface carrying bubbles of air between the abdomen and the elytra. The Gyrinidae carry a bubble of air at the tip of the abdomen, which glistens like a ball of silver when the insect is submerged. The mechanism of the air bubble supply is interesting. As the insect uses the oxygen in the bubble, the $CO_2$ given off by the insect diffuses rapidly into the surrounding water. The oxygen pressure in the remaining bubble is naturally decreased and the nitrogen pressure is increased. Oxygen therefore diffuses into the bubble from the surrounding water and the nitrogen is forced out of the bubble. Oxygen diffuses into the bubble three times as fast as the nitrogen passes out; thus the available supply of oxygen from the bubble is 13 times that originally contained in it.

A European spider, *Argyroneta aquatica,* lives among the plants at the bottom of clear, quiet ponds and breathes air which is carried down from the surface. The spider constructs a sheet of silk among the plants at the bottom of the pond and fills this silken net with air.

Fig. 370.—Diving beetles, Dytiscidae. One adult obtaining air from above the water. (*After Duncan and Pickwell.*)

This air-filled dome serves as a source of oxygen, a home, a nidus for the eggs, and a retreat for winter. As a matter of fact, the habits of this spider are little different from those of terrestrial species, except for its method of obtaining air.

Many species live just beneath the surface of the water. They obtain air by piercing the surface film with special breathing tubes. This habit occurs in adults as well as in immature forms. The rat-tailed maggots, Syrphidae, have long taillike appendages that enable the larvae to obtain air when the body is submerged beneath several inches of water or moist decayed material. This tube is telescopic and can be lengthened or shortened as needed. The tip of this tube is provided with a rosette of hairs which expand on the surface of the water and prevent the spiracle located at the tip from being submerged.

Mosquito larvae live largely at the surface of the water and descend only when alarmed. Air is drawn in through a breathing tube at the posterior end of the body. This tube varies in length in different species. A pair of spiracles located at the tip are armed with a rosette of hairs which spread out on the surface of the water when the surface film is pierced. The pupae likewise breathe at the surface of the water but in this case the anterior spiracles are developed into two earlike appendages which pierce the surface of the water. The pupae of the mosquitoes are active and descend, when disturbed, in the same way as the larvae.

The water scorpions, *Nepa* and *Ranatra*, have respiratory organs at the end of the abdomen. They consist of two long filaments which fit together to form a tube. The insects rest upon submerged vegetation and obtain air by pushing the tip of this organ through the surface film.

The larva of the mosquito, *Taeniorhynchus* (*Mansonia*) *perturbans*, has a peculiar method of obtaining air. It taps the air supply of species of sedges and cattails which grow in marshy places. The sharp spiracles are forced into the vascular roots of these plants and take the air incorporated in the large spaces between the cells of the plants. The larvae of *Donacia* take air from the roots of aquatic plants in the same manner. Incidentally, many parasitic larvae tap the air supply of the trachea of their hosts or maintain a direct connection with the outside world by protruding their spiracles through the body walls of their hosts. These larvae are actually semiaquatic.

The brown-tailed diver, *Bellura diffusa*, solves the aquatic problem in a unique manner. The young larvae are probably leaf miners but the older larvae bore downward into the petioles of pond lilies, sometimes to a depth of two feet. The larvae, although not strictly aquatic, are adapted to remain below the surface of the water while they feed. The tracheae are unusually large and probably serve as reservoirs for the storage of air for the use of the larvae when they are submerged. The posterior pair of spiracles are large and protrude because the dorsal half of the last segment is cut away. When not feeding, the larvae rest at the upper end of their burrows with the large spiracles protruding above the water.

The larva of the European moth, *Cataclysta lemnata*, constructs a diving suit. The young larva lives within a mine in the leaf which is filled with water. During this period, it absorbs oxygen through its skin. The following year, it becomes an air breather and needs protection from the water. Two pieces of leaf are cut and sewed together to form a pouch about an inch long. The edges are left open at one end of the case for a space just sufficient to allow the caterpillar to protrude its head. The opening is so arranged that, when the head is withdrawn, the sides of the opening press together, close the opening, and prevent water from entering.

**Respiration in the Truly Aquatic Species.**—The truly aquatic species live below the surface of the water and the immature forms are highly developed for an aquatic existence. To this group belong the Ephemerida, Odonata, Trichoptera, Plecoptera, and certain Lepidoptera, Diptera, Coleoptera, and Neuroptera. Oxygen is obtained in these species in one of three ways: by means of tracheal gills, by means of blood gills, or by absorption through the body wall. The majority of the species breathe by means of tracheal gills. Tracheal gills are expansions of the body wall which are filled with air and communicate with the tracheal system of the insect. Air incorporated in the water is absorbed through the thin walls of the gills. These gills are usually located along the undersides of the abdomen and are well represented in the larvae of the dobson flies and the caddis flies. In the stone flies, they are sometimes located on the head and thorax. In the dragonflies, Anisoptera, they are located in the rectum at the posterior end of the body and are known as rectal gills. In the damsel flies, Zygoptera, they take the form of three flat plates at the posterior end of the body.

Blood gills are rare in insects. They are expansions of the body wall which are filled with blood. Oxygen is taken from the water through the thin covering of the gills in much the same manner that fish and tadpoles take air from the water. Gills of this type are found in the bloodworms, Chironomidae, in certain Trichoptera, and also in exotic beetles of the genus *Pelobius*. Bloodworms live in aquatic situations where the oxygen concentration is low. The blood of these insects contains hemoglobin, which is unusual among insects. Hemoglobin absorbs 30 times as much oxygen as does water; thus oxygen is readily extracted from stagnant water.

A few insect larvae absorb oxygen through the body wall. This occurs in the naiads of May flies and in the larvae of certain Lepidoptera.

A very few larvae, as for example Nymphula, construct cases of pieces of living plants which oxygenate the water within the cases and supply the larvae with oxygen.

**Oviposition.**—Oviposition is attended with some difficulty in aquatic insects. Winged insects are often poorly fitted to enter the water to lay their eggs. Different methods have been developed to permit them to place their eggs in the water or to oviposit where the immature forms can reach the water safely.

Many of the mosquitoes, as for example *Aedes*, lay their eggs in dry places where they remain until rains or melting snows provide the necessary moisture. Most mosquitoes hibernate as eggs often in dry places and it is said that they may remain in this condition for several years before they hatch. They hatch when the moisture and temperature are satisfactory.

The Tabanidae, *Corydalis*, *Sialis*, and certain Trichoptera deposit their eggs upon the branches of trees or upon stones that overhang the water, and the young fall into the water when they hatch. The egg masses of the dobson fly, *Corydalis*, are covered with a tough, chalky material which protects them from desiccation. Those of the

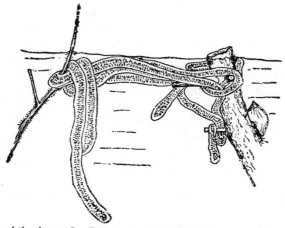

Fig. 371.—Eggs of the dragonfly, *Tetragoneuria*, hung on submerged twigs near the surface of the water. (*After J. G. Needham.*)

Tabanidae are laid in compact masses and the individual eggs have tough shells.

The net-winged midges, Blepharoceridae, and some Odonata lay their eggs in moist places at the edge of the water and the immature forms find their way to the water when they hatch. Although the habits of the net-winged midges are not fully known, it is supposed that these delicate flies could not oviposit in the swift water where the larvae live. The black flies, Simulidae, are more robust and dart at stones, as the water sprays over them, and fasten their eggs.

A number of insects drop their eggs into the water. The females of some species of dragonflies dart back and forth over the surface of the water, sweeping down at intervals and releasing one or more eggs each time the tip of the abdomen touches the water. The Plecoptera and certain short-lived Ephemerida drop their eggs in masses into the water. Individual females are often found with the two egg masses protruding from the posterior end of the body.

In some of the mosquitoes (Culicidae) and the midges (Chironomidae), the eggs often float on the surface of the water. *Chironomus meridionalis* suspends its small egg mass from a disklike mass of silk which floats on the surface of the water.

Many aquatic insects attach their eggs to submerged plants. The water pennies, Psephenidae, fasten their eggs in compact masses to aquatic plants. The whirligig beetles place them in parallel rows on plants. The damsel flies frequently rest upon floating plants, push their long abdomens completely beneath the water, and curve them beneath the plants so as to fasten their eggs on the lower surfaces of the leaves. The Hydrophilidae have diverse oviposition habits. Some lay individual eggs; others deposit them in masses. They are sometimes attached to plants but often drift freely. The *Donacia* usually lay their eggs in masses on the under surfaces of floating leaves and cover them with a gelatinous material. The Dytiscidae and a few of the Odonata deposit their eggs in slits in aquatic plants. *Haliplus ruficollis* is said to lay its eggs in dead cells of *Nitella*.

The egg masses of many insects are loosely attached to aquatic plants. The gelatinous egg masses of the May flies often have winglike expansions which entangle vegetation and prevent them from falling to the bottom of the pond where they would be prey to predacious species.

A few insects present strange egg-laying habits. The water bug, *Belostoma flumineum*, lays her eggs on the back of the male. A certain corixid lays its eggs upon crayfishes. A few parasitic Hymenoptera descend into the water to lay their eggs upon their aquatic hosts and use their wings to propel themselves through the water.

**Anchorage.**—Aquatic insects frequently face the problem of maintaining their position in the water. On the one hand, some insects thrive better if the eggs are buoyant and come to the surface; others do better if they sink to the bottom. We have already mentioned how insect eggs are attached to the surface film, to plants, to rocks, and other supports.

The larvae of some species have acquired habits or structures that solve the problem of attachment. Silken threads are used by the larvae of black flies and some aquatic caterpillars to prevent them from being washed away by the swift currents. The larvae of *Elophila* (Lepidoptera) live beneath dense silken mats. The larvae of the net-winged midges have from six to eight sucking disks which hold them firmly to the rocks in spite of the fact that they inhabit the swiftest part of the stream. Some caddis fly larvae live in dense vegetation where they have firm footing. Others live in sluggish water. The most interesting species are those that make cases of stones which act as sinkers. All case-bearing caddis fly larvae have a pair of dorsal hooks which hold them securely in their cases.   It is difficult to extricate them from their cases even after they have been killed.

Many insects use the surface film as an anchorage, some suspend themselves from it, and others rest upon it.   One of the most interesting forms resting upon the surface of the water is a small springtail, *Podura*. It attaches itself by means of the *collophore*, which is a ventral sucking disk.   It can release its hold, jump into the air, and land again on the water, on all sixes.

The water pennies, Psephenidae and the small water beetles, Dryopidae, are extremely flat and resist the force of the water.   Many of the larvae and adults that live in the water are streamlined.

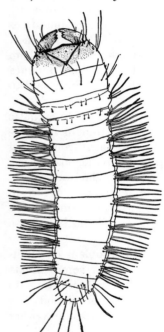

**Locomotion.**—The waterways are the great highways for insects as well as other animals.   Insects may be carried long distances by streams or ocean currents.   They move downstream as the upper tributaries dry.   They often shift their position to avoid sewage pollution.

The movements of aquatic insects consist chiefly of skipping, swimming, diving, and crawling.   Some insects are heavier than water and their problem is to maintain themselves at the surface.   The water strider stands upon the surface film which it dimples but does not break through.   If a little soap or sodium oleate is placed in a small pool inhabited by water striders, the surface tension of the water will be lowered and the insects will break through the surface film. The water springtail can jump from the water and alight unwetted.   Water fleas (Crustacea) sometimes break through the surface film and are trapped.   The Dytiscidae, Hydrophilidae, and Gyrinidae, being lighter than water, must exert great effort to descend.   To remain below they grasp some submerged object.   When they release their hold, they pop to the surface like corks.

Fig. 372.—An aquatic lepidopterous larva, *Elophila fulicalis*.   (*After Lloyd.*)

Swimming forms have a body that is smooth and compact with few or no hairs or spines.   They are able to glide through the water freely. They are usually flattened in form and have no claws.   The swimming legs have a broad fringe of hairs which serve as oars.   In the water boatmen, the body is shaped like a boat.

The jerky movements of mosquitoes, chironomids, and other aquatic larvae are produced by undulating movements of the abdomen. Both the larvae and pupae of these species are active.

The naiads of Odonata and Ephemerida are heavier than water and usually crawl upon the bottom of the pond or burrow in the mud. The dragonfly naiads expel water from the anus and propel themselves forward. Beetles of the genus *Stenus* secrete a substance that repels the surface film like camphor and drives them along the surface of the water.

**New Problems Confronting Aquatic Insects.**—There are certain problems, aside from respiration, that are peculiar to aquatic insects. Small shallow pools are inclined to dry up toward the middle of the summer. Insects have devised various means of overcoming this difficulty. Many insects develop rapidly and so mature and emerge before the pool dries. The adults of many species are winged and can migrate to new pools. Some species bury themselves in the mud or debris at the bottom of the pond and await more favorable conditions. This is also a common procedure for hibernation. Certain mosquito eggs can withstand drying for considerable periods of time.

**Emergence.**—The naiads of Plecoptera, Ephemeridae, and Odonata crawl from the water before the adults emerge. The cast skins of stone flies, May flies, and dragonflies are familiar sights about streams and ponds and indicate what has happened.

The aquatic Neuroptera leave the water and transform in cells in the ground some distance from the water.

The real problem exists in the species of Trichoptera, Lepidoptera, and Diptera that pupate in the water. These insects must make some provision to emerge without wetting their wings. The emerging habits of the Trichoptera are not well known. In the net-spinning species, the wings expand instantly when the adults reach the surface of the water. The larvae of *Psilotreta* congregate in masses, as the time for pupation occurs, and arrange their stone cases with cephalic ends directed upward so that the adults can emerge quickly. No one has observed the emergence of the delicate scaly-winged moth *Elophila*, the larva of which inhabits swift water. Emergence, however, occurs in August when the level of the streams is generally lower and the force of the water is not so great. Lloyd states that these moths were "fairly swarming as one walked through the vegetation along the water edge." He does not enlighten us as to what must be a very interesting method of emergence.

Mosquitoes and midges have pupal cases that rest upon the water and give support to the newly emerged flies until their wings expand and they are able to take flight. Many of these species also breed in quiet water where the problem of emergence is not so great. The net-winged midges emerge from swift water and, as we have described elsewhere, their wings are neatly folded within the pupa and ready for flight as soon as the flies emerge.

**Discussion of Aquatic Insects.**—More than half of the insect orders have aquatic or semiaquatic species. The Thysanura, Collembola, Dicellura, Zoraptera, Embiidina, and Dermaptera are semiaquatic in that they live in moist situations, in rotten wood, in decayed leaves or soil. The collembola, *Podura aquatica*, lives on the surface of stagnant ponds, over which these minute creatures may be seen leaping about in great numbers.

The Ephemerida, Odonata, Plecoptera and Trichoptera, with rare exceptions, are strictly aquatic. The immature forms live in the water and breathe by means of tracheal gills.

The Neuroptera, Heteroptera, Homoptera, Lepidoptera, Coleoptera, Diptera, and Hymenoptera are only partly aquatic. The immature forms of some are strictly aquatic, others are semiaquatic, but none of these orders is wholly aquatic.

**May Flies (Ephemerida).**—May flies are delicate creatures abundant about streams, ponds, and lakes. They are often attracted to lights. During the annual swarming season, vast hordes of adults appear in the air and are frequently drifted in windrows along the shores of lakes or are blown against the sides of buildings or other obstructions. *Ephemera simulans* frequently becomes a nuisance along the shores of the larger lakes. The adults accumulate in large numbers and their decaying bodies produce an offensive odor.

The adults of May flies can be recognized readily by the two or three long caudal filaments and the shape and size of the wings. There are generally two pairs of wings, the anterior pair being the larger. In the genus *Caenis*, the hind wings are absent but the adults take the same general form and one must look closely to detect that these insects have but a single pair of wings.

Several features of May flies are of special interest. The adult life is short, often lasting but a few days. Strictly speaking, they are long-lived, for the naiads live two or three years in the water. The adult takes no food. The mouth is vestigial and the alimentary canal is modified to act as a balloon which is filled with air and makes the specific gravity less. After the insect leaves the water and apparently has assumed the adult form, it moults again. These are the only insects that moult after the wings have fully expanded.

The May flies lay their eggs in the water, individually upon stems or grasses, or in gelatinous masses loosely attached to foliage by means of winglike expansions. The naiads are active creatures and are often as delicate as the adults. They usually have prominent tracheal gills along the sides of the abdomen. The abdomen is terminated by two or three caudal filaments, "tail fins." They feed upon living or dead plants. *Hexagenia, Pentagenia, Ephemera,* and *Poly-mitarcys* are burrowing May flies and the naiads live beneath sediment at the bottom of ponds. *Callibaetis, Blasturus,* and others live in slowly flowing water and climb upon vegetation. The *Heptagenia* live in swiftly flowing water and cling to the undersurfaces of stones. Their flat bodies permit them to resist the swift currents. The full-grown naiads leave the water and crawl upon some upright support to transform.

Fig. 373.—May fly naiad. (*After Needham.*)

**Stone Flies (Plecoptera).**—Stone flies are found chiefly along streams; some may be found about small swift streams, others about broad sluggish streams. Numerous factors determine the abundance and distribution of these species: the size of stream, flow of water, abundance of plant and invertebrate life, depth of water, accumulation of dead leaves and other organic material, and the type of the bottom. The adults

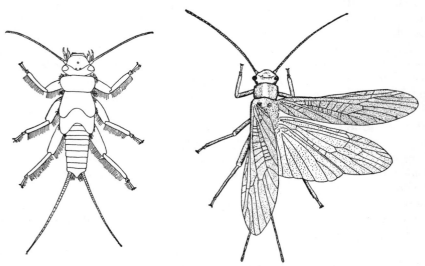

Fig. 374.—Naiad and adult of stone fly, Plecoptera.

are usually found close to the breeding grounds of the naiads. There are approximately 200 North American species. Their long antennae, comparatively short cerci, and broad hind wings distinguish these species from other aquatic insects. They have a distinct seasonal succession, as Frison (1935) remarks, "with a varied and abundant fall and winter fauna which gives us a continuous seasonal range of species from late fall to late summer." During the winter, several small black species of the genera *Allocapnia* and *Capnia* are abundant and are often found upon the snow. These are followed by species of *Isoperla*, *Hydroperla* and others. During the middle of April, gigantic species of the genus *Pteronarcys* begin to emerge. These are the largest of our stone flies. The naiadal skins are found in abundance on the trunks of trees close to the streams from which they emerge. Species of the genus *Perla* emerge during May and continue to appear during the most of the summer. *Perlista placida* is one of our common summer species, measuring approximately 9 to 11 millimeters in length. It is predominantly brown with yellow on the head and yellow costal margins to the wings.

*Acroneuria* comes forth in late May or early June and may be seen during most of the summer. *Acroneuria arida* is a common species and is principally yellow in color. Species inhabiting the larger streams drop their egg masses into the water; those in smaller streams descend into the water to lay their eggs.

The majority of the naiads cling to the lower surfaces of stones. They occur only in well-aerated water and hence are usually found in streams. Although they feed chiefly on plants, some are said to feed on aquatic insects. Most of the species possess tracheal gills which usually occur on the ventral surface of the thorax but are sometimes found on the under side of the head, on the basal abdominal segments, or on the tip of the abdomen. When full grown, the naiads leave the water to transform. The empty exuviae are commonly seen clinging to stones or other supports along the banks of streams.

**Dragonflies and Damsel Flies (Odonata).**—These beautifully colored and remarkably agile insects have attracted the attention of man for many centuries. Ancient pottery from the culture of the Mimbres, New Mexico, as well as the ancient art of China and Japan, reveals many excellent reproductions of the Odonata. The adults are common in the vicinity of streams, lakes, and ponds but are not limited to aquatic situations. They frequently visit pastures in search of insects and may be found a mile or more from water. *Anax junius* and other species congregate in the fall and spring.

There are approximately 2,000 described species of Odonata, of which about 300 occur in North America. They are divided into two suborders: the dragonflies (Anisoptera) and the damsel flies (Zygoptera). The dragonflies hold their wings outstretched when at rest; the damsel flies hold them parallel over the abdomen or uptilted. (As a matter of fact, the wings of some dragonflies droop slightly when they come to rest.) In the dragonflies, the second pair of wings are slightly broader than the first pair; in the damsel flies, the two pairs are approximately the same size. The dragonflies are powerful fliers; the damsel flies are less vigorous in their habits. The flight of the larger species of damsel flies, such as *Megaloprepus* from Panama, reminds one of an autogiro. Their wings move slowly and seem to revolve. The writer has seen them

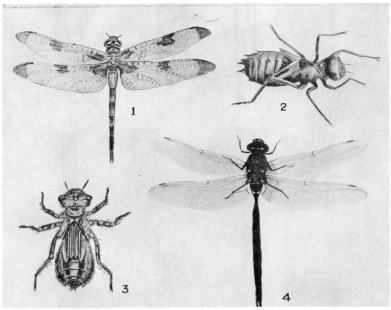

FIG. 375.—Dragonflies and naiads: (1) *Epicordulia princeps*, adult; (2) same, naiad; (3) naiad of *Sympetrum illotum*; (4) *Aeschna constricta*. (*After J. G. Needham.*)

ascend gracefully and pick off insects apparently entangled in a spider's web on a tree. Their slow movements contrast with the swift powerful flight of the dragonfly. All adults of the Odonata are predacious and take their food upon the wing.

The eggs of Odonata are laid in or near the water. Some species fly back and forth over the water, darting down at intervals to drop an egg as the abdomen touches the surface of the water. Some attach their eggs to submerged plants, others lay them in gelatinous strings which are loosely attached to the vegetation.

The naiads, with the exception of a few exotic species, are strictly aquatic. They feed upon insects and other small animals. The lower lip is greatly enlarged to form a powerful grasping weapon. When not in use, the labium is folded beneath the head and is quite inconspicuous; when in use, it is extended.

There is only one brood of dragonflies a year. Sometimes three or four years are required for the development of the naiads. The Anisop-

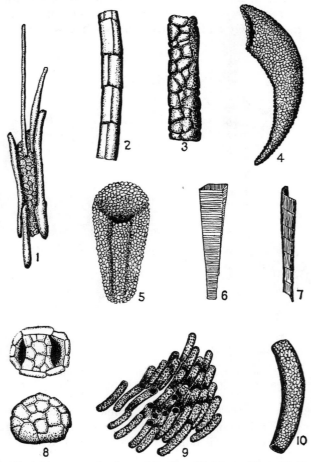

FIG. 376.—Portable cases of a few caddis flies: (1) *Limnophilus;* (2) *Neuronia vestita;* (3) *Halesus;* (4) *Leptocerus ancylus;* (5) *Molanna cinerea;* (6) *Brachycentrus nigrisoma;* (7) *Triaenodes;* (8) *Glossosoma americana,* upper and under side; (9) *Psilotreta frontalis,* a group of cases assembled for pupation; (10) a single case.

tera are usually found in the mud at the bottom of the pond and the Zygoptera usually occur under rocks in swift water. The dragonfly naiad has tracheal gills in the walls of the rectum. Water is alternately taken in and forced out through the anus. By this process the insect obtains air and propels itself through the water. The damsel fly naiad has three platelike tracheal gills at the posterior end of the body. Like the May flies and the stone flies, the naiads of Odonata crawl from the water before the adults emerge.

**Caddis Flies (Trichoptera).**—Caddis flies are mothlike insects that hold their wings over the body in a rooflike fashion. The wings are covered with hairs instead of scales. The antennae are long and slender.

There are approximately 400 North American species. With the exception of a single European genus, all caddis flies are aquatic. Some live in quiet water, others in swiftly flowing water. The majority of

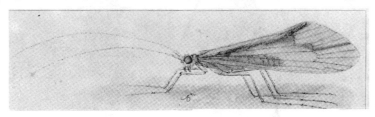

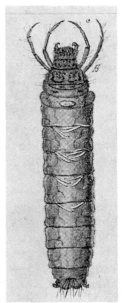

FIG. 377.—*Halesus* and its larva and pupa. (*After Needham.*)

them build portable cases, which are constructed chiefly by the Phryganeidae, Molannidae, Leptoceridae, Odontoceridae, Limnophilidae, and Sericostomatidae. All are somewhat similar in habits but differ as to habitats and materials used as coverings. The Hydropsychidae, Philopotamidae. Polycentropidae, and the Hydroptillidae do not build portable cases. Instead they live in silken tubes covered with dirt and debris. These cases are securely attached to sticks or stones. In some instances,

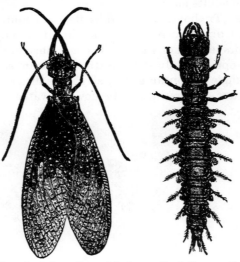

FIG. 378.—The adult male dobson fly and its larva, the hellgrammite.    (*After Comstock.*)

the cases serve as sieves to separate small organisms from the water that

FIG. 379.—The giant water bug, *Lethocerus americanus.* (*After Kellogg.*)

flows through them; in others, the larvae build delicate nets adjacent to their silken tubes to capture their food.

The larvae have only one pair of prolegs on the posterior end of the body. The prolegs bear hooks which aid the larvae in holding themselves in their cases or tubes. Most of the species possess abdominal tracheal gills. The adults apparently emerge in the water and make their way to the surface.

The Neuroptera contain four groups of aquatic species: the alder flies (Sialidae), the dobson flies (Cordyluridae), the snake flies (Raphididae), and the spongilla flies (Sisyridae). The larvae are predacious, carnivorous, or parasitic. The remaining Neuroptera are terrestrial.

The Heteroptera include a large number of aquatic or semiaquatic species. At least 15 families are closely associated with water. The water boatmen (Corixidae) and the back swimmers (Notonectidae) dive below the surface. The water scorpions (Nepidae) and the water bugs (Belostomidae) generally live submerged in grasses or vegetation at the edge of the pond. They are frequently attracted to artificial lights and may be found a considerable distance from the water. *Nepa* conceals itself in the mud or among dead leaves. *Ranatra* clings to grasses, with the respiratory tube piercing the surface film. Many of the smaller bugs —the creeping water bugs (Naucoridae), the toad-shaped bugs (Gelastocoridae), the shore bugs (Saldidae), and a few other rare species— live on the margins of streams, creep on the mud, or inhabit the grasses. The water striders (Veliidae and Gerridae), the Mesoveliidae, and the Hydrometridae walk upon the surface film. The water striders are exceedingly active creatures but the Hydrometridae are slow moving and are found mostly where plants grow in quiet water.

A few aphids, Homoptera, are found on aquatic plants but they are in no way adapted to an aquatic existence.

The Coleoptera contain many aquatic species, notably the Dytiscidae, Hydrophilidae, Gyrinidae, and Haliplidae. The larvae are predacious and truly aquatic; the adults are semiaquatic. They frequently dive beneath the surface carrying with them a supply of fresh air.

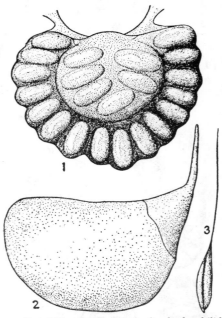

FIG. 380.—Egg cases of a few hydrophilid beetles: (1) *Helochares maculicollis* Mulsant; (2) *Hydrous triangularis* Say; (3) *Berosus peregrinus* Herbst. (*After Richmond.*)

A large proportion of the larvae of Diptera are aquatic or semiaquatic. The Dixidae, Culicidae, Chironomidae, Simulidae, and Blepharoceridae are wholly aquatic. Many of the Tabanidae, Syrphidae, Ephydridae, and Stratiomyiidae are aquatic. A few Tipulidae and Psychodidae are aquatic. Dipterous larvae that live in decayed wood, in soil, in plants, and in insects are semiaquatic.

Two groups of Lepidoptera are aquatic in the larval stage; a few species of Pyralididae and a few species of Noctuidae. The aquatic species of pyralids belong to the genera *Elophila* and *Nymphula*. *Elophila fulicalis* lives in swift water beneath sheets of silk spun across the exposed surfaces of the stones. The species of *Nymphula* live in quiet

water and feed upon various aquatic plants. The larvae have branched
gills and live completely submerged. Three North American species of

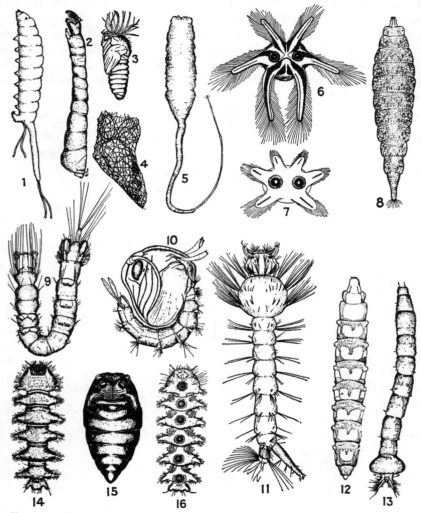

FIG. 381.—Types of aquatic dipterous larvae and pupae: (1) *Ephydra gracilis;* (2)
*Simulium pictipes;* (3) same, pupa; (4) same, pupal case; (5) *Eristalis bastardi,* dorsal view;
(6) *Prionocera fuscipennis,* spiracular disk of larva; (7) *Holorusia rubiginosa,* spiracular disk
of larva; (8) *Stratiomyia discalis;* (9) *Paradixa aliceae,* ventral view; (10) *Culex pipiens,*
pupa; (11) *Culex pipiens,* larva; (12) *Tabanus atratus;* (13) *Eriocera fultonensis;* (14)
*Blepharocera tenuipes,* upper side of larva; (15) same, pupa; (16) same, lower side of larva.
(*After Johannsen.*)

noctuids, belonging to the genus *Bellura*, are known as the divers because
the larvae burrow in lily stems and descend below the water surface

where they remain for a considerable time. The larvae are provided with large trachea which apparently serve as storage chambers for air while the larvae are beneath the water feeding. They are borers and, when not feeding, rest at the top of their burrows with the tips of the posterior spiracles protruding from the water.

None of the Hymenoptera is strictly aquatic; nevertheless leaf-mining species and parasitic species actually live in aquatic situations. Special adaptations, as previously mentioned, are necessary to obtain oxygen. There are also a few parasitic species that descend into the water to lay their eggs on aquatic hosts.

## BIBLIOGRAPHY

The references to aquatic insects have been ably summarized by C. P. Alexander (1925). Important papers appearing since that year are listed here.

ALEXANDER, C. P. (1925). An entomological survey of the salt fork of the Vermillion River in 1921, with a bibliography of aquatic insects. *Bull. Ill. Nat. Hist. Survey* **15**.

BALFOUR-BROWN, F. (1938). Systematic notes on British aquatic Coleoptera. *Ent. Mo. Mag.* (3 papers).

BARNES, T. C., and JAHN, T. L. (1934). Properties of water of biological interest. *Quart. Rev. Biol.* **9** (3). *Extensive bibliography.*

BETTEN, C., *et al.* (1934). The caddis flies or Trichoptera of New York State. *Bull. N. Y. State Mus.* **292**. *Bibliography.*

BRUES, C. T. (1927). Animal life in hot springs. *Quart. Rev. Biol.* **2** (2). *Literature.*

――― (1929). The insect fauna of thermal springs. *Trans. 4th Internat. Congress Ent.*

CARPENTER, K. (1928). Life in inland waters. Sidgwick & Jackson, Ltd., London.

CAVANAUGH, W. J., and TILDEN, J. E. (1930). Algae food, feeding and case-building habits of the larva of the midge fly, *Tanytarsus dissimilis*. *Ecology* **11** (2). *Literature cited.*

CLAUSSEN, P. W. (1931). Plecoptera nymphs of America north of Mexico. Thomas Say Foundation.

CLEGG, JOHN (1930). Aquatic insects. Marshall Press, London.

COLE, A. E. (1921). Oxygen supply of certain animals living in water containing no dissolved oxygen. *Journ. Exp. Zool.* **33** (1). *Bibliography.*

DODDS, G. S., and HISHAW, F. L. (1924–1925). Ecological studies on aquatic insects Parts I to IV. *Ecology* **5** (2), **6** (2 and 4). *Literature.*

EGGLETON, F. E. (1939). Fresh water communities. *Amer. Midl. Nat.* **21**.

FOX, D. L. (1934). Heavy water and metabolism. *Quart. Rev. Biol.* **9** (3). *Literature.*

FRISON, T. H. (1929). Fall and winter stoneflies or Plecoptera of Illinois. *Ill. Nat. Hist. Survey Bull.* **18** (2).

――― (1935). Stoneflies or Plecoptera of Illinois. *Ill. Nat. Hist. Survey Bull.* **20** (4).

GERSBACHER, W. M. (1937). Development of stream bottom communities of Illinois. *Ecology* **18** (3). *Literature cited.*

GRANT, K. (1932). Aquatic Hymenoptera. *Proc. Soc. Lond. Ent. and Nat. Hist.* (1931).

HARKNESS, W. J. K. (1935). The rôle of insect life in Ontario streams and lakes. *Can. Ent.* **67** (3).

HEADLEE, T. J. (1943). The mosquitoes of New Jersey and their control. Rutgers University Press, New Brunswick, N. J.

HEWATT, W. G. (1937). Ecological studies on selected marine intertidal communities of Monterey Bay, California. *Amer. Midl. Nat.* **18** (2). *References.*

HINMAN, E. H. (1932). The utilization of water colloids and material in solution by aquatic animals with special reference to mosquito larvae. *Quart. Rev. Biol.* **7** (2). *Literature.*

HOFFMAN, C. H. (1932). Hymenopterous parasites from the eggs of aquatic and semiaquatic insects. *Journ. Kans. Ent. Soc.* **5** (2). *Literature cited.*

IDE, F. P. (1937). Descriptions of North American species of Baetine mayflies with particular reference to the nymphal stages. *Can Ent.* **69** (11).

JEWELL, M. E. (1927). Aquatic biology of the prairie. *Ecology* **8** (3). *Literature cited.*

JOHANNSEN, O. A. (1935–1937). Aquatic Diptera. Part I, *Cornell Bull.* **164**, Part II, *Bull.* **177**, Part III, *Bull.* **205**, Part IV, *Bull.* **210**.

KARNEY, H. (1934). Biologie der Wasserinsekten. Wagner, Vienna.

KENNEDY, G. H. (1938). The present status of work on the ecology of aquatic insects as shown by the work on the Odonata. *Ohio Journ. Sci.* **38** (6). *Bibliography.*

KIMMINS, D. E. (1941). Under-water emergence of the sub-imago *Heptagenia lateralis* (Curtis) (Ephemeroptera). *Entom. Lond.* **939.**

KRECKER, F. H. (1939). A comparative study of the animal population of certain submerged plants. *Ecology* **20** (4).

LEONARD, J. W. (1939). Mortality of aquatic Diptera due to freezing. *Ent. News* **50** (4).

LLOYD, J. T. (1921). Biology of North American caddis fly larvae. *Cincinnati Bull.* **21.**

LUTZ, F. E. (1930). Caddis fly larvae as masons and builders. *Nat. Hist.* **30.**

MALOEUF, N. S. R. (1936). Quantitative studies on the respiration of aquatic arthropods and on the permeability of the outer integument to gases. *Journ. Exp. Zool.* **64** (3). *Literature cited.*

MARTIN, C. H. (1927). Biological studies of two hymenopterous parasites of aquatic insect eggs. *Ent. Amer.* **8** (3).

MELLANBY, H. (1938). Animal life in fresh water. Methuen & Co., Ltd., London.

MIALL, L. C. (1903). The natural history of aquatic insects. The Macmillan Company, New York.

MORGAN, A. H. (1930). Field book of ponds and streams. G. P. Putnam's Sons, New York.

MOULTON, F. R. (1939). Problems of lake biology. *Amer. Assoc. Adv. Sci.,* The Science Press, Lancaster, Pennsylvania.

NEEDHAM, J. G., and CHRISTENSON, R. O. (1927). Economic insects in some streams of Northern Utah. *Utah Bull.* **201.**

———, *et al.* (1935). The biology of the mayflies. Comstock Publishing Company, Inc., Ithaca, N. Y.

———, and HAYWOOD, H. B. (1939). A hand book of the dragon-flies of North America. C. A. Thomas, Springfield, Ill.

NEEDHAM, P. R. (1938). Trout streams, conditions that determine their productivity and suggestions for stream and lake management. Comstock Publishing Company, Inc., Ithaca, N. Y.

———, *et al.* (1941). A symposium on hydrobiology. *Univ. Wisconsin Press,* Madison, Wis.

RICHARDSON, R. E. (1930). Notes on the simulation of natural aquatic conditions in fresh water by the use of small non-circulating balanced aquaria. *Ecology* **11** (1).

SAUNDERS, L. G. (1928). Some marine insects of the Pacific coast of Canada. *Ann. Ent. Soc. Amer.* **21.**

SHELFORD, V. E., and EDDY, S. (1929). Methods for the study of stream communities. *Ecology* **10** (4). *Bibliography.*

STEHR, W. C., and BRANSON, J. W. (1938). An ecological study of an intermittent stream. *Ecology* **19** (2). *Literature cited.*

SULLIVAN, K. C. (1929). Notes on the aquatic life of the Niangua River, Missouri, with special reference to insects. *Ecology* **10** (3).

SURBER, E. W. (1937). Rainbow trout and bottom fauna production in one mile of stream. *Trans. Amer. Fish Soc.* **66.**

Symposium (1934). Conditions of existence of aquatic animals. *Ecol. Monogr.* **4** (4). *Bibliography.*

THORPE, W. H. (1932). Colonization of the sea by insects. *Nat. Lond.* **130.**

TOKUNAGA, M. (1940). Revision of marine craneflies (Tipulidae) with descriptions of some species. *Kontyû (Tokyo, Japan)* **14** (4).

USINGER, R. L. (1956). Aquatic insects of California. Univ. Calif. Press.

WALKER, E. M. (1933). The nymphs of the Canadian species of *Ophiogomphus* Odonata, Gomphidae. *Can. Ent.* **65** (10).

WEISS, H. B., and WEST, E. (1924). The insects and plants of a salt marsh on the coastal plain of New Jersey. *Journ. N. Y. Ent. Soc.* **32** (2).

———, and ——— (1925). Coleoptera in ocean drift. *Journ. N. Y. Ent. Soc.* **33** (1).

WELCH, PAUL S. (1935). Limnology. McGraw-Hill Book Company, Inc., New York.

WERNE, G. (1940). The soil as an ecological factor in the abundance of aquatic chironomid larvae. *Ohio Journ. Sci.* **40** (4). *Bibliography.*

ZIMMERMAN, E. C. (1842). Distribution and origin of some eastern oceanic insects. *Amer. Naturalist* **76** (764).

# CHAPTER XXII

## CASEMAKING INSECTS

The life stages of insects—eggs, larvae, and pupae—are frequently enclosed in cases. Not a few insects deposit their eggs in capsules of various kinds.* The larvae and pupae, in particular, are often protected by silk, leaves, or other materials. Perhaps one half of the species of insects make cases, known as cocoons, for pupation. Many larvae construct tubes or shelters in which they live and from which they feed. All of these species are casemakers; some of them are casebearers. The verb to bear has two distinct meanings: (1) to wear, have on, hold up, or support on one's back and (2) to carry, convey, or move about with. Thus insect cases may be attached to plants or other objects or they may be portable. The typical casebearers are those insects that make portable cases.

**Casemaking Habit.**—The casemaking habit occurs throughout practically all orders of insects with exception of a few of the more primitive orders such as Thysanura, Collembola, and Orthoptera. Insects with complete metamorphosis usually show a strong tendency toward casemaking, at least in the construction of cocoons. Casemaking may be classified according to the materials that cover the insect: (1) body secretions such as wax or silk and (2) foreign materials which may or may not be added to the body secretions. It is difficult to determine where casemaking begins. Some may criticize the definition of the waxy secretion of a coccid as a case. If the manipulation of the secreted material determines a casemaker, the covering or scale produced by a scale insect certainly fulfills the definition. When foreign materials, such as leaves, twigs, seeds, or fecula, are added to the body covering, there is little doubt that a casemaker is concerned.

The casemaking habit varies from the spinning of sheets of silk to the construction of silken tubes; from the making of cases of pure silk to the construction of cases to which materials of various kinds are added; and from the construction of cases that are permanently attached to those that are semipermanently attached or are portable. No species make cases that are completely portable, for the larvae eventually fasten their shelters for pupation. A few examples will serve to illustrate the range of the casemaking habit.

* See also the discussion of immature insects, Chap. VII.

452

The larvae of *Cryptolechia* and *Elophila* live beneath sheets of silk. The former are terrestrial and construct their tents upon the undersurfaces of the leaves of various trees; the latter are aquatic and spread their sheets of silk upon the bottom of the stream, often on the surfaces of rocks. The tents or silken shelters of these species are securely attached and, under normal conditions, the larvae do not leave them but protrude their heads to feed. Pupation takes place beneath these shelters.

Beetles of the subfamily Attelabinae also live in cases that are permanently attached but the rolls or capsules of leaves in which they live are constructed by the adult females.

FIG. 382.—The hammock moth of South America, *Perophora sanguinolenta.* The larvae construct cases of their own fecula from which they feed and in which they transform. (*After Jones.*)

FIG. 383.—Case of the apple leaf crumpler, *Mineola indigenella.* It is used at first for feeding, later for pupation.

The larvae of the leaf crumplers (Acrobasidae) attach their cases more or less permanently. The larvae construct silken tubes which are covered with dense black fecula. They feed on adjacent foliage, drawing the leaves together by means of silk to form conspicuous masses of dead leaves. With the approach of winter, the larvae of some species move their cases to the larger branches and fasten them. In the spring, the cases are loosened from their winter resting places and are moved to the opening buds where the larvae feed.

The larvae of the typical casebearers, such as the bagworms (Psychidae) and the Coleophoridae, move about freely during the summer carrying their cases wherever they go. When winter comes, these species migrate to the branches where they fasten their cases. The following spring, the Coleophoridae return to the foliage to continue their feeding.

**Silk of the Casemakers.**—The casemaking habit intergrades with the leaf-rolling habit. Both depend upon the ability to spin silk. There are two common sources of silk in insects: (1) the cephalic or modified salivary glands and (2) the caudal glands or modified Malpighian tubules. In the Embiidina, silk is spun from a gland in the metatarsus of the leg. The cephalic glands are elongate and often occupy a large part of the interior of the body. The two ducts leading from the glands unite to form a single duct which opens through the lower lip. The caudal or Malpighian glands are appendages of the alimentary canal, are arranged in pairs from two to several in number, and are often considerably branched. Malpighian glands for silk spinning usually occur in insects in which the passage between the mid- and hind intestine is closed until the final moult.

The larvae of the Lepidoptera, Trichoptera, Hymenoptera, Diptera, Siphonaptera, and Corrodentia are cephalic spinners. The Neuroptera and Coleoptera are Malpighian spinners. As far as is known, the Embiidina are the only metatarsal spinners.

**Casebearing Habit.**—Comparatively few insects construct portable cases. The larva is the casebearing stage of the insect although many species transform within cases. The habit is prominently developed in the Trichoptera and certain Lepidoptera, two orders that are somewhat closely related. It occurs to some degree in the Coleoptera. A primitive form of casebearer occurs in the Neuroptera and Heteroptera. Reference is made here to the larvae of the hemerobiids and the nymphs of certain reduviids that carry trash upon their backs. Casebearing is not universal in any order but is conspicuous in the Trichoptera where the majority of the species possess this habit. It is found in terrestrial as well as aquatic insects and is prominently a development of phytophagous species.

Fig. 384.—The masked hunter, *Reduvius personatus;* the nymphs cover themselves with particles of lint and dust. (*Charles Macnamara.*)

**Evolution of the Casebearing Habit.**—There are two general groups of casebearers: (1) the nonspinning species that cover themselves with foreign material of various kinds and (2) the silk-spinning species that construct silken cases to which leaves and other material may be added.

The waxy substance secreted upon the bodies of Coccidae and the larvae of coccinellid beetles (Hyperaspis, Scymnus, and Cryptolaemus) may represent a primitive step in the development of portable cases.

These insects carry a protective covering on their backs. Other species accidentally or intentionally cover themselves with various kinds of foreign material. May fly and dragonfly naiads often carry deposits of mud on their backs. The masked hunter, *Reduvius personatus*, carries a load of dust and lint which adheres to a viscid substance secreted by the immature stages. The larvae of *Chelymorpha* and *Cassida* deliberately place their own fecula upon their backs and the trash carriers, Hemerobiidae, place the remains of their victims upon their backs.

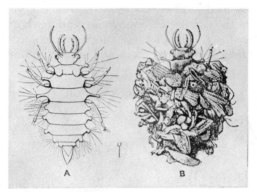

Fig. 385.—Larva of *Hemerobius*, a trash carrier: (*A*) naked larva; (*B*) larva partially covered with remains of its victims. (*After Sharp*.)

The casebearing habit has no doubt arisen independently in different groups of insects. In the silk-spinning species, this habit probably developed from the repeated use of silk. Silk is freely used for locomotion, to ensnare prey, and for other purposes. The larva probably used silk at first as a simple shelter. Later bits of foliage or other materials were added until a more substantial structure was formed.

**Discussion of the Casemaking Species.**—The Embiidina live in silken nests or galleries under stones or other objects. These silken galleries are delicate structures, very loosely constructed and permanently attached.

The Chironomidae are the outstanding casemakers in the Diptera. Many species live in delicate silken tubes to which mud and particles of sand or vegetable matter are attached. These tubes are fastened to plants or occur in profusion on the bottoms of ponds. The boot-shaped cocoons of the Simuliidae are conspicuous cases found in streams of swift, cold water.

The Lepidoptera include many casemaking forms. Both the larvae and the pupae may be enclosed in some kind of tube or case. Some of the species fasten their cases to leaves or other objects; others construct portable cases.

Examples from the Pyralididae illustrate various steps in the development of the casebearing habit. The meal moth, *Pyralis farinalis*, constructs tubes of particles of prepared food fastened together with strands of silk. The larvae of the Indian-meal moth, *Plodia interpunctella*, and the Mediterranean flour moth, *Ephestia kuehniella*, have similar habits. The Crambidae live in silken tubes composed of bits of earth or vegetable matter fastened together with silk. These occur chiefly at or below the surface of the ground at the bases of the plants on which the species feed. The wax moths, *Galleria mellonella*, live in silken galleries in the combs of honeybees and feed upon nitrogenous material in the wax. Although these species do not construct portable cases, the larvae may leave their cases and construct new ones.

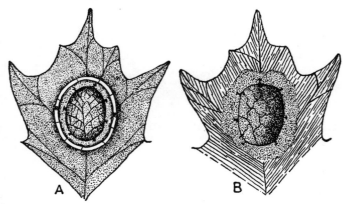

Fig. 386.—Abode of the maple casebearer, *Paraclemensia acerifoliella*: (*A*) small case showing area cut about it in preparation for enlargement; (*B*) mature case fastened to leaf with bands of silk. (*After Herrick.*)

The pyralidids of the genus *Acrobasis* make compact cornucopia-shaped cases of silk which are covered with dense black fecula. The apple leaf crumplers, *Mineola indigenella*, fasten their cases securely to the bark when winter approaches. In the spring, they loosen their cases and migrate to the opening buds to feed. Thus, the cases of this species are only semipermanently attached.

Another group of pyralidids, namely species of the genus *Nymphula*, are typical casebearers. *Nymphula maculalis* live in quiet water among yellow and white lilies. They construct cases of two pieces of lily leaf about one inch long and fasten them together with silk. These cases are filled with water and the larvae breathe by means of branched gills protruding from the sides of the body. The larvae feed upon the leafy walls of the cases. Pupation eventually takes place in these cases. In the Southern States and in protected places as far north as New York,

another species, *Nymphula obliteralis*, has similar habits but the larvae
have no gills. For oxygen they depend upon the bubbles of air that
surround them in their cases and possibly upon oxygen liberated by the
green leaves composing the cases. The cases of both
species are portable.

The Incurvariidae frequently make lenticular-shaped
cases. Certain of the species of *Incurvaria* and
*Paraclemensia* are miners at first, then construct cases.
*Paraclemensia acerifoliella* has been studied somewhat
in detail. The larva feeds upon maple, cutting an oval-
shaped piece of leaf which is attached to the leaf by means
of silk. The larva lives between the leaf and the piece
that has been fastened down. Later the larva cuts around
the edge of the oval piece of leaf and the two pieces are
fastened together to form a new case which in turn is
fastened to the surface of the leaf. When the larva needs
new feeding grounds, it cuts the supporting strands and
wanders off with its case. The larvae of *Adela* feed at first in
flowers or seeds and later construct lenticular-shaped cases.

Fig. 387.—A
pistol-type case
produced by *Col-
eophora atlantica*
on *Prunus sero-
tina.*

The *Coleophora* are the outstanding casebearers of the Lepidoptera.
There are approximately 110 North American species. The cases are
composed of silk, pubescence from leaves, and particles of
fecula. The larvae are external feeders but some mine in
a limited fashion in the leaves. They push their heads into
the leaves and feed as far as they can reach without letting
go of their cases. Two common species are the cigar
casebearer, *Coleophora fletcherella*, and the pistol casebearer,
*Coleophora malivorella*. The shapes of the two cases are
well described by their common names. They are both
pests of apple.

Fig. 388.
—A cigar-
shaped case
produced by
*Coleophora
laticornella*
on dogwood.

Two species of Lacosomidae from Northeastern America
make portable cases. *Cicinnus melsheimeri* is a well-known
casebearer. The larvae feed on the leaves of various
species of oaks and, as they grow older, make cases of pieces
of leaves.

The larvae of the Psychidae or bagworms live in
portable cases and wander about freely except as the time
for pupation arrives when they fasten their cases securely to
the host or some near-by object. The evergreen bagworm,
*Thyridopteryx ephemeraeformis*, prefers cedar and arborvitae and attaches twigs and pieces of leaves lengthwise on the
bag. Abbot's bagworm, *Oiketicus abboti*, fastens the twigs crosswise on its
case. *Solenobia walshella* makes a small case about eight millimeters long

which is covered with small grains of sand or particles of lichens and the fecula of the larva.

Several well-known casemakers belong to the Tineidae. The case-making clothes moth, *Tinea pellionella*, fashions tubes of bits of food material fastened together with silk. The carpet moth, *Trichophaga tapetzella*, makes galleries composed of silk and bits of cloth. The habits of this group are still little known and it is possible that many other casemaking forms are included.

Other common casebearing Lepidoptera are

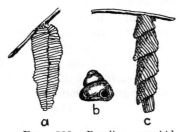

| a | b | c |

FIG. 389.— Bagworms, *Thyridopteryx ephemeraeformis*, representing portable types of cases.

FIG. 390.—Peculiar psychid cases: (a) *Amicta quadrangularis;* (b) *Apterona valvata;* (c) *Chalia hockingi.* (*After Sharp.*)

the shield bearers, *Coptodisca* and *Antispila*. These species feed during the summer as leaf miners. When fall approaches, they cut two circular pieces from the leaves and fasten them together to make neat cases. Before the leaves fall, the larvae migrate to the branches where they

FIG. 391.—Casemaking clothes moth, *Tinea pellionella*.

attach their cases for the winter. These cases are eventually permanently attached.

Many of the Olethreutidae, as for example, the eye-spotted budmoth, *Spilonota ocellana*, live in silken tubes which are permanently attached to leaf or tree. *Argyrotaenia pinatubana* lives in a neat tube of pine needles bound together by means of silk. The larva protrudes its head to feed upon the outer ends of the needles.

Many of the Trichoptera produce elaborate cases composed of particles of sand, stones, wood, bark, grass, leaves, and numerous other materials which are held together by means of silk. Their habits are discussed more in detail in Chap. XXI, Aquatic Insects. The majority of the species construct portable cases. Numerous species occur in the families Phryganeidae, Molannidae, Leptoceridae, Odontoceridae, Calamoceratidae, Limnophilidae, and Sericostomatidae. These cases are usually attached only when the larvae are ready to pupate. Some of the Hydrophilidae make portable cases, others fasten their cases firmly

Fig. 392.—Black gum showing the mines of *Antispila nyssaefoliella* and circular pieces cut by the larvae for pupal cases.

Fig. 393.—Cases of *Coptodisca splendoriferella* attached to apple twigs for hibernation.

to rocks over which the swift water flows. Grains of sand and bits of vegetable matter are attached to the cases of these species. The Hydropsychidae, Philoptamidae, and the Polycentropidae fasten their tubes and nets securely to rocks or other anchorages in rapid streams. The larvae of *Psilotreta* crawl freely over the bottom of the stream during the early part of their life, but before pupation occurs they develop a gregarious habit and assemble in great masses to transform. The species of *Glossosoma* attach their cases to stones in the stream bed where they remain during the feeding period. Before they pupate, the larvae move their cases together in large colonies.

Most of the Neuroptera spin silken cocoons. Those of the hemerobiids and the chrysopids are most familiar. The cocoon of the spongilla fly, *Climacia*, is lacelike in appearance and is a beautiful object. The hemerobiids, as mentioned previously, are primitive casebearers for they carry trash and the skins of their victims upon their backs.

In addition to the silken cocoons made by many of the Coleoptera, a few species construct interesting larval cases. The *Clythrini*, including the well-known *Coscinoptera*, make cases of chewed fragments of leaves.

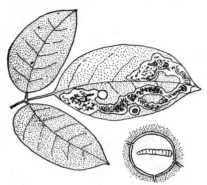

As far as is known, the larvae feed upon vegetable debris in the nests of ants. *Chlamys plicatula* live in compact cases composed of their own fecula. They carry their cases about with them by protruding the front part of their bodies through the anterior openings of the cases. These cases resemble the pellets of certain caterpillars and it is said that for this reason, birds will not pick them up.

The Attelabinae live during the larval stage in compact thimble-shaped rolls which are constructed by the females of pieces of leaves.

FIG. 394.—Leaf of *Serjania* showing mine of *Pachyschelus atroviridis* and case cut from the leaf, Panama.

The cases are firmly attached to plants. Further discussion of these species will be found in Chap. XVII, Leaf-rolling Insects.

Some of the leaf-mining Buprestidae, notably species of the genus *Pachyschelus*, make pupal cases from two circular pieces of leaves fastened together with silk.

## BIBLIOGRAPHY

See references to Trichoptera, Chap. XXI; Leaf-rolling Insects, Chap. XVII; and Immature Insects, Chap. VII.

BETTEN, C. H., *et al.* (1934). The caddis flies or Trichoptera of New York State. *N. Y. State Mus. Bull.* **292.** *Bibliography.*

CAVANAUGH, W. J., and TILDEN, J. E. (1930). Alga food, feeding and case-building of the midge fly, *Tanytarsus dissimilis*. *Ecology* **11** (2). *Literature.*

FORBES, WM. T. M. (1920). Coleophoridae by Carl Heinrich. *In*, Lepidoptera of New York and neighboring states. *Cornell Agric. Exp. Sta. Memoir* **68.**

FROST, S. W. (1927). Notes on the life history of the bud moth, *Spilonota ocellana* (D. & S.). *Journ. Agric. Res.* **35** (4).

GREEN, E. E. (1921). A note on a web-spinning Psocid. *Spolia Zeylan Colombo* **8.**

KNAB, F. (1915). Dung-bearing weevil larvae. *Proc. Ent. Soc. Wash.* **17** (4).

HOWARD, L. O., and CHITTENDEN, F. H. (1908). The bagworm. *U. S. Dept. Agric. Ent. Circ.* **97.**

MARLATT, C. L. (1915). The true clothes moth. *U. S. Dept. Agric. Farmer's Bull.* **659.**

PORTER, B, A. (1924). The bud moth. *U. S. Dept. Agric. Bull.* **1273.** *Literature.*

RIMSKY-KORSAKOV, M. (1914). Über den Bau und die Entwicklung des Spinnapparates bei Embiden. *Zeit. f. Wissenschaftliche Zool. Band* **108.**

SMITH, R. C. (1922). The biology of the Chrysopidae, *Cornell Agric. Exp. Sta. Memoir* **58.**

——— (1923). The life histories and stages of some Hemerobiids and allied species (Neuroptera). *Ann. Ent. Soc. Amer.* **16** (2).

——— (1926). The trash carrying habits of certain lace-wing larvae. *Sci. Mo.* **23.**

STAINTON, H. T. (1859–1860). Coleophora. *In,* L'histoire naturelle des Tineina. **4** (1) and **5** (2) E. S. Miller & Sohn, Berlin.

# CHAPTER XXIII

## CESSATION OF ACTIVITY

Thus far we have been discussing the activities of insects; now we come to the subject, rest or the inhibition of activity. Rest is a biological phenomenon that occurs in all forms of life, plant and animal. It is just as essential as feeding, reproduction, or any other life function and may be conspicuous, or evident only after a critical examination of the insect. There are two general forms of rest: the arrestation of development, known as *diapause*, and the arrestation of activity, which may be termed *kinetopause*. Both forms of rest may take place at the same time; for example, during hibernation, the insect ceases its activity and, during the cold portion of the winter, development may also cease. On the other hand, development may proceed while the insect is apparently inactive; for instance, when an insect is in the process of transformation, the development of legs, wings, antennae, reproductive organs, and other vital parts proceeds often at the greatest rate although the movements of the insect practically cease.

One should differentiate carefully among growth, development, and activity. Growth and development are often synonymous but growth relates principally to increase in size and may take place without development. For example, the "repletes" among honey ants increase remarkably in size due to the storage of honey; development, however, ceases when these ants pass the last moult. Development relates to the fundamental changes which the insect undergoes in passing from egg to the adult.

**Cessation of Development (Diapause).**—Henneguy used the term diapause to include the arrestation of development by internal factors; however, the term has been generally used to cover all cases of suppressed development. There are various degrees of diapause: temporary, prolonged, and complete. Complete inhibition of development and activity is death. Diapause may occur in any of the life stages of the insect but is perhaps most pronounced in the egg and the pupa. During the larval and nymphal stages, conditions that induce diapause usually force the insect into the pupal stage or into a distinct resting period.

Diapause may be imposed by external factors such as temperature, moisture, food, and oxygen or by internal factors such as heredity, enzymes, or hormones. Several theories have been propounded to explain diapause, in the true sense. Some believe that the temporary absence of hormones, necessary for growth, may check development. Others state that growth is inhibited by the accumulation of some chemical in a manner analogous to muscular fatigue in higher animals.

**Suppressed Development of Eggs.**—Eggs that are laid in the early summer generally hatch soon after they are deposited with no evident resting period. The eggs of some parasitic insects remain unhatched until they are picked up by the host. Eggs that are laid in the fall may remain dormant at least until the following spring. The eggs of the scorpion fly, *Bittacus*, and the walkingstick, *Diapheromera*, remain inactive for two years.

Temperature and moisture play the largest part in inhibiting the development of eggs. The eggs of *Sminthurus* (Collembola), which normally hatch in eight or ten days under favorable moisture conditions, have been kept dried and shriveled for 271 days. The eggs of the South American *Locusta pardalina* survive desiccation for three and a half years. The eggs of mosquitoes, laid during dry seasons, may remain unhatched for months and normally do not hatch until the following year.

Other factors may determine the hatching periods of eggs. The eggs of *Sturmia scutellata*, a tachinid parasite of the gypsy moth, are laid upon the foliage and remain unhatched until ingested with the plant food by the host caterpillar. The eggs of the horse botfly, *Gasterophilus intestinalis*, are laid upon the hairs of horses and remain inactive until they are licked by the animal and taken into the stomach. The minute spiny maggots are ready for emergence from the eggs in about seven days after they are laid but may lie dormant for several months.

Fig. 395.— Eggs of *Sturmia* (*Blepharipa*) *scutellata* on the leaf of the food plant of its host. The eggs remain on the plant until ingested by the host caterpillar. (*After Howard and Fiske*.)

Eggs may hibernate in various stages of development. In some cases, the embryo is undeveloped but, in others, it is completely formed before the winter starts. The eggs of the apple tree tent caterpillar are laid in the summer but do not hatch until the following spring. Development continues until cold weather sets in. These eggs will not hatch if they are brought indoors in the fall and kept at ordinary room temperatures between 20° and 35°C. It is essential that they be subjected to temperatures varying between 1° and 10°C. for periods of eight to ten weeks. Otherwise hatching is abnormal or does not occur at all.

Wigglesworth states that the suppression of the egg development of *Bombyx mori* may be influenced by the temperature at which the eggs of the preceding generation are incubated. Eggs incubated at 25°C. tend to produce moths that lay hibernating eggs; those incubated below

15°C. produce moths that lay nonhibernating eggs. He also gives some interesting cases of arresting development by artificial means. The eggs of *Platycnemis pennipes* and *Calliphora*, when ligated in the early embryonic stages, produce two dwarfs.

On the other hand, development is sometimes speeded up and the eggs hatch before they leave the body. This is the typical manner of reproduction among aphids during the summer. In the Hippoboscidae, nearly mature larvae are laid instead of eggs. In the Tropics or under greenhouse conditions, hibernation does not occur and the life history of certain insects may be considerably modified. Aphids continue to reproduce agamically for indefinite periods. Grain moths, clothes

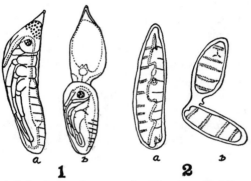

Fig. 396.—Arrested development as a result of ligation: (1) *Platycnemis pennipes;* (2) *Calliphora;* (a) normal embryos shortly before hatching; (b) embryos of similar age that have been ligated during the early stages of development.

moths, carpet beetles, and insects of similar habits breed as long as temperature, moisture, and food conditions are favorable.

**Suppressed Development in Other Immature Forms.**—After insects hatch, development usually proceeds regularly except for short periods during the moulting process or for more prolonged periods during estivation or hibernation. Adverse conditions such as lack of food, moisture, or oxygen naturally check development.

Development is usually slow in subterranean insects owing to lack of sunshine, low temperatures, and scarcity of food. The Japanese beetle spends the greater part of its life history in the ground. The white grubs (Phyllophaga) require two or three years to develop. The feeding period of the periodical cicada is prolonged over a period of 17 years.

Unfavorable conditions may cause a reversal of development. The clothes moth and *Anthrenus* continue to moult when food and moisture are unsatisfactory and become smaller in size with each moult.

**Suppressed Development in the Pupa.**—Pupal development is usually suppressed during unfavorable temperature conditions; otherwise development is rapid. The pupa of the *biston* moth is said to remain dormant for seven winters before resuming development. Hibernation is often broken by exposure to low temperatures. Moisture is an important factor in determining the emergence period of the codling moth. Mechanical injuries also tend to break hibernation. Pricks from parasites often hasten host development. Chemicals have been used successfully to break hibernation in plants and may prove useful in insects. In one case the eggs of the buck moth, *Hemileuca maia*, subjected to weak fumes of potassium cyanide, gave 100 per cent hatch in the fall.

**Cessation of Activity (Kinetopause).**—There are various degrees of the arrestation of activity known as rest, sleep, death feigning, estivation, hibernation, and death.

Activities within the egg are generally minute and inconspicuous. The embryos of some eggs have a horizontal and a vertical revolution which are visible through eggs with thin shells. The movements of jaws or mouth parts are visible especially when the insect consumes the last of the yolk. These movements may be checked during cool periods.

Naiads, nymphs, and larvae are distinguished from each other chiefly by the degrees of activity, especially in regard to transformation. Nymphs, with rare exceptions, have no definite resting period before the adults emerge. Naiads have a definite resting period but no pupae are formed. Larvae have distinct resting periods and definite pupae are formed.

Immature forms have definite periods of feeding and resting during growth and development. The most profound rest occurs at the time of moulting and during hibernation. Little has been written about sleep in caterpillars or other immature forms but it is well known that many species feed during the day and rest during the night. The apple tree tent caterpillar refrains from feeding during cold damp periods and retreats to its nest.

**Suppressed Activity in the Pupa.**—The pupal stage is characteristically a period of outward inactivity, although, during seasons of favorable temperatures, development proceeds at a greater rate than at any other time. Certain pupae, such as those of mosquitoes and chironomids, are active. Many pupae become active as time for emergence approaches. Pupae formed in the soil and many formed in wood or beneath bark work their way partly from their resting places before the adults emerge. Most pupae respond feebly to touch during the warmer seasons of the year.

**Estivation and Hibernation.**—Estivation and hibernation are somewhat similar habits but are induced by opposite conditions. Estivation takes place typically in the Tropics or in temperate areas where high temperatures or low humidity make conditions unsatisfactory for development. Certain dragonfly naiads remain alive for months in the dry sands of Australia. The eggs of mosquitoes are often laid in dry places and remain in a dormant condition until the arrival of sufficient rain enables these aquatic insects to continue their development.

Hibernation occurs in temperate or frigid areas where the temperature becomes so low that normal activities are impossible. When the temperature falls appreciably, insects usually go into hibernation. There are, however, certain exceptions. The painted lady, the monarch butterfly, and other species migrate southward to avoid unfavorable winter conditions. Other insects have modified their habits to a semidormant existence. When the temperature reaches approximately 14°C. within the hive, honeybees become active, feed, cluster about the queen, and generate sufficient heat to raise the temperature of the hive one or two degrees and thus protect her. The larvae of a species of Coleophora come out during warm periods in winter and feed upon the evergreen leaves of *Kalmia*.

Low temperatures no doubt play an important part in inducing hibernation. The cotton boll weevil begins to hibernate when the mean temperature reaches 55°F. Many insects, however, go into hibernation before cold weather sets in. One per cent of the second brood of the oriental fruit moth larvae hide away and wait until the following spring to transform. This may occur as early as June 20. As a matter of fact, these larvae really estivate until fall when they go into hibernation together with portions of the third, fourth, and fifth generations. The woolly bear caterpillars cease feeding during the latter part of summer when the temperature is high and when there is an abundance of food at hand. The codling moth likewise begins to hibernate before the temperature falls noticeably and the food fails.

Insects hibernate in all stages of development but the method is usually constant for a given species. Some insects hibernate as eggs or first instar larvae, others as full-grown larvae, and still others as pupae or adults. The rose scales may hibernate as eggs, nymphs, or as adults. The clover leaf weevil, *Hypera punctata*, hibernates in the larval, pupal, or adult stage. Groups of related insects generally show uniformity as to method of hibernation. Of the 13 North American species of *Argyrotaenia* (Eulia), all but one hibernate as pupae. Species of the genus *Cacoecia* generally hibernate as eggs. The species of the family Aegeriidae generally hibernate as larvae.

The larvae of the oriental fruit moth go into hibernation in different broods; 1 per cent of the second brood, 39 per cent of the third brood, 86.2 per cent of the fourth brood, and 100 per cent of the fifth brood hibernate. During hot dry seasons, these percentages may be increased.

**Sleep (Akinesis).**—It is well known that insects sleep or at least rest in a comatose state. Diurnal insects cease their labors at dusk and await the return of more favorable conditions to resume work. Sunshine and high temperatures, within reasonable limits, stimulate activity; low temperatures generally retard activities, and exceedingly low temperatures may cause death. Leaf-cutting ants cease their labors at remarkably regular hours in the afternoon. Those who have studied these insects say that the time of day can be told amazingly well by the hour of cessation of their activities. A passing cloud causes them to slacken their operations. The mourning-cloak butterfly and many wasps also become inactive if the sun goes under and resume their activities only when it becomes bright again.

Factors other than light or temperature may cause insects to refrain from work. Honeybees and other flower-visiting insects are governed to a large extent by the honey flow. Their enthusiasm slackens as the supply of nectar diminishes. The familiar hum of bees in a linden tree or field of clover may suddenly cease as the nectar supply is depleted and these insects may be found a mile or more away seeking more fertile feeding grounds. At the end of day, the flow of nectar is often exhausted and the bees probably stop willingly to rest.

To determine the profoundness of sleep in insects is more difficult. Many authentic records have been published to indicate that insects are in a comatose state during their evening rest. The writer has seen *Polistes* gather in small groups to spend the night away from their paper nest. This occurred so frequently that the painted surface on which they roosted became spotted with defecations. Banks found large wasps of the genus *Ammophila* and bees of the family Myzinidae asleep on grasses. They appeared between seven and eight o'clock in the evening and left before the following morning. Swartz observed 50 to 70 specimens of a species of *Priononyx* asleep on dead shrubs and others on near-by shrubs in smaller numbers in Texas. Bradley found a species of black wasp, *Priononyx atratum*, asleep on wild oats in groups of two to a couple of dozen. He collected 490 individuals in an hour's time, just before darkness, and was able to break the stems of the oats and place the specimens in a cyanide jar before they were aroused from their sleep.

It is a commonplace observation that *Vespula maculata* is aroused quite easily. Do sentinels refrain from sleep to warn the colony of approaching danger at night or are the vespulas light sleepers? Phil Rau remarks, "Some insects sleep with all the regularity of a theoretical modern infant, while others of a more unsystematic life snatch a wink when they can."

The sleeping *Heliconias* have been mentioned before. Other butterflies have frequently been observed at sleep. The milkweed butterflies, during their southern migration, congregate and sleep in great numbers upon bushes and trees.

It is well known that dragonflies rest at night and that they can be captured more easily at this time. No doubt many other diurnal insects have similar habits.

FIG. 397.—Sleeping Heliconias of Florida. The marks on the under surfaces of the wings often simulate the grasses upon which they rest. (*After Jones.*)

The resting periods of nocturnal insects have not been studied; however, many moths, especially the Noctuidae, are commonly seen resting during the day. Sometimes they crawl away to some protected place but often they remain exposed to the sunlight. These moths are oblivious to bright light but they respond to weak lights and are attracted to artificial lights at night.

Sleep has been produced artificially by Wigglesworth by removing the antennae of *Rhodnius*. After the antennae were removed, the insect settled down into a state of sleep.

**Death Feigning** (**Thanatosis**).—Death feigning is a form of sleep. The habit occurs commonly among crustaceans, reptiles, birds, mammals, and insects but is rare among fishes. From the behavior of the opossum, the common expression "playing possum" has come into everyday language. Death feigning occurs in nearly all orders of insects and has been studied especially in the beetles, the water scorpions, and the giant water bugs.

Several positions are assumed in death feints. Usually the insect draws its legs tightly against the body and remains motionless for a considerable period of time. Sometimes the legs are not drawn close to the body but are extended perpendicular to the body with the joints of the legs folded backwards. Many of the Scarabaeidae assume this position when they are thrown into a death feint. The Elateridae and the Curculionidae are easily thrown into a death feint.

Some caterpillars assume strange postures when they are alarmed: the yellow-necked caterpillar, *Datana ministra,* elevates the anterior and posterior ends of the body when disturbed. The spinx larva elevates the anterior end of the body and curls the head down so that it resembles a sphinx. The rigid attitudes assumed by many of the geometrid larvae are forms of death feints. Cutworms curl freely when disturbed and the sowbug, a crustacean, rolls itself into a ball, hence

Fig. 398.—The plum curculio, *Conotrachelus nenuphar;* above, normal position; below, one attitude assumed in a death feint.

the common name pill bug. Sawfly larvae often assume an S position when disturbed. The birch sawfly, *Croesus latitarsus,* is noted for this attitude.

The feints are generally of short duration, lasting in most cases no more than a few minutes; however, some insects assume a feigning

Fig. 399.—Yellow-necked caterpillars, *Datana ministra* (Drury), showing the position assumed when alarmed. (*After U.S. Department of Agriculture.*)

position for as long as 60 minutes. The giant water bug has been successively put in a feint for eight hours. *Belostoma* has been successively put in a feint for five hours. Experiments with the plum curculio failed to show any regularity in the duration of the feints. Fifty-three feints were produced over a period of two hours ranging from 3.16 to

6.83 minutes for each feint.  After the first few feints, they tended to show a decrease in duration until finally the curculio refused to feint longer.

The nature of the death feint is not exactly known but it is a thigmotropic response.  The nervous system is thrown into a state of abnormal inhibition by means of various mechanical stimuli such as pressure or vibration.  The muscles become tense and appendages can be removed without sign of recovery from the feint.  Decapitation or severing the body between the pro- and mesothorax broke the death feint in the curculio.

Death feigning is a protective habit.  When insects go into a feint, they usually fall to the ground, appear dead, and thus deceive their enemies.  Some insects fall so readily that it is difficult for a collector to obtain specimens unless he is aware of this habit and places a sheet or an inverted umbrella under the plant to catch them when they fall.  This habit has been somewhat successfully used to control the plum curculio. Canvas sheets are placed beneath the infested tree and the limbs jarred to bring down the beetles.  They are then gathered up and destroyed.

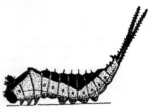

Fig. 400.—The puss caterpillar, *Cerura*, turns its tails upward when disturbed.

**Death.**—The term or span of life is limited for each species.  In protozoa, it may be a few hours; in the elephant, it is 200 years.  Mosquitoes may mature in less than a week in tropical countries.  Certain adult May flies live only for a few hours although the naiads may spend two years in the water.  Adult honeybee workers live, on the average, about six weeks.  Although the average span of life for insects is one year, many live longer.  A queen bee may live for four years and a well-fed queen ant has been known to live for 15 years.  Wood-boring larvae, white grubs, and wireworms require several years to mature.  The northern race of the periodical cicada takes 17 years to mature.  Females usually live longer than males and unmated females live longer than mated females.

The tenacity of life varies considerably.  Food is an important factor in determining the length of life.  Insects that have vestigial mouths, such as the May flies, have short adult lives.  Insects that have well-developed mouths will not live long if they do not receive food.  In rearing parasitic Hymenoptera, it is essential to provide food for the adults.  Honey or sugar water is usually employed.  The life of the oriental fruit moth adults is considerably shortened if they do not receive a small amount of food.  Unfed cotton boll weevils lived for only 10 days; properly fed weevils lived for 20 days.  Ticks have been kept for seven years without food or moisture. They shrivel and appear dead but, when food is supplied, gorge themselves and return to normal. The larvae of the museum beetle, *Trogoderma tarsalis*, are exceedingly tenacious. Newly hatched larvae that have never eaten

Fig. 401.—Darkling beetle, *Eleodes dentipes*, feigning death by standing upon its head. (*After Duncan and Pickwell.*)

live without food for four months. Full-grown larvae live without food for more than five years. *Margarodes vitium*, a scale insect of Chile, is said to remain alive for 17 years without food. There are also individual differences in vitality. Upon this fact depends a great deal of our control work. If a solution of 3 pounds of arsenate of lead to 100 gallons of water will kill one codling moth larva, theoretically it should kill all other codling moth larvae, if applied in the same manner. As a matter of fact, this is not true and it is well known that the individual resistance of a species of insect to poisons varies.

**Factors Causing Death.**—Death is the culmination of a series of activities that, as far as the insect is concerned, are directed to maturation and reproduction of its kind. It is the cessation of the vital functions without the capability of resuscitation. Few animals, insects included, die of old age. Normally death overtakes the insect after it has served its purpose—reproduction and perpetuation of its kind. Premature death may befall insects in any stage of development. The eggs and

Fig. 402.—Flies killed by an entomogenous fungus, *Empusa muscae.* A closely related species often attacks grasshoppers.

the young larva are perhaps most susceptible. Parasites, predators, and numerous diseases are responsible for the death of many insects and physical factors account for the death of many others.

Moisture and temperature are perhaps the most important physical factors causing death in insects. Within reasonable limits, insects may adapt themselves to considerable temperature variations but extreme conditions cause death. High temperatures kill by coagulating the proteins, by evaporation, and by desiccation. Low temperatures kill by the rearrangement of the cell structure and consequent destruction of the body tissues.

A few insects can withstand extremely high temperatures. The larvae of one of the stratiomyiid flies live in the hot springs of Colorado where the temperature of the water is approximately 150°F. The upper temperature limit for most insects is about 120°F. However, one out of 15 pupae of *Celerio euphorbiae* survived at a temperature of 188°F. The duration of the exposure and the humidity should be considered when discussing resistance of insects to unusual temperatures. Selecting a few examples, we find considerable variation in the resistance of insects to high temperatures: 6 to 11 minutes at 165°F will kill the larvae of the European corn borer; 5 minutes at 12 per cent humidity and 148°F. kills the cotton boll weevil; and 5 to 10 minutes at 135°F. is fatal to the clear-winged grasshopper.

Insects are better equipped to adjust themselves to low than to high temperatures but there are limits beyond which endurance fails.

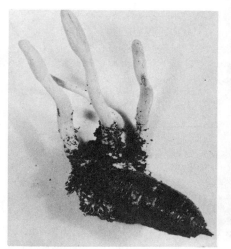

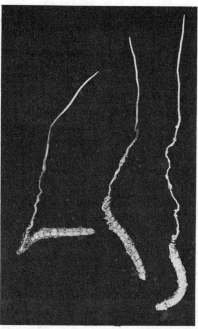

Fig. 403.—Pupa of a lepidopteron killed by *Cordyceps militaris*. (*After Overholts.*)

Fig. 404.—Wireworms attacked by *Cordyceps acicularis*. (*After Overholts.*)

The extreme for most insects is −31°F. Grain insects succumb at a temperature of 15°F. when prolonged for three hours; 20°F. for 48 hours is fatal to grasshopper adults; 25°F. for 50 minutes is fatal to the honey-bee; 20°F. for 16 hours is fatal to grasshopper eggs. The larvae of the codling moth perish when the temperature reaches −25°F.

The predators, parasites, and parasitoids responsible for the death of many insects have been discussed previously. They occur in nearly every order but are most prominent in the Neuroptera, Heteroptera, Coleoptera, Diptera, and Hymenoptera. There is no doubt that a large number of insects are destroyed by these agencies long before they reach maturity. It is generally conceded that, if it were not for these natural means of control, the earth would be overridden by insects.

We have also discussed previously such plant traps as sundew and *Utricularia* which capture many insects. *Riccia, Azolla,* or duckweed may cover water so thickly that certain insects are unable to reach the surface to obtain air. Other plants may emit secretions that are toxic to aquatic insects.

Lower forms of plant life also take their toll of insect life. *Empusa, Cordyceps, Aspergillus, Entomophthora, Sorosporella,* and other fungi attack and kill insects of many orders but are especially prevalent upon larvae that live in the ground or in damp places. Man has taken advantage of some of these organisms to combat injurious insects.

Fig. 405.—Hornworms two days after artificial inoculation with *Bacillus sphingidis.* The caterpillars are hanging by the hooks of the prolegs. (*After U.S. Department of Agriculture, Bureau of Entomology and Plant Quarantine.*)

Several species of *Aschersonia* have been used successfully in Florida to control the white fly, *Aleyrodes citri.* Other fungus diseases may work to man's disadvantage. Scab brood causes serious losses to honeybees.

Bacteria likewise play an important part in the natural control of insects. It is not uncommon to find white grubs infected with bacterial diseases. The damp situations in which they live and the moist body walls of these insects are especially favorable for infection. Bacterial diseases may find an important place in the control of injurious species that inhabit the soil. The milky disease of the Japanese beetle is caused by two species of spore-forming bacteria. Twenty five per cent of the Japanese beetle grubs were infested in 1935. The value of this disease as a means of control is still questionable. There are numerous accounts of heavy toll of armyworms, cutworms, hornworms, and grasshoppers by a bacterial disease known as septicemia. When killed by bacteria, larvae eventually turn dark brown and their contents become liquefied.

On the other hand, the European and American foul broods attack the honeybee and thus becomes pests.

Numerous viruses and protozoa attack insects. Little is known concerning the virus diseases of insects. The wilt disease of the gypsy moth and tent caterpillars is one of the best known examples. The protozoan organisms that bring about diseases of insects have not been studied as much as the bacteria and fungi. Species of *Nosema* attack the silkworm and the honeybee.

FIG. 406.—A portion of the comb of honey bee showing the effect of European foulbrood. (*A*) (*J*) (*K*) Normal sealed cells; (*B*) (*C*) (*D*) (*E*) (*G*) (*I*) (*L*) (*M*) (*P*) (*Q*) larvae affected by disease; (*R*) normal larva at age attacked by disease; (*F*) (*G*) (*N*) (*O*) dried-down larvae or scales. (*After U.S. Department of Agriculture.*)

A great number of roundworms are known to attack insects. Only a comparatively few species have been isolated and studied but it is estimated that there may be several thousands. The beetles and the Orthoptera are usually attacked by these parasites.

> The glories of our blood and state
> Are shadows, not substantial things;
> There is no armour against fate;
> Death lays his icy hands on kings:
> Scepter and crown
> Must tumble down,
> And in the dust be equal made
> With the poor crooked scythe and spade.
> —JAMES SHIRLEY
> *Death's Final Conquest*

## BIBLIOGRAPHY

**I. Longevity and Tenacity.**

ALPATOV, W. W. (1929–1930). Experimental studies on the duration of life. *Amer. Nat.* **63, 64.** *Literature.*

BAUMBERGER, J. P. (1914). Studies in the longevity of insects. *Ann. Ent. Soc. Amer.* **7** (4).

BODENHEIMER, F. S. (1938). Life intensity and longevity. *In,* Problems of animal ecology. Oxford University Press, New York.

FROST, S. W. (1936). Longevity and tenacity. *In,* Ancient artizans. Van Press, Boston. *Bibliography.*

GLASER, R. W. (1923). The effect of food on longevity and reproduction in flies. *Journ. Exp. Zool.* **38** (3).

HASE, A. (1939). Ueber die Lebenszahigkeit von *Anthrenus verbasci* L. *Angew. Ent. Berl.* **6** (1). *Bibliography.*

KOPEČ, S. (1924). Studies on the influence of inanition on the development and duration of life in insects. *Biol. Bull. Woods Hole, Mass.* **46** (1). *Bibliography.*

LABOISSIÈRE, V. (1911). Sur la longévité des insectes. *Ann. Soc. Ent. Levallvois-Perret* **15–16.**

LINSLEY, E. G. (1938). Longevity in Cerambycidae. *Pan-Pacific Ent.* **14** (4).

LUND, H. O. (1938). Studies on longevity and productivity in *Trichogramma evanescens. Journ. Agric. Res.* **56** (6). *References.*

McNEIL, F. (1886). A remarkable case of longevity in a longicorn beetle, *Eburia quadrigeminata. Amer. Nat.* **20** (1086): 1055.

MALUF, N. S. R. (1939). The longevity of insects during complete inanition. *Amer. Nat.* **73** (746). *Literature.*

PEARL, R., *et al.* (1921–1929). Experimental studies on the duration of life. *Amer. Nat.* **55** (641), **56** (643–644, 646, 655), **57** (649), **58** (654, 656), **61** (633), **63** (684). *Literature cited.*

PEARL, R., and MINER, J. R. (1935). Experimental studies on the duration of life XIV. *Quart. Rev. Biol.* **10** (1). *Literature cited.*

RILEY, C. V. (1895). Longevity in insects. *Proc. Ent. Soc. Wash.* **3** (2).

XAMBEAU, LE CAPITAINE (1907). Longévité des insectes. *Naturaliste Paris* **29.**

**II. Diapause.**—See also references to metamorphosis.

BODENHEIMER, F. S. (1938). Diapause. *In,* Problems of animal ecology. Oxford University Press, New York.

COUSIN, G. (1932). Étude éxperimentale de la diapause des insectes. *Bull. France et Belgique Suppl.* **15.**

HELLER, J. (1928). Zur Auffassung des Unterschiedes zwischen subitaner und latenter Entwicklung von Schmetterlingspuppen. *Zeit. verlag Physiol. Berlin* of

KOZHANCHIKOV, J. (1935). The rôle of anoxybiotic process in the larval diapause **8.** some Pyralidae. *C. R. Acad. Sci. Russia* **2.**

PREBBLE, M. L. (1941). The diapause and related phenomena in *Gilpina polytoma* (Hurtig). I, factors influencing the inception of diapause. II, factors influencing the breaking of diapause. *Can. Journ. Res.* **19** (10).

RICHARDS, A. G., and MILLER, A. (1937). Insect development analyzed by experimental methods, a review Parts I and II. *Journ. N. Y. Ent. Soc.* **45.**

ROUBAUD, E. (1930). Suspension evolutine et hibernation larvaire obligatoire. . . . Les diapauses vraies et les diapauses chez les insectes. *C. R. Acad. Sci. Fr. Paris* **190.**

—— (1935). Vie latente et condition hibernale provoquées par influences maternelles chez certains invertébrés. *Ann. Sci. Nat. Paris* **18.**

SQUIRE, F. A. (1937). A theory of diapause in *Platyedra gossypiella* Saund. *Trop. Agric.* **14** (10). *References.*

——— (1940). On the nature and origin of diapause in *Platyedra gossypiella.* *Bull. Ent. Res.* **31** (1).

WIGGLESWORTH, V. B. (1940). Diapause. *In,* The principles of insect physiology. E. P. Dutton & Company, Inc., New York.

III. Hibernation.

BABCOCK, K. W. (1927). The European corn borer *Pyrausta nubilalis* Hubn. I, a discussion of its dormant period. *Ecology* **8** (1). *Literature.*

BAUMBERGER, J. P. (1917). Hibernation, a periodical phenomenon. *Ann. Ent. Soc. Amer.* **10** (2).

BLAIR, K. B. (1921). Insects in winter. *Proc. Ent. Soc. Lond.* **21.**

BURGESS, E. D. (1932). A comparison of the alimentary canals of the active and hibernating adults of the Mexican bean beetle, *Epilachna corrupta* Muls. *Ohio Journ. Sci.* **32** (3).

DAWSON, R. W. (1931). The problem of voltinism and dormancy in the polyphemus moth (*Telea polyphemus* Cramer). *Journ. Exp. Zool.* **59** (1). *Literature cited.*

DIXLEY, F. A. (1896). Some aspects of hibernation. *Ent. Rec.* **7.**

DOUGLASS, J. R. (1933). Additional information on precipitation as a factor in the emergence of *Epilachna corrupta* Muls. from hibernation. *Ecology* **14** (3).

FROST, S. W. (1936). Hibernation and aestivation. *In,* Ancient artizans. Van Press, Boston. *Bibliography.*

HODSON, A. C. (1937). Some aspects of the role of water in insect hibernation. *Ecol. Monogr.* **7** (2). *Literature cited.*

HOLMQUIST, A. M. (1926). Studies in Arthropod hibernation. *Ann. Ent. Soc. Amer.* **19** (4).

——— (1931). Studies in Arthropod hibernation III, temperature in forest hibernacula. *Ecology* **12** (2). *Literature cited.*

KELLOGG, V. L. (1904). Gregarious hibernation. *Proc. Ent. Soc. Lond.* **23, 24.**

LOCHHEAD, W. (1902). The hibernation of insects. *Rept. Ent. Soc. Ontario* **32.**

McCOLLOCH, J. W., and HAYES, W. P. (1923). Soil temperature and its influence on white grub activities. *Ecology* **4** (1).

——— *et al.* (1928). Hibernation of certain scarabaeids and their *Tiphia* parasites. *Ecology* **9** (1).

MAIL, G. A. (1930). Winter soil temperatures and their relation to subterranean insect survival. *Journ. Agric. Res.* **41** (8). *Bibliography.*

MILNER, R. D. (1921). Heat production of honeybees in winter. *U. S. Dept. Agric. Bull.* **988.**

PARK, O. (1930). Studies in the ecology of forest Coleoptera with observations on certain phases of hibernation and aggregation. *Ann. Ent. Soc. Amer.* **23** (1).

PEMBRAY, M. S., and WHITE, W. H. (1896). The regulation of temperature in hybernating animals. *Journ. Physiol.* **19** (5–6).

PEUS, F., *et al.* (1937). Insekten im winter. *Mitt. detsch. ent. Ges. Berlin* **8.**

RASMUSSEN, A. T. (1916). Theories of hibernation (dealing chiefly with higher animals). *Amer. Nat.* **50** (598). *References.*

ROBINSON, WM. (1927). Water-binding capacity of colloids a definite factor in winter hardiness of insects. *Journ. Econ. Ent.* **20** (1).

SANDERSON, E. D. (1908). The relation of temperature to the hibernation of insects. *Journ. Econ. Ent.* **1** (1).

IV. Sleep.

FIEBRIG, K. (1912). Schlafende in Insekten. *Jena Zeit. Naturw.* **48.**

FROST, S. W. (1936). Sleep in insects. *In,* Ancient artizans. Van Press, Boston.

HOFFMAN, R. W. (1937). Der Inseketenschlaf als reflektoriche Immobilisation. *Naturwissenschaften, Berlin* **25**.

PICTET, A. (1904). L'instinct et le sommeil chez les insectes. *Archiv. Sci. Phys. Nat.* **4** (17).

RAU, PHIL (1938). Additional observations on the sleep of insects. *Ann. Ent. Soc. Amer.* **31** (4).

———, and RAU, NELLIE (1916). The sleep of insects; an ecological study. *Ann. Ent. Soc. Amer.* **9** (3).

YOUNG, R. T. (1935). "Sleep" aggregation in the beetle, *Altica bimarginata*. *Science* **81** (2105).

**V. Death Feigning.**

DU PORTE, E. M. (1916). Death feigning reactions in *Tychius picirostris*. *Journ. Animal Behavior* **6** (2).

——— (1917). The death feigning instinct. *Can. Ent.* **49** (7).

FABRE, J. H. (1898). La simulation de la mort. *Rev. Quest. Sci.* **2** (15).

FROST, S. W. (1936). Death feigning. *In*, Ancient artizans. Van Press, Boston.

GODGLUCK, V. (1938). Die Kataleplischen Ercheinungen beider Hemipteren. *Z. wiss Zool. Leipzig* **146**.

HOLMES, S. J. (1937). The instinct of feigning death. *Popular Sci. Mo.* **62**.

PAYNE, NELLIE (1937). Death feigning in *Sitophilus granarius* L. the granary weevil (Coleoptera; Curculionidae). *Ent. News* **48** (6).

PONGRÁCZ, S. (1917). Ueber das Todtstellen in der Insektenwelt. *Rovartani Lap.* **24**.

RABAUD, E. (1916). Le phénomène de la simulation de la mort. *C. R. Soc. Biol. Paris* **79**.

SCHMIDT, P. Y. (1935). Anabiosis in insects. Socialistic grain farmer. Sararov l. Russian.

SEVERIN, H. H. P., and H. C. (1911). Habits of *Belostoma flumineum* Say., and *Nepa apiculata* Uhler. *Journ. N. Y. Ent. Soc.* **19** (2).

STEINIGER, F. (1933). Katalepsie und visuella anpassung bei Phyllium. *Zeit. wiss. Biol.* **28**.

TURNER, C. H. (1923). The physiology of "playing possum." *Trans. Acad. Sci. St. Louis* **24**.

WADE, J. S. (1933). Beetles that stand on their heads. *Nat. Mag.* **22** (5).

WEISS, H. B. (1940). The death-feint of *Trox unistriatus* Beauv. *Journ. N. Y. Ent. Soc.* **48** (3).

——— (1940). The death-feints of *Sitophilus granarius* Linn. *Journ. N. Y. Ent. Soc.* **48** (1).

**VI. Death.**

FROST, S. W. (1936). Causes contributing to death. *In*, Ancient artizans. Van Press, Boston. *Bibliography*.

HAMILTON, JOHN (1888). Knowledge of death in insects. *Canad. Ent.* **20** (9).

LEPESCHKIN, W. W. (1931). Death and its causes. *Quart. Rev. Biol.* **6** (2). *Literature cited*.

PEARL, R. (1922). The biology of death. J. B. Lippincott Company, Philadelphia. *Bibliography*.

——— (1924). Starvation life curves. *Nat. Lond.* **113** (2850).

RABAUD, E. (1917). Essai sur la vie et la mort des espèces. *Bull. Sci. France Belg. Paris* **50**.

# APPENDIX

The Appendix includes keys to the larger groups of immature insects, general references, and material too technical for the text.

## A FIELD KEY TO THE IMMATURE FORMS OF THE ORDERS OF INSECTS EXCLUSIVE OF EGGS AND PUPAE*

1. Campodeiform, body flattened, legs more or less elongate, head usually held in a horizontal position, developing wings usually external. Forms generally known as nymphs or naiads................................................... **2**
   Eruciform, platyform, vermiform, or scarabaeiform, head usually held in a vertical position, developing wings usually internal. Forms generally known as maggots, grubs, caterpillars, or larvae............................... **21**
2. With biting mouth parts...................................................... **3**
   With sucking mouth parts..................................................... **18**
3. Abdomen with only six segments, fourth abdominal segment with a pair of appendages constituting a spring. Small forms common in decayed leaves or debris, occasionally on snow................................... *Collembola*
   Abdomen with at least nine segments, no terminal spring................. **4**
4. Terrestrial forms............................................................ **5**
   Aquatic forms................................................................ **13**
5. Cerci minute or absent....................................................... **6**
   ⋅Cerci present and usually prominent...................................... **10**
6. Species parasitic upon birds and mammals, antennae not more than five segmented, frequently only one tarsal claw..................... *Mallophaga*
   Species not parasitic, antennae with more than five segments, usually two tarsal claws................................................................. **7**
7. Antennae noticeably elongate, as long as or longer than the body. Forms resembling aphids, common on the trunks of trees, on stones or fences, or upon books or old papers............................................ *Corrodentia*
   Antennae small and often inconspicuous.................................... **8**
8. Cerci present but minute, tarsi four segmented. Nest built in wood, social insects........................................................... *Isoptera*
   Cerci absent, nonsocial insects............................................. **9**
9. Mandibles normal, of the biting type, never sickle shaped in the terrestrial species, labial palpi two segmented. (Chiefly Meloidae, Carabidae, Coccinellidea, Chrysomelidae and Staphylinidae)...................... *Coleoptera*
   Mandibles conspicuous, usually sickle shaped, labial palpi, if present, three or more segmented. (Chiefly Raphidiidae, Mantispidae, Hemerobiidae, Chrysopidae, Myrmeleonidae, and Ascalaphidae)..................... *Neuroptera*
10. Cerci short but plainly visible, compound eyes generally present.. *Orthoptera*
    Ceri long and conspicuous, compound eyes absent...................... **11**
11. Typical campodeiform larva, abdomen terminated by three segmented filaments (two cerci and a central caudal filament). Species found about old stone piles or about books and papers........................ *Thysanura*
    Abdomen terminated by only two appendages........................... **12**

* Strepsiptera, Embiidina, and Zoraptera are not included.

479

12. Minute species, body not distinctly sclerotized, eyes absent, wings never develop, cerci segmented in Campodea, forceplike in *Japyx*, found in damp places or under stone......................................... *Dicellura*
    Species 2½ millimeter or larger, body distinctly sclerotized, wings develop externally................................................. *Dermaptera*
13. Labium elongate and geniculate, gill plates in the rectum of the dragonflies (Anisoptera), as two or three plates at the end of the abdomen of the damsel flies (Zygoptera).............................................. *Odonata*
    Labium not elongate or geniculate, gills usually on the thorax or abdomen.... **14**
14. Cerci present and conspicuous........................................... **15**
    Cerci absent or inconspicuous.......................................... **16**
15. Seven or eight pairs of tracheal gills along the sides of the abdomen, usually three long segmented caudal setae, tarsal claws single.......... *Ephemerida*
    Tracheal gills usually located on the underside of the thorax, although sometimes they occur also on the basal segments of the abdomen or upon the head, only two long caudal setae, tarsal claws double................. *Plecoptera*
16. Anal segments with a pair of short appendages, often jointed and always terminated by grasping hooks, tracheal gills located on the dorsal, lateral, and ventral surfaces of the abdomen or entirely absent, mandibles of the normal biting type. The casebearing caddis flies..................... *Trichoptera*
    Anal segment with an unjointed appendage or appendages but these never bear hooks, mandibles usually sickle shaped. Free-feeding, predacious species **17**
17. Tracheal gills, if present, located on the sides of the abdomen, labial palpi two segmented. (Dytiscidae, Gyrinidae, Haliplidae, and Hydrophilidae)........
    *Coleoptera*
    Tracheal gills, if present, on the ventral surface of the abdomen, palpi if present more than two segmented. (Sialidae and Sisyridae)............ *Neuroptera*
18. Tarsi usually without claws. Species frequently found on flowers.........
    *Thysanoptera*
    Tarsi with claws....................................................... **19**
19. Tarsi each with a single claw opposed by a toothed process on the tibia. Species parasitic upon mammals............................... *Anoplura*
    Tarsi each with two claws, predacious species or plant feeders ............. **20**
20. Proboscis arising from the front margin of the head; in some aquatic species the proboscis arises farther back but in these cases the legs are fitted for grasping................................................. *Heteroptera*
    Proboscis arising from the hind margin of the head, all are plant feeders, none is truly aquatic........................................... *Homoptera*
21. Legless species, at least without thoracic legs........................... **22**
    With at least well-developed thoracic legs.............................. **25**
22. Without a distinct head capsule, head always weakly sclerotized behind and retracted into the thorax, usually only two pairs of spiracles but when more are present the anterior and posterior pairs are distinctly larger. The primitive species take many forms but the more specialized forms are maggotlike......
    *Diptera*
    With a distinct well-sclerotized head capsule........................... **23**
23. Thoracic spiracles inconspicuous if present, eyes always wanting *Siphonaptera*
    Thoracic spiracles usually present and conspicuous, eyes usually present..... **24**
24. Thoracic spiracles located on the mesothorax, prolegs absent, head never with parietals (adfrontals)......................................... *Coleoptera*
    Thoracic spiracles on the prothorax or on both the pro- and the mesothorax, ocelli absent or, only one ocellus on each side of the head, most ... *Hymenoptera*

**25.** Without distinct prolegs.................................... *Coleoptera*
   With distinct prolegs.................................................. **26**
**26.** With only one pair of prolegs on the last abdominal segment.   Noncasebearing
   caddis fly larvae........................................... *Trichoptera*
   With two or more pairs of prolegs...................................... **27**
**27.** With only one large ocellus on each side of the head, six or more pairs of prolegs without crochets (Tenthredinidae)...................... *Hymenoptera*
   With one or more small ocelli on each side of the head..................... **28**
**28.** Usually ten or more closely grouped simple eyes, sometimes absent, six to eight
   pairs of prolegs............................................ *Mecoptera*
   One to eight small ocelli on each side of the head, usually two to five pairs of
   prolegs with crochets, spiracles on the prothorax.............. *Lepidoptera*

## A FIELD KEY TO THE COMMON GROUPS OF LEPIDOPTEROUS LARVAE*

**1.** Larvae without prolegs.............................................. **2**
   Larvae with prolegs..................................................... **4**
**2.** Leaf-mining larvae..................................................
   *Tischeriidae, Heliozelidae, Gelechiidae, Nepticulidae, Gracilariidae,* and others
   Slug caterpillars, legless flattened forms.................................. **3**
**3.** Head small, usually deeply retracted into the prothorax.......... *Lycaenidae*
   Head normal and not deeply retracted into the prothorax......... *Eucleidae*
**4.** One to three pairs of prolegs including the anal pair...................... **5**
   Five pairs of prolegs (except *Notodontidae*).............................. **6**
**5.** Hooks (crochets) on prolegs uniordinal, a few................... *Noctuidae*
   Hooks on prolegs biordinal or triordinal, the loopers or measuring worms.....
                                                              *Geometridae*
**6.** Caterpillars with a horn or, rarely, a button on the dorsal surface of the last
   abdominal segment, large usually three inches long, larvae of the sphinx moths
                                                              *Sphingidae*
   Caterpillars without a horn or button on the last abdominal segment........ **7**
**7.** A distinct construction between the head and the thorax, the skippers.......
                                                              *Hesperiidae*
   No constriction between the head and thorax............................ **8**
**8.** With a Y-shaped scent gland which can be protruded from the neck of the
   caterpillar, often brightly colored, sometimes with fleshy protuberances,
   larvae of the swallowtails.................................. *Papilionidae*
   Without a scent gland as described above............................. **9**
**9.** Caterpillars with distinct fleshy protuberances........................... **10**
   Caterpillars without fleshy protuberances............................... **12**
**10.** Anal prolegs much reduced or absent...................... *Notodontidae*
   Anal prolegs as large or larger than other prolegs........................ **11**
**11.** Hooks on prolegs biordinal............................... *Citheroniidae*
   Hooks on prolegs triordinal............................... *Nymphalidae*
**12.** Caterpillars with distinct tubercles bearing short spines, usually large caterpillars, brightly colored............................................. **13**
   Caterpillars without distinct tubercles bearing short spines................. **14**
**13.** An unpaired dorsal spine or tubercle on the ninth abdominal segment, the regal,
   imperial, and other moths................................ *Citheroniidae*
   No spine on the ninth abdominal segment although there is an unpaired dorsal
   spine on the eighth segment, larvae of the giant silkworms ....... *Saturniidae*

* Advanced students should consult S. B. Fracker (1915).   *Ill. Biol. Monogr.* **2** (1),
and W. T. M. Forbes (1910).   *Ann. Ent. Soc. Amer.,* **3** (2).

**14.** Caterpillars with barbed hairs.............................................. **15**
Caterpillars without barbed hairs..................................... **18**
**15.** Poisonous caterpillars............................................... **16**
Nonpoisonous caterpillars, certain........................ *Nymphalidae*
**16.** Without prolegs, the saddleback............................. *Eucleidae*
With prolegs...................................................... **17**
**17.** With dorsal tufts of hairs on the lateral margins of the body, crochets uniordinal, the brown-tail moth (Liparidae).................... *Lymantriidae*
Without dorsal tufts of hairs, crochets biordinal, larvae of the io and maia moths.................................................... *Saturniidae*
**18.** Hairy caterpillars.................................................. **19**
Bare or inconspicuously hairy caterpillars.............................. **25**
**19.** Densely hairy caterpillars........................................... **20**
Sparsely hairy caterpillars.......................................... **24**
**20.** With distinct tufts or pencils of hair................................. **21**
Without distinct tufts or pencils of hair............................... **23**
**21.** Tufts of hair on the dorsum of the abdomen as well as pencils of hair at the anterior and posterior ends, larvae of the tussock moths........ *Lymantriidae*
Often with pencils of hair but never with dorsal tufts of hair on the abdomen.. **22**
**22.** Usually with sparse secondary hair on the head, larvae of the footman and tiger moths.............................................. *Arctiidae**
Without secondary hair on the head, cutworms and related forms, some.....
*Noctuidae**
**23.** Hooks on prolegs uniordinal................. *Arctiidae* and some *Noctuidae*
Hooks on prolegs biordinal............................... *Lasiocampidae*
**24.** Medium size caterpillars, often feeding gregariously and having the habit of throwing the anterior and posterior ends upward when disturbed, some......
*Notodontidae*
Larger caterpillars at least three inches long, larvae of the regal, imperial, and other moths........................................... *Citheroniidae*
**25.** Small greenish or yellowish caterpillars living in folded leaves or boring in stems
*Pyralididae*
Larger caterpillars about one inch long and with different habits............ **26**
**26.** Greenish or bluish caterpillars, velvety in texture, feeding upon clover or Cruciferae................................................. *Pieridae*
Caterpillars with other characteristics................................ **27**
**27.** Caterpillars frequently found in the ground, usually smooth and greasy in appearance, indistinctly marked with stripes or spots, have the habit of curling when disturbed, certain...................................... *Noctuidae*
Caterpillars not generally found in the ground......................... **28**
**28.** Boring caterpillars.................................................. **29**
Caterpillars with other habits........................................ **30**
**29.** Borers in locust, oak, or other hard woods...................... *Cossidae*
Borers in peach, iris, pine, or the stems of certain plants........ *Aegeriidae*
Borers in the stems of water lilies, certain.................... *Noctuidae*
**30.** Caterpillars living in portable cases................................. **31**
Caterpillars not living in cases....................................... **32**
**31.** Cases in the form of silken bags to which bits of leaves, twigs, or other materials are attached, bagworms, mature cases ½ inch or larger....... *Psychidae*
Cases composed of pieces of leaves, plant fuzz, silk, and fecula, mature cases seldom more than ¼ inch long............................. *Coleophoridae*

* These two groups are difficult to separate as they vary so much in form.

**32.** Leaf rollers, leaf miners, and gall markers.................... *Gelechiidae*
Leaf rollers, ugly-nest builders, and fruit borers ............. *Olethreutidae*

## A FIELD KEY TO THE COMMON GROUPS OF COLEOPTEROUS LARVAE*

**1.** Grublike larvae, body curved, thoracic legs present or absent, creamy white in color....................................................................................... **2**
Body cylindrical, flattened, or other than above........................ **6**

**2.** Large fat white grubs with well-developed legs, common in soil or rotten wood, white grubs, the larvae of June beetles, and allied forms ....... *Scarabaeidae*
Smaller grubs, thoracic legs absent or much reduced.................... **3**

**3.** Legless grubs...................................................................... **4**
Legs present but much reduced....................................... **5**

**4.** Species boring in nuts, fruits, and occasionally in stems....... *Curculionidae*
Species excavating galleries in bark or wood.................. *Scolytidae*

**5.** Species boring in dry seasoned wood.............. *Bostrichidae Anobiidae*
Species inhabiting peas and beans........................ *Bruchidae*

**6.** Body cylindrical in form, elongate, thoracic legs present................. **7**
Body of other forms...................................................... **10**

**7.** Body wall not distinctly sclerotized, a pair of prominent hooks on the dorsal aspect of the last abdominal segment, larvae live beneath bark.. *Synchroidae*
Body wall heavily sclerotized, wireworm type........................ **8**

**8.** Head flattened, epicranial suture absent or very short, mandibles with molar surfaces undeveloped, spiracles biforous or cribiform, larvae live in the soil or decayed wood............................................. *Elateridae*
Head capsule rounded, epicranial suture distinct, mandibles with molar surfaces distinctly developed.................................... **9**

**9.** Back of mandibles opposite cutting edge with sharp margins, feeders in meal and other vegetable products........................... *Tenebrionidae*
Back of mandibles opposite cutting edges without sharp margins, feeders in rotten wood............................................. *Alleculidae*

**10.** Legless species.................................................. **11**
Legs present.................................................... **12**

**11.** First thoracic segment usually flat and broad, leaf miners, borers in wood and beneath bark........................................... *Buprestidae*
First thoracic segment not distinctly flat or broad............ *Cerambycidae*

**12.** Body distinctly flattened, legs generally short.......................... **13**
Body campodeiform, legs generally long.......................... **15**

**13.** Aquatic species, larvae closely pressed to rocks, flat oval in form, dorsal plates completely obscuring the head and legs from above, larvae of water pennies............................................. *Psephenidae*
Terrestrial species, head generally visible from above, forms more elongate, dorsal plates frequently produced laterally so that they cover the sides of the body...................................................... **14**

**14.** Head much flattened, quadrangular in shape, larvae luminescent, living in the soil.................................................. *Lampyridae*
Head more rounded, larvae nonluminescent, sarcophagous, or occasionally predacious.................................................. *Silphidae*

**15.** Aquatic species................................................. **16**
Terrestrial species, rarely semiaquatic.......................... **18**

* Advanced students should consult A. G. Boving and F. C. Craighead (1931). "The Larvae of Coleoptera."

16. Mandibles distinctly toothed, not sickle shaped, legs comparatively short, head frequently upturned................................. *Hydrophilidae*
    Mandibles sickle shaped, untoothed, legs conspicuously long.............. **17**
17. Each abdominal segment with a pair of lateral filaments, caudal segment with two pairs of filaments....................................... *Gyrinidae*
    Abdominal segments without lateral filaments, caudal segment with a single pair of filaments........................................... *Dytiscidae*
18. Typical campodeiform larvae, legs noticeably long, a distinct pair of cercion the ninth abdominal segment, usually found in the ground, beneath stones, bark, etc...................................................................... **19**
    Pseudocampodeiform larvae, legs generally shorter, no cerci at the posterior end of the body, larvae usually found upon foliage...................... **20**
19. Legs six jointed, claws movable............................... *Carabidae*
    Legs five jointed, claws fused with the tarsi................. *Staphylinidae*
20. Legs and palpi comparatively short, molar areas of the mandibles well developed, leaf-feeding insects............................... *Chrysomelidae*
    Legs and labial palpi comparatively long, molar areas of the mandibles not well developed, the mandibles somewhat sickle shaped, predacious species *............................................. *Coccinellidae*

## SYNONYMY OF ORDER NAMES†

1. Thysanura, Thysanoura, Ectotrophi, Ectognatha.
2. Dicellura, Rhabdura, Diplura, Entotrophi, Entognatha, Campodeoidea, Rhabduradelphia.
3. Collembola.
4. Orthoptera, Saltatoria, Ulonata part.
5. Zoraptera, Psocoptera part.
6. Dermaptera, Euplexoptera, Coleoptera part.
7. Thysanoptera, Physopoda.
8. Embiidina, Embioptera, Embiodea, Emboidea, Oligoneura, Aetioptera.
9. Isoptera.
10. Corrodentia, Psocoptera part, Copeognatha, Pseudoneuroptera part, Platyptera part.
11. Mallophaga, Lioptera, Platyptera part, Anoplura part, Hemiptera part.
12. Anoplura, Siphunculata, Pseudorhynchota, Parasitica, Phthiraptera, Ellipoptera.
13. Homoptera, Rhynchota part, Hemiptera part, Synaptera part.
14. Heteroptera, Rhynchota part, Hemiptera part, Synaptera part.
15. Odonata, Libelluloidea, Paraneuroptera.
16. Ephemerida, Ephemeroptera, Agnatha, Plectoptera.
17. Plecoptera, Perlaria.
18. Neuroptera (Synistata, Dictyoptera, Megaloptera, Sialoidea, Raphididoidea, Raphidioidea, Emmenognatha part).
19. Mecoptera, Mecaptera, Panorpina, Panorpatae.
20. Trichoptera, Phryganoidea.
21. Lepidoptera, Glossata.
22. Diptera, Antliata, Halterata, Halteriptera, Haustellata.
23. Siphonaptera, Suctoria, Aphaniptera, Rophoteira.
24. Coleoptera, Eleuterata, Elythroptera.
25. Strepsiptera, Rhipiptera, Coleoptera part.
26. Hymenoptera.

\* With the exception of two North American species.
† Exclusive of the very early classifications such as those of Linnaeus and Brauer.

## SCHEMES OF CLASSIFICATION OF INSECT ORDERS BY VARIOUS WORKERS, EXCLUSIVE OF FOSSIL ORDERS

| | Linnaeus 1735–1768 | Brauer 1885 | Sharp 1895 | Packard 1886 | Shipley 1904 | Comstock 1912 | Folsom Wardle 1934 | Comstock 1933 | Wardle 1937 |
|---|---|---|---|---|---|---|---|---|---|
| 1 | Coleoptera | Synaptera[1] | Aptera[2] | Thysanura | Aptera[3] | Thysanura | Protura | Thysanura | Protura[4] |
| 2 | Hemiptera | Dermaptera | Orthoptera | Dermaptera | Apontoptera[5] | Ephemerida | Thysanura | Collembola | Collembola |
| 3 | Lepidoptera | Ephemeridae | Neuroptera | Orthoptera | Ellipoptera[6] | Odonata | Collembola | Orthoptera | Rhabdura[7] |
| 4 | Neuroptera | Odonata | Hymenoptera | Platyptera[8] | Aphaniptera | Plecoptera | Orthoptera | Zoraptera | Thysanura |
| 5 | Hymenoptera | Plecoptera | Coleoptera | Odonata | Orthoptera | Corrodentia | Dermaptera | Isoptera | Dictyoptera[9] |
| 6 | Diptera | Orthoptera | Lepidoptera | Plectoptera | Plecoptera | Isoptera | Plecoptera | Neuroptera | Isoptera |
| 7 | Aptera | Corrodentia | Diptera | Thysanoptera | Psocoptera[10] | Mallophaga | Isoptera | Ephemerida | Orthoptera |
| 8 | | Thysanoptera | Aphaniptera | Hemiptera | Embioptera | Euplexoptera | Embioptera | Odonata | Dermaptera |
| 9 | | Rhynchota[11] | Thysanoptera | Neuroptera | Ephemeroptera | Orthoptera | Psocoptera[12] | Plecoptera | Phylloptera[13] |
| 10 | | Neuroptera | Hemiptera | Mecaptera | Paraneuroptera[14] | Physopoda | Anoplura | Corrodentia | Plecoptera |
| 11 | | Panorpatae | | Trichoptera | Thysanoptera | Hemiptera | Mallophaga | Mallophaga | Embioptera |
| 12 | | Trichoptera | | Coleoptera | Hemiptera | Neuroptera | Siphunculata | Embiidina | Odonata |
| 13 | | Lepidoptera | | Siphonaptera | Neuroptera | Mecoptera | Ephemeroptera | Thysanoptera | Ephemeroptera |
| 14 | | Siphonaptera | | Diptera | Mecaptera | Trichoptera | Odonata | Anoplura | Homoptera |
| 15 | | Diptera | | Lepidoptera | Trichoptera | Lepidoptera | Thysanoptera | Homoptera | Hemiptera |
| 16 | | Coleoptera | | Hymenoptera | Lepidoptera | Siphonaptera | Hemiptera | Hemiptera | Psocoptera[12] |
| 17 | | Hymenoptera | | | Coleoptera | Diptera | Heteroptera | Dermaptera | Anoplura[15] |
| 18 | | | | | Strepsiptera | Coleoptera | Coleoptera | Coleoptera | Thysanoptera |
| 19 | | | | | Diptera | Hymenoptera | Strepsiptera | Strepsiptera | Mecoptera |
| 20 | | | | | Hymenoptera | | Neuroptera | Mecoptera | Trichoptera |
| 21 | | | | | | | Mecoptera | Trichoptera | Neuroptera |
| 22 | | | | | | | Trichoptera | Lepidoptera | Lepidoptera |
| 23 | | | | | | | Lepidoptera | Diptera | Diptera |
| 24 | | | | | | | Hymenoptera | Siphonaptera | Siphonaptera |
| 25 | | | | | | | Diptera | Hymenoptera | Coleoptera |
| 26 | | | | | | | Siphonaptera | | Strepsiptera |
| 27 | | | | | | | | | Hymenoptera |

[1] Hemipters of Sharp and Packard.
[2] Includes Thysanura and Collembola.
[3] Synonymous with Thysanura.
[4] Not considered as Hexapoda.
[5] Synonymous with Collembola.
[6] Synonymous with Anoplura of Comstock.
[7] Includes Campodeidae, Projapygidae, and Japygidae.
[8] Includes Phasmidae and Phyllidae.
[9] Mantidae of others.
[10] Synonymous with Corrodentia.
[11] Aptera of Sharp.
[12] Includes Zoraptera and Corrodentia.
[13] Includes Phasmidae and Phyllidae.
[14] Synonymous with Odonata.
[15] Includes Mallophaga.

ANALYSIS OF THE CONCEPTIONS OF THE ORDER PLATYPTERA OF PACKARD*

| Brauer 1885 | Packard 1886 | Sharp 1898 | Folsom 1906 | Brues and† Melander 1932 | Wardle 1937 |
|---|---|---|---|---|---|
| **Corrodentia** Termitidae Psocidae Mallophaga **Orthoptera** Embiidina and others | **Platyptera** Termitidae Mallophaga Embiidae? Psocidae? | **Neuroptera** *Mallophaga* *Pseudoneuroptera* Embiidae Termitidae Psocidae Atropidae Corrodentia **Hemiptera** *Anoplura* and others | **Platyptera** *Corrodentia* Termitidae Embiidae Psocidae **Mallophaga** **Hemiptera** Parasitica and others | **Corrodentia** Atropidae Psocidae **Embiidina** **Zoraptera** **Anoplura** **Isoptera** | **Psocoptera** *Corrodentia* Atropidae Psocidae *Zoraptera* **Embioptera** **Anoplura** *Siphunculata* *Mallophaga* **Isoptera** |

\* Orders in boldface type, suborders in italics.
† Comstock (1933) follows the same system.

**Subclasses.**—Many somewhat unsuccessful attempts have been made to divide the Hexapoda into various groups. Brauer (1885) divided the insects into two classes: Apterygogenea, or wingless insects, and Pterygogenea, or winged insects. To the former he relegated Thysanura, Collembola, and other wingless insects. All other insects were placed in the Pterygogenea. Comstock (1912) proposed simpler names for these groups, namely, Apterygota and Pterygota. In the Apterygota he placed the Collembola and the Thysanura, the primitively wingless insects; the winged species and those with the acquired wingless condition he placed in the Pterygota. Sharp (1895) proposed the name Aptera for the Apterygota. Aptera, however, was used by Linnaeus in an obscure sense to include all wingless insects as well as species of other classes. Dr. V. R. Haber has suggested that the term Pterata* be used in the place of Pterygota.

Packard (1898) attempted to divide insects on the basis of metamorphosis. In the Ametabola he placed the Thysanura, Dermaptera, Orthoptera, Platyptera, Odonata, Plecoptera, Thysanoptera, and Hemiptera. In the Metabola he placed the Neuroptera, Mecoptera, Trichoptera, Coleoptera, Siphonaptera, Diptera, Lepidoptera, and Hymenoptera.

Shipley (1904) proposed four subclasses: Apterygota, for the primitively wingless insects; Anapterygota, for the acquired wingless insects; Exopterygota, for the winged insects with incomplete metamorphosis; and Endopterygota, for the winged insects with complete metamorphosis.

Wardle (1936) proposed a classification based upon certain outstanding characteristics about which the insects of each group are associated. There are ten of these groups: The *Proturoid* insects include Protura, Collembola, Dicellura, Thysanura. The *Palaeopteroid* insects include the Dictyoptera, Isoptera, and three fossil orders. The *Orthopteroid* insects include the Orthoptera, Dermaptera, and the Phylloptera. The *Plecopteroid* insects include the Plecoptera, Embioptera, and one fossil order. The *Zygopteroid* insects include the Odonata and two fossil orders. The

\* Unpublished.

FAMILIES, GENERA, AND SPECIES OF NORTH AMERICAN LEAF-MINING INSECTS

North American Leaf-mining Lepidoptera

| Family | Genus | Approximate number of species |
|--------|-------|-------------------------------|
| Eriocraniidae | *Eriocrania* | 1 |
| Incurvariidae | *Paraclemensia* | 1 |
| Nepticulidae | *Nepticula* | 66 |
| | *Glaucolepis* | 1 |
| Tischeriidae | *Tischeria* | 20 |
| Lyonetiidae | *Proleucoptera* | 2 |
| | *Leucoptera* | 2 |
| | *Lyonetia* | 5 |
| | *Bucculatrix* | 17 |
| | *Phyllocnistis* | 11 |
| | *Bedellia* | 1 |
| Gracilariidae | *Parectopa* | 8 |
| | *Acrocercops* | 7 |
| | *Leucanthiza* | 2 |
| | *Parornix* | 9 |
| | *Marmara* | 6 |
| | *Gracilaria* | 24 |
| | *Apophthisis* | 1 |
| | *Lithocolletis* | 102 |
| | *Cremastobombycia* | 5 |
| Coleophoridae | *Coleophora* | 13 |
| Cycnodiidae | *Aphelosetia* | 15 |
| Heliozelidae | *Heliozela* | 3 |
| | *Antispila* | 9 |
| | *Coptodisca* | 11 |
| Gelechiidae | *Recurvaria* | 11 |
| | *Phthorimoea* | 6 |
| | *Gnorimoschema* | 2 |
| | *Paralechia* | 1 |
| | *Aristotelia* | 3 |
| | *Chrysopora* | 2 |
| | *Nealyda* | 3 |
| | *Gelechia* | 2 |
| | *Evippe* | 1 |
| | *Tosca* | 1 |
| Lavernidae | *Psacaphora* | 6 |
| | *Chrysopeleia* | 2 |
| | *Cosmopteryx* | 5 |
| Yponomeutidae | *Scythris* | 6 |
| | *Yponomeuta* | 1 |
| | *Argyresthia* | 3 |
| Glyphipterygidae | *Choreutis* | 1 |
| Heliodinidae | *Cycloplasis* | 1 |
| | *Litharapteryx* | 1 |

FAMILIES, GENERA, AND SPECIES OF NORTH AMERICAN LEAF-MINING INSECTS.—
(*Continued*)

| Family | Genus | Approximate number of species |
|--------|-------|-------------------------------|
| Tortricidae | *Epinotia* | 3 |
|  | *Polychrosis* | 1 |
| Pyralididae | *Melitara* | 1 |
|  | *Autocosmia* | 1 |
| Noctuidae | *Arzama* | 1 |
|  | *Nonagria* | 1 |
| Total | . . . . . . . . . . . . . . . | 408 |

North American Leaf-mining Hymenoptera

| Family | Genus | Approximate number of species |
|--------|-------|-------------------------------|
| Tenthredinidae |  |  |
|   Phyllotominae | *Phlebatrobia* | 1 |
|   Scolioneurinae | *Entodecta* | 1 |
|  | *Scolioneura* | 2 |
|  | *Metallus* | 2 |
|   Fenusinae | *Profenusa* | 1 |
|  | *Fenusa* | 2 |
|  | *Kaliofenusa* | 1 |
|   Schizocerinae | *Schizocerus* | 1 |
| Total | . . . . . . . . . . . . . . . | 11 |

North American Leaf-mining Coleoptera

| Family | Genus | Approximate number of species |
|--------|-------|-------------------------------|
| Buprestidae | *Brachys* | 1 |
|  | *Pachyschelus* | 2 |
|  | *Taphrocerus* | 1 |
| Chrysomelidae | *Zeugophera* | 2 |
|  | *Phyllotreta* | 5 |
|  | *Hippuriphila* | 1 |
|  | *Dibolia* | 1 |
|  | *Microrhopala* | 3 |
|  | *Uroplata* | 1 |
|  | *Octotoma* | 1 |
|  | *Chalepus* | 5 |
|  | *Baliosus* | 2 |
|  | *Anoplitis* | 1 |
|  | *Stenopodius* | 1 western |
|  | *Monoxia* | 1 western |
|  | *Mantura* | 1 western |
| Curculionidae | *Orchestes* | 5 largely western |
|  | *Prionomerus* | 1 |
| Total | . . . . . . . . . . . . . . . | 35 |

FAMILIES, GENERA, AND SPECIES OF NORTH AMERICAN LEAF-MINING INSECTS.—
*(Continued)*

North American Leaf-mining Diptera

| Family | Genus | Approximate number of species |
|---|---|---|
| Cecidomyiidae | *Monarthropalpus* | 1 |
| Trupaneidae | *Acidia* | 1 |
| Agromyzidae | *Agromyza* | 115 |
|  | *Phytomyza* | 42 |
|  | *Napomyza* | 1 |
| Drosophilidae | *Scaptomyza* | 2 |
| Anthomyiidae | *Pegomyia* | 5 |
|  | *Hylemyia* | 2 |
| Cordyluridae | *Cordylura* | 1 |
| Total | . . . . . . . . . . . . . . . | 170 |

*Ephemeroid* insects include the Ephemeroptera and one fossil order. The *Hemipteroid* insects include the Homoptera, Hemiptera, Psocoptera, Anoplura, Thysanoptera, and one fossil order. The *Panorpid* insects include the Mecoptera, Trichoptera, Neuroptera, Lepidoptera, Diptera, Siphonaptera, and three fossil orders. The *Coleopteroid* insects include the Coleoptera, Strepsiptera, and one fossil order. The *Hymenopteroid* insects include the Hymenoptera and one fossil order.

Other classifications have been proposed but they are confusing and do not aid in simplifying the problem.

SUMMARY OF THE MORE IMPORTANT GROUPS OF SUBTERRANEAN INSECTS

| Order | Groups | Life stages found in soil | Purpose of subterranean life |
|---|---|---|---|
| Collembola | Most springtails | Eggs, larvae, adults | Live and feed* |
| Thysanura | Machilidae | Eggs, larvae, adults | Live and feed* |
| Dicellura | Campodeidae Japygidae | Eggs, larvae, adults | Live and feed* |
| Orthoptera | Crickets Short-horned grasshoppers Cave cricket Sand cricket Mole cricket Some roaches | Eggs, nymphs, adults Eggs Eggs, nymphs, adults Eggs and nymphs Eggs, nymphs, adults Eggs and nymphs | Hide and feed* Oviposit Live and feed Live and feed Live and feed Hide* |
| Zoraptera | Zorapterans | Eggs and nymphs | Live and feed* |
| Isoptera | Some termites | Eggs, nymphs, adults | Construct nests |
| Neuroptera | Mantispids Ant lions Alder flies Ascalaphids | Pupae Larvae and pupae Pupae Larvae and pupae | Transform Construct traps Transform Feed and transform |
| Het:roptera | Shore bugs Toad-shaped bugs | Nymphs and adults Nymphs and adults | Hide Hide |
| Homoptera | Cicadas Aphids Scale insects | Nymphs Nymphs and adults Nymphs and adults | Feed and transform Reproduce, feed, and transform Reproduce, feed, and transform |
| Dermaptera | Earwigs | Eggs, nymphs, adults | Hide |
| Coleoptera | Tiger beetles Rove beetles Many beetles† Many beetles | Eggs, larvae, pupae Larvae and pupae Eggs, larvae, pupae Pupae | Live and feed in burrows Hide Feed and transform Transform |
| Lepidoptera | A few species Many species | Larvae Pupae | Feed or hide Transform |
| Diptera | Many species | Larvae and pupae | Feed and transform |
| Siphonaptera | *Tinga penetrans* | Pupae | Transform |
| Hymenoptera | Many of the Vespidae, Sphecidae, and Prosopidae Bombidae, Formicidae, and Trigonidae Many other species | Eggs, larvae, and pupae Eggs, larvae, pupae, and adults Pupae | Build solitary nests Build social nests Transformation |
| Mecoptera | Some Panorpa | Larvae | Trap and feed |

* These insects are found chiefly beneath stones or in the soil cover.
† Chiefly Carabidae, Lampyridae, Silphidae, Meloidae, Elateridae, and Scarabaeidae.

## GENERAL ENTOMOLOGICAL WORKS

ALLEE, W. C., *et al.* (1950). The principles of animal ecology. W. B. Saunders Company, Philadelphia.

BORROR, D. J. (undated). A dictionary of word roots and combining forms. Dept. Zool-Entomology, Ohio State University.

BORROR, D. J. and DELONG, M. D. (1954). Introduction to the study of insects. Rinehart and Company, New York.

BRUES, C. T. (1946). Insect dietary. Harvard Univ. Press.

BRUES, C. T., MELANDER, A. L. and CARPENTER, F. M. (1954). Classification of insects. *Bull. Mus. Comp. Zool. Harvard.* **108.**

CARPENTER, J. R. (1956). An ecological glossary. Hafner Publishing Company, New York.

CHAMBERLIN, W. J. (1952). Entomological nomenclature and literature. 3d ed. Wm. C. Brown Company.

Common names of insects (1955). *Bull. Ent. Soc. Amer.* **1** (4).

COMSTOCK, J. H. (1948). An introduction to entomology. 9th ed. Comstock Publishing Company, Inc., Ithaca, New York.

ESSIG, E. O. (1958). Insects and mites from Western North America. The Macmillan Company, New York.

———— (1942). College entomology. The Macmillan Company, New York.

FOLSOM, J. W. (1934). Entomology with special reference to its ecological aspects. 4th ed., Revised by R. A. Wardle. The Blakiston Company, Philadelphia.

FROST, S. W. (1936). Ancient artizans, the wonders of the insect world. Van Press, Boston.

———— (1942). General entomology. McGraw-Hill Book Company, New York.

———— (1952). Light traps for insect collections, survey and control. Pa. State College. *Bull.* **255.**

HENDERSON, I. F., *et al.* (1939). A dictionary of scientific terms. D. Van Nostrand Company, Inc., New York.

IMMS, A. D. (1951). A general textbook of entomology. 8th ed. E. P. Dutton, New York.

JAEGER, E. C. (1944). A source book of biological names and terms. Charles C. Thomas, Springfield, Ill.

KELLOGG, V. L. (1908). American insects. Henry Holt and Company, New York.

KENNEDY, C. H. (1948). Insect. Encyclopaedia Britanica.

LINDROTH, C. H. (1957). The faunal connections between Europe and North America. John Wiley & Sons, Inc., New York.

LITTLE, V. A. (1957). General and applied entomology. Harper Bros. Publishers, New York.

MATHESON, ROBERT (1951). Entomology for introductory courses. Comstock Publishing Company. 2d ed.

MELANDER, A. L. (1937). Source book of biological terms. The Comet Press, Brooklyn, New York.

METCALF, C. L., and FLINT, W. P. (1932). Fundamentals of insect life. McGraw-Hill Book Company, Inc., New York.

PACKARD, A. S. (1898). A text-book of entomology. The Macmillan Company, New York.

PETERSON, ALVAH (1955). A manual of entomological technique. Edwards Brothers, Ann Arbor, Mich.

Ross, H. H. (1956). A textbook of entomology. John Wiley & Sons Inc., New York.

SHARP, DAVID (1898–1899). Insects. Cambridge Natural History 5 and 6. The Macmillan Company, New York.

SMITH, R. C. (1945). Guide to the literature of the zoological sciences. Revised ed. Burgess Publishing Company, Minneapolis, Minn.

STEINHAUS, E. A., and SMITH, R. F. (1956, 1957, 1958). Annual review of entomology. Annual Reviews Inc., *Ent. Soc. Amer.* I, II, III (Issued yearly).

TORRE-BUENO, J. R. DE LA (1937). A glossary of entomology. The Science Printing Co., Lancaster, Pa.

WOOD, R. S. (1944). The Naturalist's Lexicon. Abbey Garden Press.

## RECENT REFERENCES

*Beneficial and injurious insects.*

BODENHEIMER, F. S. (1951). Insects as human food. W. Junk, 's Gravenshage.

BROWN, A. W. A. (1951). Insect control by chemicals. John Wiley & Sons Inc., New York.

CLAUSEN, C. P. (1956). Biological control of insect pests in the Continental United States. *U. S. Dept. Agric. Tech. Bull.* 1139.

CARPENTER, J. S. (1955). Mosquitoes of North America, north of Mexico. Univ. Calif. Press.

CRAIGHEAD, F. C. (1950). Insect enemies of eastern forests. *U. S. Dept. Agric. Misc. Pub.* 657.

FERNALD, H. T., and SHEPARD, H. H. (1955). Applied entomology. McGraw-Hill Book Company, Inc., New York.

FROST, S. W. (1956). United States postmarks bearing insect names. *Stamps* January 7.

FROST, S. W. (1957). Stamps of the Dutch Colonies featuring Queen Wilhelmina and sphinx moths. *Weekly Philatelic Gossip* 64 (26).

FROST, S. W. (1958). Check list of insects on stamps. *Weekly Philatelic Gossip.*

HOFFMAN, W. E. (1947). Insects as human food. *Proc. Ent. Soc. Wash.* 49 (2).

HORSFALL, W. R. (1950). Mosquitoes their bionomics and relation to disease. The Ronald Press Co., New York.

KEEN, E. P. (1952). Insect enemies of western forests. *U. S. Dept. Agric. Misc. Pub.* 273.

KENNEDY, C. H. (1943). A dragonfly nymph design on Indian pottery. *Annals Ent. Soc. Amer.* 36 (2).

LEACH, J. G. (1940). Insect transmission of plant diseases. McGraw-Hill Book Company, New York.

LLOYD, F. E. (1942). The carnivorous plants. Chronica Botanica, Waltham, Mass.

MALLIS, A. (1954). Handbook of pest control. 2d ed. Gulf Research and Development Co., Pittsburgh, Pa.

PAINTER, R. H. (1951). Insect resistance in crop plants. Macmillan Co., New York.

Ross, H. H. (1947). The mosquitoes of Illinois (Diptera, Culicidae). *Ill. Nat. Hist. Survey* V34 (1).

WILLIAMS, C. B. (1958). Insect migration. New Naturalist Series.

*Orders of insects.*

BOHART, R. M. (1941). A revision of the Strepsiptera with special reference to the species of North America. Univ. Calif. *Publs. in Ent.* **7.**

BURKS, B. D. (1953). May flies or Ephemeroptera of Illinois. *Bull. Ill. Lab. Nat. Hist.* **26.**

BYERS, G. W. (1954). Notes on North American Mecoptera. *Annals Ent. Soc. Amer.* **47** (3).

CRAMPTON, G. C., *et al.* (1942). The Diptera or true flies of Connecticut. Part VI *State Geol. & Nat. Hist. Survey Bull.* **64.**

EDWARDS, J. G. (1949). Coleoptera or beetles east of the Great Plains. Edwards Brothers, Inc.

EDWARDS, J. G. (1950). A bibliographical supplement to "Coleoptera or beetles east of the Great Plains." Edwards Brothers, Inc.

EWING, H. E. (1940). The Protura of North America. *Annals Ent. Soc. Amer.* **33.**

EWING, H. E., and FOX, I. (1943). Fleas of North America. *U. S. Nat. Museum Misc. Pub.* **500.**

FORBES, WM. T. M. (1948). Part II Lepidoptera of New York and neighboring states. *Cornell Agric. Exp. Sta. Memoir* **274.**

FORBES, WM. T.M. (1954). Part III Lepidoptera of New York and neighboring states. *Cornell Agric. Exp. Sta. Memoir* **329.**

FROST, S. W., and BROWN, J. P. (1955). A preliminary study of Pennsylvania Mecoptera. *Journ. N. Y. Ent. Soc.* **63.**

KLOTS, A. B. (1951). Field guide to the butterflies of North America, east of the Great Plains. Houghton Mifflin Co., Boston.

MACY, R. W., and SHEPARD, H. H. (1941). Butterflies, a handbook of the butterflies of the United States. Univ. Minn. Press.

NEEDHAM, J. G., and WESTFALL, M. J. (1955). A manual of the dragonflies of North America (Anisoptera). Univ. Calif. Press.

RICKER, W. E. (1952). Systematic studies in Plecoptera. Indiana Univ. *Publ. Sci. Ser. No.* **18.**

ROSS, H. H. (1944). Caddis flies or Trichoptera of Illinois. *Nat. Hist. Survey Bull.* **23** (1): 1–326.

ROSS, E. S. (1944). A revision of the Embioptera or web spinners of the New World. *U. S. Nat. Mus. Misc. Pub.* **91** (3175).

WALKER, E. M. (1937). *Grylloblatta,* a living fossil. *Trans. Roy. Soc. Can. 3rd ser., Sec. V* **31.**

# INDEX

## A

Abbot's bagworm, 457
Abbott's sawfly, 133, 148, 304
  fecula of, 148
Abdomen of insect, 171
*Abedus*, 127
Abundance of insects, 9, 35
  references to, 44
  in temperate areas, 26
  in tropics, 24–26
*Acacia cyclopis*, food of *Antheraea cytherea*, 147
*Acanthocephala femorata* attacking plants, 299
*Acanthoscelides obtectus*, 378
*Achias longividens*, sexual color forms, 184
*Achorutes*, photogenesis, 191
*Acidia fratria*, color varieties, 182
  *heraclei*, color varieties, 182
Acordulescerinae, 137
Acorn borer, 379
Acraeidae, 21
*Acraspis erinacei*, 365
Acrididae, 21
*Acripez reticulata*, sexual color forms, 183
*Acris gryllus*, song of, 199
Acrobasidae, 109
*Acrobasis caryae*, 353
  *comptoniella*, 353
  *rubrifasciella*, 353
*Acrocercops*, 333
*Acrocinus longimanus*, 43, 163
  front legs of, 163
*Acrolepia incertella*, 386
*Acroneuria arida*, 442
  rheotropism, 215
*Acrosternum*, 283, 299
  *hilare*, 284
Activity, cessation of, 462, 465
*Adalia bipunctata*, 287
Adaptations, for aquatic life, 428
  for egg laying, 125
  for flower visitation, 306, 307

Adaptations, for the gall-making habit, 362
*Adela*, 457
*Adelges abietes*, 365
*Adelops*, 420
Adjustments to swift water, 215
*Aegocera tripartitia*, sound production, 204
*Aeschna constricta*, 130, 443
Agaristidae, 21
Age of insects, 36
Aggregations, 230–239
  estivating, 232
  protective, 234
  references to, 255
  sleeping, 237
  swarming, 237
Aggressiveness in social insects, 249
*Aglossa suprealis*, 274, 277
*Agonopteryx*, 349
*Agrilus anxius*, 218, 366, 380
  geotropism in, 218
  *bilineatus*, geotropism, 218
  *ruficollis*, 366, 367, 380, 383
  geotropism in, 218
*Agrion*, 43
*Agromyza aceris*, 320, 394
  *aeniventris*, 320
  *amelanchieris*, 394
    habits, 320, 394
  *aristata*, 147, 326, 334, 337
  *borealis*, 326, 334
  *coronata*, 326
  *laterella*, 320, 356
  *melampyga*, 326, 334
  *posticata*, 326
  *pruinosa*, 320, 394
  *schineri*, habits, 320
  *simplex*, habits, 320
  *subpusilla*, 332
  *tiliae*, habits, 320
  *virens*, habits, 320
  *websteri*, habits, 32(
  *youngi*, 319

495

A CATALOGUE OF SELECTED DOVER BOOKS
IN ALL FIELDS OF INTEREST

# A CATALOGUE OF SELECTED DOVER BOOKS
# IN ALL FIELDS OF INTEREST

THE DEVIL'S DICTIONARY, Ambrose Bierce. Barbed, bitter, brilliant witticisms in the form of a dictionary. Best, most ferocious satire America has produced. 145pp.      20487-1 Pa. $1.50

ABSOLUTELY MAD INVENTIONS, A.E. Brown, H.A. Jeffcott. Hilarious, useless, or merely absurd inventions all granted patents by the U.S. Patent Office. Edible tie pin, mechanical hat tipper, etc. 57 illustrations. 125pp.      22596-8 Pa. $1.50

AMERICAN WILD FLOWERS COLORING BOOK, Paul Kennedy. Planned coverage of 48 most important wildflowers, from Rickett's collection; instructive as well as entertaining. Color versions on covers. 48pp. 8¼ x 11.      20095-7 Pa. $1.35

BIRDS OF AMERICA COLORING BOOK, John James Audubon. Rendered for coloring by Paul Kennedy. 46 of Audubon's noted illustrations: red-winged blackbird, cardinal, purple finch, towhee, etc. Original plates reproduced in full color on the covers. 48pp. 8¼ x 11.      23049-X Pa. $1.35

NORTH AMERICAN INDIAN DESIGN COLORING BOOK, Paul Kennedy. The finest examples from Indian masks, beadwork, pottery, etc. — selected and redrawn for coloring (with identifications) by well-known illustrator Paul Kennedy. 48pp. 8¼ x 11.      21125-8 Pa. $1.35

UNIFORMS OF THE AMERICAN REVOLUTION COLORING BOOK, Peter Copeland. 31 lively drawings reproduce whole panorama of military attire; each uniform has complete instructions for accurate coloring. (Not in the Pictorial Archives Series). 64pp. 8¼ x 11.      21850-3 Pa. $1.50

THE WONDERFUL WIZARD OF OZ COLORING BOOK, L. Frank Baum. Color the Yellow Brick Road and much more in 61 drawings adapted from W.W. Denslow's originals, accompanied by abridged version of text. Dorothy, Toto, Oz and the Emerald City. 61 illustrations. 64pp. 8¼ x 11.      20452-9 Pa. $1.50

CUT AND COLOR PAPER MASKS, Michael Grater. Clowns, animals, funny faces . . . simply color them in, cut them out, and put them together, and you have 9 paper masks to play with and enjoy. Complete instructions. Assembled masks shown in full color on the covers. 32pp. 8¼ x 11.      23171-2 Pa. $1.50

STAINED GLASS CHRISTMAS ORNAMENT COLORING BOOK, Carol Belanger Grafton. Brighten your Christmas season with over 100 Christmas ornaments done in a stained glass effect on translucent paper. Color them in and then hang at windows, from lights, anywhere. 32pp. 8¼ x 11.      20707-2 Pa. $1.75

SLEEPING BEAUTY, illustrated by Arthur Rackham. Perhaps the fullest, most delightful version ever, told by C.S. Evans. Rackham's best work. 49 illustrations. 110pp. 7⅞ x 10¾. 22756-1 Pa. $2.00

THE WONDERFUL WIZARD OF OZ, L. Frank Baum. Facsimile in full color of America's finest children's classic. Introduction by Martin Gardner. 143 illustrations by W.W. Denslow. 267pp. 20691-2 Pa. $2.50

GOOPS AND HOW TO BE THEM, Gelett Burgess. Classic tongue-in-cheek masquerading as etiquette book. 87 verses, 170 cartoons as Goops demonstrate virtues of table manners, neatness, courtesy, more. 88pp. 6½ x 9¼. 22233-0 Pa. $1.50

THE BROWNIES, THEIR BOOK, Palmer Cox. Small as mice, cunning as foxes, exuberant, mischievous, Brownies go to zoo, toy shop, seashore, circus, more. 24 verse adventures. 266 illustrations. 144pp. 6⅝ x 9¼. 21265-3 Pa. $1.75

BILLY WHISKERS: THE AUTOBIOGRAPHY OF A GOAT, Frances Trego Montgomery. Escapades of that rambunctious goat. Favorite from turn of the century America. 24 illustrations. 259pp. 22345-0 Pa. $2.75

THE ROCKET BOOK, Peter Newell. Fritz, janitor's kid, sets off rocket in basement of apartment house; an ingenious hole punched through every page traces course of rocket. 22 duotone drawings, verses. 48pp. 6⅞ x 8⅜. 22044-3 Pa. $1.50

PECK'S BAD BOY AND HIS PA, George W. Peck. Complete double-volume of great American childhood classic. Hennery's ingenious pranks against outraged pomposity of pa and the grocery man. 97 illustrations. Introduction by E.F. Bleiler. 347pp. 20497-9 Pa. $2.50

THE TALE OF PETER RABBIT, Beatrix Potter. The inimitable Peter's terrifying adventure in Mr. McGregor's garden, with all 27 wonderful, full-color Potter illustrations. 55pp. 4¼ x 5½. USO 22827-4 Pa. $1.00

THE TALE OF MRS. TIGGY-WINKLE, Beatrix Potter. Your child will love this story about a very special hedgehog and all 27 wonderful, full-color Potter illustrations. 57pp. 4¼ x 5½. USO 20546-0 Pa. $1.00

THE TALE OF BENJAMIN BUNNY, Beatrix Potter. Peter Rabbit's cousin coaxes him back into Mr. McGregor's garden for a whole new set of adventures. A favorite with children. All 27 full-color illustrations. 59pp. 4¼ x 5½. USO 21102-9 Pa. $1.00

THE MERRY ADVENTURES OF ROBIN HOOD, Howard Pyle. Facsimile of original (1883) edition, finest modern version of English outlaw's adventures. 23 illustrations by Pyle. 296pp. 6½ x 9¼. 22043-5 Pa. $2.75

TWO LITTLE SAVAGES, Ernest Thompson Seton. Adventures of two boys who lived as Indians; explaining Indian ways, woodlore, pioneer methods. 293 illustrations. 286pp. 20985-7 Pa. $3.00

DECORATIVE ALPHABETS AND INITIALS, edited by Alexander Nesbitt. 91 complete alphabets (medieval to modern), 3924 decorative initials, including Victorian novelty and Art Nouveau. 192pp. 7¾ x 10¾. 20544-4 Pa. $3.50

CALLIGRAPHY, Arthur Baker. Over 100 original alphabets from the hand of our greatest living calligrapher: simple, bold, fine-line, richly ornamented, etc. — all strikingly original and different, a fusion of many influences and styles. 155pp. 11⅜ x 8¼. 22895-9 Pa. $4.00

MONOGRAMS AND ALPHABETIC DEVICES, edited by Hayward and Blanche Cirker. Over 2500 combinations, names, crests in very varied styles: script engraving, ornate Victorian, simple Roman, and many others. 226pp. 8⅛ x 11. 22330-2 Pa. $4.00

THE BOOK OF SIGNS, Rudolf Koch. Famed German type designer renders 493 symbols: religious, alchemical, imperial, runes, property marks, etc. Timeless. 104pp. 6⅛ x 9¼. 20162-7 Pa. $1.50

200 DECORATIVE TITLE PAGES, edited by Alexander Nesbitt. 1478 to late 1920's. Baskerville, Dürer, Beardsley, W. Morris, Pyle, many others in most varied techniques. For posters, programs, other uses. 222pp. 8⅜ x 11¼. 21264-5 Pa. $3.50

DICTIONARY OF AMERICAN PORTRAITS, edited by Hayward and Blanche Cirker. 4000 important Americans, earliest times to 1905, mostly in clear line. Politicians, writers, soldiers, scientists, inventors, industrialists, Indians, Blacks, women, outlaws, etc. Identificatory information. 756pp. 9¼ x 12¾. 21823-6 Clothbd. $30.00

ART FORMS IN NATURE, Ernst Haeckel. Multitude of strangely beautiful natural forms: Radiolaria, Foraminifera, jellyfishes, fungi, turtles, bats, etc. All 100 plates of the 19th century evolutionist's Kunstformen der Natur (1904). 100pp. 9⅜ x 12¼. 22987-4 Pa. $4.00

DECOUPAGE: THE BIG PICTURE SOURCEBOOK, Eleanor Rawlings. Make hundreds of beautiful objects, over 550 florals, animals, letters, shells, period costumes, frames, etc. selected by foremost practitioner. Printed on one side of page. 8 color plates. Instructions. 176pp. 9³⁄₁₆ x 12¼. 23182-8 Pa. $5.00

AMERICAN FOLK DECORATION, Jean Lipman, Eve Meulendyke. Thorough coverage of all aspects of wood, tin, leather, paper, cloth decoration — scapes, humans, trees, flowers, geometrics — and how to make them. Full instructions. 233 illustrations, 5 in color. 163pp. 8⅜ x 11¼. 22217-9 Pa. $3.95

WHITTLING AND WOODCARVING, E.J. Tangerman. Best book on market; clear, full. If you can cut a potato, you can carve toys, puzzles, chains, caricatures, masks, patterns, frames, decorate surfaces, etc. Also covers serious wood sculpture. Over 200 photos. 293pp. 20965-2 Pa. $2.50

EGYPTIAN MAGIC, E.A. Wallis Budge. Foremost Egyptologist, curator at British Museum, on charms, curses, amulets, doll magic, transformations, control of demons, deific appearances, feats of great magicians. Many texts cited. 19 illustrations. 234pp. USO 22681-6 Pa. $2.50

THE LEYDEN PAPYRUS: AN EGYPTIAN MAGICAL BOOK, edited by F. Ll. Griffith, Herbert Thompson. Egyptian sorcerer's manual contains scores of spells: sex magic of various sorts, occult information, evoking visions, removing evil magic, etc. Transliteration faces translation. 207pp. 22994-7 Pa. $2.50

THE MALLEUS MALEFICARUM OF KRAMER AND SPRENGER, translated, edited by Montague Summers. Full text of most important witchhunter's "Bible," used by both Catholics and Protestants. Theory of witches, manifestations, remedies, etc. Indispensable to serious student. 278pp. 6⅝ x 10. USO 22802-9 Pa. $3.95

LOST CONTINENTS, L. Sprague de Camp. Great science-fiction author, finest, fullest study: Atlantis, Lemuria, Mu, Hyperborea, etc. Lost Tribes, Irish in pre-Columbian America, root races; in history, literature, art, occultism. Necessary to everyone concerned with theme. 17 illustrations. 348pp. 22668-9 Pa. $3.50

THE COMPLETE BOOKS OF CHARLES FORT, Charles Fort. Book of the Damned, Lo!, Wild Talents, New Lands. Greatest compilation of data: celestial appearances, flying saucers, falls of frogs, strange disappearances, inexplicable data not recognized by science. Inexhaustible, painstakingly documented. Do not confuse with modern charlatanry. Introduction by Damon Knight. Total of 1126pp.
23094-5 Clothbd. $15.00

FADS AND FALLACIES IN THE NAME OF SCIENCE, Martin Gardner. Fair, witty appraisal of cranks and quacks of science: Atlantis, Lemuria, flat earth, Velikovsky, orgone energy, Bridey Murphy, medical fads, etc. 373pp. 20394-8 Pa. $3.00

HOAXES, Curtis D. MacDougall. Unbelievably rich account of great hoaxes: Locke's moon hoax, Shakespearean forgeries, Loch Ness monster, Disumbrationist school of art, dozens more; also psychology of hoaxing. 54 illustrations. 338pp. 20465-0 Pa. $3.50

THE GENTLE ART OF MAKING ENEMIES, James A.M. Whistler. Greatest wit of his day deflates Wilde, Ruskin, Swinburne; strikes back at inane critics, exhibitions. Highly readable classic of impressionist revolution by great painter. Introduction by Alfred Werner. 334pp. 21875-9 Pa. $4.00

THE BOOK OF TEA, Kakuzo Okakura. Minor classic of the Orient: entertaining, charming explanation, interpretation of traditional Japanese culture in terms of tea ceremony. Edited by E.F. Bleiler. Total of 94pp. 20070-1 Pa. $1.25

*Prices subject to change without notice.*
Available at your book dealer or write for free catalogue to Dept. GI, Dover Publications, Inc., 180 Varick St., N.Y., N.Y. 10014. Dover publishes more than 150 books each year on science, elementary and advanced mathematics, biology, music, art, literary history, social sciences and other areas.